Springer Series in Statistics

Perspectives in Statistics

Springer
New York
Berlin
Heidelberg
Barcelona
Budapest
Hong Kong
London
Milan
Paris
Santa Clara
Singapore
Tokyo

Springer Series in Statistics

Andersen/Borgan/Gill/Keiding: Statistical Models Based on Counting Processes.
Andrews/Herzberg: Data: A Collection of Problems from Many Fields for the Student and Research Worker.
Anscombe: Computing in Statistical Science through APL.
Berger: Statistical Decision Theory and Bayesian Analysis, 2nd edition.
Bolfarine/Zacks: Prediction Theory for Finite Populations.
Borg/Groenen: Modern Multidimensional Scaling: Theory and Applications
Brémaud: Point Processes and Queues: Martingale Dynamics.
Brockwell/Davis: Time Series: Theory and Methods, 2nd edition.
Daley/Vere-Jones: An Introduction to the Theory of Point Processes.
Dzhaparidze: Parameter Estimation and Hypothesis Testing in Spectral Analysis of Stationary Time Series.
Fahrmeir/Tutz: Multivariate Statistical Modelling Based on Generalized Linear Models.
Farrell: Multivariate Calculation.
Federer: Statistical Design and Analysis for Intercropping Experiments.
Fienberg/Hoaglin/Kruskal/Tanur (Eds.): A Statistical Model: Frederick Mosteller's Contributions to Statistics, Science and Public Policy.
Fisher/Sen: The Collected Works of Wassily Hoeffding.
Good: Permutation Tests: A Practical Guide to Resampling Methods for Testing Hypotheses.
Goodman/Kruskal: Measures of Association for Cross Classifications.
Gouriéroux: ARCH Models and Financial Applications.
Grandell: Aspects of Risk Theory.
Haberman: Advanced Statistics, Volume I: Description of Populations.
Hall: The Bootstrap and Edgeworth Expansion.
Härdle: Smoothing Techniques: With Implementation in S.
Hart: Nonparametric Smoothing and Lack-of-Fit Tests.
Hartigan: Bayes Theory.
Heyde: Quasi-Likelihood And Its Application: A General Approach to Optimal Parameter Estimation.
Heyer: Theory of Statistical Experiments.
Huet/Bouvier/Gruet/Jolivet: Statistical Tools for Nonlinear Regression: A Practical Guide with S-PLUS Examples.
Jolliffe: Principal Component Analysis.
Kolen/Brennan: Test Equating: Methods and Practices.
Kotz/Johnson (Eds.): Breakthroughs in Statistics Volume I.
Kotz/Johnson (Eds.): Breakthroughs in Statistics Volume II.
Kotz/Johnson (Eds.): Breakthroughs in Statistics Volume III.
Kres: Statistical Tables for Multivariate Analysis.
Küchler/Sørensen: Exponential Families of Stochastic Processes.
Le Cam: Asymptotic Methods in Statistical Decision Theory.

(continued after index)

Emanuel Parzen Kunio Tanabe
Genshiro Kitagawa
Editors

Selected Papers of Hirotugu Akaike

Springer

Emanuel Parzen
Department of Statistics
Texas A&M University
College Station, TX 77843
USA

Kunio Tanabe
Genshiro Kitagawa
The Institute of Statistical Mathematics
4-6-7 Minami-Azabu, Minato-ku
Tokyo 106
Japan

Library of Congress Cataloging-in-Publication Data
Akaike, Hirotugu, 1927–
[Selections. 1998]
Selected papers of Hirotugu Akaike / Emanuel Parzen, Kunio Tanabe, Genshiro Kitagawa, editors.
p. cm. — (Springer series in statistics)
Includes bibliographical references and index.
ISBN 0-387-98355-4 (hardcover : alk. paper)
1. Time–series analysis. 2. Mathematical statistics. I. Parzen, Emanuel, 1929– . II. Tanabe, Kunio, 1941– . III. Kitagawa, G. (Genshiro), 1948– . IV. Title. V. Series.
QA280.A33 1998
519.5—dc21 97-34592

Printed on acid-free paper.

Production managed by Allan Abrams; manufacturing supervised by Jacqui Ashri.
Camera-ready copy prepared by the editors.
Printed and bound by Edwards Brothers, Inc., Ann Arbor, MI.
Printed in the United States of America.

9 8 7 6 5 4 3 2 1

ISBN 0-387-98355-4 Springer-Verlag New York Berlin Heidelberg SPIN 10646832

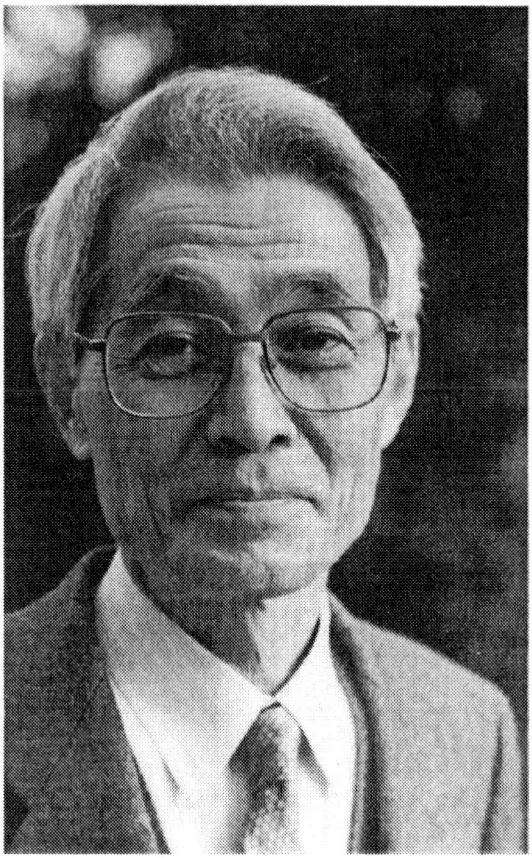

Hirotugu Akaike

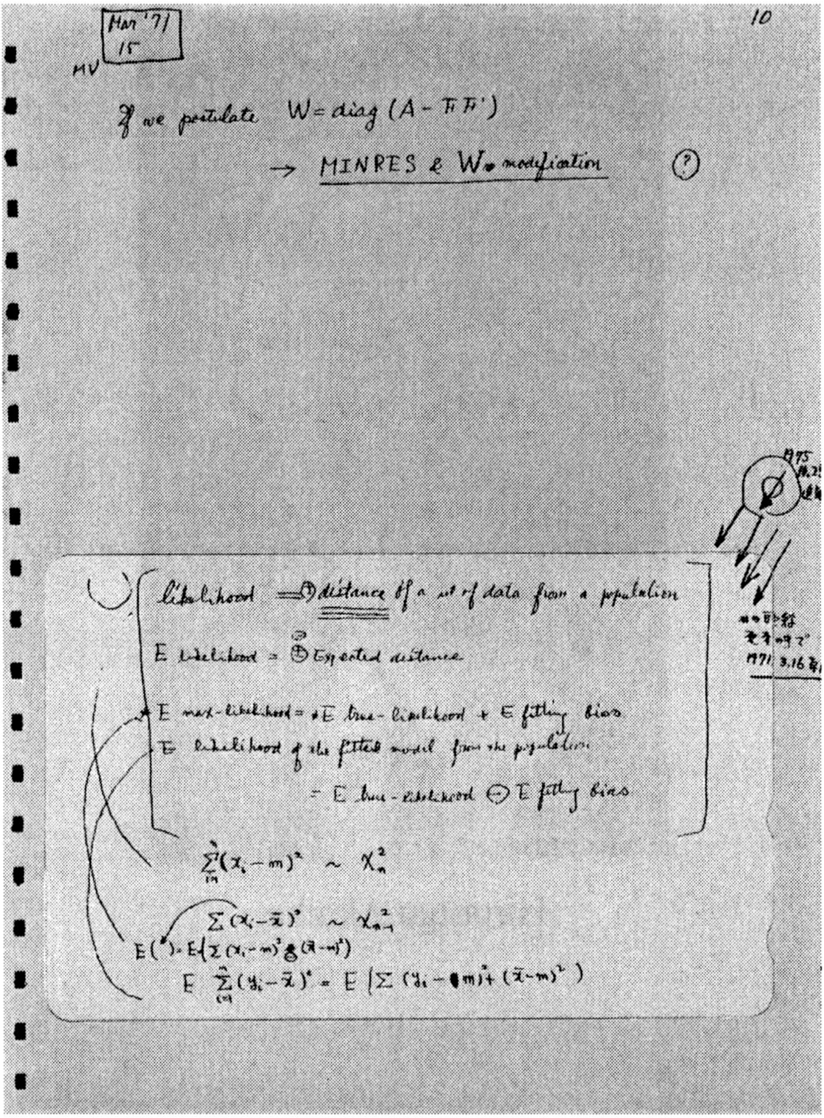

Mar '71 15

MV

If we postulate W = diag (A − ΓΓ')

→ MINRES & W modification

likelihood = distance of a set of data from a population

E likelihood = Expected distance

E max-likelihood = E true-likelihood + E fitting bias

E likelihood of the fitted model from the population

= E true-likelihood − E fitting bias

$\sum_{i=1}^{n}(x_i - m)^2 \sim \chi_n^2$

$\sum (x_i - \bar{x})^2 \sim \chi_{n-1}^2$

A photocopy of the page which indicates the birth of AIC.

Contents

Foreword

The pioneering research of Hirotugu Akaike has an international reputation for profoundly affecting how data and time series are analyzed and modelled and is highly regarded by the statistical and technological communities of Japan and the world. His 1974 paper "A new look at the statistical model identification" (IEEE Trans Automatic Control, AC-19, 716-723) is one of the most frequently cited papers in the area of engineering, technology, and applied sciences (according to a 1981 Citation Classic of the Institute of Scientific Information). It introduced the broad scientific community to model identification using the methods of Akaike's criterion AIC. The AIC method is cited and applied in almost every area of physical and social science.

The best way to learn about the seminal ideas of pioneering researchers is to read their original papers. This book reprints 29 papers of Akaike's more than 140 papers. This book of papers by Akaike is a tribute to his outstanding career and a service to provide students and researchers with access to Akaike's innovative and influential ideas and applications.

To provide a commentary on the career of Akaike, the motivations of his ideas, and his many remarkable honors and prizes, this book reprints "A Conversation with Hirotugu Akaike" by David F. Findley and Emanuel Parzen, published in 1995 in the journal *Statistical Science.* This survey of Akaike's career provides each of us with a role model for how to have an impact on society by stimulating applied researchers to implement new statistical methods.

The significant contributions of Akaike to society, stimulating applications of his time series methods by his engineering friends, were recognized in Japan by significant prizes and honors: the Ishikawa prize in 1972, the Okochi prize in 1980, the Purple Ribbon Medal given by the Emperor of Japan in 1989, the Asahi Prize in 1989, and the First Japan Statistical Society Prize in 1996.

The selected papers in this book are divided into six groups, representing successive phases of Akaike's research interests during his more than 40 years of work at the prestigious Institute of Statistical Mathematics in Tokyo (from 1952 until his retirement in 1994). He was Director General of the Institute from 1986 to 1994. The Institute provided Akaike with a unique environment which he felt protected his freedom of choice of subject and way of developing research.

1. Two papers, called Precursors, represent the years 1952–1960 which were a launching period; a paper on gap (zero-one valued) processes and a paper on methods of optimization and numerical analysis.

2. Three papers are reprinted from the years 1960–1965, emphasizing Frequency Domain Time Series Analysis; they are concerned with power-spectrum analysis and frequency response analysis. Applications of these methods by his engineering friends was reported in Akaike (1962), (1964).

3. Six papers describe Akaike's innovative research after 1965 emphasizing Time Domain Time Series Analysis with influential (and prize winning) applications of these methods to analysis of feedback systems and autoregressive model fitting for control of cement rotary kilns and power plants.

4. Six papers describe Akaike's research after 1970 which developed his world famous criterion AIC for model identification. Fundamental research on using state space models to identify parsimonious models for multivariate time series is presented in the 1974 paper on Markovian representation of stochastic processes.

5. Nine papers represent Akaike's research after 1977 on Bayesian statistical analysis and entropy methods of statistical inference whose goal is the practical application of Bayesian models by developing criteria to compare competing models and priors.

6. Three papers written since 1985 exposit Akaike's philosophy of statistical thinking. The review paper "Prediction and entropy", published in *A Celebration of Statistics*, summarize the statistical literacy requirements of a modern statistician who wants to benefit from Akaike's research strategy: AIC, Bayes procedure, entropy, entropy maximization principle, information, likelihood, model selection, predictive distribution.

Many researchers at the Institute of Statistical Mathematics and throughout the world profitably apply information methods of statistical inference developed by Akaike as the result of many years of successfully applying statistics to important real problems. We hope that these "Selected Papers of Hirotugu Akaike" will stimulate future generations of statisticians and applied researchers to learn and teach Akaike's philosophy of statistical research.

Emanuel Parzen

Reprinted by permission of
Institute of Mathematical Statistics

Statistical Science
1995, Vol. 10, No. 1, 104–117

A Conversation with Hirotugu Akaike

David F. Findley and Emanuel Parzen

Abstract. Hirotugu Akaike was born in Fujinomiya City, Shizuoka Prefecture, Japan on the fifth of November 1927. He studied at the Naval Academy of Japan, the First Higher School and the University of Tokyo, where he earned his B.S. degree and his external Doctor of Science degree, both in mathematics.

After receiving his bachelor's degree in 1952, he was hired by the Institute of Statistical Mathematics, which had been founded eight years earlier by the Japanese government. He was Director of the institute's Fifth Division, concerned with time series analysis and control, from 1973 until 1985. When the institute was reorganized as an interuniversity research institute in 1986, he became a Professor and Director of the Department of Prediction and Control. In 1987, he became Director General of the Institute, the position from which he retired on March 31, 1994. He was also Professor and Head of the Department of Statistical Science of the Graduate University for Advanced Studies, an independent university whose departments are distributed among the 11 interuniversity research institutes, from 1988 until 1994.

He has held visiting positions at a number of universities: Princeton (1966–1967), Stanford (1967, 1979), Hawaii (1972), the University of Manchester Institute of Science and Technology (1973), Harvard (Vinton Hayes Senior Fellow in Engineering and Applied Physics, 1976), Wisconsin–Madison (Mathematics Research Center, 1982) and several Japanese universities.

His honors include two major technology prizes, each shared with one or more collaborating engineers: with Toichiro Nakagawa, he was awarded the 1972 Ishikawa Prize for modernization of production management by the Ishikawa Prize Committee of the Japan Union of Scientists and Engineers; and, with Hideo Nakamura and others, he received the 1980 Okochi Prize of the Okochi Memorial Foundation for contributions to production engineering. In 1989, he was the recipient of two of Japan's most respected culture and science awards, the Purple Ribbon Medal given by the Emperor of Japan and the Asahi Prize of the Asahi Shimbun Foundation, awards which recognize writers and artists and other citizens as well as inventors and scientists for distinguished contributions to Japanese society. He was a member of the Science Council of Japan from 1988 to 1991.

He has published more than 140 papers and several monographs and textbooks. His 1972 monograph with T. Nakagawa on the statistical analysis and control of dynamic systems has been republished in English translation (Akaike and Nakagawa, 1988). To indicate the magnitude of the impact of the methods described in this book, Professor Genshiro Kitagawa kindly provided us with a table from an article published in Japan in February 1994 listing the outputs of electric power plants in Japan that were built to be controlled by statistical models based on these methods. The table shows these plants generated approximately 12% of Japan's electrical power obtained from nonnuclear and nonhydroelectric sources.

The initial conversation, in which David Findley and Emanuel Parzen spoke with Professor Akaike, took place in May 1992 at the University of Tennessee in Knoxville during the "First U.S.–Japan Conference on

Note added by the editors of this volume:
In lines 13 and 14, 1986 and 1987 should
read 1985 and 1986, respectively.

the Frontiers of Statistical Modeling: An Information Approach." Findley later obtained clarifications and amplifications of some points from Professor Akaike during visits to the Institute of Statistical Mathematics in Tokyo in March 1993 and February 1994.

EDUCATION

Findley: Hiro, Manny Parzen and I are very pleased to have this opportunity to explore some aspects of your scientific development and career. The success of your methods in applications in many fields has helped us to understand more fully what statistical science and statisticians can contribute to other fields. To begin, could you tell us something about your education and the decisions that led to your becoming a statistician and time series analyst?

Akaike: The part of my education that is closely related to my present career started perhaps when I was a student at the Naval Academy during the wartime, because I learned something about statistics there, for example, fitting a straight line by using the method of least squares. After the war I was interested in going to the University of Tokyo to learn about atomic physics, but I got the impression that research on atomic energy was being discouraged in Japan. So I planned to approach the subject indirectly through electrical engineering. But in the year when I wanted to begin, age restrictions prevented me from taking the entrance examination to the university. So I followed a general science curriculum at the First Higher School (Daiichi Kōtō Gakkō) and afterwards went to the University of Tokyo to study mathematics. At that time I was very much interested in learning how to solve nonlinear differential equations and related problems so that I could help engineers. But the mathematicians were more interested in finding conditions for the existence of solutions and similar abstract issues, so I looked outside pure mathematics and started learning probability to prepare for statistics.

I came to the Institute of Statistical Mathematics in 1952. Since I was very much interested in radio receivers and the related electronics, it was natural that I developed an interest in the subject of time series. Consequently, from the very beginning of my career, I was doing research in time series.

David F. Findley is Principal Researcher at the Statistical Research Division of the Bureau of the Census, Washington, DC 20233-9100. Emanuel Parzen is Distinguished Professor of Statistics at Texas A&M University, College Station, Texas 77843-3143.

Findley: What led you to the institute?

Akaike: I was trying to get a job with the government as a statistician. Through an interview with the National Personnel Authority I almost went to the Ministry of International Trade and Industry, but then Professor Zyoiti Suetuna suggested that I go to the institute and meet Dr. Chikio Hayashi (who later became its director general). I knew a little bit about the institute through their publications. Also, Dr. Hayashi's group was making significant contributions to the field of social surveys and sample surveys. Since I saw they were doing practical applications, I thought I could learn something there.

Parzen: At that time, was anyone at the institute doing work in time series analysis?

Akaike: Not specifically, but Dr. Masami Sugawara, who later became Director of the Research Center for Disaster Prevention, was doing an analysis of river flooding using a kind of simulator with fluid. That study was very interesting to me. Probably he was the only person working with actual time series. Of course there were people who were interested in some theoretical aspects.

Findley: Did you have any training in time series analysis before you went to the institute?

Akaike: No. I had read Cramér's book on random variables and probability distributions in a seminar with Professor Suetuna during my last year at the university and had attended probability courses given by Professor Yukiyoshi Kawada in my first and last years. That was the extent of my education related to statistics.

Findley: So you learned the techniques of time series analysis on your own, perhaps through reading some of the available literature?

Akaike: I read everything I could find. There were chapters on time series analysis in some Japanese books on statistics, and I had a chance to look at Wold's book.

FIRST EXPERIENCES WITH MODELING

Findley: In your writings, you have argued forcefully for the important role played by models in the development of statistical solutions to applied problems. Could you tell us something about your early experiences with statistical models, experiences which might have brought you to this view?

Akaike: I'm not quite sure what the main reason is for my interest in modeling, but maybe I have some instinctive tendency to treat problems through modeling because I am very engineering-oriented. In the time series literature I read then, only such things as the estimation of autocovariances were discussed. I was more interested in developing statistical models which could be manipulated. Particularly when I did an analysis of stock price series, I felt ordinary time series analysis was almost useless. You need to have very detailed knowledge about the time series itself before you apply any established technique. I was really looking for the structure of individual time series.

This was the state of mind or psychological situation in which my work on modeling began. My first opportunity to apply modeling to a practical problem came when Mr. Akinori Shimazaki of the Sericultural Experiment Station of the Ministry of Agriculture (now a professor at Shinshu University) visited the colleague I shared an office with. I overheard that he was having trouble applying ordinary control chart techniques to the process of winding filaments of silk from bunches of cocoons into a thread and onto a reel. I felt that a model which I had developed for the analysis of traffic flow in the street could be applied to this problem, and I went to Mr. Shimazaki to propose that he try this model. This was quite successful.

Findley: So modeling traffic flow was the first successful modeling experience that you had?

Akaike: In a sense. I did this just to check my idea that there must be some structure in the flow, which I treated as a very simple stationary time series of ones and zeroes: one car in the sampling interval or no cars. By assuming the independence of the lengths of time intervals between cars, I was able to derive the structure of the series, which I called a gap process (Akaike, 1956). It is a stationary renewal process in discrete time. Essentially the same structure was applicable in the silk reeling process to the dropping ends of exhausted silkworm cocoons. From the test reeling carried out to determine the proper boiling condition for the cocoons, the distribution of filament lengths and the structure of the gap process were estimated. The time series of the number of dropping ends under normal operating conditions is represented by the sum of as many gap processes as there are cocoons. The resulting process provides a basis or reference process to use for the detection of abnormalities in the actual reeling process (Akaike, 1959). Shimazaki tested and implemented this procedure with so much success that it enabled him to earn the first doctorate in agriculture awarded for work in the area of sericultural engineering by the University of Tokyo, and it eventually led to further research which changed the method of silk production in Japan. Afterward, he became Japan's leading researcher in the area of silk manufacturing and received several prizes. This important real proof of practical applicability gave me confidence in the power of models.

Findley: If I have understood you correctly, the work you did modeling traffic flow was a project you chose for yourself as an exercise in statistical modeling. This suggests that you had considerable freedom in the choice of projects you worked on.

Akaike: Definitely. I think this was a very fortunate situation.

CONTACTS WITH ENGINEERS VIA THE FREQUENCY DOMAIN

Findley: Did the institute have many contacts then with industrial researchers?

Akaike: No. When I entered the institute I was in the division mainly concerned with the analysis of social phenomena. But the director of the division, Dr. Hayashi, didn't force me to work in only that area. Then I was put in charge of a newly created section for the study of time series in another division. Afterward, I was able to develop contacts with people in industry, many of whom are now my friends. But it took almost 10 years for me to develop substantial contacts, and this happened largely by coincidence.

Actually, when I entered the institute, there were several people from outside the institute I would get together with to study statistics, at the house of a friend, Ichiro Kaneshige. After Kaneshige began to work for the Isuzu Motor Company, I was always suggesting to him that he try some statistical techniques. Then, at his request, Isuzu sent him to do research for a year at the Tokyo Institute of Technology, and he told me that the subject he had chosen was the analysis of random vibration by spectral techniques.

Findley: And you worked with him on this topic, and this led to a successful procedure for analyzing suspension systems?

Akaike: Yes. He got a prize from the Automobile Technology Association for his work on the analysis and design of a suspension system by statistical methods. This was the first practical application of a procedure for the estimation of frequency response functions of linear systems that I developed with Y. Yamanouchi of the Transportation Technical Research Institute of the Ministry of Transportation (Akaike and Yamanouchi, 1962; Akaike and Kaneshige, 1962). The original, nonstatistical approach to estimating this function requires using many sinusoidal inputs to the system at different frequencies. But this is often not practical, for example, when you are steering a car on the road! Then you need to work with the natural input, from normal driving. N. Goodman had published a statistical estimation approach in 1957, but in his paper little attention was given to the effect of the window chosen for the smoothing required for spectrum estimation. In a 1960 paper, I showed that this effect was the source of the severe underestimates of the gain function obtained by Darzell and Yamanouchi when they used Goodman's procedure, and I proposed a solution. The 1962 paper with Yamanouchi gives a full theoretical analysis.

Findley: Did you have to know anything about automotive engineering to participate in this research?

Akaike: No. But I thought I needed to have some kind of direct feeling for random vibration. So I bought a motor scooter and drove around Mount Fuji.

Findley: And that was useful?

Akaike: Yes. I could easily see how the surface of the road changes when it is used by heavy trucks, and, of course, I could see the interaction between the frequency characteristics of the suspension system and the wavy pattern of the road, because these roads are unpaved.

Findley: Do you feel that this was essential for your success with the project?

Akaike: Not in this case. But still, it's very important to have some direct feeling for the subject, as I did in the case of silk reeling, due to my father's raising silkworms.

FEEDBACK ANALYSIS REQUIRES BOTH FREQUENCY AND TIME DOMAINS

Findley: You continued to work on frequency domain estimation techniques for a while, but when you started to develop diagnostics for systems with feedback, you found it necessary to return to modeling in the time domain. This work on feedback systems was very fruitful. It ultimately led to your various FPE (final prediction error) order selection criteria, which in turn stimulated you to develop your general model selection criterion AIC (an information criterion). But this research first produced your relative spectral power feedback diagnostic and your impulse response estimates for feedback, powerful tools that are hardly known to statisticians outside Japan. The only place I have seen them in a western journal read by statisticians is in your recent partially tutorial paper with Takao Wada and others in *Computers and Mathematics with Applications* (Wada et al., 1988). So it would be useful if you could talk at some length about this work.

Akaike: We had a special research project on the practical application of frequency response function estimation techniques. I was able to get some support for this from the Ministry of Education, which funded over and oversaw the institute then as now. I asked engineers in various fields to come try our method of frequency response function estimation with their own data, which they could bring on an analog tape. I had developed a switching circuit to feed a two-variate time series into our computer from such a tape through an analog-to-digital converter.

The main results were reported in a supplement of the *Annals* of our institute in 1964. I think these reports convincingly showed the practical applicability of statistical techniques for the estimation of frequency response functions. But I recognized that there are many important problems where there is feedback from the output to the input. Then the basic assumption for the application of these techniques breaks down. I tried to find some resolution of this difficulty, but it took a long time.

I reported my first analysis of the failure of the frequency domain methods at the 1966 symposium at Madison, Wisconsin, on "Spectral analysis of time series" (Akaike, 1967). At that time I wasn't successful yet. It was early in the next year, after I finished my visit at Stanford, where you had invited me, Manny, that I finally developed an idea how to treat this problem (Akaike, 1968). I was forced to come back to the time domain because the condition of physical realizability, concerned with the fact that effects propagate forward but not backward in time, was not easy to implement with frequency domain techniques. This finally led me to the general use of multivariate autoregressive modeling for both analysis and control in this type of problem.

Findley: Did you have any practical experience with feedback?

Akaike: Yes. By the time I developed this analysis technique with multivariate time series models, I was already designing my own audio amplifier with multiple feedback, and I experienced much difficulty in designing the proper feedback.

Findley: Did you have a specific application in mind at the time you developed these techniques?

Akaike: I had already begun working with Dr. Toichiro Nakagawa from Chichibu Cement Company to find a method for the control of an industrial cement rotary kiln. The basic model I used for a feedback system includes a model for each variable of the system in terms of its own past and the contemporary and past values of the other variables, together with a driving noise. Each variable's driving noise process is presumed to follow an autoregressive model and is assumed to be uncorrelated with the noise processes of the other variables. This is a crucial assumption which makes it possible to represent the spectrum of each system variable as a sum of contributions, one from each of the noise processes. This is the source of the diagnostic you mentioned. It enables you to see which variable plays the dominant role in generating the fluctuations of another variable in a particular frequency band.

This idea has been useful in developing an understanding of the behavior of a variety of feedback systems. One successful application was made by Dr. Kohyu Fukunishi at Hitachi Ltd.'s Atomic Energy Research Laboratory. He did an analysis of the abnormal behavior of a nuclear power plant by comparing the relative spectral power contributions for normal and abnormal operation. He was able to confirm the correctness of a guess about the source of the abnormal fluctuations. His paper was published in *Nuclear Science and Engineering* (Fukunishi, 1977), and I heard that it led to his receiving an honor as the author of the best paper published in this area in the journal in the preceding three years.

In a very recent application, Wada and other medical scientists (Wada, Sato and Matsuo, 1993) were able to use the spectrum diagnostic and the impulse response functions calculated from the feedback model to solve a long-standing problem of differentiating the roles of chloride and potassium in the development of metabolic alkalosis (above-normal concentrations of bicarbonate in the blood and in other body fluids). The analysis showed a significant contribution of chloride concentration to bicarbonate concentration but none in the opposite direction.

I think this technique can be applied with little effort once you have an appropriate record of observations. You then get good insight into the system. I keep insisting that analysis before control is very important. Some people think only of control and immediately apply control methods, but it's dangerous. Unless you really know the structure of the fluctuations and introduce appropriate improvements to the system itself before implementing the control, you might not get good results. Yes, before you take your medicine, you had better adjust your physiology by adhering to regular activity! I believe that these diagnostics could also be applied to determine some of the feedback structure of economic time series.

Parzen: When you did this work on feedback and control did you have a staff to help you? Did you have assistants to do numerical work for you and things like that?

Akaike: Yes. At the time we were always helped by assistant researchers who were quite capable at doing programming and statistical calculations. I was already being helped by Miss Arahata, who has assisted me for many years.

FINAL PREDICTION ERROR ORDER SELECTION CRITERIA

Findley: It was in the course of this work that you developed two of your final prediction error criteria, FPE and MFPE, for selecting the order of scalar and vector autoregressions. What motivated this development?

Akaike: In the feedback model, a model order, meaning the maximum time lag of past values used, must be specified both for the system variables and for the noise autoregressions; that is, two orders are needed to specify the basic model. (It became one order later with the use of multivariate autoregressive models.) I suggested the use of this basic model to the people at Chichibu Cement Company, where Dr. Nakagawa was working. They got some interesting results, but were always having trouble fixing these orders. They kept asking me by telephone how to handle this problem, and it was very difficult even for me to make clear decisions.

I decided that the simpler situation is the modeling of the noise itself. This is just an ordinary stationary autoregressive model. But if you have a procedure for the choice of the model order, then you get an estimate of the power spectrum automatically. That means you don't need a statistician to get an estimate of the power spectrum. This is quite serious, you know! (laughter) So I was a little bit dubious about the idea of developing a definite procedure for the choice of model order. But I remembered from my visit to Princeton that people in the U.S. are pragmatic in the sense that if they can get a reasonable result, they think it's okay.

So rather than sticking to a strict attitude requiring precise justifications, I decided if I could produce a fairly reasonable answer, then that would be sufficient. I tried various possible criteria and did an enormous amount of numerical computation. Eventually, I recognized that the final use of the model is for prediction, and I thought that the expected value of the mean square error of prediction of an independent realization would give me a kind of basic criterion for the choice of the order. So I produced an almost unbiased estimate of this criterion. By watching the behavior of this estimated value of expected one-step-ahead-prediction-error variance in simulations first, I could see that there was a dip at the proper order and after that the output became very stable. Other criteria I had considered before this one had worked well with some particular model but not well with others. This one worked well with various choices of models, so I thought I had finally found a good criterion.

When I extended the idea to multivariate time series, I recognized that there are several measures of one-step prediction error, and I chose as basic criterion the generalized variance, the value of the determinant of the covariance matrix of the one-step-ahead prediction error. When I thought more about the estimated value of this covariance matrix, I recognized that the log of the estimated generalized variance appeared in a paper by Whittle in the formula for the asymptotic maximized log-likelihood of the Gaussian model. That was when I first recognized that there was some connection with the maximum likelihood concept. After I developed the appropriate FPE criterion, the engineers at Chichibu used it for order selection without my help.

FIRST CONTACTS OUTSIDE JAPAN

Findley: A few moments ago you mentioned that in the United States you found a more practical attitude towards statistical methods, which helped you. Could you elaborate?

Akaike: Yes. This may not be the general attitude of American statisticians, but when I visited Princeton University in 1966 and 1967 at Professor John Tukey's invitation, I attended a lecture by him. He was talking about how to handle abnormal observations in a time series, and his attitude toward this type of problem was quite pragmatic. He wasn't following any particular model of outlier but he was developing reasonable procedures to eliminate irregular observations. Even though this was not completely theory-based, it was very reasonable. My position at the time was either the rather theoretical one that unless we have a definite reason to eliminate these observations, we shouldn't eliminate them, or an extremely practical one. I remember saying when asked how to handle this type of problem that you only have to use a digital-to-analog converter with an audio amplifier and listen to the sound: if there are abnormal observations you will hear a clicking sound. I recall that John Hartigan enjoyed this answer. So I was alternating between the two extremes, very practical or very strict, but Professor Tukey was in-between, developing reasonable analyses and also producing very reasonable ideas about how to handle such problems. I thought maybe this is a good attitude. Consequently, during my later work on autoregressive model selection, I thought that even if I couldn't really get a definitive solution to the problem of the choice of order, if I could find a solution which produced a reasonable answer in many situations, then that would be sufficient, even if I couldn't prove any kind of optimality. This attitude helped me very much.

Parzen: This story really begins in 1965. There was an event called the U.S.–Japan Joint Seminar on engineering applications of stochastic processes, which I want to ask about next. But first I want to mention that, as a consequence of that seminar, I invited you to come to the United States. Then John Tukey heard about your planned visit to the U.S. and asked you to spend the first part of the trip in Princeton, which you did. Now let's go back to the U.S.–Japan Joint Seminar. Could you tell us something about the origin of that? What I know is that Frank Kozin received a grant to conduct a U.S.–Japan joint seminar but didn't know what topic to select. Will Gersch is a friend of Frank Kozin, and Will had read a review I wrote in *Mathematical Reviews* of your work on spectral estimation. On the basis of that, they decided you were the person in Japan to contact to organize the Japanese side of the seminar, and you were contacted by Frank Kozin. Take it from there.

Akaike: I received a letter from Frank Kozin. He was visiting Taiwan at the time and asked me to be something like the chairman for the seminar. But since I was too junior, I had to find someone else who could do this. I talked to my friends in engineering and they decided to ask Professor Takashi Isobe of the University of Tokyo to organize this activity. He and some other engineers proposed this seminar to the Japan Society for the Promotion of Science, JSPS. It was accepted, and the U.S.–Japan Joint Seminar took place in 1965. Manny came, and Professor Drennick from Brooklyn Polytechnic Institute and Professor Ho from Harvard.

This was the beginning of my contact with the outside world. Before that time I was only interested in developing statistical techniques based on the demand within Japan, because unless you have a real problem close at hand, you cannot develop any good ideas.

Parzen: When I invited you to come to the United States, what was your feeling about that?

Akaike: Well, we had already met and I had found that there were many good colleagues in the time series area. I felt this was an opportunity for me, because in Japan I was rather isolated. There were not many time series analysts, particularly time series analysts interested in engineering, an extremely difficult situation for me. Sometimes it is very stressful when you are exploring a new area and you feel that you are alone.

STATISTICAL CONTROL OF COMPLEX INDUSTRIAL PROCESSES

Findley: Your feedback diagnostics and FPE criteria evolved as powerful new tools around which you and Dr. Nakagawa developed a statistical-model-based approach for controlling industrial processes. This approach has been successfully applied to many processes too complex for classical, differential-equation-based control procedures. How did this work with Dr. Nakagawa begin?

Akaike: We met around 1961 at a meeting of the Society of Instrument and Control Engineers where I insisted to the speaker that he did not have enough data to resolve different peaks in a spectrum the way he wanted to. Doctor Nakagawa was sitting near me and decided to ask me to help him to clarify the characteristics of a measurement device he had developed which used a kind of nonlinear noise reduction technique. He then came to Tokyo from Chichibu to talk with me. After we began to work together I asked him to provide me with a problem related to control. I wanted very much to demonstrate the usefulness of statistical methods for controlling a modern industrial process. But it took several years until he brought me the record of operation of a cement kiln. The paper on feedback I presented in Wisconsin in 1966 was a report about our initial work with such data. This work finally gave me the complete idea of a multivariate system with feedback, which we implemented into a linear quadratic Gaussian control approach.

Findley: This is described in your 1972 book with Dr. Nakagawa, which is now available in English (Akaike and Nakagawa, 1988).

Akaike: Yes. By that time we had a complete success proving the effectiveness of this approach, because the control realized by this approach showed a significant reduction of the power spectrum, compared with human operator control, in the very low frequency range. This means that long-lasting movements away from the desired state of the system were reduced.

Findley: I believe that your work with Dr. Nakagawa is of historical significance for statistics. Some of the applications of your book's methods and software have been extraordinary. At the conference this week, we heard another of your collaborators, Dr. Nakamura, mention that already nineteen 500-megawatt and larger thermoelectric power plants, 16 in Japan, one in Canada and two in China, have been built to be controlled by such statistical models, with more plants still under construction in Japan.

Akaike: I was extremely pleased when I saw the operation of the first power plant that used a controller designed by this technique, which was built by the Kyushu Electric Power Company, the employer of Dr. Nakamura at that time. The plant engineer there told me that the quality of control achieved was very good, and they were quite appreciative of the controller's performance. I was particularly delighted to learn, after the plant had been in operation for about a year and they had to perform some maintenance work, that even after this work the controller still performed quite well without much adjustment. This meant that the whole system was quite robust. I think that this point was also mentioned by Dr. Nakamura in his talk (Nakamura, 1994). You could very easily see that when the control was on, the whole system became stable, and the plant's ability to track the changing electricity load demand was increased.

Findley: Doctor Nakamura also mentioned that the controller is easy enough to use and so effective that someone can successfully manage the operation of the plant without having a detailed engineering knowledge of the plant. Doctor Nakamura feels that this property of the controller insures that it can be used for many years.

Akaike: Doctor Nakamura spent perhaps four or five years just to implement this type of control on a power plant. He even refused a promotion in order to be able to continue the research on this.

Findley: In 1972, you and Dr. Nakagawa were honored for your joint work with the Ishikawa prize given by the Ishikawa Prize Committee of the Japanese Union of Scientists and Engineers. Could you tell us about this prize?

Akaike: I think the prize is normally given to a research group, or sometimes to institutions, who develop significant contributions to the modernization of production management. I believe that our work on the control of cement kiln production and particularly its general formulation for other possible applications was considered to be a significant contribution in this area.

Findley: Were you the first statistician to receive this honor?

Akaike: I'm not sure; but until that time not many individuals received the prize. Of course there were two of us, Dr. Nakagawa and myself.

Findley: And in 1980, Dr. Nakamura and you were awarded the Okochi prize of the Okochi Memorial Foundation for the successful work on power plant control described in your joint paper in *Automatica* (Nakamura and Akaike, 1981). What kind of prize is this?

Akaike: Someone connected to the awarding of this prize told me that it is something like a Japan Olympic medal for production engineers. So, in this case, I was considered an engineer.

THE PATH TO AIC

Findley: How were you led to discover AIC?

Akaike: I was interested in the factor analysis problem. I don't know why, but maybe this was because of some numerical aspect of the model. Factor analysis requires numerical optimization of the likelihood function to get the solution, and I have always been interested in numerical procedures. I have one paper in this area.

Parzen: In what year?

Akaike: In 1959.

Parzen: When you came to Stanford, Professor Forsythe knew your name very well. Is this the paper that he knew about?

Akaike: Yes. The paper gave a mathematical proof of the convergence of the error term of a gradient "hill climbing" optimization method, irrespective of the dimension of the function's argument, into a two-dimensional region, a peculiar phenomenon. Professor Forsythe had a paper in the *Pacific Journal of Mathematics* verifying this behavior for the special case of a three-dimensional optimization and commenting that even this special case required a very complicated mathematical treatment. I got interested, and one day the idea came to me that this method could be interpreted as a transformation of a probability distribution on the eigenvectors of the quadratic form associated with the locally quadratic approximation to the function. This is the kind of approximation used in my derivation of AIC also.

While I was thinking about factor analysis, I got an invitation to the Second International Symposium on Information Theory from Dr. Tsybakov, who was in Moscow. I recognized that there was a similarity between the choice of the number of factors in factor analysis and the determination of the order of an autoregression, but I couldn't really find any definite connection between the two. For example, in the case of autoregressive models, the idea of prediction is clear. You just compute the predicted value of the next time point. But what are you doing when you fit the factor analysis model to data? What is the prediction in that case? I wasn't sure about this analogy, and since there was a time limit for the submission of the manuscript to the symposium, my attention was split by two separate efforts: preparation of a manuscript for the symposium concerning my experience with autoregressive models for spectral estimation; and finding a good concept for the similarity between the autoregressive modeling and factor analysis problems. I was under very much pressure, psychological pressure, so during the nighttime I often woke up thinking and during the daytime I was almost sleeping.

One day I recognized that the factor analysis people are maximizing the likelihood and this maximization is trying to get a good distributional model for the purpose of prediction. However, in this case the prediction is not a value, but is the fitted distribution itself, which is applied to understand the next similar observation. For the next similar problem you use the present model, and the accuracy criterion for this prediction is given by the log-likelihood function. Then once I got this far, it was just one step to recognize that the expected log-likelihood is related to the Kullback information. This idea came when I was standing on the train from my home to the institute. I still have the page of my notebook where I wrote down one or two lines to explain this. That was the solution.

Findley: This was in 1971?

Akaike: In March of 1971. This experience is described in the historical note that the Institute for Scientific Information asked me to produce (This Week's Citation Classic, 1981).

Parzen: Let's go over those references at this point. I see in your vita that you have two short papers in the proceedings of the Fifth Hawaii International Conference on System Sciences, one entitled "Uses of an information theoretic quantity for statistical model identification" and the other "Automatic data structure search by the maximum likelihood." So you announced some of these ideas in Hawaii?

Akaike: Yes. Actually I first announced the idea in the annual meeting of Japan Statistical Society in July 1971. Then I went to Armenia to the Second International Symposium on Information Theory in September and gave a talk on this subject. Professor Tom Kailath from Stanford was there, and he became interested in this idea and asked me to contribute to a special volume on time series analysis of the *IEEE Transactions on Automatic Control* (Akaike, 1974a), where you also have a paper, Manny. After coming back from the symposium, I visited Hawaii for a year, so I prepared a talk on this subject and presented the results of several applications there. The paper that I presented at Tsahkadsor in Armenia was eventually published in 1973, so there was a delay in the publication.

Parzen: This paper was previously very inaccessible but is now available with an introduction by Jan de Leeuw in the first volume of *Breakthroughs in Statistics* (Johnson and Kotz, 1991, pages 599–624). I should also point out that the *IEEE*

Transactions paper was listed as one of the most frequently cited papers in the area of engineering technology and applied sciences, and, of course, it's cited and applied in almost every area of physical and social science. How did you become aware of all the different fields in which people were applying AIC?

Akaike: I received papers from many people reporting the results of their applications of AIC in various areas. After I published the paper in the *IEEE Transactions*, I also wrote a paper for the popular Japanese magazine *Mathematical Science*. This explained how the idea was developed and how one can apply it to practical problems. These two articles were the source for the spread of the criterion. People developed their own applications just from reading these papers. The idea is so simple.

Findley: From looking at citation indexes, it appears that the use of AIC by statisticians developed more slowly than its use by other researchers. How did you feel about the negative reactions of some statisticians to AIC?

Akaike: My main concern was the fact that there were so few statisticians with sufficient experience in handling real problems. Lacking such experience, they could not check the validity of the basic idea of AIC on their own problems. Many had such a dogmatic attitude that they did not even question the use of a maximum likelihood estimate when nothing was known about the "true" form of the distribution, yet they criticized the use of a minimum AIC estimate. People with serious statistical problems, like applied engineers, could easily appreciate the contribution of AIC simply by getting useful answers to problems which could not be handled by a conventional statistical approach.

Findley: The first paper I know of in a statistical journal to use AIC is the one Dick Jones published in 1975 in the *Journal of Statistical Computation and Simulation*. This paper is concerned with an application to biostatistics (Jones, 1975a). He also applied AIC to meteorology (Jones, 1975b). He must have learned about it from you during your visit to Hawaii.

Akaike: Yes, in the first version of his paper, he called it something like ASC for "Akaike selection criterion." I thought this was not very appropriate. I just used IC, information criterion, in the paper on Markovian representations of time series (Akaike, 1974b), but when I was preparing my manuscript for the *IEEE Transactions* my assistant Miss Arahata was doing some programming for me. I asked her to calculate some values of the IC criterion. She was programming in FORTRAN and needed to put a different letter at the beginning of IC since it has a noninteger value, so she put the A in front of it. I thought this might be a good idea as it still suggested an information criterion. So in the paper for the *IEEE Transactions* I used AIC to denote this criterion, and also suggested the name minimum AIC estimate, MAICE. Of course, I was aware that there would be a succession of criteria, AIC, BIC, . . . , and minimum BIC would be MBICE, etc.

OTHER CRITERIA

Findley: Around 1976, you developed a criterion for regressor selection you called BIC which gives consistent estimates of the correct regressor in overparameterized linear regression situations.

Akaike: That BIC criterion, which I presented in Dayton, Ohio, in 1976 (Akaike, 1977b), was derived by a Bayesian argument with Gaussian likelihoods. For regression, it is asymptotically equivalent to the criterion independently obtained by Gideon Schwarz. He used a Bayesian argument in the more general situation of models from a Koopman–Darmois exponential family to derive his criterion, in which the term added to minus two times the log maximum likelihood is the number of parameters multiplied by the logarithm of the sample size, instead of by the number 2 as in AIC. Earlier, I had considered a modification of FPE to obtain a consistent estimator of the order of a finite-order autoregression, the consistency property that Schwarz's criterion has (Akaike, 1970).

Parzen: What about Mallow's C_P? Could you discuss the connection with C_P?

Akaike: I must confess that I was not a good reader of journals, so I didn't know at first about Mallow's C_P criterion. Only after I wrote the original paper on AIC and was preparing the manuscript for the *IEEE Transactions* did I suddenly realize that there were two papers whose ideas were very close to my own. Mallow's 1973 paper was one, and the other was a 1966 paper by Davisson (Davisson, 1966). Davisson discussed the problem of order selection, but proposed no definite procedure. And, of course, Mallow's criterion is just for scalar regression problems.

Parzen: Davisson was the first person to give the formula for the effect of parameter estimation on one-step prediction mean square error.

Akaike: Yes. Since I was only interested in practical applications, when forced to do some theoretical thinking I usually developed my own ideas from scratch. Of course, there are similarities or connections with the ideas in these two papers, but the idea of using log-likelihood as a general criterion was quite new.

STATE SPACE AND SOFTWARE

Parzen: I'd like to bring in one other visit that seems to have been significant. In 1973, just after the development of AIC, you visited the University of Manchester Institute of Science and Technology, invited by Professor Maurice Priestley, and you wrote two very important papers there. In them you used state space ideas to solve the identifiability problem for representing general vector ARMA processes, and you developed a minimal representation. These papers are cited in leading textbooks and monographs. Please talk about this period.

Akaike: Actually, Professor Priestley and his group were interested in my work on feedback system analysis. He already had a paper in *Automatica* on this subject. When I visited UMIST, they were surprised to find me working in the time domain. They expected a frequency domain person. While I was in Honolulu, I had almost completed the basic framework for the state space Markovian representation of time series based on the concept of predictor space. I had already finished my work on the canonical correlation basis interpretation, but I had to write up the final results for the multivariate ARMA model. When I was in Manchester, I produced two technical reports, one on Markovian representations of stationary time series and the other on stochastic realizations. These were motivated by an earlier contact with Professor Rudolf Kalman (Akaike, 1977c).

Findley: Your TIMSAC 74 programs implementing these procedures were the first widely distributed programs for multivariate autoregressive-moving-average modeling. It must be said that the large number of programs published by you and your collaborators is a very impressive body of research software. There are the four TIMSAC, or "time series analysis and control," packages, and there are programs on other topics such as density estimation and contingency table analysis. The University of Tulsa, where I once taught, has distributed over 700 copies of the TIMSAC software. The development of this software clearly took a great deal of your time.

Akaike: Right. It also involved many younger members of the institute as coauthors. The programs were developed initially to check the applicability of our ideas to actual problems. I recognized that such programs might be useful to people who read the papers describing our new methods and wished to try them. Finally, I decided to distribute them to as many people as possible, because I thought that through this process people can understand how powerful statistics is.

There is another area of research and software development at the institute which has been very fruitful. That is point process modeling. This was stimulated by a visit by Professor David Vere-Jones. I had asked the Japan Society for the Promotion of Science to invite him. This led to contacts between the institute and researchers in the area of seismology and to his collaboration with Yosihiko Ogata, who has become the institute's main researcher in point-process modeling and its application to earthquake modeling and prediction. Professor Vere-Jones was giving a series of lectures on point processes at the institute when I came back from Harvard, which Ogata and Tohru Ozaki were attending. I asked them what is the most basic component of point process models, and they answered that it is intensity function, which describes the intensity of occurrence of an event at a particular time point. I then encouraged them to develop and fit intensity function models for earthquake data. This has led to very interesting modeling work for various kinds of earthquakes and for their distribution over the Earth's surface. Ogata has concentrated mainly on earthquake occurrences under or near the Japan mainland. He has used smoothness priors to obtain a smooth spatial image of the intensity distribution.

BAYESIAN MODELING

Findley: You have published more than a dozen papers concerned with Bayesian modeling and Bayesian philosophical issues, and most of your modeling examples and applications use the smoothness priors you just mentioned. Please tell us how your Bayesian modeling ideas developed.

Akaike: The idea of using Bayesian modeling was originally motivated by the desire to use the full information that a set of competing models can provide. I also wanted, in this way, to overcome the inadmissibility problems identified so clearly by Professor Stanley Sclove in situations in which a criterion such as AIC is used, for example, to select the order of a polynomial. This led to the discussion of combining models given in my 1979 *Biometrika* paper (Akaike, 1979). When I visited Harvard, I concentrated on Bayesian statistics. The most bothersome aspect was the split between the objectivist and subjectivist statisticians. If subjective information is so different from objective information, how can you combine both kinds? I spent a lot of time researching the historical development of this type of discussion and eventually ended up with the resolution that a Bayesian model is just another type of statistical model for extracting the information provided by the data, in order to obtain a good distribution applicable to a future observation. Then everything became quite transparent. Since this is just a statistical model, you can use the concept of likelihood, and also expected log-likelihood, as the basic criterion for the evaluation of the model. So, if there is an undetermined hyperparameter in the prior, for example, then you can adjust it by the method of maximum likelihood applied to the marginal density. As I discussed in my paper on the objective use of Bayesian models (Akaike, 1977a), this is formally the same as I. J. Good's type II maximum likelihood procedure. But he, by contrast, regards the procedure as a compromise between Bayesian and frequentist approaches with no particular justification.

Once I got this far, it was just one more step to the idea that we need a systematic approach to the construction of priors. It seemed to me it is only our lack of understanding of the nature of the prior that impedes our developing ideas for practical applications of Bayesian models. I was familiar with procedures for ill-posed problems, as described in Tikhonov's paper, for example (Tikhonov, 1965). Usually, somewhat artificial smoothness constraints are introduced. In the seasonal adjustment situation I was considering, this led to constrained least squares. I put an additive quadratic term into an exponent with a minus sign in front and, with the proper normalizing constant, this became a Gaussian prior density. Estimation of the variance ratios in this prior by maximum likelihood led immediately to a seasonal adjustment procedure whose simplicity seemed remarkable to me since it's just a minor modification of generalized least squares. The first output, for an artificial time series published by the Economic Planning Agency, was very encouraging. So I brought this to the First International Bayesian Meeting in Valencia, Spain. That was the beginning of my practical application of Bayesian models. In the paper I presented there (Akaike, 1980), I also showed how parameters for Stein-type shrinkage estimators, ridge regression and Shiller's smoothness-prior-based distributed-lag estimator can be determined by this approach, as well as O'Hagan's localized regression. Of course, in applications, you might need some criterion like the ABIC (A Bayesian information criterion) that I obtained for these models in order to compare competing models and priors.

I don't say that a smoothness prior is necessarily reasonable as a prior. Only that it is a useful prior if it helps you construct a model that extracts information of interest to you from the data.

Findley: Smoothness prior models with hyperparameters estimated by maximum likelihood have been used in a variety of applications by Kitagawa and Gersch, who are writing a book on this topic, and by others, particularly Andrew Harvey and his collaborators, with what they call structural time series models, which seem to have been stimulated to an extent by a Kitagawa and Gersch paper they reference.

Akaike: Actually those later models you mention are based on state space models, which are usually a kind of product of the imagination. Once you have the Bayesian interpretation, the motivation for such models becomes very clear.

Findley: We find your BAYSEA smoothness prior seasonal adjustment program useful at the Census Bureau, especially with time series which are too short for other seasonal adjustment methods. How did you become interested in seasonal adjustment?

Akaike: Because I have expertise in spectral analysis, people in the Economic Planning Agency thought I would be an appropriate person to comment on an extensive study they did of seasonal adjustment procedures. Actually the person in charge of producing this report was attending a course given by our institute for the public. I think I was talking on time series analysis, particularly frequency domain analysis, at that time in the 1960's, so I only gave comments from the point of view of the frequency domain properties of the filters used for seasonal adjustments.

Findley: This was a course offered by your institute for any person from the public who wished to attend?

Akaike: Right. My contact with Dr. Nakamura came through a course of this type.

PREDICTION AND ENTROPY

Parzen: You have given a very rich summary of your ideas on prediction and entropy in a paper with this title in the 1985 ISI Centenary volume *A Celebration of Statistics*, edited by Atkinson and Fienberg and published by Springer-Verlag. Please give us a quick introduction to what you think the importance is to statistics of the ideas of entropy and prediction.

Akaike: Unless we have some reasonable prediction for the future, we cannot choose appropriate actions for the present. So this uncertainty about the future is the source, I guess, for our activities directed toward statistical understanding and analysis of data. In that sense, the predictive point of view is a prototypical point of view to explain the basic activity of statistical analysis. Within statistics, the entropy criterion is quite general and also quite natural. Of course, I don't have any mathematical proof that this is a uniquely natural criterion, but if you look through the history of statistics, you can see, as I discussed in this paper, that when people were coming close to the concept of entropy, they were producing very successful results and when they were far away from this concept, they were not so productive, at least from my point of view. So this is some kind of historical evidence for the productivity of the concept of information or entropy.

Findley: The cost of the entropy or information-theoretic point of view seems to be that it's necessary to construct a model from which a likelihood function can be obtained. If you can do that, then you have criteria with which you can compare competing models and this is, of course, very important. Science is, in part, an arena in which models compete, and the comparison of models is fundamental. But a significant part of the current development of statistics is focused on nonparametric methods which are often not very closely related to likelihood functions. Do you see any possibilities for modifying your conceptual framework so that it can also be used with more nonparametric models?

Akaike: In the past history of science people were always producing results by using some parameterization. For example, Newton's work on mechanics. So this is a very popular and useful point of view. If we can find a very powerful procedure which does not use this kind of structure, then, of course, I can easily change my point of view any time. I am, in that sense, very flexible. When I was young, I was interested in things like nonparametric or distribution-free statistics and the nonparametric approach. But when I looked through the development of science, particularly natural science, I found that many significant developments were related to the putting forward of hypotheses which clearly depend on parameters. So I thought perhaps this is the right way. It's just a subjective judgment.

Also, through the development of specific, appropriate types of models (and by insisting on their development) the accumulation of experience can be accomplished very quickly, and this accumulation becomes a kind of common resource for future development. Otherwise someone develops some arbitrary idea which cannot be used by anyone else. This communicability is very important. So is the portability and the compact parameterization when you are implementing the model for something like on-line control.

FISHER, INDUCTION AND ABDUCTION

Findley: In the first paper you presented at this conference (Akaike, 1994), you compared your views of likelihood with those of R. A. Fisher. It would be interesting for our readers if you could summarize your understanding of the differences.

Akaike: I didn't have much time to study historical developments, but when I read through some of the work of Fisher I recognized that his definition of likelihood was for a fixed model, a distribution whose functional form was given, known, and only the parameters were unknown. As I began to do autoregressive modeling, especially multivariate modeling, I became very aware of the problem of comparing different models. Also I had an impression that the division of theory into estimation and testing was somewhat unproductive in the area of practical applications. If you come to the choice of one particular model from among several alternatives by the conventional approach, you probably have to depend on some kind of test procedure rather than estimation. But to me the problem of determining the best model for the data from a collection of models was quite a natural extension of estimation.

When I was preparing my manuscript for this conference, I remembered a paper by the American scientist and philosopher Charles Saunders Peirce on the topic of induction and "abduction," meaning the generation of hypothesis (Peirce, 1955). As I reread it, I saw clearly that in scientific explorations it is the generation and comparison of hypotheses which is most important. Then it became quite clear to me why people outside of statistics showed such interest in applying AIC as soon as it was published. Their main interest was developing hypotheses based on data. So in that sense, model selection is more closely connected to what Peirce calls abductive logic. Fisher emphasized the use of likelihood in the inductive phase of inference, obtaining information about the population from the available sample under the assumption of a hypothesis, most specifically information about the parameters of the model that determines the likelihood (Fisher, 1935). In actual situations, however, you have to develop various possible hypotheses and compare them based on observations; AIC is directed toward comparison of different models. So I recognized that Fisher and I have different ways of interpreting the likelihood function. But as everyone knows, Fisher had very deep experience and insight into practical aspects of statistics. From that common experience, many of my attitudes are similar to Fisher's.

Findley: Some of your experience as a time series analyst is a kind of experience that Fisher did not have. For example, he did not work with complex dynamic physical systems as far as I know. Do you think that there is a special perspective that comes from time series analysis which has helped you find new approaches to some fundamental questions of statistics?

Akaike: If the models are clearly specified by your own past experience, then there's no need to develop a particular selection criterion or evaluation criterion. However, in time series analysis, there are many applications of linear system theory which are in a sense quite nonparametric. Of course the models are parametric, but, through different choices of the model order or dimension of the system, you get broad flexibility from the models, just like polynomial fitting. So this is a typical situation where you have to make some kind of decision, where you're going to choose one model in a practical application. I think this forced me in the direction of model selection, so I was lucky. Originally I was not very interested in this type of model. It's too general. If it's too general you cannot easily bring your own ideas or insight into the modeling. But it had very wide applicability, so I changed my mind.

DIRECTOR GENERAL OF THE INSTITUTE WITH NEW RESPONSIBILITY FOR GRADUATE STATISTICAL EDUCATION

Findley: You became Director General of the Institute of Statistical Mathematics in 1987, just after it changed from being almost exclusively a research institute to also being an institution of graduate education, the first nonuniversity institution in Japan to grant a Ph.D. in statistics. Could you please share with us some of your ideas concerning the training of statisticians?

Akaike: Statistics is a very difficult subject in the sense that it is essentially related to information, and information has no physical form. This means statistics is related to a subject lacking form, so it's very difficult to explain to society how it is important. From that viewpoint, I think the only solution is to get people from various disciplines to obtain some training in statistics and bring the knowledge back to their own fields. We need to train some professional statisticians, of course, but a very important part of our activity and mission should be directed toward the dissemination of statistical knowledge to other disciplines.

Our applicants come with a Master's degree, or equivalent work experience, from engineering and mathematics and medicine and physics and so forth, so we have to provide some basic training in statistics. But they can start their research in a particular area from the very beginning if they wish. The staff of our institute has such a broad spectrum of backgrounds and interests that we can accommodate the interests of almost any applicant. Some already have previous contacts with the institute. This year two students finished their Ph.D.'s. One got a job at the University of Tokyo in the Department of Mathematical Engineering and Information Physics. Another student, a woman, got a job at the Central Research Laboratory of Hitachi, Ltd. So they both have very respectable places where they can continue their research. So I think we are very fortunate.

Findley: In other words, your students will not receive general training in statistics as much as training that's relevant to the area of their interest. And they will not go back to a university necessarily, certainly not to a statistics department as they might in the U.S.A., because there aren't any statistics departments in Japan. Rather they will go back to departments which have some connection with the background that they had, or to departments or companies that value the background that they've gotten from your institute.

Akaike: This is exactly what we hoped for and it seems that our hope is being realized.

Findley: So you are training statistical scientists rather than statisticians! (Laughter.)

Akaike: Maybe that's a good expression. Yes, of course.

THE INFLUENCE OF THE SECOND WORLD WAR

Findley: Before we close this interview, could you tell us if there is some important way in which the experiences you had during the Second World War influenced your career.

Akaike: I think there are actually two aspects, one quite personal. I developed an interest in mathematics because of an uncle who was killed during the war. He was a Navy pilot and was interested in mathematics. He sometimes talked to me about calculus and other topics in mathematics. The other aspect is that after the end of the war, as the Naval Academy was being closed, the person in charge of the academy stressed the importance of our role in postwar society, that when Japan was recovering from this damage we must do our best to rebuild the country.

I think this was accepted by students at that time without any question. After I started my study of mathematics, eventually I thought I could serve society or help people more directly by doing statistics. Later, when my friends from the Naval Academy would meet occasionally, there was a spread of political attitudes from right to left, but we would just get together and discuss each one's opinion and develop understanding and there was no antagonistic feeling there. They just wanted to be of service to the society I guess. This was a kind of generation-dependent attitude that maybe the people who are students today have moved away from and might call an obsolete attitude. But still I think it gave us a quite stable feeling and strength. There is no serious split among us, because we have some common concern for the society that we can discuss with each other. Maybe this explanation is quite incomplete.

Parzen: I am very happy that I had the opportunity to meet you in 1965 and that we have been in contact all these years. I never thought of us as representing the U.S. and Japan, just two good friends who are working together. I think this is important. International cooperation is based on the personal friendship of the scientists, and we learn from each other in many ways besides our common interest in science. Thank you for this conversation.

Akaike: Thank you. Maybe I have to add just one more point, a point we discussed partially. One reason I could develop this kind of contact in my own research is the type of protection provided by the institute in maintaining my complete freedom of choice of subject and way of developing research. I think this is very important and sometimes very difficult to maintain, this freedom of research, without any concern about promotion and so forth. This is a very unique environment.

Findley: I also want to thank you for this conversation.

Akaike: Thank you very much. Thank you.

REFERENCES

AKAIKE, H. (1956). On a zero-one process and some of its applications. *Ann. Inst. Statist. Math.* **8** 87–94.

AKAIKE, H. (1959). On the statistical control of the gap process. *Ann. Inst. Statist. Math.* **10** 233–259.

AKAIKE, H. (1967). Some problems in the application of the cross spectral method. In *Spectral Analysis of Time Series* (B. Harris, ed.) 81–107. Wiley, New York.

AKAIKE, H. (1968). On the use of a linear model for the identification of feedback systems. *Ann. Inst. Statist. Math.* **20** 425–429.

AKAIKE, H. (1970). On a semi-automatic power spectrum estimation procedure. In *Proceedings of the Third Hawaii International Conference on System Sciences* 974–977. Western Periodicals Company, Honolulu.

AKAIKE, H. (1974a). A new look at the statistical model identification. *IEEE Trans. Automat. Control* **AC-19** 716–723.

AKAIKE, H. (1974b). Markovian representation of stochastic processes and its application to the analysis of autoregressive moving average processes. *Ann. Inst. Statist. Math.* **26** 363-387.

AKAIKE, H. (1977a). An objective use of Bayesian models. *Ann. Inst. Statist. Math.* **29** 9–20.

AKAIKE, H. (1977b). On entropy maximization principle. In *Applications of Statistics* (P. R. Krishnaiah, ed.) 27–41. North-Holland, Amsterdam.

AKAIKE, H. (1977c). Canonical correlation analysis of time series and the use of an information criterion. In *System Identification: Advances and Case Studies* (R. K. Mehra and D. G. Laniotis, eds.) 29–76. Academic, New York.

AKAIKE, H. (1978). A Bayesian analysis of the minimum AIC procedure. *Ann. Inst. Statist. Math.* **30** 9–14.

AKAIKE, H. (1979). A Bayesian extension of the minimum AIC procedure of autoregressive model fitting. *Biometrika* **66** 237–242.

AKAIKE, H. (1980). Likelihood and the Bayes procedure. In *Bayesian Statistics* (J. M. Bernardo, M. H. DeGroot, D. V. Lindley and A. F. M. Smith, eds.) 143–166. Univ. Press Valencia.

AKAIKE, H. (1994). Implications of informational point of view on the development of statistical science. In *Proceedings of the First US/JAPAN Conference on The Frontiers of Statistical Modeling: An Informational Approach* **3** (H. Bozdogan, ed.) 27–38. Kluwer, Dordrecht.

AKAIKE, H. and KANESHIGE, I. (1962). Some estimation of vehicle suspension system's frequency response by cross-spectral method. In *Proceedings of the 12th Japan National Congress for Applied Mechanics* 241–244. Univ. Tokyo Press.

AKAIKE, H. and NAKAGAWA, T. (1988). *Statistical Analysis and Control of Dynamic Systems.* Kluwer, Dordrecht.

AKAIKE, H. and YAMANOUCHI, Y. (1962). On the statistical estimation of frequency response function. *Ann. Inst. Statist. Math.* **14** 23–56.

DAVISSON, L. D. (1966). The prediction error of stationary Gaussian time series of unknown covariance. *IEEE Trans. Inform. Theory* **IT-11** 527–532.

FISHER, R. A. (1935). The logic of inductive inference. *J. Roy. Statist. Soc. Ser. A* **98** 39–54.

FUKUNISHI, K. (1977). Diagnostic analyses of a nuclear power plant using multivariate autoregressive processes. *Nuclear Science and Engineering* **62** 215–225.

JOHNSON, N. L. and KOTZ, S., eds. (1991). *Breakthroughs in Statistics. Volume I: Foundations and Basic Theory.* Springer, New York.

JONES, R. H. (1975a). Probability estimation using a multinomial logistic function. *J. Statist. Comput. Simulation* **3** 315–329.

JONES, R. H. (1975b). Estimation of spatial wave number spectra and falloff rate with unequally spaced observations. *J. Atmospheric Sci.* **32** 260–268.

NAKAMURA, H. (1994). Statistical identification and optimal control of thermal power plants. In *Proceedings of the First US/Japan Conference on the Frontiers of Statistical Modeling: An Informational Approach 3* (H. Bozdogan, ed.) 57–79. Kluwer, Dordrecht.

NAKAMURA, H. and AKAIKE, H. (1981). Statistical identification for optimal control of supercritical thermal power plants. *Automatica* **17** 143–155.

PEIRCE, C. S. (1955). Abduction and induction. In *Philosphical Writings of Peirce* (J. Buchler, ed.), 150–156. Dover, New York.

THIS WEEK'S CITATION CLASSIC (1981). *Current Contents* **51** 22 (December 21). [Also included in (1986) *Contemporary Classics in Engineering and Applied Science* 42 (A. Thackary, ed.) ISI Press, Philadelphia.

TIKHNOV, A. N. (1965). Incorrect problems of linear algebra and a stable method for their solution *Soviet Math. Dokl.* **6** 989–991.

WADA, T., AKAIKE, H., YAMADA, Y. and UDAGAWA, E. (1988). Application of multivariate autoregressive modelling for analysis of immunologic networks in man. *Comput. Math. Appl.* 15 713–722.

WADA, T., SATA, S. and MATSUO, N. (1993). Application of multivariate autoregressive modeling for analyzing chloride/potassium/bicarbonate relationship in the body. *Medical and Biological Engineering and Computers* **31** S99–S107.

List of Publications of Hirotugu Akaike

Papers written in English

1. Note on the decision problem. (with K. Matusita) *Ann. Inst. Statist. Math.*, Vol. 4, (1952) 11–14.
2. On a matching problem. (with C. Hayashi) *Ann. Inst. Statist. Math.*, Vol. 4, (1953) 55–64.
3. An approximation to the density function. *Ann. Inst. Statist. Math.*, Vol. 6, (1954) 127–132.
4. Decision rules, based on the distance, for the problems of independence, invariance and two samples. (with K. Matusita) *Ann. Inst. Statist. Math.*, Vol. 7, (1955) 67–80.
5. Monte Carlo method applied to the solution of simultaneous linear equations. *Ann. Inst. Statist. Math.*, Vol. 7, (1955) 107–113.
6. On the distribution of the product of two Γ-distributed variables. *Ann. Inst. Statist. Math.*, Vol. 8, (1956) 53–54.
7. On a zero-one process and some of its applications. *Ann. Inst. Statist. Math.*, Vol. 8, (1956) 87–94.
8. On optimum character on von Neumann's Monte Carlo model. *Ann. Inst. Statist. Math.*, Vol. 7 (1956) 183–193.
9. On ergodic property of a tandem type queueing process. *Ann. Inst. Statist. Math.*, Vol. 9, (1957) 13–21.
10. On a computation method for eigenvalue problems and its application to statistical analysis. *Ann. Inst. Statist. Math.*, Vol. 10, (1958) 1–20.
11. On a successive transformation of probability distribution and its application to the analysis of the optimum gradient method. *Ann. Inst. Statist. Math.*, Vol. 11, (1959) 1–16.
12. On the statistical control of the gap process. *Ann. Inst. Statist. Math.*, Vol. 10, (1959) 233–259.
13. Analysis of the effect of timing-error on the frequency characteristics of sampled-data. *Proceedings of the 10th Japan National Congress for Appl. Mech.*, (1960) 387–389.
14. Effect of timing-error on the power spectrum of sampled data. *Ann. Inst. Statist. Math.*, Vol. 11, (1960) 145–165.
15. On a min-max theorem and some of its application. (with Y. Saigusa) *Ann. Inst. Statist. Math.*, Vol. 12, (1960) 1–5.

16. On a limiting process which asymptotically produces f^{-2} spectral density. *Ann. Inst. Statist. Math.*, Vol. 12, (1960) 7–11.

17. Undamped oscillation of the sample autocovariance function and the effect of prewhitening operation. *Ann. Inst. Statist. Math.*, Vol. 13, (1961) 127–143.

18. Analytical studies on fluctuations found in time series of daily milk yield. (with H. Matsumoto and Y. Saigusa), *National Institute of Animal Health Quarterly*, Vol. 2, (1962) 161–171.

19. On the design of lag window for the estimation of spectra. *Ann. Inst. Statist. Math.*, Vol. 14, (1962) 1–21.

20. On the statistical estimation of frequency response function. (with Y. Yamanouchi) *Ann. Inst. Statist. Math.*, Vol. 14, (1962) 23–56.

21. Some estimation of vehicle suspension system's frequency response by cross-spectral method. (with I. Kaneshige) *Proceedings of the 12th Japan National Congress for Appl. Mech.*, (1962) 241–244. Theoretical and Applied Mechanics, Japan National Committee for Theoretical and Applied Mechanics Science Council of Japan.

22. Statistical measurement of frequency response function. *Supplement III, Ann. Inst. Statist. Math.*, (1964) 5–17.

23. An analysis of statistical response of backlash. (with I. Kaneshige) *Ann. Inst. Statist. Math.*, Supplement III, (1964) 99–102.

24. On the statistical estimation of the frequency response function of a system having multiple input. *Ann. Inst. Statist. Math.*, Vol. 17, (1965) 185–210.

25. Note on the higher order spectra. *Ann. Inst. Statist. Math.*, Vol. 18, (1966) 123–126.

26. On the use of non-Gaussian process in the identification of a linear dynamic system. *Ann. Inst. Statist. Math.*, Vol. 18, (1966) 269–276.

27. Some problems in the application of the cross-spectral method. *Spectral Analysis of Time Series* (B. Harris ed.) John Wiley, (1967) 81–107.

28. On the use of an index of bias in the estimation of power spectra. *Ann. Inst. Statist. Math.*, Vol. 20, (1968) 55–69.

29. Low pass filter design. *Ann. Inst. Statist. Math.*, Vol. 20, (1968) 271–297.

30. On the use of a linear model for the identification of feedback systems. *Ann. Inst. Statist. Math.*, Vol. 20, (1968) 425–439.

31. A method of statistical identification of discrete time parameter linear systems. *Ann. Inst. Statist. Math.*, Vol. 21, (1969) 225–242.

32. Fitting autoregressive models for prediction. *Ann. Inst. Statist. Math.*, Vol. 21, (1969) 243–247.

33. Implementation of computer control of a cement rotary kiln through data analysis. (with T. Otomo and T. Nakagawa) *Preprints, Tech. Session 66, Fourth Congress of IFAC*, Warszawa, June (1969) 115–140.

34. Load history effects on structural fatigue. (with S. R. Swanson) *Proc. Institute of Environmental Sciences*, (1969) 66–77.

35. Power spectrum estimation through autoregressive model fitting. *Ann. Inst. Statist. Math.*, Vol. 21, (1969) 407–419.

36. Statistical predictor identification. *Ann. Inst. Statist. Math.*, Vol. 22, (1970) 203–217.

37. A fundamental relation between predictor identification and power spectrum estimation. *Ann. Inst. Statist. Math.*, Vol. 22, (1970) 219–223.

38. On a decision procedure for system identification. *Preprints, IFAC Kyoto Symposium on System Engineering Approach to Computer Control*, (1970) 485–490.

39. On a semi-automatic power spectrum estimation procedure. *Proc. 3rd Hawaii International Conference on System Sciences*, (1970) 974–977.

40. Autoregressive model fitting for control. *Ann. Inst. Statist. Math.*, Vol. 23, (1971) 163–180.

41. Automatic data structure search by the maximum likelihood. *Computers in Biomedicine, Supplement to Proc. 5th Hawaii International Conference on System Sciences*, Western Periodicals Company (1972) 99–101.

42. Statistical approach to computer control of cement rotary kilns. (with T. Otomo and T. Nakagawa) *Automatica*, Vol. 8, (1972) 35–48.

43. Use of an information theoretic quantity for statistical model identification. *Proc. 5th Hawaii International Conference on System Sciences*, Western Periodicals Co., (1972) 249–250.

44. Block Toeplitz matrix inversion. *SIAM J. Appl. Math.*, Vol. 24, (1973) 234–241.

45. Information theory and an extension of the maximum likelihood principle. *Proc. 2nd International Symposium on Information Theory* (B. N. Petrov and F. Csaki eds.) Akademiai Kiado, Budapest, (1973) 267–281. (Reproduced in Breakthroughs in Statistics, Vol. I, Foundations and Basic Theory, S. Kotz and N.L. Johnson eds., Springer-Verlag, New York, (1992) 610–624.)

46. Maximum likelihood estimation of structural parameters from random vibration data. (with W. Gersch and N.N. Nielsen), *Journal of Sound and Vibration*, (1973) Vol. 31, No. 3, 295–308.

47. Maximum likelihood identification of Gaussian autoregressive moving average models. *Biometrika*, Vol. 60, (1973) 255–265.

48. Stochastic theory of minimal realization. *IEEE Trans. Automat. Contrl.*, AC-19, (1974) 667–674.

49. A new look at the statistical model identification. *IEEE Trans. Automat. Contrl.*, AC-19, No. 6, (1974) 716–723.

50. Markovian representation of stochastic processes and its application to the analysis of autoregressive moving average processes. *Ann. Inst. Statist. Math.*, Vol. 26, (1974) 363–387.

51. Markovian representation of stochastic processes by canonical variables. *SIAM J. Control*, 13, (1975) 162–173.

52. TIMSAC-74, A time series analysis and control program package (1), (with E. Arahata and T. Ozaki). *Computer Sciences Monographs*, No. 5, The Institute of Statistical Mathematics, Tokyo, (1975).

53. TIMSAC-74, A time series analysis and control program package (2). (with E. Arahata and T. Ozaki). *Computer Science Monographs*, No. 6, The Institute of Statistical Mathematics, Tokyo, (1976).

54. An objective use of Bayesian models. *Ann. Inst. Statist. Math.*, Vol. 29, (1977) 9–20.

55. An extension of the method of maximum likelihood and the Stein's problem. *Ann. Inst. Math.*, Vol. 29, (1977) 153–164.

56. Canonical correlation analysis of time series and the use of an information criterion. *System identification: Advances and Case Studies*, D. G. Lainiotis and R. K. Mehra eds., Academic Press, (1977) 27–96.

57. Information and statistical model building. *Towards a Plan of Actions for Mankind*, Vol. 4, M. Marois, ed., Pergamon Press, Oxford, (1977) 147–151.

58. On entropy maximization principle. P. R. Krishnaiah, ed., *Applications of Statistics*, North-Holland Publishing Company, (1977) 27–41.

59. On the statistical model of the Chandler Wobble. (with M. Ooe and Y. Kaneko), *Publications of the International Latitude Observatory*, Vol. 9, No. 1, (1977).

60. Spectrum estimation through parametric model fitting. *Stochastic Problems in Dynamics*, B. L. Clarkson, ed., Pitman Publishing, London, (1977) 348–363.

61. A Bayesian analysis of the minimum AIC procedure. *Ann. Inst. Statist. Math.*, Vol. 30, A, (1978) 9–14.

62. Analysis of cross classified data by AIC. (with Y. Sakamoto) *Ann. Inst. Statist. Math.*, Vol. 30, B, (1978) 185–197.

63. A new look at the Bayes Procedure. *Biometrika*, Vol. 65, (1978) 53–59.

64. A procedure for the modeling of non-stationary time series. (with G. Kitagawa) *Ann. Inst. Statist. Math.*, Vol. 30, No. 2, B, (1978) 351–363.

65. Comments on model structure testing in system identification. *Int. J. Control*, Vol. 27, (1978) 323–324.

66. Covariance matrix computation of the state variable of a stationary Gaussian process. *Ann. Inst. Statist. Math.* 30, (1978) Part B, 499–504.

67. GALTHY, a probability density estimator. (with E. Arahata) *Computer Science Monographs*, No. 9, The Institute of Statistical Mathematics, Tokyo, (1978).

68. On newer statistical approaches to parameter estimation and structure determination. *A Link Between Science and Applications of Automatic Control*, Vol. 3 (A.Niemi, ed.), Pergamon Press, Oxford, (1978) 1877–1884.

69. On the likelihood of a time series model. *The Statistician*, Vol. 27, (1978) 217–235.

70. Time series analysis and control through parametric models. *Applied Time Series Analysis*, D. F. Findley, ed., Academic press, New York, (1978) 1–23.

71. Robot data screening of cross-classified data by an information criterion. (with Y. Sakamoto) *Proceedings of the international conference on cybernetics and society*, Vol. 1, IEEE, (1978) 398–403.

72. A Bayesian extension of the minimum AIC procedure of autoregressive model fitting. *Biometrika*, 66, 2, (1979) 237–242.

73. Application of optimal control system to a supercritical thermal power plant. (with H. Nakamura, M. Uchida and T. Kitami) 1979 Control of Power Systems Conference Record, *IEEE*, New York, (1979) 10–14.

74. TIMSAC-78 (with G. Kitagawa, E. Arahata and F. Tada) *Computer Science Monographs*, No. 11, The Institute of Statistical Mathematics, Tokyo, (1979).

75. On the construction of composite time series models. *Proceedings of the 42nd Session of the International Statistical Institute*, Vol. 1, (1979) 411–422.

76. Use of statistical identification for optimal control of a supercritical thermal power plant. (with H. Nakamura) *Identification and System Parameter Identification*, Edited by R. Isermann, Oxford and New York, (1979) 221–232.

77. A Bayesian approach to the trading-day adjustment of monthly data. (with M. Ishiguro) *Time Series Analysis*, O.D. Anderson and M.R. Perryman, eds., North-Holland, Amsterdam, (1980) 213–226.

78. BAYSEA, A Bayesian seasonal adjustment program. (with M. Ishiguro) *Computer Science Monographs*, No. 13, The Institute of Statistical Mathematics, Tokyo, (1980).

79. Seasonal adjustment by a Bayesian modeling. *Journal of Time Series Analysis*, Vol. 1, No. 1, (1980) 1–13.

80. The interpretation of improper prior distributions as limits of data dependent proper prior distributions. *J. R. Statist. Soc. B*, Vol. 42 (1980) 46–52.

81. Ignorance prior distribution of a hyperparameter and Stein's estimator. *Ann. Inst. Statist. Math.*, Vol. 32, A (1980) 171–179.

82. Likelihood and the Bayes procedure. *Bayesian Statistics*, J. M. Bernardo, M. H. De Groot, D. V. Lindley and A. F. M. Smith, eds., Valencia, Spain; University Press, (1980) 143–166, (discussion 185–203).

83. On the identification of state space models and their use in control. *Directions in Time Series*, D. R. Brillinger and G. C. Tiao, eds., The Institute of Mathematical Statistics, California, (1980) 175–187.

84. On the use of the predictive likelihood of a Gaussian model, *Ann. Inst. Statist. Math.*, Vol. 32, A (1980), 311–324.

85. Trend estimation with missing observation. (with M. Ishiguro) *Ann. Inst. Statist. Math.*, Vol. 32, B (1980) 481–488.

86. Likelihood of a model and information criteria. *Journal of Econometrics*, Vol. 16, (1981) 3–14.

87. Modern development of statistical methods. *Trends and Progress in System Identification,* P. Eykhoff, ed., Pergamon Press, Oxford, (1981) 169–184.

88. On the fallacy of the likelihood principle. *Statistics and Probability Letters*, Vol. 1, (1981) 75–78.

89. On TIMSAC-78. (with G. Kitagawa) *Applied Time Series Analysis II*, D.F. Findley, ed., (1981) 499–547.

90. Recent development of statistical methods for spectrum estimation. *Recent Advances in EEG and EMG Data Processing*, N. Yamaguchi and K. Fujisawa, eds., Elsevier/North-Holland Biomedical Press, Amsterdam, (1981) 63–78.

91. Statistical identification for optimal control of supercritical thermal power plants. (with H. Nakamura), *Automatica*, Vol. 17, No. 1, (1981) 143–155.

92. Statistical information processing system for prediction and control. *Scientific Information Systems in Japan*, H. Inoue, ed., North-Holland, Amsterdam, (1981) 237–241.

93. On linear intensity models for mixed doubly stochastic Poisson and self-exciting point processes. (with Y. Ogata) *J. R. Statist. Soc. B*, 44, (1982) 102–107.

94. The application of linear intensity models to the investigation of causal relations between a point process and another stochastic process. (with Y. Ogata and K. Katsura) *Ann. Inst. Statist. Math.*, Vol. 34, B, (1982) 373–387.

95. A quasi Bayesian approach to outlier detection. (with G. Kitagawa) *Ann. Inst. Statist. Math.*, Vol. 34, B, (1982) 389–398.

96. A Bayesian approach to the analysis of earth tides. (with M. Ishiguro, H. Ooe and S. Nakai) *Proceedings of the Ninth International Symposium on Earth Tides,* J. T. Kuo, ed., (1983) 283–292.

97. Comparative study of the X-11 and BAYSEA procedures of seasonal adjustment. *Applied Time Series Analysis of Economic Data,* Arnold Zellner, ed., Economic Research Report ER-5, U. S. Department of Commerce, Bureau of the Census, (1983) 17–30.

98. Information measures and model selection. *Proc. 44th Session of the International Statistical Institute,* Vol. 1, (1983) 277–291.

99. On minimum information prior distribution. *Ann. Inst. Statist. Math.*, Vol. 35, A, (1983) 139–149.

100. Statistical inference and measurement of entropy. *Scientific Inference, Data Analysis, and Robustness,* Academic Press, Ins., (1983) 165–189.

101. Comment, *Journal of Business and Economic Statistics,* Vol. 2, No. 4, (1984) 321–322.

102. On the use of Bayesian models in time series analysis. *Robust and Nonlinear Time Series Analysis,* J. Franke, W. Härdle and D. Martin, eds., Springer-Verlag, New York, (1984) 1–16.

103. Prediction and entropy. *A Celebration of Statistics,* A.C. Atkinson and E. Fienberg, eds., Springer-Verlag, New York, (1985) 1–24.

104. TIMSAC-84 Part 1 and 2. (with T. Ozaki *et al.*) *Computer Science Monographs,* No. 22 and 23, The Institute of Statistical Mathematics, Tokyo, (1985).

105. Autoregressive models provide stochastic descriptions of homeostatic processes in the body. (with T. Wada and E. Kato) *Japanese Journal of Nephrology,* Vol. 28, (1986) 263–268.

106. Frequency dependency of causal factors for hypertension in hemodialysis patients. (with T. Wada, S. Sudoh and E. Kato) *Japanese Journal of Nephrology,* Vol. 28, (1986) 1237–1243.

107. The selection of smoothness priors for distributed lag estimation. *Bayesian Inference and Decision Techniques with Applications: Essays in Honor of Bruno de Finetti,* P. K. Goel and A. Zellner, eds., North-Holland, Amsterdam, (1986) 109–118.

108. Use of statistical models for time series analysis. *Proceedings of the International Conference on Acoustics, Speech and Signal Processing,* ICASSP 86, Tokyo, IEEE, (1986) 3147–3155.

109. Some reflections on the modeling of time series. *Time Series and Econometric Modeling,* I. B. MacNeill and G. J. Umphrey, eds., Reidel, Dordrecht, (1987) 13–28.

110. Comment on "Prediction of future observations in growth curve models" by C. R. Rao. *Statistical Science*, Vol. 2, (1987) 464–465.

111. Factor analysis and AIC. *Psychometrika*, Vol. 3, (1987) 317–332.

112. Applications of multivariate autoregressive modeling for an analysis of immunologic networks in man. (with T. Wada, Y. Yamada and E. Udagawa) *Computers & Mathematics with Applications*, Vol. 15, (1988) 713–722.

113. Application of the multivariate autoregressive model. *Advances in Statistical Analysis and Statistical Computing*, Vol. 2, R. S. Mariano, ed., JAI Press Inc., Greenwich, Connecticut, (1989) 43–58.

114. Bayesian modeling for time series analysis. *Advances in Statistical Analysis and Statistical Computing*, Vol. 2, R.S. Mariano, ed., JAI Press Inc., Greenwich, Connecticut, (1989) 59–69.

115. DALL: Davidon's algorithm for log likelihood maximization – A Fortran subroutine for statistical model builders –. (with M. Ishiguro) *Computer Science Monographs*, No. 25, The Institute of Statistical Mathematics, Tokyo, (1989).

116. Comment on "The unity and diversity of probability" by Glenn Shafer. *Statistical Science*, Vol. 5, (1990) 444–446.

117. Experiences on the development of time series models. *Proceedings of the First US/JAPAN Conference on The Frontiers of Statistical Modeling: An Informational Approach*, Vol. 1, H. Bozdogan, ed., Kluwer Academic Publishers, Dordrecht, (1994) 33–42.

118. Implications of informational point of view on the development of statistical science. *Proceedings of the First US/JAPAN Conference on The Frontiers of Statistical Modeling: An Informational Approach*, Vol. 3, H. Bozdogan, ed., Kluwer Academic Publishers, Dordrecht, (1994) 27–38.

Papers written in Japanese

1. 傳播現象の統計数理的解析 –I – マイクロウェイブに於けるフェイディングの一分析 – 統計数理研究輯報, 第 11 号, (1953), 1–90. (Statistical analysis of propagation phenomena–I, An analysis on the fading of microwave. *The Research Report of the Institute of Statistical Mathematics,* Vol. 11.)

2. 系列現象の統計的解析 –II – 株価変動の統計的解析 – 統計数理研究所彙報, 第 1 巻, 第 2 号, (1954) 47–62, and 57–58. (Statistical analysis of serial phenomena–II, Statistical analysis of stock price. *The Research Report of the Institute of Statistical Mathematics,* Vol. 1, No. 2.)

3. カルナップ 確率の論理学的基礎. 科学基礎論研究, Vol. 1, No. 4, (1955) 36–41. (Review on Carnap, Logical foundation of probability, The University of Chicago Press, 1950, *Journal of the Japan Society of Philosophy of Science*, Vol. 1, No. 4.)

4. 系列現象の統計的解析 –III – 株価と新聞内容の統計的解析 – 統計数理研究所彙報, 第 3 巻, 第 1 号, (1955) 3–26. (Statistical analysis of serial phenomena–III, Stock price and content of newspaper. *The Research Report of the Institute of Statistical Mathematics*, Vol. 3, No. 1.)

5. 系列現象の統計的解析 –IV, – モンテカルロ法へのリレー計算機の利用について – (三枝八重子と共著), 統計数理研究所彙報, 第 5 巻, 第 1 号, (1957) 58–65. (Statistical analysis of serial phenomena–IV, Monte Carlo methods performed with relay computer. (with Y. Saigusa) *The Research Report of the Institute of Statistical Mathematics*, Vol. 5, No. 1.)

6. 系列現象の統計的解析 –V–(1) – 間隔過程と繰糸工程管理 –, 統計数理研究所彙報, 第 5 巻, 第 2 号, (1958) 133–139. (Statistical analysis of serial phenomena–V–(I), Gap process and quality control in filature industry. *The Research Report of the Institute of Statistical Mathematics*, Vol. 5. No. 2.)

7. 確率過程に関する統計理論の発展の方向について. 統計数理研究所彙報, 第 7 巻, 第 1 号, (1959) 65–80. (On the trend in statistical theory about stochastic processes. *The Research Report of the Institute of Statistical Mathematics*, Vol. 7, No. 1.)

8. 密度関数の統計的推定について. 統計数理研究所彙報, 第 12 巻, 第 1 号, (1964) 117–131. (On the statistical estimation of density functions. *The Research Report of the Institute of Statistical Mathematics*, Vol. 12, No. 1.)

9. 製糸工程の統計的管理法に関する研究 –IV-, 自動繰糸工程における煮熟繭移行の管理に関する研究 (嶋崎昭典と共著). 蚕糸試験場報告, 第 20 巻, 第 2 号, (1966) 71–186. (Studies on the statistical control of the raw silk production process–IV, Control methods of the flow of cooked cocoons in the automatic reeling process. (with A. Shimazaki) *The Bulletin of the Sericultural Station*, Tokyo, Vol. 20, No. 2.)

10. スペクトル解析, 相関関数およびスペクトル – その測定と応用 –, 磯部 孝編, 東京大学出版会, (1968) 28–46. (Spectrum Analysis, Correlation Function and Spectrum–Measurement and Application–, T. Isobe ed., University of Tokyo Press.)

11. 時系列解析の現況. 計測と制御, 第 8 巻, 第 3 号, (1969) 176–182. (Present status of time series analysis, *Journal of the Society of Instrument and Control Engineers*, Vol. 8, No. 3.)

12. 時系列の解析と予測と制御. 科学基礎論研究, 第 10 巻, 第 2 号, (1971) 73–77. (Time series analysis and prediction and control, *Journal of the Japan Society of Philosophy of Science*, Vol. 10, No. 2.)

13. 統計的モデルの決定のための新しい方法について. Computation & Analysis Seminar, Japan, Vol. 6, No. 1, (1974) 43–51. (On a new method for statistical model selection, *Computation & Analysis Seminar, Japan*, Vol. 6, No. 1.)

14. 統計的システムの表現. 電気学会雑誌, 95 巻 2 号, (1975) 25–31. 117–123. (Representation of statistical system, *Trans. IEE of Japan*, Vol. 95, No. 2.

15. 情報量規準 AIC とは何か. 数理科学, 第 14 巻, 第 3 号, (1976) 5–11. (What is the information criterion AIC?, *Suri Kagaku (Mathematical Sciences)*, Vol. 14, No. 3.)

16. 統計的情報とシステム理論. 数学, 第 29 巻, 第 2 号, (1977) 97–109. (Statistical information and system theory, *Sugaku (Mathematics)*, Vol. 29, No. 2.)

17. 火力発電プラントの最適制御のための統計的アプローチ. (中村秀雄, 平野敏彦と共著) 電気学会論文誌, 第 98 巻 B　第 7 号, (1979) 601–608. (Statistical approach to the optimal control of a thermal-power plant, (with H. Nakamura and T. Hirano) *Trans. IEE of Japan*, Vol. 98–B, No. 7.)

18. 確率系の実現問題. 計測と制御, 第 17 巻, 第 12 号, (1978) 891–898. (Problem of realization of stochastic system, *Journal of the Society of Instrument and Control Engineers*, Vol. 17, No. 12.)

19. 統計的検定の新しい考え方. 数理科学, 第 17 巻　第 12 号, (1979) 51–57. (New look on statistical test, *Suri Kagaku (Mathematical Sciences)*, Vol. 17, No. 12.)

20. エントロピーとモデルの尤度. 日本物理学会誌, 第 35 巻, 第 7 号, (1980) 608–614. (Entropy and the likelihood of a model, *Butsuri (Physics)*, Vol. 35, No. 7.)

21. 統計的推論のパラダイムの変遷について. 統計数理研究所彙報, 第 27 巻 第 1 号, (1980). (On the transition of the paradigm of statistical inference, *The Research Report of the Institute of Statistical Mathematics*, Vol. 27, No. 1.)

22. モデルによってデータを測る. 数理科学, 第 19 巻, 第 3 号, (1981) 7–10. (Measuring data by models, *Suri Kagaku (Mathematical Sciences)*, Vol. 19, No. 3.)

23. 統計とエントロピー. 数学セミナー, 第 21 巻, 第 12 号, (1982) 2–12. (Statistics and entropy, *Sugaku Seminar (Mathematics Seminar)*, Vol. 21, No. 12.)

24. 比較代表制の確率論的分析. 統計数理研究所彙報, 第 31 巻, 第 2 号, (1983) 129–132. (A probabilistic analysis of the electoral system of proportional representation. *Proceedings of the Institute of Statistical Mathematics*, Vol. 31, No. 2.)

25. 確率の解釈における困難について. 統計数理研究所彙報, 第 32 巻, 第 2 号, (1984) 117–128. (On the difficulty in the interpretation of probabilities. *Proceedings of the Institute of Statistical Mathematics*, Vol. 32, No. 2.)

26. エントロピーを巡る混乱. 数理科学, 第 23 巻, 第 1 号, (1985) 53–57. (Confusion of entropy, *Suri Kagaku (Mathematical Sciences)*, Vol. 23, No. 1.)

27. AIC による推論. 科学基礎論研究, 第 19 巻, 第 2 号, (1989) 73–79. (Inference by AIC, *Journal of the Japan Association for Philosophy of Science*, Vol. 19, No. 2.)

28. 知識の科学としての統計学. 科学, Vol. 59, No. 7, (1989) 446–454. (Statistics as a science of knowledge, *Kagaku (Science)*, Vol. 59, No. 7.)

29. 統計学研究の方策について. 日本統計学会誌, 第 19 巻, 第 2 号, (1989) 123–128. (On the strategy for the research of statistics, *Journal of the Japan Statistical Society*, Vol. 19, No. 2)

30. 事前分布の選択とその応用. ベイズ統計学とその応用, 鈴木雪夫・国友直人編, 東京大学出版会, (1989) 81–98. (Selection of prior distribution and its application, *Bayesian Statistics and Its Applications*, Y. Suzuki and N. Kunitomo, eds.)

31. 計算機社会と統計的データ処理. 日本統計学会誌, 第 21 巻, 第 3 号 (増刊号), (1992) 323–327. (Computer society and statistical data processing, *Journal of the Japan Statistical Society*, Vol. 21, No. 3.)

32. 統計モデルによるデータ解析. 脳と発達, 第 24 巻, 第 2 号, (1992) 127–133. (Data analysis by statistical models, *No to Hattatsu (Brain and Development)*, Vol. 24, No. 2.)

33. 統計科学とは何だろう. 統計数理, Vol. 42, No. 1, (1994) iii–ix. (Statistical Science, What It Means, *Proceedings of the Institute of Statistical Mathematics*, Vol. 42, No. 1.)

34. モデルを通してデータを読む. トライボロジスト, Vol. 40, No. 7, (1995) 573–578. (Reading data through models, *Journal of Japanese Society of Tribologists*, Vol. 40, No. 7.)

35. AIC と MDL と BIC. オペレーションズ・リサーチ, Vol. 41, No. 7, (1996) 375–378. (AIC, MDL and BIC, *Communications of the Operations Research Society of Japan*, Vol. 41, No. 7.)

Books

1. 現代社会とマスコミュニケーション　マスコミュニケーション講座, 第五巻, （日高六郎他と共著）河出書房, (1955). (Modern Society and Mass Communication, Lectures on Mass Communication, Vol. 5, (coauthored with R. Hidaka) Kawade Shobo)

2. 世論に関する考え方（蝋山政道他と共著）　新日本教育協会, (1955). (Views on Public Opinion (coauthored with M. Royama *et al.*), Shin Nihon Kyoiku Kyokai))

3. ダイナミックシステムの統計的解析と制御（中川東一郎と共著）サイエンス社, (1972). (Statistical Analysis and Control of Dynamic Systems (coauthored with T. Nakagawa))

4. 確率論・統計学（林知己夫他と共著）放送大学教育振興会, (1985). (Probability and Statistics (coauthored with C. Hayashi *et al.*), The Society for the Promotion of the University of the Air)

5. 統計学特論（林知己夫, 鈴木達三と共著）放送大学教育振興会, (1986). (Special Topics of Statistics (coauthored with C. Hayashi and T. Suzuki), The Society for the Promotion of the University of the Air)

6. 時系列論（尾崎 統他と共著）放送大学教育振興会, (1988). (Theory of Time Series (coauthored with T. Ozaki *et al.*), The Society for the Promotion of the University of the Air)

7. 時系列解析の実際 II（北川源四郎と共編著）朝倉書店, (1995). (Practice of Time Series Analysis II (coauthored with G. Kitagawa *el al.*), Asakura Publishing Company, Tokyo, (1995).

8. Statistical Analysis and Control of Dynamic Systems. (coauthored with T. Nakagawa) Kluwer Academic Publishers, Dordrecht, (1988).

Reprinted from *Annals of the Institute of Statistical Mathematics*, Vol. 8, No. 2, 1956, 87-94 by permission of the Institute of Statistical Mathematics

ON A ZERO-ONE PROCESS AND SOME OF ITS APPLICATIONS

By HIROTUGU AKAIKE

(Received Nov. 15, 1956)

0. Introduction and summary

We have many occasions to analyse the pattern of occurrences of serial random events, for example, automobile flows, dropping ends of cocoon filaments. Many works have been done for the case when the lengthes of gaps independently follow one and the same negative exponential distribution. In this case, as is well known, the number of occurrences in one time interval is entirely independent of that in another time interval which does not overlap the former, and follows the Poisson distribution. Nevertheless, we sometimes observe random events concerning which the numbers of occurrences in some disjoint time intervals are not entirely independent, that is, they show wave-like movement. In some cases, this is due to the fact that the distribution of gaps are not necessarily of negative exponential type. In this paper we shall treat a discrete parameter processes for the purpose to analyse practically the processes of the type just stated.

The process we treat here will be a powerfull tool for analysis of the fundamental strucutre of the process representing the dropping ends of cocoon filaments. Some examples of its application will also be given.

1. Definition of gap-process

Consider a stochastic process $\{x_n(\omega)\,;\ n=0, 1, 2, \cdots\}$, where $x_0(\omega)\equiv 1$ and $x_n(\omega)=1$ or 0, and which is governed by the gap distribution $\text{Prob}\,\{x_{n+1}=0,\ x_{n+2}=0,\ \cdots x_{n+\nu-1}=0,\ x_{n+\nu}=1|x_n=1\}\equiv p_\nu$ $(\nu=1, 2, \cdots)$.

Then we have $P_n\equiv\text{Prob}\,\{x_n(\omega)=1\}=\sum_{\nu=1}^{n} p_\nu P_{n-\nu}$ where $P_{n-\nu}$ is given by the recurrence relation

$$P_0\equiv 1, \qquad P_\mu=\sum_{\nu=1}^{\mu} p_\nu P_{\mu-\nu}\,.$$

As for the joint distribution of $x_n(\omega)$'s, from the relation

$$\text{Prob}\{x_{n+\nu}(\omega)=1|x_n(\omega)=1,\ x_{n-1}(\omega),\ \cdots,\ x_0(\omega)\}$$
$$=\text{Prob}\{x_{n+\nu}(\omega)=1|x_n(\omega)=1\}=P_\nu$$

we have

$$\text{Prob}\{x_n(\omega)=1,\ x_{n+\nu_1}(\omega)=1,\ x_{n+\nu_1+\nu_2}(\omega)=1,\ \cdots\ x_{n+\nu_1+\nu_2+\cdots+\nu_k}(\omega)=1\}$$
$$=P_n P_{\nu_1} P_{\nu_2}\cdots P_{\nu_k}{}^*.$$

Now, as is well known, by the recurrent event theory, when the greatest common divisor of those ν for which $p_\nu>0$ is one, we have

$$\lim_{n\to\infty} P_n=P=\frac{1}{L} \quad \text{where} \quad L=\sum \nu p_\nu$$

(see [1]). Hence, when we are only concerned with the steady state of this process after a long time has passed the process may be represented by another strictly stationary zero-one process which is defined as follows.

We call the stochastic process $\{x_n(\omega),\ -\infty<n<\infty\}$ a gap process where for almost all ω $x_n(\omega)=0$ or 1 and

$$\text{Prob}\{x_n(\omega)=1,\ x_{n+\nu_1}(\omega)=1,\ \cdots,\ x_{n+\nu_1+\cdots+\nu_k}(\omega)=1\}=PP_{\nu_1}\cdots P_{\nu_k}$$

where P_ν is given from some $\{p_\nu\}$ with $\sum_{\nu\geq 1} p_\nu=1$ and $p_\nu\geq 0$ by the recurrence relation $P_0=1$, $P_\nu=\sum_{\mu=1}^{\nu} p_\mu P_{\nu-\mu}$ and $P=\lim_{\nu\to\infty} P_\nu$. We shall call $\{p_\nu\}$ the corresponding gap distribution. Its meaning will be obvious from the former consideration.

2. Spectral function of bounded gap-process

In this section we shall study some of the spectral character of the gap process whose gap lengthes are bounded by some fixed value. The restriction of boundedness for the gap lengthes is not a serious one for practical problems. Now, let the gap distribution be $\{p_\nu;\ \nu=1,2,\cdots,k.\ \sum_{\nu=1}^{k} p_\nu=1\}$. Namely, consider the gaps not longer than k. Then we have the following:

THEOREM. *When* $p_{\nu_i}\neq 0\ (i=1,\cdots,\mu)$ *and* $\sum_{i=1}^{\mu} p_{\nu_i}=1$ *hold and the*

* To determine the joint distribution of zero-one variables only the values of the type $\text{Prob}\{x_{n_1}(\omega)=1,\ x_{n_2}(\omega)=1\ \cdots\ x_{n_k}(\omega)=1\}$ are needed.

For example, to get the value of $\text{Prob}\{x_{n_1}(\omega)=1,\ x_{n_2}(\omega)=0,\ x_{n_3}(\omega)=1\}$ the following relation can be used.

$$\text{Prob}\{x_{n_1}(\omega)=1,\ x_{n_2}(\omega)=0,\ x_{n_3}(\omega)=1\}$$
$$=\text{Prob}\{x_{n_1}(\omega)=1,\ x_{n_2}(\omega)=1\}-\text{Prob}\{x_{n_1}(\omega)=1,\ x_{n_2}(\omega)=1,\ x_{n_3}(\omega)=1\}.$$

greatest common divisor of $(\nu_1, \nu_2, \nu_3, \cdots, \nu_\mu)$ *is* 1, *the* $\{x_n(\omega)-P\}$ *process has the continuous spectral density function*

$$G'(\lambda)=2R(0)+4\sum_{n=1}^{\infty}R(n)\cos 2\pi n\lambda$$

where
$$R(n)=P(P_n-P)\,.$$

PROOF. As is easily seen, the covariance function of the $\{x_n(\omega)-P\}$ process is given by $R(n)=E\{x_{n+m}(\omega)-P\}\{x_m(\omega)-P\}=P(P_n-P)$. Thus the process is stationary in the wide sense. Now, as the P_n's are the solution for the difference equation

$$P_n=p_1P_{n-1}+p_2P_{n-2}+\cdots+p_kP_{n-k}$$

under the initial condition

$$P_0=1,\ P_1=p_1,\ P_2=p_1P_1+p_2,\ \cdots,\ P_{k-1}=p_1P_{k-2}+\cdots+p_{k-1},$$

it can be represented in the form

$$P_n=c_1z_1^n+c_2z_2^n+\cdots+c_kz_k^n$$

where z_i's are the roots of characteristic equation $z^k-p_1z^{k-1}-p_2z^{k-2}-\cdots-p_k=0$ and c_i's are complex numbers. We can take $z_1=1$, as 1 is clearly a root of the characteristic equation. Moreover, it is shown from the following inequality that 1 is a simple root of the characteristic equation:

$$\frac{d}{dz}(z^k-p_1z^{k-1}-p_2z^{k-2}-\cdots-p_k)\Big|_{z=1}$$
$$=k\Big(1-\frac{k-1}{k}p_1-\frac{k-2}{k}p_2-\cdots-\frac{1}{k}p_{n-1}-\frac{0}{k}p_k\Big)=\sum_{\nu=1}^{k}\nu p_\nu>0\,.$$

If $|z_j|=1$ holds for some j, we can represent it by $z_i=e^{i\theta}$, and this θ must satisfy the following equation

$$1=p_1e^{-i\theta}+p_2e^{-i2\theta}+\cdots+p_ke^{-ik\theta}$$
$$=p_{\nu_1}e^{-i\nu_1\theta}+p_{\nu_2}e^{-i\nu_2\theta}+\cdots+p_{\nu_\mu}e^{-i\nu_\mu\theta}\,.$$

Hence we must have the following equations

$$e^{-i\nu_j\theta}=1\qquad j=1,2,\cdots,\mu\,,$$

that is,

$$\nu_j\theta=\kappa_j2\pi\qquad(\kappa_j\ \text{integer})\,.$$

Now, as the greatest common divisor of $\nu_1, \nu_2, \cdots, \nu_\mu$ is unity, we have for some integral values $\alpha_1, \alpha_2, \cdots, \alpha_\mu$

$$1=\alpha_1\nu_1+\alpha_2\nu_2+\cdots+\alpha_\mu\nu_\mu .$$

Thus we have

$$\theta=(\alpha_1\kappa_1+\alpha_2\kappa_2+\cdots+\alpha_\mu\kappa_\mu)2\pi$$

or $z_i=1$. From these facts we get a conclusion that $|z_i|<1$ for $i\neq 1$. Under the condition of the theorem we must have

$$\lim_{n\to\infty} P_n=P=\frac{1}{L}$$

where $L=\sum \nu p_\nu$, and from this it can be seen that $c_1=P$ and

$$P_n-P=\sum_{i=2}^{k} c_i z_i^n \qquad |z_i|<1 .$$

Therefore, we have

$$\sum_{n=0}^{\infty} |R(n)|=P\sum_{n=0}^{\infty}|P_n-P|\leq P\sum_{n=0}^{\infty}\sum_{i=2}^{k}|c_i||z_i|^n<+\infty$$

and the existence of the continuous spectral density function

$$G'(\lambda)=2R(0)+4\sum_{1}^{\infty} R(n)\cos 2\pi n\lambda$$

is assured [2].

From this theorem it follows that almost all gap-processes we observe have continuous spectral functions and that some of them show wave-like movements. In such cases, according to the theory of stationary process, the ordinary periodogram analysis is of less use, and the structure of gap distribution must be considered first. When $R(n)$'s $(n=0, 1\cdots, k)$ are given, we can easily obtain the values $\{p_\nu ; \nu=1, 2, \cdots k\}$ by using the recurrence relation $p_\nu=P_\nu-\sum_{\mu=1}^{\nu-1} p_\mu P_{\nu-\mu}$.

3. Examples. (Practical application of gap process theory).

The examples we treat in this section are essentially continuous parameter processes. But as the treatment of such processes are not easy we have applied discrete approximations to them which we believe to have sufficient accuracy for practical purposes. Original processes are continuous parameter processes of which almost all sample functions are step-functions with jumps of unit saltus, and the time lengthes τ between jumps are governed independently of each other by one and the same continuous probability distribution $p(\tau)d\tau$.

Let τ_0 and ε be positive numbers such that

$$\int_0^{\tau_0} p(t)dt \leq \varepsilon \qquad \frac{T}{\tau_0}\varepsilon \ll 1$$

where T is the whole length of the time for observation. When we put $\varepsilon(\tau) \equiv \int_0^{\tau} p(t)dt$, we have for continuous $p(t)$ a positive h such that

$$\varepsilon(\tau) = \tau p(h\tau) \qquad 0 < h < 1\,,$$

so, whenever $\lim_{t\to 0} p(t) = 0$, we can find a pair of ε and τ_0 which satisfies the above stated conditions. This means that for continuous $p(\tau)$ with $\lim_{t\to 0} p(t) = 0$, there exists such τ_0 that there are few intervals containing more than two occurrences of events in some finite number of observations each with the duration of time T. To treat such processes the following procedure will always be a pertinent one.

Given a record of occurrences of events, we take as the time unit for digitization such τ' that there is no time interval of length τ' containing more than two records of occurrences.

1. *Flow of automobiles.*

We have observed the flow of automobiles at some points in Tokyo. The distribution of the number of cars per unit time interval seems to be fairly well fitted by the Poisson distribution (Fig. 1). Nevertheless, when we analyse the dependence between the numbers of cars in succesive time intervals we can recognize the existence of serial correlations. This fact contradicts to the fundamental assumption that the numbers of cars in mutually disjoint time intervals are independent of each other for a simple Poisson process. Therefore, we have digitized the data by taking τ' as half the shortest record of time interval between the cars, and analysed the distribution of the length of time intervals between

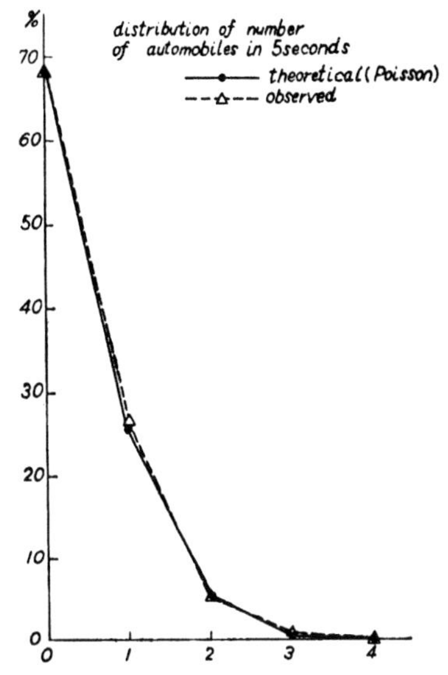

Fig. 1.

successive cars and got the result as shown in Fig. 2.

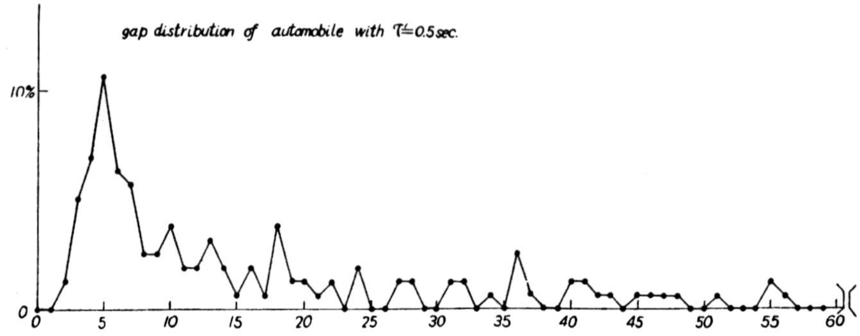

Fig. 2.

This figure suggests that the above hypothesis does not hold true, that is, the length of time interval between successive cars does not follow the negative exponential distribution. We have taken the Fig. 2 as giving the gap distribution $\{p_\nu\}$ and calculated P_ν's under the hypothesis that the process is a gap one. The result is graphically shown in Fig. 3. This shows a fairly good fit and the existence of serial correlation between successive data can be considered as due to the narrowness of the road. We observed 156 cars in this experiment.

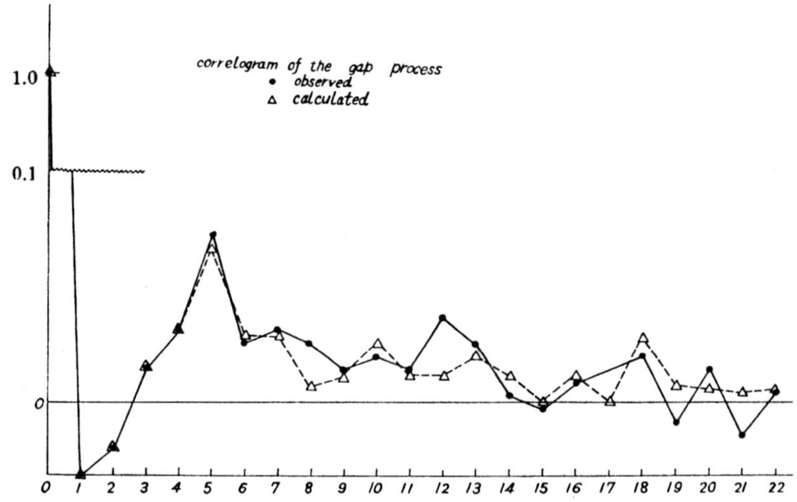

Fig. 3.

2. *Dropping ends of cocoon filament.*

When we operate the silk-reel we sometimes observe the wave-like pattern of stoppings by dropping ends of cocoon. This seems due to the fact that the distribution of length between successive ends does not follow the negative exponential distribution. Experimental data show that the distribution of nonbreaking length of a bave is such as shown in Fig. 4*.

Taking this figure as giving a gap distribution $\{p_\nu\}$ we get the sequence $\{P_\nu\}$. The values $(P_\nu - P)/(1-P)$ are considered as showing the correlogram of the process** (Fig. 5).

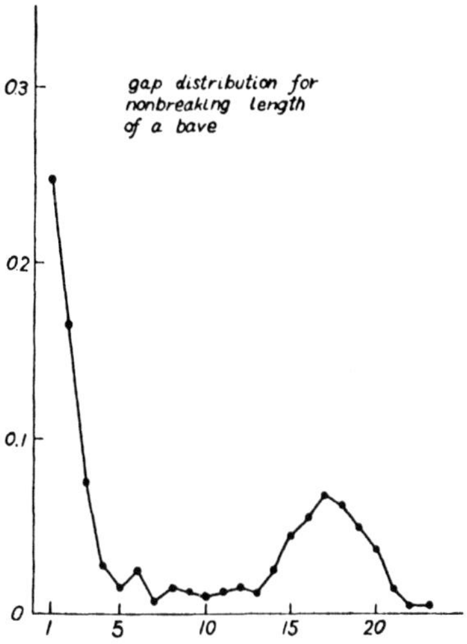

Fig. 4.

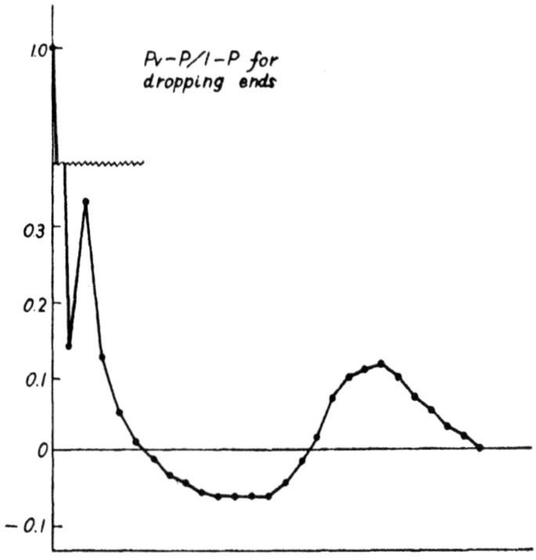

Fig. 5.

4. Acknowledgement

The author expresses his thanks to Mr. K. Morita, member of Electrical Communication Laboratory, Nippon Telegraph and Telephone Public Corporation, who kindly provided us with the convenience to use a recording apparatus for our experiments on automobile flows and to Mr. A. Shimazaki* for calling the author's attention to the problem of dropping ends of cocoon filaments.

* The data is offered to the author from Mr. A. Shimazaki of Sericultural Experiment Station, Tokyo.

** This line of approach seems to give some light on the development of the production of raw silk, and is now being put forward by Mr. A. Shimazaki.

Thanks are also due to Mr. M. Sibuya of our Institute for kind discussion on the dropping ends problem and to Misses Y. Saigusa and T. Kageyama for their recording and arranging work of data.

THE INSTITUTE OF STATISTICAL MATHEMATICS

REFERENCES

[1] W. Feller, *An Introduction to Probability Theory and its Applications.* John Wiley & Sons, Inc., New York, (1950), pp. 243–248.

[2] J. L. Doob, *Stochastic Processes.* John Wiley & Sons, Inc., New York, (1953), p. 476.

Reprinted from *Annals of the Institute of Statistical Mathematics*, Vol. 11, No. 1, 1959, 1-16 by permission of the Institute of Statistical Mathematics

ON A SUCCESSIVE TRANSFORMATION OF PROBABILITY DISTRIBUTION AND ITS APPLICATION TO THE ANALYSIS OF THE OPTIMUM GRADIENT METHOD

BY HIROTUGU AKAIKE

(Received July 25, 1959)

§ 0. Introduction and summary.

In this paper we define a type of transformation of probability distribution and analyze the limiting behavior of the result of successive applications of the transformation to some initial probability distribution. By using the results of this analysis we can get a fairly general insight into the so-called optimum-gradient method in numerical analysis. We can prove the conjecture which was stated by Forsythe and Motzkin [7] and was used as the logical basis of an acceleration procedure for the optimum gradient method [4] [5] [6]. It was stated by Forsythe [4] that this conjecture seems to be hard to prove as the related transformation is rather complicated. But our present proof is rather simple. Further, we can see the relation between the condition-number of the related matrix and the convergence rate of the optimum gradient method. By using the relation which according to [5] is first proved by Kantrovich [8], we can say that when the matrix is ill-conditioned the convergence rate tends near to its worst possible value. Using the same data as those treated by Forsythes in paper [5], we give some numerical examples.

There are many important problems, not necessarily of linear type, where the gradient method is applicable [2], [3]. Even in these non-linear case we can expect that when the approximation proceeds the problem will be reduced essentially to the linear one. One of such examples was discussed in the former paper [1] of the present author. Thus, though the results of this paper are concerned with the solution of the simultaneous linear equation $Ax = b$, these results will be generally useful for the analysis of limiting behavior of the approximate solutions in the optimum gradient method.

§ 1. A successive transformation of probability distribution.

In this section we shall treat probability distributions over a set of mutually different real numbers $\lambda_1, \lambda_2, \cdots, \lambda_n$. First we shall give the definition of our transformation T. For a probability distribution $\boldsymbol{P} \equiv \{(\boldsymbol{P})_\nu ;\ \nu=1, 2, \cdots, n\}$ where $(\boldsymbol{P})_\nu$'s represent the probabilities attached to λ_ν's respectively the image $\boldsymbol{TP} \equiv \{(\boldsymbol{TP})_\nu ; \nu=1, 2, \cdots, n\}$ of $\boldsymbol{P}$ by T is given by

$$(\boldsymbol{TP})_\nu = \frac{(\boldsymbol{P})_\nu(\lambda_\nu - \bar{\lambda}(\boldsymbol{P}))^2}{\sum_{\mu=1}^{n}(\boldsymbol{P})_\mu(\lambda_\mu - \bar{\lambda}(\boldsymbol{P}))^2} \qquad \nu=1, 2, \cdots, n$$

where $\bar{\lambda}(\boldsymbol{P}) = \sum_{\mu=1}^{n} \lambda_\mu(\boldsymbol{P})_\mu$ and T has as its domain the set of $\boldsymbol{P}$'s with $\sum_{\mu=1}^{n}(\boldsymbol{P})_\mu(\lambda_\mu - \bar{\lambda}(\boldsymbol{P}))^2 > 0$. Here we shall give some preliminary lemmas. We shall hereafter use the notation $\overline{f(\lambda)}(\boldsymbol{P})$ for $\sum_{\mu=1}^{n} f(\lambda_\mu)(\boldsymbol{P})_\mu$.

LEMMA 1. *We have*

$$\bar{\lambda}(\boldsymbol{TP}) = \bar{\lambda}(\boldsymbol{P}) + \frac{\overline{(\lambda - \bar{\lambda}(\boldsymbol{P}))^3}(\boldsymbol{P})}{\overline{(\lambda - \bar{\lambda}(\boldsymbol{P}))^2}(\boldsymbol{P})} .$$

PROOF.

$$\begin{aligned}
\bar{\lambda}(\boldsymbol{TP}) &= \sum_{\nu=1}^{n} \lambda_\nu(\boldsymbol{TP})_\nu = \frac{\sum_{\nu=1}^{n} \lambda_\nu(\lambda_\nu - \bar{\lambda}(\boldsymbol{P}))^2(\boldsymbol{P})_\nu}{\overline{(\lambda - \bar{\lambda}(\boldsymbol{P}))^2}(\boldsymbol{P})} \\
&= \frac{\sum_{\nu=1}^{n}(\lambda_\nu - \bar{\lambda}(\boldsymbol{P}))^3(\boldsymbol{P})_\nu + \bar{\lambda}(\boldsymbol{P})\sum_{\nu=1}^{n}(\lambda_\nu - \bar{\lambda}(\boldsymbol{P}))^2(\boldsymbol{P})_\nu}{\overline{(\lambda - \bar{\lambda}(\boldsymbol{P}))^2}(\boldsymbol{P})} \\
&= \frac{\overline{(\lambda - \lambda(\boldsymbol{P}))^3}(\boldsymbol{P})}{\overline{(\lambda - \bar{\lambda}(\boldsymbol{P}))^2}(\boldsymbol{P})} + \bar{\lambda}(\boldsymbol{P}) .
\end{aligned}$$

LEMMA 2. *We have*

$$\overline{(\lambda - \bar{\lambda}(\boldsymbol{TP}))^2}(\boldsymbol{TP}) \geqq \overline{(\lambda - \bar{\lambda}(\boldsymbol{P}))^2}(\boldsymbol{P})$$

where $=$ *holds only when there are only two* ν*'s with* $(\boldsymbol{P})_\nu > 0$.

Note; We are considering only $\boldsymbol{P}$'s with $\overline{(\lambda - \bar{\lambda})(\boldsymbol{P}))^2}(\boldsymbol{P}) > 0$, and the case where there is only one ν with $(\boldsymbol{P})_\nu > 0$ is out of our consideration. The lemma states that by transformation T the variance of the distribution increases except for the special type of distributions stated in the lemma. Thus for $\boldsymbol{P}$ in the domain of T, $\boldsymbol{TP}$ is again in the domain of T.

PROOF.

By using the result of lemma 1 we have

$$\lambda_\nu-\bar{\lambda}(TP)=\lambda_\nu-\bar{\lambda}(P)+\{\bar{\lambda}(P)-\bar{\lambda}(TP)\}=\lambda_\nu-\bar{\lambda}(P)-\frac{\overline{(\lambda-\bar{\lambda}(P))^3}(P)}{\overline{(\lambda-\bar{\lambda}(P))^2}(P)}\,.$$

Thus we get

$$\overline{(\lambda-\bar{\lambda}(TP))^2}(TP)-\overline{(\lambda-\bar{\lambda}(P))^2}(P)$$

$$=\frac{\sum_{\nu=1}^{n}(\lambda_\nu-\bar{\lambda}(TP))^2(\lambda_\nu-\bar{\lambda}(P))^2(P)_\nu-\{\overline{(\lambda-\bar{\lambda}(P))^2}(P)\}^2}{\overline{(\lambda-\bar{\lambda}(P))^2}(P)}$$

$$=\frac{\{\overline{(\lambda-\bar{\lambda}(P))^4}(P)\}\{\overline{(\lambda-\bar{\lambda}(P))^2}(P)\}-\{\overline{(\lambda-\bar{\lambda}(P))^3}(P)\}^2-\{\overline{(\lambda-\bar{\lambda}(P))^2}(P)\}^3}{\{\overline{(\lambda-\bar{\lambda}(P))^2}(P)\}^2}\,.$$

For the sake of simplicity we shall use here the abbreviated notation $M_k(P)$ in place of $\overline{(\lambda-\bar{\lambda}(P))^k}(P)$. Then we have

$$\{\overline{(\lambda-\bar{\lambda}(P))^4}(P)\}\{\overline{(\lambda-\bar{\lambda}(P))^2}(P)\}-\{\overline{(\lambda-\bar{\lambda}(P))^3}(P)\}^2-\{\overline{(\lambda-\bar{\lambda}(P))^2}(P)\}^3$$

$$=\begin{vmatrix} 1 & 0 & M_2(P) \\ 0 & M_2(P) & M_3(P) \\ M_2(P) & M_3(P) & M_4(P) \end{vmatrix}\,.$$

If we represent by λ the random variable which follows the probability distribution P i. e. prob $\{\lambda=\lambda_\nu\}=(P)_\nu$ $\nu=1, 2, \cdots, n$, the above stated determinant is the determinant of the product moment matrix of the random variables $(\lambda-E(\lambda))^0\equiv 1$, $\lambda-E(\lambda)$ and $(\lambda-E(\lambda))^2$. The product moment matrix is positive semi-definite and thus the value of the above determinant is non-negative. The value of the determinant is equal to zero if and only if there is non trivial linear relation between 1, $\lambda-E(\lambda)$ and $(\lambda-E(\lambda))^2$. Suppose that there exist constants α_0, α_1, α_2 not all equal to zero satisfying the relation $\alpha_0 1+\alpha_1(\lambda-E(\lambda))+\alpha_2(\lambda-E(\lambda))^2=0$ with probability 1. Such a relation can hold only when there are not more than two values of ν for which prob $\{\lambda=\lambda_\nu\}=(P)_\nu>0$. Taking into account the fact that we have excluded the case where λ is identically equal to a fixed constant, we get the proof of the final statement of the lemma.

Using the results of these lemmas we want to analyze the limiting behavior of $P^{(k)}\equiv T^kP^{(0)}$ as k tends to infinity where $T^kP^{(0)}$ denotes the result of k successive applications of the transformation T to $P^{(0)}$ and the initial probability distribution $P^{(0)}$ is supposed to satisfy the condition $M_2(P^{(0)})>0$.

It is obvious that any infinite subsequence of the sequence $\{P^{(k)};$ $k=1, 2, \cdots\}$ contains its own convergent subsequence. For a convergent subsequence $\{P^{(\alpha_j)}; j=1, 2, \cdots\}$ of original $\{P^{(k)}\}$ we shall represent by $P_\alpha^{(\infty)}$ its limiting distribution.

Now as $\overline{(\lambda-\bar{\lambda}(P^{(\alpha_j)}))^2}(P^{(\alpha_j)})=\sum_{\nu=1}^{n}(\lambda_\nu-\bar{\lambda}(P^{(\alpha_j)}))^2(P^{(\alpha_j)})$ and $\bar{\lambda}(P^{(\alpha_j)})=\sum_{\nu=1}^{n}\lambda_\nu(P^{(\alpha_j)})_\nu$ it can be seen that

$$\lim_{j\to\infty}\bar{\lambda}(P^{(\alpha_j)})=\bar{\lambda}(P_\alpha^{(\infty)})$$

$$\lim_{j\to\infty}\overline{(\lambda-\bar{\lambda}(P^{(\alpha_j)}))^2}(P^{(\alpha_j)})=\overline{(\lambda-\bar{\lambda}(P_\alpha^{(\infty)}))^2}(P_\alpha^{(\infty)}) .$$

Further by taking into account the result of lemma 2 we can see that $\overline{(\lambda-\bar{\lambda}(P)^{(\alpha_j)}))^2}(P^{(\alpha_j)})$ tends monotone-increasingly to $\overline{(\lambda-\bar{\lambda}(P_\alpha^{(\infty)})^2}(P_\alpha^{(\infty)})$ and also that

$$\overline{(\lambda-\bar{\lambda}(P^{(\infty)}))^2}(P_\alpha^{(\infty)})=\lim_{k\to\infty}\overline{(\lambda-\bar{\lambda}(P^{(k)}))^2}(P^{(k)})=\overline{(\lambda-\bar{\lambda}(TP_\alpha^{(\infty)}))^2}(TP_\alpha^{(\infty)})$$

holds.

Now as $M_2(P^{(0)})>0$, we have $M_2(P_\alpha^{(\infty)})>0$, and we have seen that the equality $M_2(TP_\alpha^{(\infty)})=M_2(P_\alpha^{(\infty)})$ holds. Thus from the result of lemma 2 we can see that there are just two numbers $\nu(\alpha, 1)$ and $\nu(\alpha, 2)$ for which $(P_\alpha^{(\infty)})_{\nu(\alpha,1)}>0$, $(P_\alpha^{(\infty)})_{\nu(\alpha,2)}>0$ and $(P_\alpha^{(\infty)})_{\nu(\alpha,1)}+(P_\alpha^{(\infty)})_{\nu(\alpha,2)}=1$ hold. Thus it has been proved that the limit points of the sequence $\{P^{(k)}\}$ are contained in the set of points or distributions of the type analogous to the $P_\alpha^{(\infty)}$ which we have obtained above.

Now we shall further prove that the sequence $\{P^{(k)}\}$ is itself oscillatorily convergent in the sense which will be described bellow.

Taking into account of the monotone-increasing property of the sequence $M_2(P^{(k)})$ we have seen that any $P_\alpha^{(\infty)}$ has one and the same variance $M_2(P_\alpha^{(\infty)})=\lim_{k\to\infty}M_2(P^{(k)})$. Now the distribution $P_\alpha^{(\infty)}$ is characterized by the quantities $\nu(\alpha, 1)$, $\nu(\alpha, 2)$, $(P_\alpha^{(\infty)})_{\nu(\alpha,1)}$, $(P_\alpha^{(\infty)})_{\nu(\alpha,2)}$ and by the relation $M_2(P_\alpha^{(\infty)})=(P_\alpha^{(\infty)})_{\nu(\alpha,1)}(P_\alpha^{(\infty)})_{\nu(\alpha,2)}(\lambda_{\nu(\alpha,1)}-\lambda_{\nu(\alpha,2)})^2=\lim_{k\to\infty}M_2(P^{(k)})$.

Thus the set $\prod_\infty$ of all $P_\alpha^{(\infty)}$'s are composed of at most finitely many distributions. We shall denote the element of $\prod_\infty$ by P_i $i=1, 2, \cdots,$ N. P_i is characterized by the quantities $(P_i)_{\nu(i,1)}>0$ $(P_i)_{\nu(i,2)}>0$.

Now we can see that if there is some P_i with $(P_i)_{\nu(i,1)}\neq(P_i)_{\nu(i,2)}$ then the distribution P_i^* with $(P_i^*)_{\nu(i,1)}=(P_i)_{\nu(i,2)}$ and $(P_i^*)_{\nu(i,2)}=(P_i)_{\nu(i,1)}$ is also contained in $\prod_\infty$. This is proved by taking into account of the fact that if $P^{(i_j)}\to P_i(j\to\infty)$ then $TP^{(i_j)}\to TP_i(j\to\infty)$ and the fact that $TP_i=P_i^*$.

It is obvious that if $(P_i)_{\nu(i,1)}=(P_i)_{\nu(i,2)}=1/2$ we have $TP_i=P_i$. Hereafter, we shall sometimes use the notation $P(\nu(i, 1), \nu(i, 2))$ to represent the P_i to clarify the character of the distribution. We shall here observe the domain of T in its natural relative topology generated by the ordinary topology of n-dimensional Euclidean space. Now as is easily seen T is continuous in its domain, and for any neighborhood $U\{P(\nu(i, 1), \nu(i, 2))\}$ of $P(\nu(i, 1), \nu(i, 2))$ there exists a proper neighborhood $V\{P(\nu(i, 2), \nu(i, 1)\}$ of $P(\nu(i, 2), \nu(i, 1))$ with the property that for any P in $V\{P(\nu(i, 2), \nu(i, 1))\}$ its transform TP is contained in $U\{P(\nu(i, 1), \nu(i, 2))\}$. Here we take the neighborboods $U\{P(\nu(i, 1), \nu(i, 2))\}$ of P_i's so as to be mutually disjoint. We define the neighborhood $W\{P_i\}$ of P_i as the intersection of $U\{P_i\}$ and $V\{P_i\}$ where V is defined as above. Now there exists a number M such that $P^{(k)}$ with k greater than M are all contained in $\sum_{i=1}^{N} W\{P_i\}$, otherwise there must be some limit point outside Π_∞.

We shall represent by $\{P^{(i_j)}, j=1, 2, \cdots\}$ the subsequence of $\{P^{(k)}; k=M+1, M+2, \cdots\}$ contained in $W\{P_i\}$. Then it is seen that $\{P^{(k)}; k=M+1, M+2, \cdots\}=\sum_{i=1}^{N}\{P^{(i_j)}; j=1, 2, \cdots\}$ holds as the relation between the sets of points in the sequences. Now from the definition of M and $P^{(i_j)}$ we can see that $TP^{(i_j)} \in \sum_{i=1}^{N} W\{P_i\}$ holds for any i and j and further by the definition of W and V we can see that $\{TP^{(i_j)}; \{j=1, 2, \cdots\} \subset U\{TP_i\}=U\{P(\nu(i, 2), \nu(i, 1))\}$. As the $U\{P_i\}$'s are taken to be mutually disjoint, so $TP^{(i_j)} \notin W\{P_i\}$ for P_i different from TP_i. Thus we have $\{TP^{(i_j)}; j=1, 2, \cdots\} \subset W\{TP_i\}$. From this last relation we can also see that $\{T^2P^{(i_j)}; j=1, 2, \cdots\} \subset W\{P_i\}$, consequently, that $T^{2r+1}P^{(i_j)} \in W\{TP_i\}$ and $T^{2r+2}P^{(i_j)} \in W\{P_i\}$ hold for $r=0, 1, 2, \cdots$ and that if $P^{(i_j)}$ is identical to some $P^{(k)}$ of the original sequence $\{P^{(k)}\}$, then $P^{(k+s)} \in W\{P_i\} \cup W\{TP_i\}$ holds for $s=0, 1, 2, \cdots$. From these relations it follows that Π_∞ is composed of at most two points $P^{(\infty)}$ and $P^{*(\infty)}$ satisfying the relation $P^{*(\infty)}=TP^{(\infty)}$. When $TP^{(\infty)}=P^{(\infty)}$, Π_∞ is composed of only one point $P^{(\infty)}=P(\nu(\infty, 1), \nu(\infty, 2))$ with $(P^{(\infty)})_{\nu(\infty,1)} = (P^{(\infty)})_{\nu(\infty,2)}=1/2$. Thus we have proved the following;

THEOREM 1: *For any* $P^{(0)}$ *with* $\overline{(\lambda-\lambda(P^{(0)}))^2}(P^{(0)})>0$ *the sequences of distributions* $\{T^{2r}P^{(0)}; r=1, 2, \cdots\}$ *and* $\{T^{2r+1}P^{(0)}; r=0, 1, 2, \cdots\}$ *are convergent to some limiting distributions* $P^{(\infty)}$ *and* $P^{*(\infty)}$, *respectively. These distributions are characterized by* $(P^{(\infty)})_{\nu(\infty,1)}>0$ *and* $(P^{(\infty)})_{\nu(\infty,2)}>0$ *satisfying the relation* $(P^{(\infty)})_{\nu(\infty,1)}+(P^{(\infty)})_{\nu(\infty,2)}=1$ *and* $(P^{*(\infty)})_{\nu(\infty,1)} =$

$(\boldsymbol{P}^{(\infty)})_{\nu(\infty,2)}$, $(\boldsymbol{P}^{*(\infty)})_{\nu(\infty,2)}=(\boldsymbol{P}^{(\infty)})_{\nu(\infty,1)}$. *We have* $T\boldsymbol{P}^{(\infty)}=\boldsymbol{P}^{*(\infty)}$ *and* $\boldsymbol{P}^{*(\infty)}$ *is identical to* $\boldsymbol{P}^{(\infty)}$ *if and only if* $(\boldsymbol{P}^{(\infty)})_{\nu(\infty,1)}=(\boldsymbol{P}^{(\infty)})_{\nu(\infty,2)}=1/2$ *holds.*

Note; We can see that λ_i's with $(\boldsymbol{P}^{(0)})_i=0$ should have been entirely discarded at the very out-set of our whole discussion, and we shall hereafter assume that $(\boldsymbol{P}^{(0)})_i>0$ for $i=1, 2, \cdots, n$.

We shall here discuss some of the effects of the initial distribution $\boldsymbol{P}^{(0)}$ to the limiting distributions $\boldsymbol{P}^{(\infty)}$ and $\boldsymbol{P}^{*(\infty)}$.

THEOREM 2: *If* $\lambda_1<\lambda_i<\lambda_n (i=2, 3, \cdots, n-1)$, then we have

$$\{\nu(\infty, 1), \nu(\infty, 2)\}=\{1, n\}$$

that is, the limiting distributions have their total probability attached to both extremal points.

PROOF; It is obvious that $\lambda_1<\bar{\lambda}(\boldsymbol{P})<\lambda_n$ holds when $(\boldsymbol{P})_1>0$ and $(\boldsymbol{P})_n>0$ hold. Thus from the definition of T it is seen that $(\boldsymbol{P}^{(k)})_1>0$ and $(\boldsymbol{P}^{(k)})_n>0$ hold when $(\boldsymbol{P}^{(k-1)})_1>0$ and $(\boldsymbol{P}^{(k-1)})_n>0$ hold. From this fact, under the condition of the theorem, we have $(\boldsymbol{P}^{(k)})_1>0$ and $(\boldsymbol{P}^{(k)})_n>0$ for $k=0, 1, 2, \cdots$. Suppose $\nu(\infty, 2)<n$ holds. Then we have $\lambda_{\nu(\infty,1)}<\bar{\lambda}(\boldsymbol{P}^{(\infty)}), \bar{\lambda}(\boldsymbol{P}^{*(\infty)})<\lambda_{\nu(\infty,2)}<\lambda_n$ and we can find a K and $\bar{\lambda}, \underline{\lambda}$ such that, $\lambda_{\nu(\infty,1)}<\underline{\lambda}<\bar{\lambda}(\boldsymbol{P}^{(k)})<\bar{\lambda}<\lambda_{\nu(\infty,2)}$ holds for all $k\geqq K$. Now taking into account of the relation

$$\begin{aligned}
&(\boldsymbol{P}^{(k+j)})_n/(\boldsymbol{P}^{(k+j)})_{\nu(\infty,2)} \\
&\quad=\{(\boldsymbol{P}^{(k)})_n/(\boldsymbol{P}^{(k)})_{\nu(\infty,2)}\}\prod_{i=1}^{j}\{(\lambda_n-\bar{\lambda}(\boldsymbol{P}^{(k+i)}))^2/(\lambda_{\nu(\infty,2)}-\bar{\lambda}(\boldsymbol{P}^{(k+i)}))^2\} \\
&\quad\geqq\{(\boldsymbol{P}^{(k)})_n/(\boldsymbol{P}^{(k)})_{\nu(\infty,2)}\}\prod_{i=1}^{j}\{(\lambda_n-\underline{\lambda})^2/(\lambda_{\nu(\infty,2)}-\underline{\lambda})^2\}\ ,
\end{aligned}$$

we can see that $(\boldsymbol{P}^{(k)})_n$ cannot tend to 0, this contradicts to the definition of $\boldsymbol{P}^{(\infty)}$. The case where it is supposed that $1<\nu(\infty, 1)$ holds is treated in the same manner.

TREOREM 3: *Under the condition of theorem 2 if there is a* λ_i $(i\neq 1, n)$ *for which* $\bar{\lambda}(\boldsymbol{P}^{(k)})\neq\lambda_i$ *for all* $k=0, 1, 2, \cdots$ *then the following inequality holds*

$$\left(\frac{\lambda_n-\lambda_1}{2}\right)^2+\left(\lambda_i-\frac{\lambda_n+\lambda_1}{2}\right)^2\geqq 2\left(\bar{\lambda}(\boldsymbol{P}^{(\infty)})-\frac{\lambda_n+\lambda_1}{2}\right)^2.$$

PROOF: We have

$$
\begin{aligned}
&(\boldsymbol{P}^{(k+2)})_i/(\boldsymbol{P}^{(k+2)})_1 \\
&\quad = \{(\boldsymbol{P}^{(k)})_i/(\boldsymbol{P}^{(k)})_1\}\,\{(\lambda_i-\bar{\lambda}(\boldsymbol{P}^{(k)}))^2(\lambda_i-\bar{\lambda}(\boldsymbol{P}^{(k+1)}))^2\} \\
&\qquad /\{(\lambda_1-\bar{\lambda}(\boldsymbol{P}^{(k)}))^2(\lambda_1-\bar{\lambda}(\boldsymbol{P}^{(k+1)}))^2\}\ .
\end{aligned}
$$

For any $\delta>0$ there is a $K(\varepsilon)$ such that for $k\geq K(\varepsilon)$ we have

$$
\begin{aligned}
&\{(\lambda_i-\bar{\lambda}(\boldsymbol{P}^{(k)}))^2(\lambda_i-\bar{\lambda}(\boldsymbol{P}^{(k+1)}))^2\}/\{\lambda_1-\bar{\lambda}(\boldsymbol{P}^{(k)}))^2(\lambda_1-\bar{\lambda}(\boldsymbol{P}^{(k+1)}))^2\} \\
&> \{(\lambda_i-\bar{\lambda}(\boldsymbol{P}^{(\infty)}))^2(\lambda_i-\bar{\lambda}(\boldsymbol{P}^{*(\infty)}))^2\}/(\lambda_1-\bar{\lambda}(\boldsymbol{P}^{(\infty)}))^2(\lambda_1-\bar{\lambda}(\boldsymbol{P}^{*(\infty)}))^2\}-\varepsilon\ .
\end{aligned}
$$

From this relation we have

$$
(\lambda_i-\bar{\lambda}(\boldsymbol{P}^{(\infty)}))^2(\lambda_i-\bar{\lambda}(\boldsymbol{P}^{*(\infty)}))^2/(\lambda_1-\bar{\lambda}(\boldsymbol{P}^{(\infty)}))^2(\lambda_1-\bar{\lambda}(\boldsymbol{P}^{*(\infty)}))^2\leqq 1
$$

and

$$
\begin{aligned}
&\left(\lambda_i-\frac{\lambda_n+\lambda_1}{2}-\left(\bar{\lambda}(\boldsymbol{P}^{(\infty)})-\frac{\lambda_n+\lambda_1}{2}\right)\right)^2\left(\lambda_i-\frac{\lambda_n+\lambda_1}{2}-\left(\bar{\lambda}(\boldsymbol{P}^{*(\infty)})-\frac{\lambda_n+\lambda_1}{2}\right)\right)^2 \\
&\leqq\left(\lambda_1-\frac{\lambda_n+\lambda_1}{2}-\left(\bar{\lambda}(\boldsymbol{P}^{(\infty)})-\frac{\lambda_n+\lambda_1}{2}\right)\right)^2\left(\lambda_1-\frac{\lambda_n+\lambda_1}{2}-\left(\bar{\lambda}(\boldsymbol{P}^{*(\infty)})-\frac{\lambda_n+\lambda_1}{2}\right)\right)^2 .
\end{aligned}
$$

By using the relation $\bar{\lambda}(\boldsymbol{P}^{*(\infty)})=\lambda_n+\lambda_1-\bar{\lambda}(\boldsymbol{P}^{(\infty)})$ we have

$$
\begin{aligned}
&\left(\lambda_i-\frac{\lambda_n+\lambda_1}{2}-\left(\bar{\lambda}(\boldsymbol{P}^{(\infty)})-\frac{\lambda_n+\lambda_1}{2}\right)\right)^2\left(\lambda_i-\frac{\lambda_n+\lambda_1}{2}+\left(\bar{\lambda}(\boldsymbol{P}^{(\infty)})-\frac{\lambda_n+\lambda_1}{2}\right)\right)^2 \\
&\leqq\left(\frac{\lambda_1-\lambda_n}{2}-\left(\bar{\lambda}(\boldsymbol{P}^{(\infty)})-\frac{\lambda_n+\lambda_1}{2}\right)\right)^2\left(\frac{\lambda_1-\lambda_n}{2}+\left(\bar{\lambda}(\boldsymbol{P}^{(\infty)})-\frac{\lambda_n+\lambda_1}{2}\right)\right)^2
\end{aligned}
$$

or

$$
\begin{aligned}
&\left(\left(\lambda_i-\frac{\lambda_n+\lambda_1}{2}\right)^2-\left(\bar{\lambda}(\boldsymbol{P}^{(\infty)})-\frac{\lambda_n+\lambda_1}{2}\right)^2\right)^2 \\
&\qquad\qquad\leqq\left(\left(\frac{\lambda_1-\lambda_n}{2}\right)^2-\left(\bar{\lambda}(\boldsymbol{P}^{(\infty)})-\frac{\lambda_n+\lambda_1}{2}\right)^2\right)^2 .
\end{aligned}
$$

From this last inequality we can get the desired result.

§ 2. Application to the optimum gradient method

In this section we shall discuss the application of the results of the former section to the so-called optimum gradient method. Let A be an n-by-n matrix, not singular, and let x, b denote the n-rowed column vectors. The optimum gradient method for the solution of simultaneous linear equation $Ax=b$ with respect to the metric $\|\ \|_P$ is defined as follows*); Take a positive definite symmetric matrix P. Then the gradient of the error function $\| Az-b \|_P^2=(Az-b, P(Az-b))$ at z is given by $2(A'P(Az-b))$. Given the k-th approximate solution $x_k=x+\varepsilon_k$ where x represents the desired solution the optimum gradient method proceeds by finding the $(k+1)$-th approximate solution $x_{k+1}=x_k-\gamma_k\zeta_k$ where

$$\begin{aligned} \operatorname*{Min}_{\gamma} \| A(x_k-\gamma\zeta_k)-b \|_P^2 &= \|A(x_k-\gamma_k\zeta_k)-b\|_P^2 \\ \gamma_k &= (A\varepsilon_k, A\zeta_k)_P/\|A\zeta_k\|_P^2 \\ \zeta_k &= \frac{1}{2} \text{ gradient at } x_k = A'PA\varepsilon_k . \end{aligned}$$

We shall represent by $(0<)\lambda_1 \leqq \lambda_2 \leqq \cdots \leqq \lambda_n$ the eigenvalues of $A'PA$ and by $\xi_1, \xi_2, \cdots, \xi_n$ the corresponding eigenvectors which are supposed to be orthonormal. We exclude the trivial case where $\lambda_1=\lambda_2=\cdots=\lambda_n$. Then we have the following.

THEOREM 4; *In the optimum gradient method with respect to the metric* $\| \ \|_P$

i) ε_k $(=x_k-x)$ *tends to be approximated by a linear combination of two fixed eigenvectors of* $A'PA$ *with the eigenvalues equal to* $\operatorname{Max}(\lambda_i ; (\varepsilon_0, \xi_i)\neq 0)$ *and* $\operatorname{Min}(\lambda_i ; (\varepsilon_0, \xi_i)\neq 0)$, *respectively, and*

ii) ε_k *alternates asymptotically in two fixed directions.*

PROOF: Here we prove the theorem for the case where $\lambda_1<\lambda_2<\cdots<\lambda_n$ and $(\varepsilon_0, \xi_1)\neq 0$, $(\varepsilon_0, \xi_n)\neq 0$ hold. Modifications necessary for the proofs of other cases are obvious and are omitted here.

Suppose that the error ε_k of the k-th approximate solution is represented as $\varepsilon_k=\sum_{i=1}^n \alpha_i^{(k)}\lambda_i^{-1}\xi_i$. Then ε_{k+1} is given by the following;

$$\begin{aligned} \varepsilon_k &= \sum_{i=1}^n \alpha_i^{(k+1)}\lambda_i^{-1}\xi_i=\varepsilon_k-\gamma_k\zeta_k=\varepsilon_k-\gamma_k A'PA\varepsilon_k \\ &= \sum_{i=1}^n \alpha_i^{(k)}(1-\gamma_k\lambda_i)\lambda_i^{-1}\xi_i=\gamma_k \sum_{i=1}^n \alpha_i^{(k)}(\gamma_k^{-1}-\lambda_i)\lambda_i^{-1}\xi_i \end{aligned}$$

where $\gamma_k^{-1}=\| A\zeta_k \|_P^2/(A\varepsilon_k, A\zeta_k)_P=\sum_{i=1}^n (\alpha_i^{(k)})^2\lambda_i/\sum_{i=1}^n (\alpha_i^{(k)})^2>0$.

If we consider the set of values $((\alpha_1^{(k)})^2/\sum_{i=1}^n(\alpha_i^{(k)})^2, (\alpha_2^{(k)})^2/\sum_{i=1}^n(\alpha_i^{(k)})^2, \cdots, (\alpha_n^{(k)})^2/\sum_{i=1}^n(\alpha_i^{(k)})^2)$ as a probability distribution $P^{(k)}$ over $(\lambda_1, \lambda_2, \cdots, \lambda_n)$, then

*) Here we use the notations $\| x \|_P^2$ and $(x, y)_P$ to represent the quantities (x, Px) and (x, Py), respectively, where (x, y) denotes the ordinary inner product of vectors x and y. A' denotes the transposition of A.

$\boldsymbol{P}^{(k+1)}=((\alpha_1^{(k+1)})^2/\sum_{i=1}^n(\alpha_i^{(k+1)})^2, (\alpha_2^{(k+1)})^2/\sum_{i=1}^n(\alpha_i^{(k+1)})^2, \cdots, (\alpha_n^{(k+1)})^2)/\sum_{i=1}^n(\alpha_i^{(k+1)})^2)$ is represented as $\boldsymbol{P}^{(k+1)}=T\boldsymbol{P}^{(k)}$, transformation T being defined in § 1. Thus the results of the former section are applicable to the present sequence of $\boldsymbol{P}^{(k)}$'s and we have for some non-zero c

$$\lim_{h\to\infty}\boldsymbol{P}^{(2h)}=\left(\frac{1}{1+c^2}, 0, \cdots, 0, \frac{c^2}{1+c^2}\right)\equiv\boldsymbol{P}^{(\infty)}$$

$$\lim_{h\to\infty}\boldsymbol{P}^{(2h+1)}\left(\frac{c^2}{1+c^2}, 0, \cdots, 0, \frac{1}{1+c^2}\right)\equiv\boldsymbol{P}^{*(\infty)}$$

and the limit points of the sequence of direction-cosines

$$\left(\frac{\alpha_1^{(2h)}}{\sqrt{\sum_{i=1}^n(\alpha_i^{(2h)})^2}}, \frac{\alpha_2^{(2h)}}{\sqrt{\sum_{i=1}^n(\alpha_i^{(2h)})^2}}, \cdots, \frac{\alpha_n^{(2h)}}{\sqrt{\sum_{i=1}^n(\alpha_i^{(2h)})^2}}\right)$$

are limited to the set of points of the type

$$\left(\pm\frac{1}{\sqrt{1+c^2}}, 0, \cdots, 0, \pm\frac{c}{\sqrt{1+c^2}}\right).$$

Now consider the transformation $\hat{T}$ defined over the set of points $\{(\delta_1, \delta_2, \cdots, \delta_n); \sum_{i=1}^n \delta_i^2=1, \delta_i^2<1 (i=1, 2, \cdots, n)\}$ by the following;

$$\hat{T}(\delta_1, \delta_2, \cdots, \delta_n)$$
$$=\left(\frac{\delta_1(\bar{\lambda}-\lambda_1)}{\sqrt{\sum_{i=1}^n\delta_i^2(\bar{\lambda}-\lambda_i)^2}}, \frac{\delta_2(\bar{\lambda}-\lambda_2)}{\sqrt{\sum_{i=1}^n\delta_i^2(\bar{\lambda}-\lambda_i)^2}}, \cdots, \frac{\delta_n(\bar{\lambda}-\lambda_n)}{\sqrt{\sum_{i=1}^n\delta_i^2(\bar{\lambda}-\lambda_i)^2}}\right)$$

where $\bar{\lambda}=\sum_{i=1}^n\delta_i^2\lambda_i$. Then we have

$$\left(\frac{\alpha_1^{(k+1)}}{\sqrt{\sum_{i=1}^n(\alpha_i^{(k+1)})^2}}, \frac{\alpha_2^{(k+1)}}{\sqrt{\sum_{i=1}^n(\alpha_i^{(k+1)})^2}}, \cdots, \frac{\alpha_n^{(k+1)}}{\sqrt{\sum_{i=1}^n(\alpha_i^{(k+1)})^2}}\right)$$
$$=\hat{T}\left(\frac{\alpha_1^{(k)}}{\sqrt{\sum_{i=1}^n(\alpha_i^{(k)})^2}}, \frac{\alpha_2^{(k)}}{\sqrt{\sum_{i=1}^n(\alpha_i^{(k)})^2}}, \cdots, \frac{\alpha_n^{(k)}}{\sqrt{\sum_{i=1}^n(\alpha_i^{(k)})^2}}\right)$$

and relations such as

$$\hat{T}\left(\frac{1}{\sqrt{1+c^2}}, 0, \cdots, 0, \frac{c}{\sqrt{1+c^2}}\right)=\left(\frac{|c|}{\sqrt{1+c^2}}, 0, \cdots, 0, \frac{-c}{|c|\sqrt{1+c^2}}\right)$$

$$\hat{T}\left(\frac{|c|}{\sqrt{1+c^2}}, 0, \cdots, 0, \frac{-c}{|c|\sqrt{1+c^2}}\right)=\left(\frac{1}{\sqrt{1+c^2}}, 0, \cdots, 0, \frac{c}{\sqrt{1+c^2}}\right).$$

Obviously $\hat{T}$ is continuous in its domain and by applying the eintirely similar arguments in the proof of theorem 1 of § 1, which were used to show the oscillatory convergence of $P^{(k)}$, we can get the conclusions i) and ii) of the present theorem.

Note: It may be of some use to remark here the relations $(\zeta_k, \zeta_{k+1})=0$,

$$\frac{\|\zeta_{k+1}\|_I^2}{\|\zeta_k\|_I^2}=\frac{\overline{(\lambda-\bar{\lambda}(P^{(k)}))^2}(P^{(k)})}{\{\bar{\lambda}(P^{(k)})\}^2}=(\text{coefficient of variation of } P^{(k)})^2 .$$

Thus taking into account the fact that $\gamma_k^{-1}=\bar{\lambda}(P^{(k)})$ and lemma 2 of § 1, we can see that the ratio $\|\zeta_{k+1}\|_I^2/\|\gamma_k\zeta_k\|_I^2=\overline{(\lambda-\bar{\lambda}(P^{(k)}))^2}(P^{(k)})$ tends monotone-increasingly to the limit $(\lambda_n-\lambda_1)^2c^2(1+c^2)^{-2}$.*)

This theorem provides the theoretical foundations of the following acceleration procedure which is a slightly modified version of the acceleration procedure proposed by Forsythe and Motzkin [5].

Acceleration procedure for the optimum gradient method:

In the optimum gradient method when the direction of vector ζ_k is nearly the same as that of ζ_{k-2}, it is recommendable to insert the step defined by the following;

$$x_{k+1}=x_k-\hat{\gamma}_k(x_{k-2}-x_k)$$

where

$$\hat{\gamma}_k=(A\varepsilon_k, A(x_{k-2}-x_k))_P/\|A(x_{k-2}-x_k)\|_P^2 .$$

The rationale for this procedure is as follows; The fact that the direction of ζ_k is nearly the same of that of ζ_{k-2} means that $\zeta_{k-2}\doteqdot\|\zeta_{k-2}\|_I\|\zeta_k\|_I^{-1}\zeta_k$ where the sign $\doteqdot$ is used in place of the description " is approximately equal to ". Then we can see that

$$\begin{aligned} x_{k-2}-x_k &= \varepsilon_{k-2}-\varepsilon_k=(A'PA)^{-1}(\zeta_{k-2}-\zeta_k) \\ &\doteqdot (\|\zeta_{k-2}\|_I\|\zeta_k\|_I^{-1}-1)(A'PA)^{-1}\zeta_k \\ &= (\|\zeta_{k-2}\|_I\|\zeta_k\|_I^{-1}-1)\varepsilon_k \end{aligned}$$

holds and it is obvious that in this case $x_{k-2}-x_k$ will be a good candidate for the correcting term of the $x_k=x+\varepsilon_k$. Now if we can suppose that $\varepsilon_{k-2}\doteqdot\beta(\lambda_1^{-1}\xi_1+c\lambda_n^{-1}\xi_n)$ then we have

---

*) It is supposed here that $(P^{(0)})_1>0$ and $(P^{(0)})_n>0$ hold.

$$\zeta_{k-2} \doteqdot \beta(\xi_1 + c\xi_n)$$
$$\gamma_{k-2} = (A\varepsilon_{k-2}, A\zeta_{k-2})_P / \| A\zeta_{k-2} \|_P^2 \doteqdot (1+c^2)/(\lambda_1 + c^2\lambda_n)$$
$$\varepsilon_{k-1} = \varepsilon_{k-2} - \gamma_{k-2}\zeta_{k-2} \doteqdot \beta(\lambda_1 + c^2\lambda_n)^{-1}(\lambda_n - \lambda_1)c^2 \{\lambda_1^{-1}\xi_1 - c^{-1}\lambda_n^{-1}\xi_n\}$$
$$\| A\varepsilon_{k-2} \|_P^2 = (\varepsilon_{k-2}, \zeta_{k-2}) \doteqdot \beta^2(\lambda_1^{-1} + c^2\lambda_n^{-1})$$
$$\| A\varepsilon_{k-1} \|_P^2 = (\varepsilon_{k-1}, \zeta_{k-1}) \doteqdot \beta^2(\lambda_1 + c^2\lambda_n)^{-2}(\lambda_n - \lambda_1)^2 c^4(\lambda_1^{-1} + c^{-2}\lambda_n^{-1})$$
$$\varepsilon_k \doteqdot \beta(\lambda_n - \lambda_1)^2(\lambda_1 + c^2\lambda_n)^{-1}(\lambda_1 + c^{-2}\lambda_n)^{-1}\{\lambda_1^{-1}\xi_1 + c\lambda_n^{-1}\xi_n\}$$
$$\doteqdot \| A\varepsilon_{k-1} \|_P^2 \| A\varepsilon_{k-2} \|_P^{-2} \varepsilon_{k-2} .$$

It is obvious that in this case the direction of ζ_k is nearly the same as that of ζ_{k-2}. This fact and the result of theorem 1 which assures that we can expect that ε_{k-2} takes the form just stated when k becomes large, give the theoretical basis of the recommendation of the present acceleration procedure.

The rate of convergence of the optimum gradient method:

By using the results of calculations in the former paragraph we can see that when the distribution $P^{(k)}$ corresponding to ε_k tends alternatingly to the limiting distributions $P^{(\infty)} = (1/(1+c^2), 0, \cdots, 0, c^2/(1+c^2))$ and $P^{*(\infty)} = (c^2/(1+c^2), 0, \cdots 0, 1/(1+c^2))$ i.e. ε_k tends to be approximated by $\beta_k\{\lambda_1^{-1}\xi_1 + c\lambda_n^{-1}\xi_n\}$ with some scalar β_k and c, the rate of convergence $\| A\varepsilon_{k+1} \|_P^2 / \| A\varepsilon_k \|_P^2$ tends to the value $(\lambda_n - \lambda_1)^2\{(\lambda_1 + \lambda_n)^2 + (c - c^{-1})^2\lambda_1\lambda_n\}^{-1}$. Thus the rate of convergence of the optimum gradient method is eventually determined by the value of c^2 which is inherited from the initial vector x_0 or $P^{(0)}$. The values $(\lambda_n - \lambda_1)^2\{(\lambda_1 + \lambda_n)^2 + (c - c^{-1})^2\lambda_1\lambda_n\}^{-1}$ attains its maximum value $(\lambda_n - \lambda_1)^2(\lambda_1 + \lambda_n)^{-2}$ when $c^2 = 1$ holds. Now $(\lambda_n - \lambda_1)^2/(\lambda_n + \lambda_1)^2 = (t-1)^2/(t+1)^2$ where $t \equiv \lambda_n/\lambda_1$ is a so-called condition-number of the matrix of $A'PA$.

We shall here make a slight digression for investigation of the meaning of the condition-number. Now define the function $f(\hat{x}, c) \equiv \| A(\hat{x} - c\hat{\zeta}) - b \|_P^2 / \| A\hat{x} - b \|_P^2$ of a vector $\hat{x}(\neq x)$ and a scaler c where $\hat{\zeta} = A'PA\hat{\varepsilon}$, $\hat{\varepsilon} = \hat{x} - x$ (x is the solution of $Ax = b$). If $\hat{\varepsilon} = \sum_{i=1}^{n} \beta_i \xi_i$, we have $f(\hat{x}, c) = \sum_{i=1}^{n} \beta_i^2\lambda_i(1 - c\lambda_i)^2 / \sum_{i=1}^{n} \beta_i^2\lambda_i$ and thus $\underset{c}{\mathrm{Min}}\, \underset{\hat{x}}{\mathrm{Max}} f(\hat{x}, c) = \underset{c}{\mathrm{Min}}\, (\mathrm{Max}\, ((1 - c\lambda_1)^2, (1 - c\lambda_n)^2)) = (1 - 2\lambda_1/(\lambda_1 + \lambda_n))^2 = (\lambda_n - \lambda_1)^2/(\lambda_n + \lambda_1)^2$ for $c = 2/(\lambda_1 + \lambda_n)$. Obviously $\underset{c}{\mathrm{Min}}\, \underset{\hat{x}}{\mathrm{Max}} f(\hat{x}, c) \geqq \underset{\hat{x}}{\mathrm{Max}}\, \underset{c}{\mathrm{Min}} f(\hat{x}, c)$ holds and $\mathrm{Min}\, f(\hat{x}, c) = f(\hat{x}, \hat{\gamma})$ where $\hat{\gamma} = (A\hat{\varepsilon}, A\hat{\zeta})_P / \| A\hat{\zeta} \|_P^2$ is given by the optimum gradient method. Thus we get $f(\hat{x}, \hat{\gamma}) \leqq (\lambda_n - \lambda_1)^2/(\lambda_n + \lambda_1)^2$. We have already seen that $f(\hat{x}, \hat{\gamma}) = (\lambda_n - \lambda_1)^2/(\lambda_n + \lambda_1)^2$ holds for $\hat{x}$ with corresponding probability distribution $\hat{P}=(1/2, 0, \cdots, 0, 1/2)$ and we can see that the value $(\lambda_n-\lambda_1)^2/(\lambda_n+\lambda_1)^2=(t-1)^2/(t+1)^2$ gives the least upper bound of the convergence rate $f(\hat{x}, \hat{\gamma})$ of the optimum gradient method. It is stated in [5] that a proof of this fact was given by Kantrovitch [8] but as the Kantrovitch's paper was not available, we gave a proof to it for the sake of completeness of the following discussion.

Here we use the result of theorem 3 of § 1 to see why the optimum gradient method often converges with the convergence rate nearly equal to its worst possible value $(\lambda_n-\lambda_1)^2/(\lambda_n+\lambda_1)^2$. This fact was also noticed by Forsythes [5]. For $P^{(\infty)}=(1/(1+c^2), 0, \cdots, 0, c^2/(1+c^2))$ and $P^{*(\infty)}=(c^2/(1+c^2), 0, \cdots, 0, 1/(1+c^2))$ we have $\bar{\lambda}(P^{(\infty)})=(\lambda_1+c^2\lambda_n)/(1+c^2)$ and $\bar{\lambda}(P^{*(\infty)})=(c^2\lambda_1+\lambda_n)/(1+c^2)$ and thus we have $(\bar{\lambda}(P^{(\infty)})-(\lambda_n+\lambda_1)/2)^2=(\bar{\lambda}(P^{*(\infty)})-(\lambda_n+\lambda_1)/2)^2=(\lambda_n-\lambda_1)^2(1-c^2)^2/4(1+c^2)^2$. Thus under the condition which assumes that the point λ_i is not discarded during the course of approximation procedure, we have from the result of theorem 3 of § 1

$$\left(\frac{\lambda_n-\lambda_1}{2}\right)^2+\left(\lambda_i-\frac{\lambda_n+\lambda_1}{2}\right)^2\geqq\frac{(1-c^2)^2}{2(1+c^2)^2}(\lambda_n-\lambda_1)^2 .$$

By putting $\delta_i\equiv(\lambda_i-(\lambda_n+\lambda_1)/2)/((\lambda_n-\lambda_1)/2)$ we get from the above inequality the following

$$4\left\{\frac{1+\delta_i^2}{1-\delta_i^2}\right\}\geqq(c-c^{-1})^2 .$$

Thus, for example, when there exists some λ_i which satisfies the condition of theorem 3 and with $|\delta_i|\ll 1$ we can expect that $(c-c^{-1})^2$ is near or less than 4. Now the convergence rate at the point corresponding to the $P^{(\infty)}=(1/(1+c^2), 0, \cdots, 0, c^2/(1+c^2))$ is given by $(\lambda_n-\lambda_1)^2(\lambda_n+\lambda_1)^{-2}\{1+(c-c^{-1})^2(\lambda_n\lambda_1^{-1}+\lambda_1\lambda_n^{-1}+2)^{-1}\}^{-1}$ and by putting $\varepsilon=1-(\lambda_n-\lambda_1)/(\lambda_n+\lambda_1)$ this is represented as $(\lambda_n-\lambda_1)^2(\lambda_n+\lambda_1)^{-2}\{1+(c-c^{-1})^2(2\varepsilon^{-1}+1+\varepsilon(2-\varepsilon)^{-1})^{-1}\}^{-1}$. Thus we can see that the convergence rate tends to be greater than $(\lambda_n-\lambda_1)^2(\lambda_n+\lambda_1)^{-2}\{1+(\varepsilon/2)(c-c^{-1})^2\}^{-1}$ or a fortiori than $(\lambda_n-\lambda_1)^2(\lambda_n+\lambda_1)^{-2}\{1+2\varepsilon(1+\delta_i^2)(1-\delta_i^2)^{-1}\}^{-1}$. From this last relation we can see that when we use the optimum gradient method for ill-conditioned ($\varepsilon\ll 1$) matrix $A'PA$, then the rate of convergence tends near to its worst possible value, especially when there is some λ_i near the value $(\lambda_n+\lambda_1)/2$ i.e. $|\delta_i|\ll 1$.

As to one of the numerical examples described by Forsythes [5], and which will be treated fully in the following paragraph of this section, we have $\varepsilon=0.01073$ and

$$\begin{aligned}
&(\lambda_n-\lambda_1)^2(\lambda_n+\lambda_1)^{-2}\{1+2\varepsilon(1+\delta_i^2)(1-\delta_i^2)^{-1}\}^{-1} \\
&=(\lambda_n-\lambda_1)^2(\lambda_n+\lambda_1)^{-2}\times 0.9789=0.9580 \quad \text{for} \quad \delta_5^2=0.001321 \\
&=(\lambda_n-\lambda_1)^2(\lambda_n+\lambda_1)^{-2}\times 0.7171=0.7018 \quad \text{for} \quad \delta_2^2=0.896854\,.
\end{aligned}$$

We can clearly see in this example the effect of the small value of ε making the convergence rate tend near to its worst possible value. Thus, it seems that our present analysis gives fairly general theoretical explanation to the fact that the optimum gradient method converges slowly when the matrix is ill-conditioned.

Numerical example

Here we shall present some numerical results obtained by using the same A and b as those treated by Forsythes [5] and by putting $P=I$ (identity matrix). These results are obtained by using a FACOM-128 relay computer of our institute. We have used the computation scheme of optimum gradient method where the acceleration step is inserted automatically when the condition $(\zeta_{k-2}, \zeta_k)_I/(\|\zeta_{k-2}\|_I\|\zeta_k\|_I)>\delta$ is satisfied for some preassigned value of $\delta\,(1>\delta>0)$.

We have

$$A=\begin{bmatrix}
\sqrt{0.00268704} & & & & & 0 \\
& \sqrt{0.01581310} & & & & \\
& & \sqrt{0.08234830} & & & \\
& & & \sqrt{0.17590130} & & \\
0 & & & & \sqrt{0.25946632} & \\
& & & & & \sqrt{0.49823436}
\end{bmatrix}$$

$b=[0,\ 0,\ 0,\ 0,\ 0,\ 0,]$ and

$$A'PA=\begin{bmatrix}
0.00268704\ (=\lambda_1) & & & & & 0 \\
& 0.01581310\ (=\lambda_2) & & & & \\
& & 0.08234830\ (=\lambda_3) & & & \\
& & & 0.17590130\ (=\lambda_4) & & \\
0 & & & & 0.25946632\ (=\lambda_5) & \\
& & & & & 0.49823436\ (=\lambda_6)
\end{bmatrix}.$$

We have used three values 0.99, 0.999 and 0.9999 as δ. In fig's. 1, 2 and

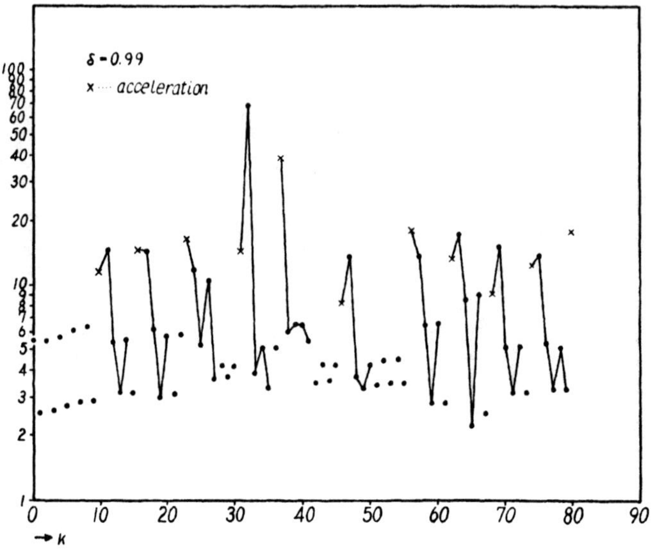
δ = 0.99
× ····· acceleration
100
90
80
70
60
50
40
30
20
10
9
8
7
6
5
4
3
2
1
0
10
20
30
40
50
60
70
80
90
→ k

Fig. 1

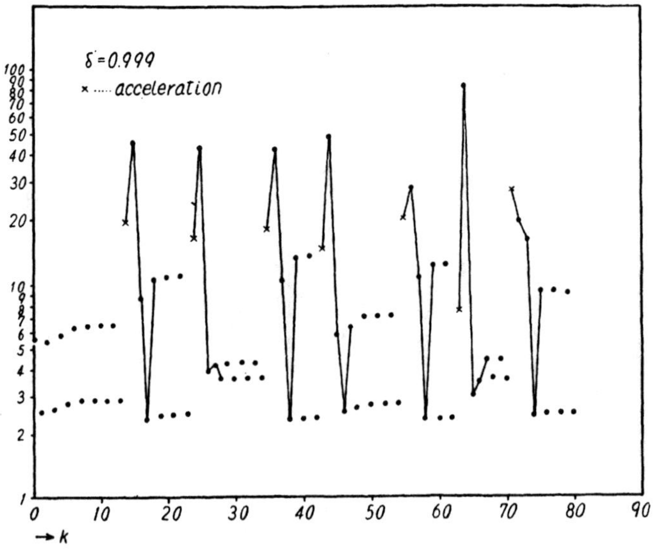
δ = 0.999
× ····· acceleration
100
90
80
70
60
50
40
30
20
10
9
8
7
6
5
4
3
2
1
0
10
20
30
40
50
60
70
80
90
→ k

Fig. 2

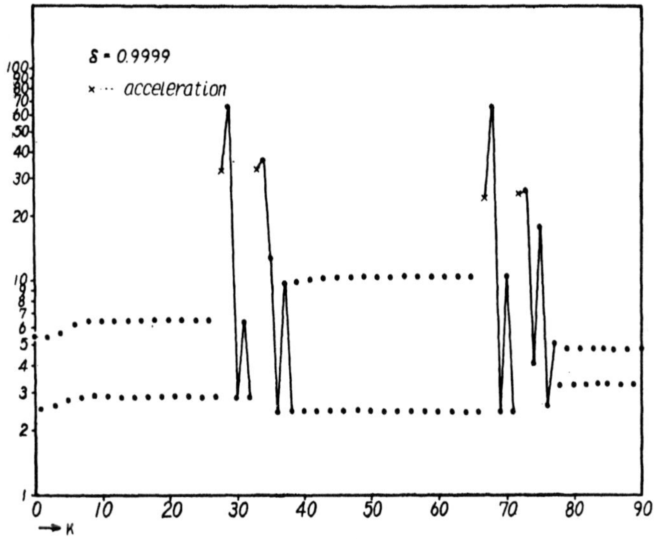

Fig. 3

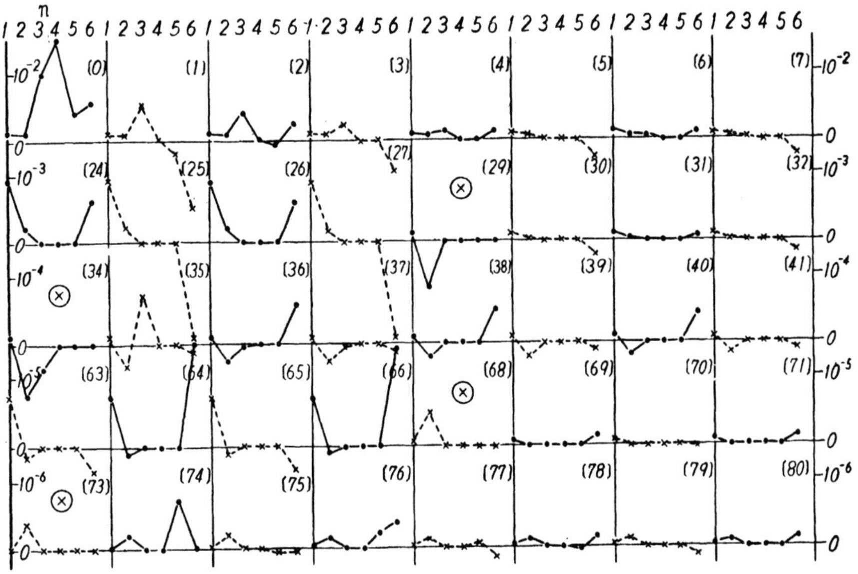

Fig. 4

3 the sequence of γ_k's are represented by • and $\hat{\gamma}_k$'s (acceleration step) are represented by ×. These three figures correspond to the above mentioned three values of δ. In fig. 4 part of the values of ζ_k's are illustrated for the case $\delta=0.9999$. The numbers in the blacket in fig. 4 correspond to the k's of fig. 3 and the symbols ⊗ signify the result of acceleration procedure.

We can see in these examples a good agreement between the results of our theoretical analysis made in this paper and the results of practical computations.

Acknowledgement

I wish to express my thanks to Mr. K. Isii for reading the manuscript of § 1 of this paper. Thanks are also due to Miss Y. Saigusa who performed the programmings and operations of FACOM-128 automatic relay computer to prepare the numerical results illustrated in § 2.

THE INSTITUTE OF STATISTICAL MATHEMATICS

REFERENCES

[1] Akaike, H. "On a computation method for eigenvalue problems and its application to statistical analysis," *Annals of the Institute of Statistical Mathematics* 10, 1-20 (1958).

[2] Chernoff, H. and Divinsky, N. "The computation of maximum-likelihood estimates of linear structural equations," Chapter 10 in *Studies in Ecnometric Method, Cowles Commission Monograph* 14, Wm. C. Hood and T. C. Koopmans editors, John Wiley and Sons, Inc. New York, (1953).

[3] Crockett, J. B. and Chernoff, H. "Gradient methods of maximization," *Pacific J. Math.*, 5, 33-50 (1955).

[4] Forsythe, G. E. "Solving linear algebraic equations can be interesting," *Bull. Amer. Math. Soc.*, 59, 299-329 (1953).

[5] Forsythe, A. I. and Forsythe, G. E. "Punched-card experiments with accelerated gradient methods for linear eqations," *Contributions to the Solution of systems of Linear Equations and the Determination of Eigenvalues, N.B.S. Applied Mathematics Series* 39, Olga Taussky editor, 55-69 (1954).

[6] Forsythe, G. E. and Motzkin, T. S. "Asymptotic properties of the optimum gradient method," *Bul. Am. Math. Soc.*, 57, 183 (1951) (Abstract).

[7] Forsythe, G. E. and Motzkin, T. S. "Acceleration of the optimum gradient method," Preliminary report, *Bul. Am. Math. Soc.*, 57, 304-305 (1951) (Abstract).

[8] Kantrovich, L. V. "Functional analysis and applied mathematics," *Uspekhi Mathematicheskikh Nauk* (NS), 3, No. 6, 89-185 (1951).

Reprinted from *Annals of the Institute of Statistical Mathematics*, Vol. 11, No. 3, 1960, 145-165 by permission of the Institute of Statistical Mathematics

EFFECT OF TIMING-ERROR ON THE POWER SPECTRUM OF SAMPLED-DATA

By HIROTUGU AKAIKE

(Received Feb. 15, 1960)

0. Introduction and summary

Nowadays it is very common to use time-sampled data to analyse or control an object fluctuating continuously in time. This seems to be motivated by the recent developments of digital methods serving for these purposes. In relation to such procedures there are many papers which treat the noise or error due to quantization ([4], [5]). The effect of time-sampling on the spectral properties is also well-known as folding or aliasing for the case where the fluctuation of object is represented by a stationary stochastic process and the timings are performed without error. As to the cases where timing-errors are present we have seen yet little quantitative description of their effects on the spectral properties of the time-sampled data [3]. In the present paper we treat this problem for the case where the timing is independent of the original process and the intervals between sampling-time points form a stationary process. After the general discussion of this case we treat two special types of time-sampling in more details. The first corresponds to the case where, though it is intended to sample the record at the time points which are the integral multiples of a constant time Δt, the deviations of the sampling-time points from the preassigned ones are present and form a purely random process, i.e., they are random variables which are mutually independent and follow one and the same probability distribution. The second corresponds to the case where the sampling-time points form a renewal process, i.e., the interval lengthes between successive sampling-time points form a purely random process. The results of our analyses show clearly how the timing-error affects the power spectral distribution function of the time-sampled data. The effects are essentially non-linear but the time-sampling of the first type may be described as a low-pass filter with an inner white noise source. We can see further that in practical applications even the time-sampling procedures of the second type may sometimes act approximately as a low-pass filter with an inner white noise source. These results seem to give a mathematical explanation of the fact, mentioned by D. T. Ross in [3], that the time-sampling was sometimes vaguely (and incorrectly) considered to be a filtering. By using the continuous records of the outputs of accelerometers mounted on the frame and on the front axle of an automobile we illustrate these theoretical results by numerical examples. These records were obtained by the members of the research department of the Isuzu Motor Company and were presented to the author by Mr. Itiro Kanesige of the department for the purpose to develop the statistical research of the relations between vehicular oscillations and road surfaces. When we were reading these records using a rule we had to face two main sources of error. The one was the error in the horizontal position of the rule and the other was that due to quantization. The results of our investigation described in this paper enable us to make a quantitative evaluation of the effect of the reading-errors on the forms of the power spectral distribution functions of these time-sampled data.

1. Spectral properties of sampled-data.

We shall here consider a strictly stationary real stochastic process $\{x(t, \omega);\ -\infty<t<\infty\}$ with continuous time parameter t. It is assumed that

a.1. the process has zero-mean and finite second order moments,

a.2. almost all sample functions of the process have finite limit from the right,

$$x(t+, \omega)=\lim_{s \downarrow t} x(s, \omega)$$

for all t, and

a.3. the process is continuous in the sense of mean square,

$$\lim_{s-t\to 0} E\{|x(s, \omega)-x(t, \omega)|^2\}=0 .$$

It is fairly obvious that in almost every physical realizations of stochastic processes these conditions are satisfied. To see further that the condition a.3. really does not seriously restrict the generality of the process, the reader is recommended to consult the book [1] by J. L. Doob. Here we take another strictly stationary stochastic process $\{\Delta\tau_n(\omega);\ -\infty<n<\infty\}$ defined on the same ω space as $x(t, \omega)$ process and with discrete time parameter n. We define $\tau_n(\omega)$ by the followings;

$$\tau_0(\omega)=\varepsilon(\omega)$$
$$\tau_n(\omega)-\tau_{n-1}(\omega)=\Delta\tau_{n-1}(\omega)$$

where $\varepsilon(\omega)$ is a random variable. Now we define a sequence of ω functions $\{x_{\tau,n}(\omega);\ -\infty<n<\infty\}$ as follows;

$$\begin{aligned} x_{\tau,n}(\omega)&=x(\tau_n(\omega),\omega) && \text{if Prob }\{\omega';\ \tau_n(\omega')=\tau_n(\omega)\}>0\\ &=x(\tau_n(\omega)+,\omega) && \text{otherwise.}\end{aligned}$$

We can see that the function $x_{\tau,n}(\omega)$ is well defined by the assumption a.2. To see that $\{x_{\tau,n}(\omega);\ -\infty<n<\infty\}$ forms a stochastic process we adopt the following discrete approximation procedure described in [2]. Denote by S_τ the set of values which some τ_n takes with positive probability. Obviously S_τ is at most enumerably infinite. For each positive integer q choose finitely many points

$$a_1^{(q)}<a_2^{(q)}<\cdots<a_{n_q}^{(q)}$$

in such a way that first q point of S_τ enumerated in some order are $a_j^{(q)}$'s and that every points in the interval $[-q,q]$ is within distance $1/q$ of some $a_j^{(q)}$ and $\pm\infty$ and 0 are some $a_j^{(q)}$'s. Define the stochastic process $\{\tau_n^{(q)}(\omega);\ -\infty<n<\infty\}$ by

$$\begin{aligned}\tau_n^{(q)}(\omega)&=a_1^{(q)} && \text{if } \tau_n(\omega)\le a_1^{(q)}\\ &=a_j^{(q)} && \text{if } a_{j-1}^{(q)}<\tau_n(\omega)\le a_j^{(q)}\\ &=0 && \text{if } \tau_n(\omega)>a_{n_q}^{(q)}.\end{aligned}$$

If we define

$$x_{\tau^{(q)},n}(\omega)=x(\tau_n^{(q)}(\omega),\omega)\ ,$$

it is obvious that $x_{\tau^{(q)},n}(\omega)$ is a measurable ω function, i.e., a random variable and

$$\lim_{q\to\infty} x_{\tau^{(q)},n}(\omega)=x_{\tau,n}(\omega)$$

holds with probability 1. Hence $x_{\tau,n}(\omega)$ is a random variable. Hereafter we shall sometimes omit the variable ω in the expression of random variables. Now consider a set $\{m_\nu;\ \nu=1,2,\cdots,k\}$ of arbitrary finite number of integers satisfying the relation $m_1<m_2<\cdots<m_k$ and a set $\{\xi_{m_\nu};\ \nu=1,2,\cdots,k\}$ of real numbers. Then we have

$$\begin{aligned}&\text{Prob }\{x_{\tau^{(q)},m_1}\le\xi_{m_1},\ x_{\tau^{(q)},m_2}\le\xi_{m_2},\ \cdots,\ x_{\tau^{(q)},m_k}\le\xi_{m_k}\}\\ &\quad=\sum_{(j_1,j_2,\cdots,j_k)}\text{Prob }\{x(a_{j_1})\le\xi_{m_1},\ x(a_{j_2})\le\xi_{m_2},\ \cdots,\ x(a_{j_k})\le\xi_{m_k}\ \text{and}\\ &\qquad\qquad \tau_{m_1}^{(q)}=a_{j_1},\ \tau_{m_2}^{(q)}=a_{j_2},\ \cdots,\ \tau_{m_k}^{(q)}=a_{j_k}\}\end{aligned}$$

where a_{j_ν} denotes $a_{j_\nu}^{(q)}$ for $\nu=1, 2, \cdots, k$ and the summation is taken over all the possible arrangements $(j_1, j_2, \cdots, j_k)$ where j_ν is one of $(1, 2, \cdots, n_q)$. Denote by $\phi^{(q)}(\lambda_1, \lambda_2, \cdots, \lambda_k; m_1, m_2, \cdots, m_k)$ the characteristic function of $(x_{\tau^{(q)}.m_1}, x_{\tau^{(q)}.m_2}, \cdots, x_{\tau^{(q)}.m_k})$ and by $\phi^{(q)}(\lambda_1, \lambda_2, \cdots, \lambda_k \mid a_{j_1}, a_{j_2}, \cdots, a_{j_k}; m_1, m_2, \cdots, m_k)$ the conditional characteristic function of $(x_{\tau^{(q)}.m_1}, x_{\tau^{(q)}.m_2}, \cdots, x_{\tau^{(q)}.m_k})$ conditioned by the random variables $(\tau_{m_1}^{(q)}, \tau_{m_2}^{(q)}, \cdots, \tau_{m_k}^{(q)})$. Then we have

$$\begin{aligned}&\phi^{(q)}(\lambda_1, \lambda_2, \cdots, \lambda_k; m_1, m_2, \cdots, m_k)\\&\quad=\sum_{(j_1, j_2, \cdots, j_k)} \phi^{(q)}(\lambda_1, \lambda_2, \cdots, \lambda_k \mid a_{j_1}, a_{j_2}, \cdots, a_{j_k}; m_1, m_2, \cdots, m_k)\\&\quad\quad\times \text{Prob}\,(\tau_{m_1}^{(q)}=a_{j_1}, \tau_{m_2}^{(q)}=a_{j_2}, \cdots, \tau_{m_k}^{(q)}=a_{j_k}) .\end{aligned}$$

Denote by $\phi(\lambda_1, \lambda_2, \cdots, \lambda_k; m_1, m_2, \cdots, m_k)$ the characteristic function of $(x_{\tau.m_1}, x_{\tau.m_2}, \cdots, x_{\tau.m_k})$. Then we have

$$\phi(\lambda_1, \lambda_2, \cdots, \lambda_k; m_1, m_2, \cdots, m_k)=\lim_{q\to\infty} \phi^{(q)}(\lambda_1, \lambda_2, \cdots, \lambda_k; m_1, m_2, \cdots, m_k) .$$

Hereafter we shall restrict our attention to the case where $\{\tau_n(\omega)\}$ and $\{x(t, \omega)\}$ are mutually independent. For this case we can put

$$\begin{aligned}&\phi^{(q)}(\lambda_1, \lambda_2, \cdots, \lambda_k \mid a_{j_1}, a_{j_2}, \cdots, a_{j_k}; m_1, m_2, \cdots, m_k)\\&\qquad\qquad=\psi(\lambda_1, \lambda_2, \cdots, \lambda_k; a_{j_1}, a_{j_2}, \cdots, a_{j_k})\end{aligned}$$

where $\psi(\lambda_1, \lambda_2, \cdots, \lambda_k; a_{j_1}, a_{j_2}, \cdots, a_{j_k})$ denotes the characteristic function of $(x(a_{j_1}, \omega), x(a_{j_2}, \omega), \cdots, x(a_{j_k}, \omega))$. From the continuity assumption a.3. for the $x(t)$ process it follows that $\psi(\lambda_1, \lambda_2, \cdots, \lambda_k; a_{j_1}, a_{j_2}, \cdots, a_{j_k})$ is continuous in $(a_{j_1}, a_{j_2}, \cdots, a_{j_k})$. Obviously $\lim_{q\to\infty} \tau_n^{(q)}(\omega)=\tau_n(\omega)$ with probability 1 and thus the finite dimensional distribution of $(\tau_{m_1}^{(q)}, \tau_{m_2}^{(q)}, \cdots, \tau_{m_k}^{(q)})$ converges to that of $(\tau_{m_1}, \tau_{m_2}, \cdots, \tau_{m_k})$ as $q\to\infty$. Thus taking into account the boundedness of $\psi(\lambda_1, \lambda_2, \cdots, \lambda_k; a_{j_1}, a_{j_2}, \cdots, a_{j_k})$ we have

$$\begin{aligned}&\phi(\lambda_1, \lambda_2, \cdots, \lambda_k; m_1, m_2, \cdots, m_k)\\&\quad=\int\psi(\lambda_1, \lambda_2, \cdots, \lambda_k; a_{j_1}, a_{j_2}, \cdots, a_{j_k})dP_{m_1, m_2, \cdots, m_k}(a_{j_1}, a_{j_2}, \cdots, a_{j_k})\end{aligned}$$

where $P_{m_1, m_2, \cdots, m_k}(a_{j_1}, a_{j_2}, \cdots, a_{j_k})=\text{Prob}\,(\tau_{m_1}\le a_{j_1}, \tau_{m_2}\le a_{j_2}, \cdots, \tau_{m_k}\le a_{j_k})$. From the strict stationarity of the $x(t)$ process

$$\psi(\lambda_1, \lambda_2, \cdots, \lambda_k; a_{j_1}, a_{j_2}, \cdots, a_{j_k})$$

can be represented in the form

$$\hat{\psi}(\lambda_1, \lambda_2, \cdots, \lambda_k; a_{j_2}-a_{j_1}, a_{j_3}-a_{j_2}, \cdots, a_{j_k}-a_{j_{k-1}})$$

and we have

$$\phi(\lambda_1, \lambda_2, \cdots, \lambda_k; m_1, m_2, \cdots, m_k)$$
$$=\int \hat{\psi}(\lambda_1, \lambda_2, \cdots, \lambda_k; b_2, b_3, \cdots, b_k) dP_{(m_2-m_1, m_3-m_2, \cdots, m_k-m_{k-1})}(b_2, b_3, \cdots, b_k)$$

where

$$P_{(m_2-m_1, m_3-m_2, \cdots, m_k-m_{k-1})}(b_2, b_3, \cdots, b_k)$$
$$=\text{Prob}\,\{\tau_{m_2}-\tau_{m_1}\leq b_2, \tau_{m_3}-\tau_{m_2}\leq b_3, \cdots, \tau_{m_k}-\tau_{m_{k-1}}\leq b_k\} .$$

Taking into account of the strict stationarity of the $\Delta\tau_n$ process we can see from this equation that $P_{(m_2-m_1, m_3-m_2, \cdots, m_k-m_{k-1})}$, and so the characteristic function $\phi(\lambda_1, \lambda_2, \cdots, \lambda_k; m_1, m_2, \cdots, m_k)$, is completely determined by the differences m_j-m_{j-1} ($j=2, 3, \cdots, k$). Thus the process $\{x_{\tau,n}(\omega)\}$ is seen to be strictly stationary. Taking into account the inequalities

$$\left|\frac{\partial}{\partial\lambda_\nu}\psi(\lambda_1, \lambda_2, \cdots, \lambda_k; a_{j_1}, a_{j_2}, \cdots, a_{j_k})\right|\leq E\,|\,x(a_{j_\nu})\,|\leq\sigma$$

$$\left|\frac{\partial^2}{\partial\lambda_\mu\partial\lambda_\nu}\psi(\lambda_1, \lambda_2, \cdots, \lambda_k; a_{j_1}, a_{j_2}, \cdots, a_{j_k})\right|\leq E\,|\,x(a_{j_\mu})x(a_{j_\nu})\,|\leq\sigma^2$$

where $\sigma^2=E\,|\,x(t)\,|^2$, we get

$$\frac{\partial}{\partial\lambda_\nu}\phi(\lambda_1, \lambda_2, \cdots, \lambda_k; a_{j_1}, a_{j_2}, \cdots, a_{j_k})$$
$$=\int\frac{\partial}{\partial\lambda_\nu}\psi(\lambda_1, \lambda_2, \cdots, \lambda_k; a_{j_1}, a_{j_2}, \cdots, a_{j_k})dP_{m_1, m_2, \cdots, m_k}(a_{j_1}, a_{j_2}, \cdots, a_{j_k})$$

$$\frac{\partial^2}{\partial\lambda_\mu\partial\lambda_\nu}\phi(\lambda_1, \lambda_2, \cdots, \lambda_k; a_{j_1}, a_{j_2}, \cdots, a_{j_k})$$
$$=\int\frac{\partial^2}{\partial\lambda_\mu\partial\lambda_\nu}\psi(\lambda_1, \lambda_2, \cdots, \lambda_k; a_{j_1}, a_{j_2}, \cdots, a_{j_k})dP_{m_1, m_2, \cdots m_k}(a_{j_1}, a_{j_2}, \cdots, a_{j_k})$$

and

$$E\{x_{\tau,n}(\omega)\}=0 ,$$

$$\rho(k)\equiv E\{x_{\tau,n+k}(\omega)x_{\tau,n}(\omega)\}=\int R(a_{n+k}-a_n)dP_{n,n+k}(a_n, a_{n+k})$$
$$=\int R(\tau)dP_{(k)}(\tau)$$

where $R(\tau)=E\{x(t+\tau, \omega)x(t, \omega)\}$ and $P_{(k)}(\tau)=\text{Prob}\,(\tau_k-\tau_0\leq\tau)$. Clearly this last relation holds for any k positive or negative and we have

$$\rho(0)=\sigma^2 .$$

Thus $x_{\tau,n}$ process is stationary also in the wide sense, i.e., has finite second order moments.

Obviously this $x_{\tau,n}$ process is a mathematical representation of the sequence of data which are time-sampled by using the timing impulses situated at the time points τ_n. The purpose of the present paper is to study the spectral properties of this $x_{\tau,n}$ process.

Now we have

$$\begin{aligned}\rho(k)&=\int R(\tau)dP_{(k)}(\tau)=\int_{(\tau)}\left[\int_{(f)}e^{2\pi if\tau}dP(f)\right]dP_{(k)}(\tau)\\&=\int_{(f)}\left[\int_{(\tau)}e^{2\pi if\tau}dP_{(k)}(\tau)\right]dP(f)\\&=\int\phi_k(f)dP(f)\end{aligned}$$

where $P(f)$ is the power spectral distribution function of the $x(t)$ process, continuous from the right and with $p(-\infty)=0$, $P(\infty)=\sigma^2$, and

$$\phi_k(f)=\int e^{2\pi if\tau}dP_{(k)}(\tau)\ .$$

As the $x(t)$ process is real we can assume that $P(-f)=\sigma^2-P(f)$ holds at the continuity points of $P(f)$ and we have

$$\rho(k)=\int_{(f)}\phi_k(f)dP(f)=2\int_{0+}^{\infty}Re(\phi_k(f))dP(f)+P(0)-P(0-)\ .^{*)}$$

This shows that $\rho(k)$ is obtainable as an output power of some (imaginary) filter with signed power transfer function $2Re(\phi_k(f))$ under the input $\{x(t)\}$. Now we can evaluate the power spectrum of the $x_{\tau,n}$ process by using some smoothing process or a filter. Take a convergence factor or a sequence of real numbers c_k such as

$$c_0=1\ ,$$

$$c_{-k}=c_k$$

and

$$\sum_{k=-\infty}^{\infty}|c_k|<\infty\ .$$

We define the smoothing function corresponding to the sequence $\{c_k\}$

$$h(f)=\sum_{k=-\infty}^{+\infty}c_ke^{2\pi ikf}\ .$$

Obviously $h(f)$ is a real continuous even periodic function with period 1 and

$$\int_{-1/2}^{1/2}h(f)df=1\ .$$

*) $Re(\phi_k(f))$ denotes the real part of $\phi_k(f)$.

Here we further assume that $h(f)\geq 0$, i.e., the sequence c_k is positive definite. Now we shall represent by $P_\tau(f)$ the power spectral distribution function of the $x_{\tau,n}$ process, which is continuous to the right, monoton non-decreasing and with $\lim_{f\to -1/2} P_\tau(f)=0$, $P_\tau(\frac{1}{2})=\sigma^2$. Then we have

$$\begin{aligned}\sum_{k=-\infty}^{\infty} c_k\rho(k)e^{-2\pi ikf} &= \sum_{k=-\infty}^{+\infty} e^{-2\pi ikf}c_k\int_{-1/2}^{1/2} e^{2\pi ikf'}dP_\tau(f') \\ &=\int_{-1/2}^{1/2}\left[\sum_{k=-\infty}^{\infty} c_k e^{-2\pi ik(f-f')}\right]dP_\tau(f') \\ &=\int_{-1/2}^{1/2} h(f-f')dP_\tau(f') .\end{aligned}$$

Define

$$h*P_\tau(f)=\int_{-1/2}^{1/2} h(f-f')dP_\tau(f') ,$$

then $h*P(f)$ is continuous with respect to f and

$$\int_{-1/2}^{1/2} h*P_\tau(f)df=\int_{-1/2}^{1/2} dP_\tau(f')=\sigma^2 .$$

Now consider a set $[\{c_k^{(n)};\ -\infty<k<\infty\};\ n=1, 2, 3, \cdots]$ of convergence factors $\{c_k^{(n)}\}$ with the corresponding smoothing functions $h^{(n)}(f)$ for which

$$\lim_{n\to\infty}\int_\alpha^\beta h^{(n)}(f)df=1$$

holds for any α and β satisfying $-\frac{1}{2}\leq\alpha<0<\beta\leq\frac{1}{2}$. Then we have from the eveness of $h^{(n)}(f)$

$$\lim_{n\to\infty}\int_0^\beta h^{(n)}(f)df=\lim_{n\to\infty}\int_\alpha^0 h^{(n)}(f)df=\tfrac{1}{2}$$

and for α' and β' such that $0\notin[\alpha', \beta']$ and $-\frac{1}{2}\leq\alpha'<\beta'\leq\frac{1}{2}$ we have

$$\lim_{n\to\infty}\int_{\alpha'}^{\beta'} h^{(n)}df=0 .$$

Now for x and y satisfying $-\frac{1}{2}\leq x<y<\frac{1}{2}$ we have

$$\begin{aligned}\lim_{n\to\infty}\int_x^y h^{(n)}*P_\tau(f)df &=\lim_{n\to\infty}\int_x^y\left[\int_{-1/2}^{1/2} h^{(n)}(f-f')dP_\tau(f')\right]df \\ &=\int_{-1/2}^{1/2}\left[\lim_{n\to\infty}\int_x^y h^{(n)}(f-f')df\right]dP_\tau(f') \\ &=\tfrac{1}{2}[(P_\tau(y)-P_\tau(y-))+(P_\tau(x)-P_\tau(x-))]+P_\tau(y-)-P_\tau(x)\end{aligned}$$

where $P_\tau(x-)=P_\tau(x)=0$ by definition for $x=-\frac{1}{2}$, and

$$\lim_{n\to\infty}\int_{-1/2}^{1/2} h^{(n)}*P_\tau(f)df=\sigma^2=P_\tau(\tfrac{1}{2}) .$$

Thus we have

$$\lim_{y\uparrow 1/2}\lim_{n\to\infty}\left(\int_{-1/2}^{1/2} dP_\tau(f)-\int_{-1/2}^{y} h^{(n)}*P_\tau(f)df\right)=\lim_{y\uparrow 1/2}\lim_{n\to\infty}\int_{y}^{1/2} h^{(n)}*P_\tau(f)df$$
$$=P_\tau(\tfrac{1}{2})-P_\tau(\tfrac{1}{2}-) .$$

For the continuity interval $[x, y]$ of $P_\tau(x)$ we have

$$\lim_{n\to\infty}\int_{x}^{y} h^{(n)}*P_\tau(f)df=P_\tau(y)-P_\tau(x) ,$$

and thus for z in $(-\frac{1}{2}, \frac{1}{2})$

$$\lim_{\substack{y\downarrow z\\ x\uparrow z}}\lim_{n\to\infty}\int_{x}^{y} h^{(n)}*P_\tau(f)df=P_\tau(z)-P_\tau(z-) .^{*)}$$

Further if

$$\lim_{n\to\infty} h^{(n)}*P_\tau(f)=\hat{p}(f)$$

almost everywhere and

$$\int_{-1/2}^{1/2}\hat{p}_\tau(f)df=\sigma^2$$

then we have $dP_\tau(f)=\hat{p}_\tau(f)df$.**)

Here we shall derive concrete results for two types of τ_n process.

1. First we shall consider the case where absolute clock pulses are available and timing-errors (deviations of sampling-time points from the corresponding true clock pulses) form a purely random process. In this case we have

$$\tau_n=n\Delta t+\varepsilon_n \quad\text{and}\quad \Delta\tau_n=\varepsilon_{n+1}-\varepsilon_n+\Delta t$$

where Δt is a fixed non-negative constant and ε_n's are mutually independent random variables following one and the same distribution. We shall call this procedure time-sampling of purely random type.

As

$$\tau_{k+n}-\tau_n=\varepsilon_{k+n}-\varepsilon_n+k\Delta t$$

holds, we have for $k\neq 0$

---

*) The symbol $\lim_{\substack{y\downarrow z\\ x\uparrow z}}$ denotes $\lim_{y-x\to 0}$ where x and y are taken to satisfy $x<z<y$.

**) Hereafter we shall use the notation $dP(f)$ to denote the measure function determined by $P(f)$. When $P(f)$ is absolutely continuous and with density function $p(f)$ then we write $dP(f)=p(f)df$.

$$\phi_k(f)=\int e^{2\pi i f\tau}dP_{(k)}(\tau)=e^{2\pi i k\Delta t f}|\phi(f)|^2$$

where

$$\phi(f)=E\{\exp(2\pi i f\varepsilon_n)\} .$$

Thus

$$\rho(0)=\sigma^2=\int_{-\infty}^{\infty}dP(f)$$

and for $k\neq 0$

$$\rho(k)=\int_{-\infty}^{\infty}e^{2\pi i k\Delta t f}|\phi(f)|^2dP(f) .$$

Here we shall disregards the trivial case where $\Delta t=0$ and $\varepsilon_n=0$ with probability 1. If we represent by $d|\phi|^2P_A(f)$ the aliased form of $d|\phi(f)|^2P(f)$ for the folding frequency $1/(2\Delta t)$, i.e.,

$$d|\phi|^2P_A(f)=\sum_{\nu=-\infty}^{\infty}d\left|\phi\left(\frac{\nu}{\Delta t}+f\right)\right|^2P\left(\frac{\nu}{\Delta t}+f\right)$$

then we have for $k\neq 0$

$$\rho(k)=\int_{-1/(2\Delta t)}^{1/(2\Delta t)}e^{2\pi i k\Delta t f}d|\phi|^2P_A(f) .$$

From this expression we can at once see that $dP_\tau(f)$ is given by the following

$$dP_\tau(f)=d|\phi|^2P_A\left(\frac{f}{\Delta t}\right)+\left[\int_{-\infty}^{\infty}(1-|\phi(f')|^2)dP(f')\right]df .$$

Of course we can easily obtain this results by exactly following the smoothing procedure described in the preceding section. For practical applications expressions such as

$$\rho(k)=\int_{-1/2\Delta t}^{1/2\Delta t}e^{2\pi i k\Delta t f}dP_\tau^{\Delta t}(f) \qquad k=\cdots,-2,-1,0,1,2,\cdots,$$

$$dP_\tau^{\Delta t}(f)=d|\phi|^2P_A(f)+\Delta t\left[\int_{-\infty}^{\infty}(1-|\phi(f')|^2)dP(f')\right]df ,$$

for $-1/(2\Delta t)\leq f\leq 1/(2\Delta t)$, will be better suited. This last relation clearly shows the effect of timing error on the power spectrum. If there is no timing error we have $\phi(f)=1$ and $dP_\tau^{\Delta t}(f)=dP_A(f)$. Thus time-sampling causes aliasing.*) When the timing-errors are present and are not lattice

*) As to the use of the word "aliasing" see [1]. From the above expression of aliased form of a spectral function we suppose that it will be more natural to consider that the aliased form is obtained by "piling up" the sliced spectrum rather than by "folding".

valued we have $|\phi(f)|^2<1$ for $f\neq 0$. Further when the distribution function of error ε_n has probability density function we have $\lim_{|f|\to\infty}|\phi(f)|=0$. Thus we can see that the time-sampling of purely random type usually acts as a low-pass filter with an inner white noise source. The power of this white noise is the same as that of the higher frequency component excluded by the filter from the original $x(t)$ process.

Estimation of the term $\int_{-\infty}^{\infty}(1-|\phi(f)|^2)dP(f)$.

It will be desirable to get an estimate of the term

$$\Delta t\int_{-\infty}^{\infty}(1-|\phi(f)|^2)dP(f)\ .$$

If such an estimate is available, by subtracting it from the estimate of $dP_\tau^{\Delta t}(f)$ we can estimate $d|\phi|^2P_\Delta(f)$ which will be a good estimate of $dP(f)$ for f near zero, for proper Δt and $\phi(f)$. Now consider two mutually independent readings of the same $x(t)$ process. We shall represent them as $x_{\tau_{1,n}}$ and $x_{\tau_{2,n}}$ with

$$\tau_{1,n}=n\Delta t+\varepsilon_{1,n}\ ,$$

$$\tau_{2,n}=n\Delta t+\varepsilon_{2,n}$$

where $\{\varepsilon_{1,n}\}$ and $\{\varepsilon_{2,n}\}$ represent timing-errors which form mutually independent purely random processes with one and the same finite dimensional distribution. Then we can see by using the result for time-sampling of purely random type with $\Delta t=0$ that

$$E|x_{\tau_{1,n}}-x_{\tau_{2,n}}|^2=2(\sigma^2-ER(\varepsilon_{1,n}-\varepsilon_{2,n}))$$

holds where

$$ER(\varepsilon_{1,n}-\varepsilon_{2,n})=E\int_{-\infty}^{\infty}\exp\{2\pi i(\varepsilon_{1,n}-\varepsilon_{2,n})f\}dP(f)=\int_{-\infty}^{\infty}|\phi(f)|^2dP(f)\ .$$

Thus $\frac{1}{2}(x_{\tau_{1,n}}-x_{\tau_{2,n}})^2$ is an unbiased estimate of

$$\int_{-\infty}^{\infty}(1-|\phi(f)|^2)dP(f)$$

and by using the sample mean of the variable for sufficiently large number of n's we can practically estimate the desired quantity.

2. Next we shall consider the case where the interval lengthes between successive sampling time points form a purely random process.

This is the case where only relative clock pulses are available. Interval length from the former sampling time point τ_{n-1} is strictly measured and when it reaches preassigned value Δt next observation is made about $x(t)$ but with timing-error ε_n. We shall call this procedure the time-sampling of renewal type. In this case $\tau_n=\tau_{n-1}+\Delta t+\varepsilon_n$ and ε_n's are assumed to form a purely random process. Notice that when we are not considering the operation in real-time $\Delta t+\varepsilon_n$ may take negative values.

Now if we define

$$\phi(f)=E\{\exp[2\pi if(\Delta t+\varepsilon_n)]\}$$

we have

$$\begin{aligned}\phi_k(f)&=\phi^k(f) && \text{for } k\geq 0\\ &=\phi^{-k}(-f) && \text{for } k<0\,,\end{aligned}$$

and

$$\rho(k)=\int_{-\infty}^{\infty}\phi_k(f)dP(f)\,.$$

To evaluate the power spectral distribution function of $x_{\tau,n}$ process we shall use the convergence factors $\{c_k^{(n)}\}$ defined by

$$c_k^{(n)}=c_n^{|k|}$$

where $0<c_n<1$ and $\lim_{n\to\infty} c_n=1$.

Here we want to mention a theorem which is well known in the theory of functions and stated as follows; suppose $u(f)$ is a function defined for f in $-\frac{1}{2}\leq f\leq\frac{1}{2}$ and is integrable $[-\frac{1}{2},\frac{1}{2}]$. Then if $u(f)$ is continuous at $f=f_0$ the function

$$u(r,f')=\int_{-1/2}^{1/2}u(f)\frac{1-r^2}{|1-re^{-2\pi i(f-f')}|^2}df\,,$$

defined for r and f' satisfying $0<r<1$, $-\frac{1}{2}\leq f'\leq\frac{1}{2}$, converges to $u(f_0)$ as $re^{2\pi if'}$ tends to $e^{2\pi if_0}$. A direct consequence of this theorem is that by the present definition of our $\{c_k^{(n)}\}$ the functions

$$h^{(n)}(f)=\sum_{k=-\infty}^{\infty}c_k^{(n)}e^{-2\pi ikf}=\frac{1-|c_n|^2}{|1-e^{-2\pi if}c_n|^2}$$

have the properties which we have postulated in § 1 as necessary for $h^{(n)}(f)$'s to serve for our present purpose. Now we define

$$K_n(\phi, f, f') = \sum_{k=-\infty}^{\infty} c_k^{(n)} \phi_k(f') e^{-2\pi i k f}$$
$$= \frac{1 - |c_n \phi(f')|^2}{|1 - e^{-2\pi i f} c_n \phi(f')|^2} .$$

Then if we use the representation

$$\phi(f') = r' e^{2\pi i s'} \qquad (0 \leq r' \leq 1 , \quad -\tfrac{1}{2} < s' \leq \tfrac{1}{2})$$

we have

$$K_n(\phi, f, f') = \sum_{k=-\infty}^{\infty} e^{-2\pi i k (f-s')} (c_n r')^{|k|} = \frac{1 - |c_n r'|^2}{|1 - e^{-2\pi i (f-s')} c_n r'|^2} ,$$

$$K_n(\phi, f, f') \geq 0$$

and

$$\int_{-1/2}^{1/2} K_n(\phi, f, f') df = (c_n r')^0 = 1 .$$

Now we shall define for $\phi(f') = r' e^{2\pi i s'}$ $(0 \leq r' \leq 1, -\tfrac{1}{2} < s' \leq \tfrac{1}{2})$ and f $(-\tfrac{1}{2} \leq f \leq \tfrac{1}{2})$

$$K(\phi, f, f') = \frac{1 - |r'|^2}{|1 - e^{-2\pi i (f-s')} r'|^2} \qquad \text{when} \quad r' < 1,$$
$$= 0 \qquad \text{when} \quad r' = 1$$

and for x and y satisfying $-\tfrac{1}{2} \leq x < y < \tfrac{1}{2}$

$$\tilde{\chi}_{[x,y],\phi}(f') = 1 \qquad \text{when } r' = 1 \text{ and } x < s' < y$$
$$= \tfrac{1}{2} \qquad \text{when } r' = 1 \text{ and } s' = x \text{ or } s' = y$$
$$= 0 \qquad \text{otherwise.}$$

Then by taking into account the results of the above-mentioned theorem we can get for x and y satisfying $-\tfrac{1}{2} < x < y < \tfrac{1}{2}$

$$\lim_{n\to\infty} \int_x^y K_n(\phi, f, f') df = \tilde{\chi}_{[x,y],\phi}(f') + \int_x^y K(\phi, f, f') df ,$$

and for x and y satisfying $-\tfrac{1}{2} \leq x < y < \tfrac{1}{2}$

$$\lim_{n\to\infty} \int_x^y h^{(n)} * P_\tau(f) df = \lim_{n\to\infty} \int_x^y \left[\int_{-\infty}^{\infty} K_n(\phi, f, f') dP(f') \right] df$$
$$= \lim_{n\to\infty} \int_{-\infty}^{\infty} \left[\int_x^y K_n(\phi, f, f') df \right] dP(f')$$
$$= \int_{-\infty}^{\infty} \tilde{\chi}_{[x,y],\phi}(f') dP(f') + \int_x^y \left[\int_{-\infty}^{\infty} K(\phi, f, f') dP(f') \right] df .$$

Thus we have

$$P_\tau(\tfrac{1}{2})-P_\tau(\tfrac{1}{2}-)=\sigma^2-\lim_{\nu\uparrow 1/2}\lim_{n\to\infty}\int_{-1/2}^{\nu} h^{(n)}*P_\tau(f)df$$
$$=\sigma^2-\int_{-\infty}^{\infty}\Big[\lim_{\nu\uparrow 1/2}\tilde{\chi}_{[-1/2,\nu],\phi}(f')+\int_{-1/2}^{1/2}K(\phi,f,f')df\Big]dP(f')$$
$$=\int_{-\infty}^{\infty}\tilde{\chi}_{1/2,\phi}(f')dP(f')$$

where $\tilde{\chi}_{1/2,\phi}(f')=1$ if and only if $\phi(f')=e^{\pi i}$ and $=0$ otherwise. If we further define

$$\tilde{\chi}_{s,\phi}(f')=\lim_{\substack{\nu\downarrow s\\ x\uparrow s}}\tilde{\chi}_{[x,\nu],\phi}(f') \qquad \text{for } s\neq\pm\tfrac{1}{2}$$
$$=\lim_{\nu\downarrow -1/2}\tilde{\chi}_{[-1/2,\nu],\phi}(f') \qquad \text{for } s=-\tfrac{1}{2}$$

we have for s in $[-\frac{1}{2},\frac{1}{2}]$

$$P_\tau(s)-P_\tau(s-)=\int_{-\infty}^{\infty}\tilde{\chi}_{s,\phi}(f')dP(f')\ .$$

We shall hereafter analyse these results in more details. There are three classes of $\phi(f')$. The first is composed of those $\phi(f')$ for which $|\phi(f')|=1$ holds for all f'. The second is composed of those $\phi(f')$ for which the minimum of the absolute values of those f' $(\neq 0)$, for which $|\phi(f')|=1$ hold, takes some positive value f_0 which depends on ϕ. The third is composed of those $\phi(f')$ for which $|\phi(f')|=1$ holds only at $f'=0$. When $\phi(f')$ is of the first class it can be represented in the form

$$\phi(f')=e^{2\pi i f'\Delta\tau}$$

by some real constant $\Delta\tau$ and corresponds to the time-sampling with the length $\Delta\tau$ of sampling interval and without timing-error. Hereafter we shall disregard the trivial case where $\Delta\tau=0$ holds. Now we have

$$K(\phi,f,f')=0$$

and for x and y satisfying $-\frac{1}{2}<x<y<\frac{1}{2}$

$$\lim_{n\to\infty}\int_x^y h^{(n)}*P_\tau(f)df=\int_{-\infty}^{\infty}\tilde{\chi}_{[x,y],\phi}(f')dP(f')$$
$$=\sum_{\nu=-\infty}^{\infty}\left[\frac{P\Big(\frac{y+\nu}{\Delta\tau}\Big)+P\Big(\frac{y+\nu}{\Delta\tau}-\Big)}{2}-\frac{P\Big(\frac{x+\nu}{\Delta\tau}\Big)+P\Big(\frac{x+\nu}{\Delta\tau}-\Big)}{2}\right].$$

This is the formula showing the folding or alaising. The line spectra of this case are given by

$$P_\tau(s)-P_\tau(s-)=\sum_{\nu=-\infty}^{\infty}\left\{P\left(\frac{s+\nu}{\Delta\tau}\right)-P\left(\frac{s+\nu}{\Delta\tau}-\right)\right\} \qquad \text{for } -\tfrac{1}{2}<s\leq\tfrac{1}{2}.$$

If the original $P(f)$ has a density function $p(f)$ then we have

$$dP_\tau(f)=\left[\sum_{\nu=-\infty}^{\infty}p\left(\frac{s+\nu}{\Delta\tau}\right)\right]df\,.$$

As for $\phi(f')$ of the second class we can express it in the form

$$\phi(f')=\exp\left\{2\pi i\left(\frac{f'}{f_0}\right)s_0\right\}\left[\sum_{k=-\infty}^{\infty}p_k\exp\left\{2\pi ik\left(\frac{f'}{f_0}\right)\right\}\right]$$

where s_0 is such that $-\frac{1}{2}<s_0\leq\frac{1}{2}$ and $p_k=\text{Prob}\,\{\Delta t+\varepsilon_n=f_0^{-1}(s_0+k)\}$. In this case $|\phi(f')|=1$ holds only at $f'=\nu f_0$ $(\nu=\cdots,-1,0,1,\cdots)$ where $\phi(f')=e^{2\pi i\nu s_0}$. When $s_0\neq 0$ the line spectra are obtained by properly re-scaling the ordinates of the spectra obtained by piling up the line spectra at $f=\nu f_0$ $(\nu=\cdots,-1,0,1,\cdots)$ of $P(f)$ sliced at the frequencies $(\mu+\frac{1}{2})(f_0/s_0)$ $(\mu=\cdots,-1,0,1,\cdots)$, i.e., $P_\tau(s)-P_\tau(s-)=$sum of line spectra of $P(f)$ at νf_0's where $\nu f_0=(\mu+s)(f_0/s_0)$ holds for some integer μ.*) When $s_0=0$ the line spectrum is present only at the origin, or the total power of line spectra at $f=\nu f_0$'s of the original $P(f)$ is transformed into continuous of the d.c. (direct current) component of $x_{\tau,n}$. Now the power part of the $P_\tau(f)$ is seen to have a density function

$$\int_{-\infty}^{\infty}K(\phi,f,f')dP(f')\,.$$

Thus when $P(f)$ has a density function $p(f)$ we have

$$dP_\tau(f)=\left[\int_{-\infty}^{\infty}\frac{1-|\phi(f')|^2}{|1-e^{-2\pi if}\phi(f')|^2}p(f')df'\right]df\,.$$

When $\phi(f')$ belongs to the third class there may be a line spectrum in $P_\tau(f)$ only at the origin and it is equal to that of the line spectrum of the original $P(f')$ at $f'=0$. Thus in this case only the power of the d.c. component of the original process is preserved as line spectrum and becomes the power of the d.c. component of the time-sampled process. Thus if only the d.c. component is absent in the original process we always have absolutely continuous spectrum given by

$$dP_\tau(f)=\left[\int_{-\infty}^{\infty}\frac{1-|\phi(f')|^2}{|1-e^{-2\pi if}\phi(f')|^2}dP(f')\right]df\,.$$

---

*) When there is line spectrum in $P(f)$ at $f=(\mu+\frac{1}{2})(f_0/s_0)$ it must be piled up at $s=\frac{1}{2}$ in $P_\tau(f)$.

From these results we can see that the present sampling procedure distributes the power $dP(f')$ at f', of the original process, over the range $[-\frac{1}{2}\leq f\leq \frac{1}{2}]$ following the distribution function given by $\{(1-|\phi(f')|^2)/|1-e^{-2\pi if}\phi(f')|^2\}df$. The distribution function given by $\{(1-|\phi(f')|^2)/|1-e^{-2\pi if}\phi(f')|^2\}df$ should be interpreted as $\delta(f-s')df$ for $\phi(f')=1\cdot e^{2\pi is'}$ $(-\frac{1}{2}<s'<\frac{1}{2})$ where $\delta(f-s')$ denotes the Dirac's δ-function and for this frequency f' our time-sampling procedure acts as if there were no timing-errors.*) The distribution given by $\{(1-|\phi(f')|^2)/|1-e^{-2\pi if}\phi(f')|^2\}df$ gives the spectral distribution of randomly phase modulated sinusoidal sequence $\left\{\exp\left[2\pi if'\left(n\Delta t+\sum_{j=1}^{n}\varepsilon_j(\omega)\right)\right];\ n=\cdots,-1,0,1,\cdots\right\}$. An analogous interpretation is also possible for the case of the time-sampling of purely random type and we can see that our present sampling procedures are essentially non-linear. Taking into account the fact that $\{(1-|\phi(f')|^2)/|1-e^{-2\pi if}\phi(f')|^2\}df$ tends to the uniform distribution as $|\phi(f')|\to 0$ and tends to the Dirac's δ-function as $|\phi(f')|\to 1$, we can see that the power $dP(f')$ is conserved near the s' when r' of $\phi(f')=r'e^{2\pi is'}$ is nearly equal to 1 and spread all over the range when r' is nearly equal to 0. When the distribution function of the sampling-time interval is absolutely continuous we have

$$|\phi(f')|\to 0 \qquad (|f'|\to\infty)$$

and we can see that the power at the higher frequencies is spread nearly uniformly all over the frequency range of $P_r(f)$, while the power near the zero frequency is conserved near the zero frequency.

Thus from the results in this and the preceeding paragraphs we can see that in practical applications of time-sampling procedures of these two types, if in the original process there is some power at some separated very high frequency band, the time-sampling may appear as a low-pass filter with an inner white noise source. This fact will show why time-sampling was sometimes considered to be a filtering while it is essentially a folding which is non-linear.

2. Numerical example**)

Here we shall illustrate the results in the preceeding section by some numerical examples. The estimates of the spectral density functions

*) Obvious modification is necessary for $s'=+\frac{1}{2}$.

**) In this section we shall sometimes use the notations of random variables to represent one of their realizations so long as it does not introduce serious ambiguities.

illustrated in this section were obtained by the following numerical procedure;

a) given a time sampled data $\{x_{\tau,n};\ n=1, 2, \cdots, N\}$ we computed $C(k)$'s $(k=0, 1, 2, \cdots, h)$

$$C(k)=\frac{1}{N}\left(\sum_{n=1}^{N-k} \tilde{x}_{\tau,n+k}\tilde{x}_{\tau,n}\right)$$

where $\tilde{x}_{\tau,n}=x_{\tau,n}-\bar{x}_\tau$ and $\bar{x}_\tau=\frac{1}{N}\sum_{n=1}^{N} x_{\tau,n}$,

b) then the transforms $\tilde{p}(f)$ of this $\{C(k)\}$ where computed for $f=(j/h)(1/2)$ $(j=0, 1, 2, \cdots, h)$

$$\tilde{p}\left(\frac{j}{h}\cdot\frac{1}{2}\right)=C(0)+2\sum_{k=1}^{h-1} C(k)\cos\left(\frac{jk}{h}\pi\right)+C(h)\cos(j\pi)\ ,$$

c) these $\tilde{p}(f)$'s were then further smoothed to give our estimate $p(f)$ for $f=(j/h)(1/2)$ $(j=0, 1, 2, \cdots, h)$

$$p\left(\frac{j}{h}\cdot\frac{1}{2}\right)=0.23\tilde{p}\left(\frac{j-1}{h}\cdot\frac{1}{2}\right)+0.54\tilde{p}\left(\frac{j}{h}\cdot\frac{1}{2}\right)+0.23\tilde{p}\left(\frac{j+1}{h}\cdot\frac{1}{2}\right)$$

where

$$\tilde{p}\left(-\frac{1}{h}\cdot\frac{1}{2}\right)=\tilde{p}\left(\frac{1}{j}\cdot\frac{1}{2}\right) \text{ and } \tilde{p}\left(\frac{h+1}{h}\cdot\frac{1}{2}\right)=\tilde{p}\left(\frac{h-1}{h}\cdot\frac{1}{2}\right).$$

Taking into account of the symmetricity of the present $p(f)$ we have considered the values of $P(f)$ only for positive f. As to the analytical details of the present numerical procedure the reader is recommended to consult the paper [1] by Blackman and Tukey. In the following we shall denote the value of $p(f)/C(0)$ simply as $p(f)$. In Fig. 1 the $p(f)$ of a time-sampled data $\{x_{\tau,n};\ n=1, 2, \cdots, N\}$ is shown where $N=530$ and $h=60$. The data was read from a continuous record of a typical oscillation of the frame of an automobile running over a gravel road. Here $\tau_n=\tau_0+n\Delta t$ and Δt was taken to be 1/50 sec. In this data there may be some errors in τ_n but we shall disregard it now as our concern here is with the comparison of this $p(f)$ with other $p(f)$'s which were obtained from the present data by some artificial random sampling procedures which will be described in the following.

Fig. 2 shows the effect of timing-error of purely random type. The crosses show the $p(f)$ of the data $\{x'_{\tau,\nu}\}$ which was time-sampled from the primary data $\{x_{\tau,\nu}\}$ and

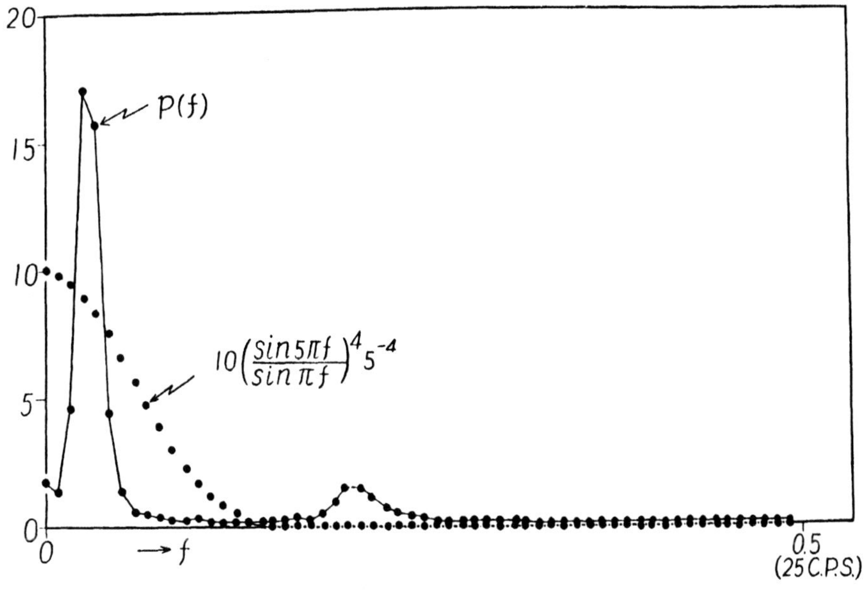

Fig. 1.

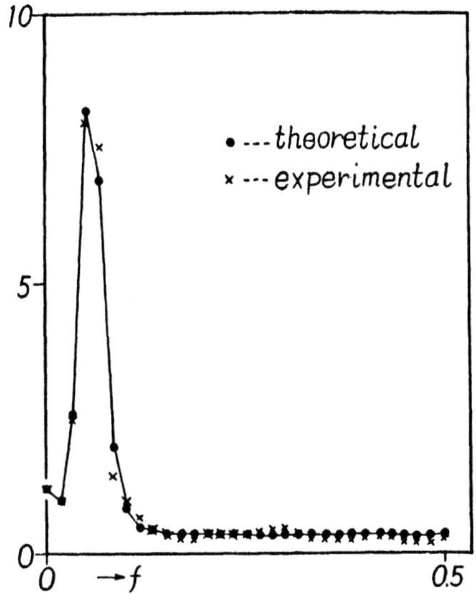

Fig. 2. Effect of time-sampling of purely random type.

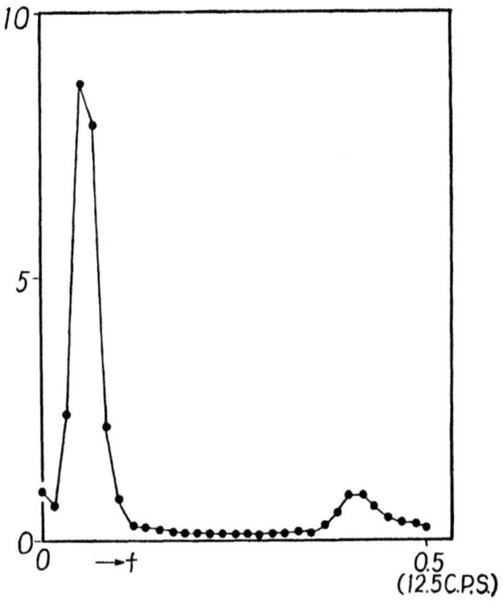

Fig. 3. Effect of pure folding.

$$x^I_{\tau,\nu}=x_{\tau,2\nu+\varepsilon_\nu}$$

where $\{\varepsilon_\nu\}$ is a purely random process such that

$$\Pr\{\varepsilon_\nu=\mu\}=\frac{5-|\mu|}{25}$$

for integer μ in the range $-4\leq\mu\leq4$. Here $N=263$ and $h=30$. In Fig. 2 the dots show the theoretically expected values of $p(f)$ which were obtained by using the results of the preceding section and the $p(f)$ of Fig. 1 in place of the true value of $p(f)$ of $\{x_{\tau,n}\}$. We can see a fairly good agreement. In the present example we have

$$|\phi(f)|^2=\left(\frac{\sin 5\pi f}{\sin \pi f}\right)^4 5^{-4}$$

and its values are plotted, being multiplied by a constant factor 10, for $f=j/2h=j/120$ $(j=0, 1, 2, \cdots, 60)$ in Fig. 1.

Fig. 3 shows the $p(f)$ which corresponds to the case where $\varepsilon_\nu\equiv0$ and $x^I_{\tau,\nu}=x_{\tau,2\nu}$ and illustrates the pure folding. By comparing Fig. 2 with Fig. 3 we can clearly see the effect of timing-error. We have further made an experiment of the estimation of the term

$$\int_{-\infty}^{\infty}(1-|\phi(f)|^2)dP(f)\,.$$

By another independent reading we obtained $\{x^{I'}_{\tau,\nu}\}$ and got

$$\frac{1}{2}\left[\frac{1}{263}\sum_{\nu=1}^{263}(x^I_{\tau,\nu}-x^{I'}_{\tau,\nu})^2\right]=0.308\times C(0) \text{ of } x_{\tau,n}\,.$$

Now we can see

$$\frac{1}{2}E(x^I_{\tau,\nu}-x^{I'}_{\tau,\nu})^2=R(0)-\sum_k R(k)\sum_\mu \Pr\{\varepsilon_\nu=\mu+k\}\Pr\{\varepsilon_\nu=\mu\}$$

holds where $R(k)=E\{x_{\tau,n}x_{\tau,n+k}\}$. We computed another estimate of $\frac{1}{2}E(x^I_{\tau,\nu}-x^{I'}_{\tau,\nu})^2$ by putting $C(k)$ of $\{x_{\tau,n}\}$ in place of $R(k)$ in the above formula and it was found to be $0.327\times C(0)$. This last value was used to draw the doted curve of Fig. 2. Thus the present result suggests that for the time-sampling of purely random type, if there is some power at some separated very high frequency band in the original process and the timing-errors are continuously distributed and their range is sufficiently small compared with the wave length of the lower frequency band but sufficiently big compared with the wave length of that high frequency band, then by using the estimate of

$$\int_{-\infty}^{\infty}(1-|\phi(f)|^2)dP(f)$$

described in the former section we may obtain a better estimate of $p(f)$ of the original process. Fig. 4 and 5 show the effect of this correction procedure. There are also presented the order $(1/12)(C(0)^{-1})$ of quantization noise which is assumed approximately to be a white noise. $p(f)$'s in Figs. 4 and 5 were obtained from the data which were read by using a rule, at each timing mark which were 1/100 sec. apart each other, from the continuous records of the outputs of an accelerometer of strain gauge type mounted on the front axle of an automobile running at the speed of 30 km/h and 60 km/h respectively. Here $N=500$, $h=50$ for Fig. 4 and $N=250$, $h=50$ for Fig. 5. We have felt some

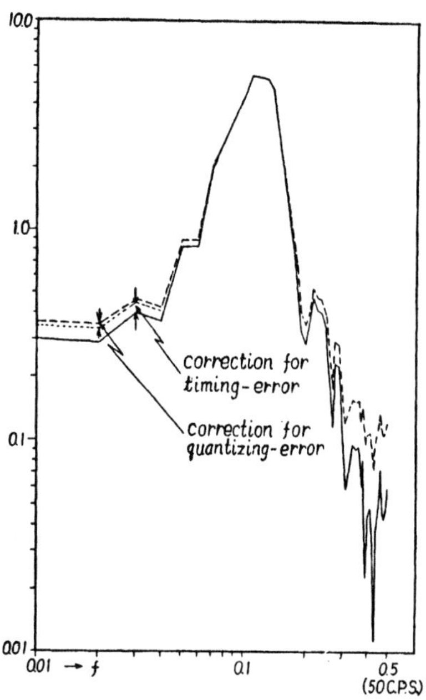

Fig. 4. Spectrum of vertical acceleration of a front axle (at 30km/h).

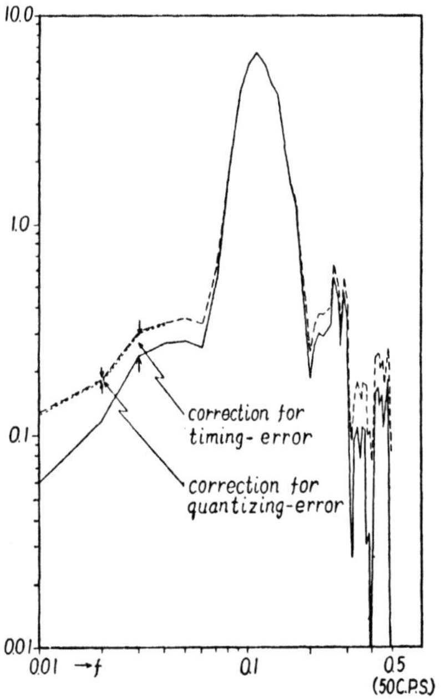

Fig. 5. Sepectrum of vertical aceeleration of a front axle (at 60 km/h).

uncertainty in measuring these data due to the existence of components of very high frequency which might be 100 or 400 cycle per second or higher. We have considered that the uncertainty is mainly due to the fluctuations of the horizontal position of the eye or the rule. As the precise timing marks were available at each 1/100 second such readings will correspond to the time-sampling of purely random type. We can see from the present results that there are more power at higher frequencies in the case of Fig. 5, and this has increased the difficulty in reading the corresponding data. We can see further that timing-error causes little effect on the estimates of $p(f)$ in absolute value. But taking into account of the fact that the present estimate $p(f)$ keeps nearly the same relative accuracy all over the range of f, we have to pay attension, for $p(f)$ at low levels, to the bias of white noise type due to timing-error besides that due to quantization.

In Fig. 6 is illustrated a $p(f)$ of a time-sampled data obtained from the former $\{x_{\tau,\nu}\}$ by a time-sampling procedure of renewal type. The crosses show the values of $p(f)$ of $\{x_{\tau,\nu}^{II}\}$

$$x_{\tau,\nu}^{II}=x_{\tau,\varepsilon_1+\varepsilon_2+\cdots+\varepsilon_\nu} \qquad \nu=1, 2, \cdots, N$$

where $\{\varepsilon_\nu\}$ is a purely random process and

$$\Pr\{\varepsilon_\nu=1\}=\Pr\{\varepsilon_\nu=3\}=\tfrac{1}{4}$$
$$\Pr\{\varepsilon_\nu=2\}=\tfrac{1}{2}\ .$$

Here $N=268$ and $h=30$. The dots represent approximations to the theoretically expected values of $p(f)$ and were obtained by approximately applying the theoretical result of the preceding section to the $p(f)$ of $\{x_{\tau,n}\}$. We can see a fairly good agreement in this case too.

In Fig. 7 are illustrated the values of $C(k)/C(0)$ which were used for the computations of $p(f)$'s of Figs. 1, 2 and 6. The Figs. 1, 2, 3 and 6 show how the present time-sampling procedures act like low-pass filters.

In the present section we have not discussed the sampling variations of our estimates. We did so as our main concern in this section was with the analysis of the biases of our estimates and not of the variances. The discussion of the sampling fluctuations of our estimates is possible at least for the Gaussian case and the reader is recommended to consult the paper [1] for that purpose.

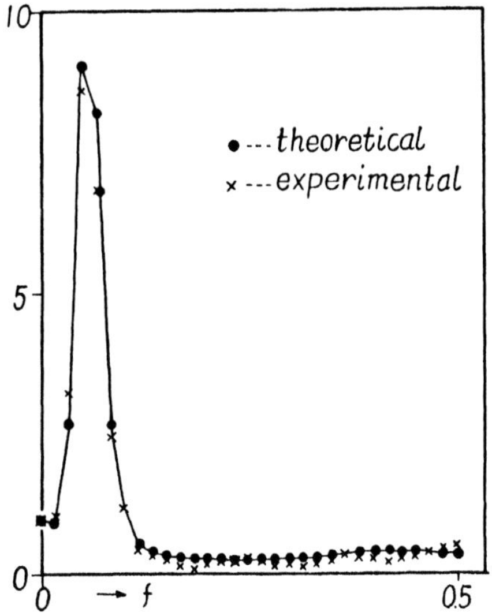

Fig. 6. Effect of time-sampling of renewal type.

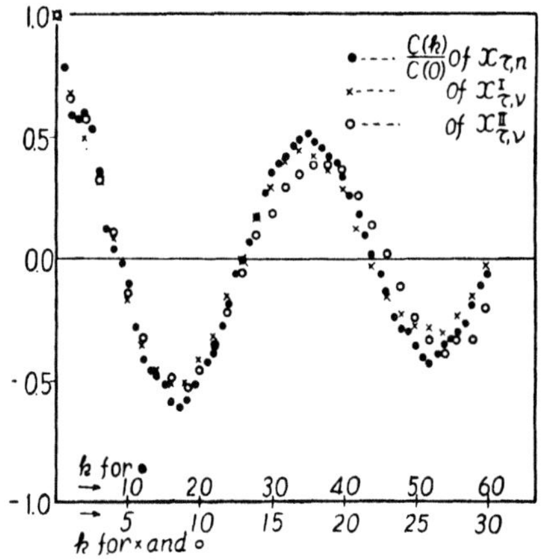

Fig. 7. Correlograms of $\{x_{\tau,n}\}$, $\{x^{I}_{\tau,\nu}\}$ and $\{x^{II}_{\tau,\nu}\}$.

Acknowledgement

I am indebted to the Isuzu Motor Company for providing me with the records of outputs of acceleromers and to Mr. I. Kanesige for stimulating me to the present investigation. I am grateful to Mr. M. Motoo for reading the manuscript of this paper. I am also grateful to Miss Y. Saigusa for programming the computations on FACOM-128 relay computer of our institute and to Mrs. T. Isii for preparing the numerical inputs to the machine and co-operating with Miss Saigusa in computations.

THE INSTITUTE OF STATISTICAL MATHEMATICS

REFERENCES

[1] R. B. Blackman and J. W. Tukey, "The measurement of power spectra from the point of view of communications engineering", *Bell System Technical Journal* 37, 185–282, 485–569 (1958).

[2] J. L. Doob, "*Stochastic Process*" John Wiley & Sons, Inc. (1953).

[3] D. T. Ross, "Sampling and quantizing", Chaper II in *notes on Analog-Digital Conversion Techniques*, Alfred K. Susskind editor, Technology Press of M.I.T. and John Wiley & Sons, Inc. New York, (1957).

[4] J. W. Tukey, and R. W. Hamming, "Measuring noise color" *Bell Telephon Lab. Memo.*, MM-49-110-119 (1949).

[5] B. Widrow, "A study of rough amplitude quantization by means of Nyquist sampling theory", *Trans. I.R.E.*, PGCT-3, 266–276 (1956)

Reprinted from *Annals of the Institute of Statistical Mathematics*, Vol. 12, No. 1, 1960, 7-11 by permission of the Institute of Statistical Mathematics

ON A LIMITING PROCESS WHICH ASYMPTOTICALLY PRODUCES f^{-2} SPECTRAL DENSITY*

By HIROTUGU AKAIKE

(Received July 20, 1960)

1. Introduction and summary

In the recent papers in which the results of the spectral analyses of roughnesses of runways or roadways are reported [1, 2, 3,4] the power spectral densities of approximately of the form f^{-2} (f: frequency) are often treated. This fact directed the present author to the investigation of the limiting process which will provide the f^{-2} form under fairly general assumptions. In this paper a very simple model is given which explains a way how the f^{-2} form is obtained asymptotically. Our fundamental model is that the stochastic process, which might be considered to represent the roughness of the runway, is obtained by alternative repetitions of roughening and smoothing. We can easily get the limiting form of the spectrum for this model. Further, by taking into account the physical meaning of roughening and smoothing we can formulate the conditions under which this general result assures that the f^{-2} form will eventually take place.

The derivation of the present result is entirely formal, and it is hoped that further discussion of the model will be given from the standpoint of road engineering as to whether or not it can be adopted as a starting point of the approximation to the above stated f^{-2} phenomena.

While giving the mathematical reasoning of the present model we discuss very briefly about the numerical results reported in the paper by Wilbur E. Thompson to draw the reader's attension to the necessity of a trend elimination procedure in the spectral analysis of runway or roadway elevations. This discussion will be of some help to those who have occasion to check our model in some practical problems in the future.

2. Derivation of the f^{-2} form

We shall consider the sequence of stationary stochastic processes

* A part of the results of the present paper was announced in July 7, 1960 at the 28th annual meeting of the Japan Statistical Association.

$\varepsilon_j(t)$ $(j=1, 2, \cdots)$ where $\varepsilon_j(t)$'s are supposed to be mutually independent and to follow the same finite dimensional distributions. We shall assume that the process $\varepsilon_j(t)$ has zero-mean, finite variance, and spectral density function $p_\varepsilon(f)$. Consider the initial process $x_0(t)$ which may be considered as representing the initial state of roughness. First we smooth this $x_0(t)$ by an integrable smoothing function $s(t)$ and then roughen by adding a stationary stochastic process $\varepsilon_1(t)$ to it. Thus, denoting by $x_1(t)$, the result of one time application of the smoothing and roughening operation, we have

$$x_1(t) = s*x_0(t) + \varepsilon_1(t)$$

where by definition

$$s*x_0(t) = \int_{-\infty}^{\infty} x_0(t-\tau)s(\tau)d\tau \ ,$$

$$s(t) \geqq 0 \ ,$$

$$\int_{-\infty}^{\infty} s(t)dt = 1 \ .$$

We shall assume that $x_0(t)$ satisfies some regularity condition which makes the operations $s*s*\cdots*s*x_0(t)$ meaningfull. As such a condition we have the bounded measurability of $x_0(t)$. If we represent by $x_n(t)$ the result of n successive applications of alternative smoothing and roughening operations, we get

$$x_n(t) = s*x_{n-1}(t) + \varepsilon_n(t) \ ,$$

and successively,

$$x_n(t) = \varepsilon_n(t) + s*\varepsilon_{n-1}(t) + s*s*\varepsilon_{n-2}(t) + \cdots + s*s*\cdots*s*\varepsilon_1(t) + s*s*\cdots*s*x_0(t) \ .$$

If we put

$$y_n(t) = \varepsilon_n(t) + s*\varepsilon_{n-1}(t) + s*s*\varepsilon_{n-2}(t) + \cdots + s*s*\cdots*s*\varepsilon_1(t)$$

it can be seen that the process $y_n(t)$ is stationary and has power spectral density $p_n(f)$ which is given as

$$p_n(f) = (1 + |\sigma(f)|^2 + |\sigma(f)|^{2\times 2} + \cdots + |\sigma(f)|^{2(n-1)})p_\varepsilon(f)$$
$$= \frac{1-|\sigma(f)|^{2n}}{1-|\sigma(f)|^2} p_\varepsilon(f)$$

where by definition

$$\sigma(f) = \int_{-\infty}^{\infty} e^{-2\pi i f t} s(t)dt \ .$$

Thus we get the general limiting formula

$$\lim_{n\to\infty} p_n(f) = \frac{1}{1-|\sigma(f)|^2} p_\epsilon(f) \qquad \text{for } f \neq 0 .$$

From this result we can see that if we carefully separate the "trend" or the residual effect $s*s\cdots*s*x_0(t)$, if any, which can not be filtered out by the repeated applications of the smoothing function $s(t)$, and if we restrict our attension to some range $f \geqq f_0$ ($f_0>0$, arbitrarily small but fixed), we can expect that after a sufficient number of repetitions of alternative smoothing and roughening the form of the power spectrum will become stable. Thus, it can be said that our model gives a limiting form of the spectrum. Of course, the restriction of our attension to the range $f \geqq f_0$ ($f_0>0$ arbitrarily small but fixed) is necessary as the power around the zero frequency grows indefinitely as n tends to infinity.

Our limiting form is given to the process $y_n(t)$ and it is obvious that when we try to apply the present result to some practical phenomena care should be taken to eliminate the trend completely before applying the ordinary spectral analysis to observed data. The spectral densities of runway roughness reported in the paper by Thompson [3] are at once seen to be quite-misleading as the trend elimination procedure adopted there was quite insufficient. The effect of such a trend is quite serious at the lower frequencies and the variations of root-mean-square values, σ's, of runway elevations reported therein can almost entirely be attributed to the "trends" of the runways. Taking into consideration this fact we can see that f^{-2} is a fairly good approximation to the results.

Neglecting the term $s*s*\cdots*s*x_0(t)$ we shall hereafter concern ourselves only with $y_n(t)$ and its power spectral density $p_n(f)$. From the general result just obtained we shall now try to deduce the f^{-2} form.

The term roughening means that the power spectrum of $\varepsilon_j(t)$ is widely spread compared with the form of $\sigma(f)$, or the term smoothing means that $|\sigma(f)|^2$ when multiplied with $p_\epsilon(f)$ cuts the power of $p_\epsilon(f)$ substantially. Further, in the cases of runways the effective range of smoothing must be short compared with the wave length which is in the range of our concern, i.e., the smoothing is not so sufficient that when it is applied once it substantially suppresses the roughness of the runway or roadway. These physical considerations lead us to the following assumptions.

A.1. $\varepsilon_j(t)$ is a white noise, i.e., $p_\epsilon(f)=c$ (>0) in the range of our concern.

A.2. $\int_{-\infty}^{\infty} t^2 s(t)dt < +\infty$, i.e., taking into account the relation $\int s(t)dt = 1$, $|\sigma(f)|^2$ permits the expression

$$|\sigma(f)|^2 = 1 - bf^2 + o(f^2) \qquad (|f| \to 0) ,$$

where

$$b = \left(\int_{-\infty}^{\infty} t^2 s(t)dt - \left(\int ts(t)dt\right)^2\right)(2\pi)^2 .$$

A.3. $|\sigma(f)|^2 - (1 - bf^2)$ is sufficiently small in magnitude in the range of f of our concern.

Then we get the following

THEOREM. *Under the above stated assumptions* A.1 *and* 2 *we have*

$$\lim_{n\to\infty} p_n(f) = \frac{c}{bf^2} + o(f^2) \qquad (|f| \to 0) .$$

Thus if the assumption A.3 holds and if we neglect the very low frequency component we observe that the f^{-2} law is valid. We can interpret the result of the theorem as saying that when the intensity of roughening, c, is large or the range of smoothing, b, is small the power of roughness increases, and in the opposite case it decreases. If we plot the spectral density on the (log, log) paper the line representing the spectral density of slope -2 goes upward for intensive roughening and poor smoothing and goes downward in the opposite case. At least qualitatively this fact seems to be in agreement with the observations reported in the paper by Walls and others [4]. If we further consider the approximation of the form

$$|\sigma(f)|^2 = 1 - bf^2 + df^4 + o(f^4) \qquad (d > 0)$$

we have

$$\lim_{n\to\infty} p_n(f) = \frac{c}{bf^2(1 - df^2/b)} + o(f^4) \qquad (|f| \to 0) .$$

Such an approximation may become necessary if the observed spectral density shows concavity upwards at higher frequency besides the effects of holding and other noises in the course of computation which have tendencies to whiten the spectrum. It seems that in practical cases $s(t)$ may be taken nearly as Gaussian. This is supported by the reasoning that in some cases $s(t)$ itself may be taken to be a result of repeated convolutions of smoothing functions and thus the central limit theorem applies. In any case $|\sigma(f)|^2$ may be considered to be the characteristic function of the difference of two random variables which are mutually independent and follow one and the same distribution and thus will be close to the Gaussian error function. These considerations will be of some use to those who want to use our present result for the understanding of some practical f^{-2} phenomena.

3. Acknowledgements

I wish to express my hearty thanks to Mr. I. Kaneshige of the research department of the Isuzu Motor Company, who drew my attension to the present phenomena. Thanks are also due to Mr. S. Takeda of the Transportation Technical Research Institute of the Ministry of Transportation for kind discussions of the present results.

THE INSTITUTE OF STATISTICAL MATHEMATICS

REFERENCES

[1] Ichiro Kaneshige, "Measurements of power spectra of roadways," *Isuzu Motor Company Technical Report* 3379, 1960. (in Japanese)

[2] Shun Takeda, "An investigation of airplane loads during taxiing," *Monthly Reports of Transportation Technical Research Institute.* Vol. 10, No. 5, 1960. (in Japanese)

[3] Wilbur E. Thompson, "Measurements and power spectra of runway roughness at airports in countries of the North Atlantic Treaty Organization," *NACA TN* 4303, 1958.

[4] James H. Walls, John, C. Houbolt, and Harry Press, "Some measurements and power spectra of runway roughness," *NACA TN* 3305, 1954.

Reprinted from *Annals of the Institute of Statistical Mathematics*, Vol. 14, No. 1, 1962, 23-56 by permission of the Institute of Statistical Mathematics

ON THE STATISTICAL ESTIMATION OF FREQUENCY RESPONSE FUNCTION

BY HIROTUGU AKAIKE and YASUFUMI YAMANOUCHI

(Received July 16, 1962)

1. Introduction and summary

At present, the spectral method is used very commonly for the analysis of an electrical or mechanical system. The spectral method is used not only for the estimation of the individual spectral density functions of the input and output of the system but also for the estimation of the frequency characteristics of the system.

The statistical method of estimation of the frequency characteristics of a system is best suited for this purpose as it can be applied without disturbing the normal operation of the system and even under the existence of the additive disturbance of extraneous noises. Statistical method for the estimation of the power spectral density has been brought to a considerable development by the valuable contributions of many statisticians, for example, those of J. W. Tukey [4, 9]. As for the estimation of the frequency response function of a linear and time-invariant system we have a paper by N. R. Goodman [8]. The method described by Goodman was a direct application of the method of estimation of the spectral density to that of the crosspectral density, but some experimenters who applied this kind of method to their numerical data experienced the very low coherency of their estimates. As far as we know, this fact was first recognized experimentally and announced by J. F. Darzell and Y. Yamanouchi [6].

The fruitful results of the method of estimation of the spectral density are mainly due to its success in reducing the variance by using proper smoothing operations. Methods of smoothing or averaging such as those named "hanning" and "hamming" and so forth were derived as those which have desirable properties of concentration of the effective range of smoothing. The smoothing operation is carried out by taking the product of the sample covariance function and a smoothing kernel or a lag window. It is well known that the autocovariance function does not contain any information about the phase of each frequency component contained in the original data. However, the crosscovariance function contains information of the phase, and the alignment of the phase shift of the frequency response function at frequencies in the effective range of the smoothing is most important to get a valid estimate of the amplitude gain. In Goodman's paper little attention was paid to this fact. Some experimenters who applied the same kind of method directly to their numerical data observed a very low coherency of their estimates at frequencies near the peak of the amplitude gain.

In 1960 we gave a proof that this low coherency is due to the inappropriate use of the lag window [3] and showed that, in the case of Darzell and Yamanouchi [6], the shifting of the time axis of one of the data greatly improves the coherency [11]. Almost at the same time Darzell and W. J. Pierson Jr. published an interesting paper in which the proper use of the time shift operation was also recommended to avoid the reduction in coherency [7]. Because of the reasons stated in the preceding paper [1, p. 2], discussions in this paper will be restricted to the lag windows of trigonometric sum type which are defined in section 2. First the sampling variability of the estimate of a frequency response function obtained by using a smoothing operation will be evaluated and then the bias caused by use of a lag window will be analysed. Even if the amplitude gain of the system under investigation is assumed to be locally constant or linear in the range of the smoothing operation, rapid change of the phase shift may still occur, and it is shown that this tends to reduce the value of the estimate of the amplitude gain. When we recall that in the physically realizable minimum-phase system the rate of change of the phase shift with respect to the increase of frequency is large at the peak of the amplitude gain [10, p. 45], we can see that this observation is of general meaning. Further the bias due to the use of the window is observed to be usually small in the case of the estimation of the phase shift. Taking into account that for the input of white noise the overall reduction in coherency is kept minimum when the integral of the square of a crosscovariance function multiplied by the lag window is maximum, it is recommended, to obtain a good overall view, in the practical application of the present estimation procedure, first to shift the time axis of the crosscovariance function so as to situate the origin at the maximum of the envelope of the sample crosscovariance function. The result of the analysis of the sampling variability of the estimate shows that at frequencies where the values of the coherency are high the lag window of which the Fourier transform, or the spectral window, has narrow bandwidth should be used. Numerical comparison of the windows are made and the use of the window Q, which has been recommended for estimation of the power spectral density in the preceding paper, is recommended for this case too. Sampling variability of an estimate of the coherency is discussed and the mean and the variance of the estimate are numerically given for some typical cases. Formulae for approximate evaluation of these two quantities are also given. Construction of the confidence region for the frequency response function is tentatively suggested. In the final section some numerical examples are given to show a practical meaning of the contents of this paper.

2. Sampling variability of the estimate

In this section an estimation procedure of the frequency response function of a linear time-invariant system will be discussed and the sampling variability of the estimate will be evaluated. The evaluation proceeds entirely analogously to that of the estimate of the spectral density function treated in the preceding paper [1]. All the stochastic processes treated in this parer are assumed to be Gaussian. Therefore the results of the evaluation of the sampling variability of the estimate have only the meaning of rough approximation to non-Gaussian cases. Even for the Gaussian case the evaluation can be carried out only approximately, so in practical applications of the method it is most recommended to check the sampling variability of the estimate numerically by some artificial experiments.

We observe here two stationary stochastic processes $x(t)$ and $y(t)$ which are real and are related by

$$x(t)=\int_{-\infty}^{\infty}\exp(2\pi ift)dZ(f)^{*}$$

$$y(t)=\int_{-\infty}^{\infty}\exp(2\pi ift)A(f)dZ(f)+n(t)$$

where $A(f)$ is the frequency response function or the gain of the system under consideration and $n(t)$ is a Gaussian process which is independent of the process $x(t)$, and which represents the additive disturbance of the extraneous noise. $x(t)$ and $n(t)$ are assumed to have absolutely continuous power spectral distributions with power spectral density functions $p_x(f)$ and $p_n(f)$ respectively, and to have zero-means.

We shall assume $p_x(f)$, $p_n(f)$ and $|A(f)|$ to be bounded continuous functions of f.

Under the present assumption the process $y(t)$ has an absolutely continuous power spectral distribution with power spectral density $p_y(f)$ given by

$$p_y(f)=|A(f)|^2p_x(f)+p_n(f).$$

We shall also use the Fourier representation of $n(t)$

* This is the Fourier representation of $x(t)$ and $Z(f)$ is a complex orthogonal process with $E|dZ(f)|^2=p_x(f)df$.

$$n(t)=\int_{-\infty}^{\infty}\exp(2\pi ift)dN(f).$$

It is known that when $y(t)$ represents a stationary output of a linear time-invariant system, of which the impulsive response function is $h(t)$, and which is distorted with an additive random noise $n(t)$, i.e.,

$$y(t)=\int_0^{\infty}x(t-\tau)h(\tau)d\tau+n(t),$$

we have

$$A(f)=\int_0^{\infty}\exp(-2\pi if\tau)h(\tau)d\tau.$$

$|A(f)|$ is called the amplitude gain of the system and $\phi(f)=\arg A(f)$ the phase shift of the system. Now, suppose data are given and are represented by

$$\{(x(t),\, y(t));\, -T\leqq t\leqq T\}.$$

After the preceding paper [1, p. 3] we define for integral ν's

$$X\left(\frac{\nu}{2T}\right)=\frac{1}{\sqrt{2T}}\int_{-T}^{T}\exp\left(-2\pi i\frac{\nu}{2T}t\right)x(t)dt$$

$$Y\left(\frac{\nu}{2T}\right)=\frac{1}{\sqrt{2T}}\int_{-T}^{T}\exp\left(-2\pi i\frac{\nu}{2T}t\right)y(t)dt$$

$$N\left(\frac{\nu}{2T}\right)=\frac{1}{\sqrt{2T}}\int_{-T}^{T}\exp\left(-2\pi i\frac{\nu}{2T}t\right)n(t)dt.$$

Then

$$X\left(\frac{\nu}{2T}\right)=\int_{-\infty}^{\infty}W_T\left(\frac{\nu}{2T}-f\right)dZ(f)$$

$$Y\left(\frac{\nu}{2T}\right)=\int_{-\infty}^{\infty}W_T\left(\frac{\nu}{2T}-f\right)A(f)dZ(f)+N\left(\frac{\nu}{2T}\right)$$

$$N\left(\frac{\nu}{2T}\right)=\int_{-\infty}^{\infty}W_T\left(\frac{\nu}{2T}-f\right)dN(f)$$

where

$$W_T(f)=\frac{1}{\sqrt{2T}}\int_{-T}^{T}\exp(-2\pi ift)dt\,.$$

Throughout the present paper it is assumed that T is sufficiently large so that, the real and the imaginary parts of $X(\nu/2T)$'s or $N(\mu/2T)$'s can be considered to be mutually independent Gaussian random variables with zero means and variances $p_x(\nu/2T)/2$ or $p_n(\mu/2T)/2$, respectively, for positive ν and μ, as was observed in the preceding paper, [1. p. 5].

Obviously, from the assumption of this paper it follows that $X(\nu/2T)$'s and $N(\mu/2T)$'s are mutually independent.

Hereafter, in this section, we shall adopt the extremely simplified assumption of constant $A(f)$. Then

$$\int_{-\infty}^{\infty} W_T\left(\frac{\nu}{2T}-f\right)A(f)dZ(f)\approx A\left(\frac{\nu}{2T}\right)\int_{-\infty}^{\infty} W_T\left(\frac{\nu}{2T}-f\right)dZ(f)$$

and

$$Y\left(\frac{\nu}{2T}\right)\overline{X\left(\frac{\nu}{2T}\right)}\approx A\left(\frac{\nu}{2T}\right)\left|X\left(\frac{\nu}{2T}\right)\right|^2+N\left(\frac{\nu}{2T}\right)\overline{X\left(\frac{\nu}{2T}\right)}$$

where $\approx$ means "equal to" under the present assumption. From this result one may imagine that the statistic

$$\frac{Y\left(\frac{\nu}{2T}\right)\overline{X\left(\frac{\nu}{2T}\right)}}{\left|X\left(\frac{\nu}{2T}\right)\right|^2}\approx A\left(\frac{\nu}{2T}\right)+\frac{N\left(\frac{\nu}{2T}\right)}{X\left(\frac{\nu}{2T}\right)}$$

will form an estimate of $A(\nu/2T)$. However, taking into account that $(p_x(\nu/2T))^{-1}|X(\nu/2T)|^2$ is distributed approximately as a χ^2 with d.f. 2, we can see that the square of the absolute value of the error term $X(\nu/2T)^{-1}\,N(\nu/2T)$ will not have finite expectation. This fact suggests that the statistic of this type will not be an appropriate one for the estimation of $A(f)$, and that it will be necessary to increase the d.f. of the denominator to obtain a reliable estimate. Using proper weight $\{w_\nu\}$ (satisfying $\sum_\nu w_\nu=1$) we now introduce our estimate $A(f_\mu)$:

$$\hat{A}(f_\mu)=\frac{\sum_\nu w_\nu Y(f_\mu-f_\nu)\overline{X(f_\mu-f_\nu)}}{\sum_\nu w_\nu|X(f_\mu-f_\nu)|^2}\ *$$

where $f_\mu=\mu/2T$. For this estimate we have

$$\hat{A}(f_\mu)\approx A(f_\mu)+\frac{\sum_\nu w_\nu N(f_\mu-f_\nu)\overline{X(f_\mu-f_\nu)}}{\sum_\nu w_\nu|X(f_\mu-f_\nu)|^2}.$$

Taking into account that we have almost surely $\sum_\nu|X(f_\nu)|^2=\int_{-T}^{T}|x(t)|^2dt$ and $\sum_\nu|N(f_\nu)|^2=\int_{-T}^{T}|n(t)|^2dt$, we get, in case w_ν's $(\nu\geqq\mu)$ are small enough and the last quantity in the following is finite,

---

*In this paper we shall use the notation $\sum_\nu$ in place of $\sum_{\nu=-\infty}^{\infty}$ and $\sum_{\nu\neq\mu}$ in place of $\sum_{\nu=-\infty}^{\mu-1}+\sum_{\nu=\mu+1}^{\infty}$.

$$E\left|\frac{\sum_\nu w_\nu N(f_\mu-f_\nu)\overline{X(f_\mu-f_\nu)}}{\sum_\nu w_\nu|X(f_\mu-f_\nu)|^2}\right|^2$$

$$=\sum_\nu\sum_{\nu'} w_\nu w_{\nu'} E\left[\frac{N(f_\mu-f_\nu)\overline{N(f_\mu-f_{\nu'})}\,\overline{X(f_\mu-f_\nu)}X(f_\mu-f_{\nu'})}{|\sum_\nu w_\nu|X(f_\mu-f_\nu)|^2|^2}\right]$$

$$\sim E\left[\frac{\sum_\nu |w_\nu|^2|X(f_\mu-f_\nu)|^2}{|\sum_\nu w_\nu|X(f_\mu-f_\nu)|^2|^2}\right]p_n(f_\mu)$$

$$\leqq \operatorname*{Max}_{\nu'}|w_{\nu'}|E\left[\frac{\sum_\nu |w_\nu||X(f_\mu-f_\nu)|^2}{|\sum_\nu w_\nu|X(f_\mu-f_\nu)|^2|^2}\right]p_n(f_\mu)$$

where the symbol $\sim$ means "approximately equal to" when T is sufficiently large so that $p_n(f_\mu-f_\nu)$ can be considered to be a constant $p_n(f_\mu)$ in evaluating the quantity under consideration. For almost all the lag windows which will be used practially it holds that $\operatorname*{Max}_{\nu'}|w_{\nu'}|=w_0$ and $w_\nu=|w_\nu|$ at those ν's where $|w_\nu|$ is large. Therefore, in this case, the last term may be replaced by

$$w_0E\left[\frac{1}{\sum_\nu w_\nu|X(f_\mu-f_\nu)|^2}\right]p_n(f_\mu).$$

In this paper we approximately regard the distribution of variable $\sum_\nu w_\nu|X(f_\mu-f_\nu)|^2p_x(f_\mu)^{-1}$ as a Γ-distribution, or more roughly as a χ^2-distribution, of which characteristics are given by $E(\sum_\nu w_\nu|X(f_\mu-f_\nu)|^2)$ and $D^2(\sum_\nu w_\nu|X(f_\mu-f_\nu)|^2)$. If a χ^2-distribution is adopted, its d.f. h is given by

$$h=\text{the nearest integer to }\frac{2(E\sum_\nu w_\nu|X(f_\mu-f_\nu)|^2)^2}{D^2(\sum_\nu w_\nu|X(f_\mu-f_\nu)|^2)}\sim\frac{2}{\sum_\nu|w_\nu|^2},$$

and when $h>2$

$$E\left[\frac{1}{\sum_\nu w_\nu|X(f_\mu-f_\nu)|^2}\right]\sim\frac{h}{h-2}p_x(f_\mu)^{-1}.$$

Thus an evaluation formula of the magnitude of the relative error of our estimate $\hat{A}(f_\mu)$ is given by

$$E\left|\frac{\hat{A}(f_\mu)-A(f_\mu)}{A(f_\mu)}\right|^2\sim w_0\frac{h}{h-2}\frac{p_n(f_\mu)}{|A(f_\mu)|^2p_x(f_\mu)}.$$

This evaluation may seem to be somewhat conservative. However, when we compare numerically the present result with a more optimistic one, which is obtained by assuming $\sum_\nu |w_\nu|^2 |X(f_\mu - f_\nu)|^2 / \sum_\nu w_\nu |X(f_\mu - f_\nu)|^2 \doteqdot \sum_\nu w_\nu^2 / \sum_\nu w_\nu$ and is given by replacing w_0 in the above formula by $\sum_\nu w_\nu^2$, we can see that the difference is small, say of the order of 10 to 20% in the sense of root mean square, for the windows of ordinary use. From this result we may conclude that we can conveniently use any of these two formulae for the evaluation and analysis of the sampling variability of our estimate.

Following the definition in the preceding paper [1] let us call the weight $\{w_\nu\}$ a spectral window corresponding to a lag window $W(t)$ when it satisfies

$$w_\nu = \frac{1}{2T}\int_{-T}^{T} \exp\left(-2\pi i \frac{\nu}{2T} t\right) W(t)dt.$$

The lag window of trigonometric sum type of the kth order is defined by

$$\begin{aligned} W(t) &= \sum_{n=-k}^{k} a_n \exp\left(2\pi i \frac{n}{2T_m} t\right) & |t| < T_m{}^* \\ &= \frac{1}{2}\sum_{n=-k}^{k} a_n \exp\left(2\pi i \frac{n}{2T_m} t\right) & |t| = T_m \\ &= 0 & |t| > T_m \end{aligned}$$

where a_n's are real and $a_{-n} = a_n$.

For the lag windows of trigonometric sum type

$$w_0 = \frac{1}{2T}\int_{-T}^{T} W(t)dt = \frac{2T_m}{2T} a_0$$

$$\sum_n |w_n|^2 = \frac{1}{2T}\int_{-T}^{T} |W(t)|^2 dt = \frac{2T_m}{2T} \sum_{n=-k}^{k} |a_n|^2$$

and

$$h = \text{the nearest integer to } 2\left(\frac{2T}{T_m}\right)\frac{1}{2\sum_{n=-k}^{k} |a_n|^2}.$$

The coherency $\gamma^2(f)$ at the frequency f is defined by

$$\begin{aligned} \gamma^2(f) &= \frac{|A(f)|^2 p_x(f)}{p_y(f)} \\ &= 1 - \frac{p_n(f)}{p_y(f)}. \end{aligned}$$

---

* If $\sum_\nu |w_\nu| < +\infty$ is necessary, we have only to modify this $W(t)$ to be smooth at $t = \pm T_m$. See also [1, p. 9].

When we assume that h is fairly large so that $h(h-2)^{-1}$ can be taken to be nearly equal to 1, the final evaluation formula of the relative accuracy of our estimate of the frequency response function will be given by

$$E\left|\frac{\hat{A}(f_\mu)-A(f_\mu)}{A(f_\mu)}\right|^2 \sim \frac{T_m}{2T} 2a_0 \frac{1-\gamma^2(f_\mu)}{\gamma^2(f_\mu)} .$$

If the above stated second evaluation formula is adopted,

$$E\left|\frac{\hat{A}(f_\mu)-A(f_\mu)}{A(f_\mu)}\right|^2 \sim \frac{T_m}{2T} 2 \sum_{n=-k}^{k} a_n^2 \frac{1-\gamma^2(f_\mu)}{\gamma^2(f_\mu)} .$$

By using this result and that of the analysis of bias, which will be given in the following section 3, the design principle of the optimum lag window of trigonometric sum type may be given entirely analogously as in the preceding paper. Therefore, we can see that the results of the analysis of the windows obtained in the preceding paper are all useful for this case too.

We have derived the present evaluation formulae under the assumption that $A(f)$ can be considered to be a constant in the range of the smoothing. The assumption of this kind caused little trouble in the case of estimation of the power spectral density, but in the case of estimation of the frequency response function the situation is different.

The lag windows of trigonometric sum type that have been used hitherto are all symmetric with respect to the y-axis hence are free from the biases of odd orders and they caused little trouble in the estimation of the power spectral density even in the case where the density function showed large linear variation. As to the estimation at the peaks and valleys of the power spectral density function, statisticians say that they are estimating the "averaged power spectra". In the estimation of the frequency response function the situation is the same and one must satisfy himself with an estimate of the "averaged frequency response function". However, there exists a difference in that the total power remains unchanged by averaging while the frequency response function may even "disappear" by averaging when the variation of phase shift exists. The apparent reduction in the amplitude gain is thus introduced by averaging and in case where the main concern of the experimenter is in the analysis of the coherency or of the linearity of the system such an estimate may be quite useless or even misleading.

In the following section the bias introduced by averaging in the estimation of the frequency response function will be discussed.

3. Bias due to smoothing

In this section we adopt the approximation to regard

$$Y(f_\mu)=A(f_\nu)X(f_\mu)+N(f_\mu) \qquad \mu=0,\ \pm 1,\ \pm 2,\ \cdots$$

which will hold strictly only when the $x(t)$ is circulating with period $2T$ and will hold approximately in the sense of mean square, as T tends to infinity. Then assuming the existence of $E[\sum_\nu |w_\nu||X(f_\mu-f_\nu)|^2/\sum_\nu w_\nu|X(f_\mu-f_\nu)|^2]$ we have

$$E(\hat{A}(f_\mu))\sim \sum_\nu A(f_\mu-f_\nu)E\left[\frac{w_\nu|X(f_\mu-f_\nu)|^2}{\sum_\lambda w_\lambda|X(f_\mu-f_\lambda)|^2}\right].$$

For the sake of simplicity put $x=w_\nu|X(f_\mu-f_\nu)|^2$ and $y=\sum_{\lambda\neq\nu} w_\lambda|X(f_\mu-f_\lambda)|^2$. Even under present assumption that $|X(f_\mu-f_\nu)|$'s are mutually independent and follow distributions of χ^2-type with d.f. 2, the quantity $x/(x+y)$ may not have any finite expectation when some of the w_λ's take negative values. Obviously this is due to the positive probability of $x+y$ being in the neighbourhood of zero, and if this probability is very small we may practically restrict our consideration to the case where $x+y$ is greater than some positive constant. For evaluation of the expectation of $x/(x+y)$ in this restricted sense we shall here approximate $x/(x+y)$ by $(x/(x+y))_\Delta$ which is defined as

$$\left(\frac{x}{x+y}\right)_\Delta=\frac{1}{x_0+y_0}(x_0+\Delta x)\left(1-\frac{\Delta x+\Delta y}{x_0+y_0}+\left(\frac{\Delta x+\Delta y}{x_0+y_0}\right)^2\right)$$

where $x_0=Ex$, $y_0=Ey$, $\Delta x=x-x_0$ add $\Delta y=y-y_0$. When the coefficient of variation of $x+y$, which is approximately equal to $\sqrt{2/h}$, is, fairly small, say less than 1/3, this approximation will give an estimate of the bias of $x/(x+y)$ from $x_0/(x_0+y_0)$.

The expectation of $(x/(x+y))_\Delta$ is

$$E\left(\frac{x}{x+y}\right)_\Delta=\frac{x_0}{x_0+y_0}-\frac{E(\Delta x)^2}{(x_0+y_0)^2}+\frac{x_0E(\Delta x+\Delta y)^2}{(x_0+y_0)^3}+\frac{E(\Delta x)^3}{(x_0+y_0)^3}.$$

Under the assumption of this paper

$$x_0=w_\nu p_x(f_\mu-f_\nu)$$

$$x_0+y_0=\sum_\lambda w_\lambda p_x(f_\mu-f_\lambda)$$

$$E(\Delta x)^2=w_\nu^2p_x^2(f_\mu-f_\nu)$$

$$E(\Delta x+\Delta y)^2=\sum_\lambda w_\lambda^2p_x^2(f_\mu-f_\lambda)$$

$$E(\Delta x)^3=2(w_\nu p_x(f_\mu-f_\nu))^3$$

when $p_x(f)$ is considered to be nearly a constant in the neighourhood of f_μ, and it can be seen that the ratio of the last term of the right hand side of $E(x/(x+y))_\Delta$ to the third term is nearly equal to $2w_\nu^2/\sum_\lambda w_\lambda^2$. Taking into account the relations

$$w_0 = \frac{2T_m}{2T} a_0$$

$$\sum_\lambda w_\lambda^2 = \frac{2T_m}{2T} \sum_n a_n^2$$

for the lag windows of trigonometric sum type, we can see that when h defined in section 2 or T/T_m is fairly large this last term usually takes a very small value for the windws used ordinarily. For the evaluation, therefore $\bar{E}(x/(x+y))_\Delta$ will be used in place of $E(x/(x+y))_\Delta$ where $\bar{E}(x/(x+y))_\Delta$ is defined as

$$\bar{E}\left(\frac{x}{x+y}\right)_\Delta = \frac{x_0}{x_0+y_0}\left(1+\frac{E(\Delta x+\Delta y)^2}{(x_0+y_0)^2}\right) - \frac{E(\Delta x)^2}{(x_0+y_0)^2}\ .$$

This $\bar{E}(x/(x+y))_\Delta$ gives the accurate result of $E(\sum_\nu w_\nu|X(f_\mu-f_\nu)|^2/\sum_\nu w_\nu|X(f_\mu-f_\nu)|^2)=1$ when it is formully applied to the evaluation of individual $E(w_\mu|X(f_\mu-f_\nu)|^2/\sum_\nu w_\nu|X(f_\mu-f_\nu)|^2)$, which is a main reason why we have adopted this $\bar{E}(x/(x+y))_\Delta$ in the present investigation.*

When we assume that $p_x(f)$ is locally flat around f_μ we have

$$E(\hat{A}(f_\mu)) = \sum_\nu A(f_\mu - f_\nu)\{w_\nu(1+\sum w_\lambda^2) - w_\nu^2\}$$

where the expectation should be understood in the restricted sense as mentioned above. When the assumption of local flatness of $p_x(f)$ is not satisfied, we have only to replace w_λ's in this formula of $E(\hat{A}(f_\mu))$ by w_λ''s which are proportional to $w_\lambda p_x(f_\mu - f_\lambda)$.

We analyse here the effect of smoothing operation when the assumption of constant $A(f_\mu - f_\nu)$ is not admitted. Besides the bias of $E(\hat{A}(f_\mu))$ from $A(f_\mu)$, the variation of $A(f_\mu - f_\nu)$ causes the sampling variability of the term $\sum_\nu w_\nu A(f_\mu - f_\nu)|X(f_\mu-f_\nu)|^2/\sum_\nu w_\nu|X(f_\mu-f_\nu)|^2$. However, taking into account that there usually exists inevitable sampling variability due to the additive noise, we shall limit our attension to the analysis of the bias of $E(\hat{A}(f_\mu))$ from $A(f_\mu)$ assuming that $p_x(f)$ is locally flat around f_μ. It should be noted that when $A(f)$ can be con-

* Strict evaluation of $E(x/(x+y))$ is possible if $w_\lambda > 0$ and $w_\lambda p_x(f_\mu - f_\lambda) \neq w_{\lambda'} p_x(f_\mu - f_{\lambda'})$ $(\lambda \neq \lambda')$ hold, but the result is not directly applicable to the present analysis.

sidered to be a constant in the effective range of the smoothing operation the assumption of local flatness of $p_{\Delta}(f)$ is unnecessary to keep our estimate unbiased. This shows that the local flatness of the amplitude gain and that of phase shift have great influence on the estimation of the frequency response function. Therefore, we can guess that in this case some sort of "preflatening" operation will play the important role which was played by the prewhitening operation in the estimation of the power spectral density function.

Now let us consider the case where $A(f)$ admits the local approximation

$$A(f_\mu - f_\nu) = \alpha(1 + \beta f_\nu + \gamma f_\nu^2) \exp(2\pi i f_\nu T_\mu)^*$$

where β, γ and T_μ are real constants and $\alpha = A(f_\mu)$. Then

$$E(\hat{A}(f_\mu)) \sim \sum_\nu A(f_\mu - f_\nu)\{w_\nu(1 + \sum_\lambda w_\lambda^2) - w_\nu^2\}$$

$$= \alpha(\sum_\nu w_\nu \exp(2\pi i f_\nu T_\mu) + \beta \sum_\nu w_\nu f_\nu \exp(2\pi i f_\nu T_\mu)$$

$$+ \gamma \sum_\nu w_\nu f_\nu^2 \exp(2\pi i f_\nu T_\mu))(1 + \sum_\lambda w_\lambda^2)$$

$$- \alpha(\sum_\nu w_\nu^2 \exp(2\pi i f_\nu T_\mu) + \beta \sum_\nu w_\nu^2 f_\nu \exp(2\pi i f_\nu T_\mu) + \gamma \sum_\nu w_\nu^2 f_\nu^2 \exp(2\pi i f_\nu T_\mu))$$

$$= \alpha(W(T_\mu) + \frac{\beta}{2\pi i} \dot{W}(T_\mu) + \frac{\gamma}{(2\pi i)^2} \ddot{W}(T_\mu))$$

$$+ (\sum_\lambda w_\lambda^2)\alpha \left\{ W(T_\mu) - W^{*2}(T_\mu) + \frac{\beta}{2\mu i}(\dot{W}(T_\mu) - \dot{W}^{*2}(T_\mu)) \right.$$

$$\left. + \frac{\gamma}{(2\pi i)^2}(\ddot{W}(T_\mu) - \ddot{W}^{*2}(T_\mu)) \right\} (C.\ 3)$$

where $(C.\ 3)$ shows summability by the 3rd Cesàro mean [1. p. 7],

$$W^{*2}(t) = (\sum_\lambda w_\lambda^2)^{-1} \int_{-\infty}^{\infty} W(t-s) W(s) ds,$$

$$\dot{W}(T_\mu) = \frac{d}{dt} W(t) \Big|_{t=T_\mu},$$

$$\ddot{W}(T_\mu) = \frac{d^2}{dt^2} W(t) \Big|_{t=T_\mu},$$

$$\dot{W}^{*2}(T_\mu) = \frac{d}{dt} W^{*2}(t) \Big|_{t=T_\mu},$$

$$\ddot{W}^{*2}(T_\mu) = \frac{d^2}{dt^2} W^{*2}(t) \Big|_{t=T_\mu},$$

* This local approximation of the frequency response function constitutes the basis of our present analysis. Its extension to the cases with the polynomial factor of higher degrees is straightforward.

and the existence of these quantities is assumed. The second term of this last expression of $E(\hat{A}(f_\mu))$ represents the contribution of $E(x/(x+y)) - E(x)/E(x+y)$ and usually is very small while the first term represents the main effect of the smoothing operation. Hereafter, we restrict our attension to this first term. This term is analogous to that obtained in the preceding paper [1, p. 7] by the analysis of the bias in the estimation of the power spectral density. However, there is one important difference in that $W(0)$, $\dot{W}(0)$ and $\ddot{W}(0)$ of the formers are replaced by $W(T_\mu)$, $\dot{W}(T_\mu)$ and $\ddot{W}(T_\mu)$ in the present formulae, respectively.

Taking into account that $2\pi T_\mu$ represents $-d[\arg A(f)]/df|_{f=f_\mu}$, we can see that this appearance of T_μ in the formula is showing that the allignment of the phase shift between the output and input is essential for the estimation of $A(f)$. This fact was entirely disregarded in the paper by Goodman [8]. As to the lag windows of trigonometric sum type hitherto used, $W(T_\mu)$ is less than one except for the case where $T_\mu=0$ holds, hence even if the contributions of the term βf_ν, βf_ν^2 are small the estimate may suffer a significant decrease in the amplitude gain by the contribution of T_μ.

Taking into account the very rapid change of phase shift of the frequency response function of the ship model treated by Darzell and Yamonouchi [6], this gives a theoretical explanation to the apparently very low amplitude gain, accordingly to the apparently very low coherency, observed there at the frequencies where the amplitude gain showed a significant peak in the analysis of the response of the ship model to the waves.

The present representation of the bias suggests a method how to remedy this defect of the window. For we can see that if we replace w_ν by $w_\nu \exp(-2\pi i f_\nu T_\mu)$ or $W(t)$ by $W(t-T_\mu)$ then we have

$$\sum_\nu w_\nu \exp(-2\pi i f_\nu T_\mu) A(f_\mu - f_\nu) = \alpha\left(W(0) + \frac{\beta}{2\pi i}\dot{W}(0) + \frac{\gamma}{(2\pi i)^2}\ddot{W}(0)\right) (C,\ 3),$$

and if we use the window $W_k(2, *)$ which was defined in the preceding paper and for which $\dot{W}(0) = \ddot{W}(0) = 0$ holds, we have

$$\sum_\nu w_\nu \exp(-2\pi i f_\nu T_\mu) A(f_\mu - f_\nu) = \alpha\ (C.\ 3).$$

However, if we try to use this procedure practically, we have to put an estimate of T_μ in place of T_μ itself, and this suggests the necessity of the use of lag windows with somewhat flat tops. It is known that in a physically realizable minium-phase system there exists functional relationship between the amplitude gain and the phase shift, so to know of T_μ is actually to know of the shape of the amplitude gain near the frequency f_μ, which we are struggling to obtain. Fortunately, however, it can be seen, by inserting some concrete values into the present evaluation formula of the bias, that even when $|T_\mu|$ is fairly large $W(T_\mu) - (\gamma/(2\pi))^2 \ddot{W}(T_\mu)$ may sometimes be much greater then $|(\beta/2\pi)\dot{W}(T_\mu)|$ and the bias of the phase shift due to the smoothing is relatively small. This fact and the following consideration of the overall reduction in the gain shows that we can usually get a reliable estimate of the phase shift and thus of T_μ, and the estimate is used to obtain the final estimate of $A(f_\mu)$.

To get an overall view of the effect of the smoothing on the frequency response function we make an analysis in the time domain [3]. It is assumed that $A(f)$ is the Fourier transform of some function $h(t)$ and is represented as

$$A(f)=\int_{-T}^{T} \exp(-2\pi i f t)h(t)dt .$$

Then we have

$$\sum_{\nu} w_\nu A(f_\mu - f_\nu)=\int_{-T}^{T} \exp(-2\pi i f_\mu t)W(t)h(t)dt .$$

and

$$\left(\frac{1}{2T}\right) \sum_{\mu} \Big|\sum_{\nu} w_\nu A(f_\mu - f_\nu)\Big|^2=\int_{-T}^{T} |W(t)h(t)|^2 dt$$

$$\left(\frac{1}{2T}\right) \sum_{\mu} |A(f_\mu)|^2=\int_{-T}^{T} |h(t)|^2 dt .$$

From these formulae and the fact that $W(t)$ is usually less than unity, it can be seen that, to keep minimum the overall reduction in the gain, $W(t)$ which gives the maximum of $\int_{-T}^{T} |W(t)h(t)|^2 dt$ should be used. In the practical application of our estimation procedure we can observe the sample crosscovariance function which is a convolution of $h(t)$ with the sample autocovariance function of $x(t)$ distorted by the existence of some additive disturbance $n(t)$, and the sample crosscovariance function provides some information of the shape of $h(t)$, especially when some type of overall prewhitening operation is applied to the input.* Therefore in case where the input is considered to be fairly white in the range of our concern, we can obtain a fairly good overall estimate of $A(f)$ by shifting the center of the lag window or the origin of the time axis of the sample crosscovariance function to that time point t where the maximum of the absolute value of the sample crosscovariance function occurs. From the result of the foregoing analysis, it can be seen that

* As to the overall prewhitening operation, some numerical examples are illustrated in the former paper by one of the authors [2].

the local whiteness of the input is desired to keep the estimate of the frequency response function unbiased and overall whiteness is only necessary to get a good primary estimate.

We shall here analyse the effect of the smoothing operation for a certain concrete type of $A(f)$. We assume

$$h(t)=\frac{\omega_n}{\sqrt{1-\zeta^2}}\exp(-\zeta\omega_n t)\sin\sqrt{1-\zeta^2}\,\omega_n t) \qquad t\geqq 0$$

$$=0 \qquad t<0.$$

Then we have

$$A(f)=\frac{\omega_n^2}{(\omega_n^2-(2\pi f)^2)+2\zeta\omega_n i(2\pi f)}=|A(f)|\exp(i\phi)$$

$$\phi=\tan^{-1}\left(\frac{-2\zeta\omega_n 2\pi f}{\omega_n^2-(2\pi f)^2}\right).$$

This $A(f)$ is the frequency response function of the system of which behavior is described by the second order differential equation

$$\left(\tilde{T}\frac{d^2}{dt^2}+\frac{d}{dt}+K\right)\theta_O(t)=K\theta_I(t)$$

where $\theta_I(t)$ and $\theta_O(t)$ represent the input and output of the system and

$$\omega_n^2=\frac{K}{\tilde{T}}$$

$$\zeta=\frac{2}{2\sqrt{K\tilde{T}}}.$$

Consider the case where $\zeta\ll 1$ or very lightly damped oscillating system with one degree of freedom. In this case $|A(f)|$ shows its peak near its natural resonant frequency $f_n=\omega_n/2\pi$, and a value of T_μ at $f_\mu=f_n$, which appeared in the local approximation formula of $A(f)$, will be given by

$$T_\mu=-\frac{1}{2\pi}\frac{d\phi}{df}\bigg|_{f=f_n}=\frac{1}{\zeta\omega_n}.$$

Further, if we define $\Delta f_+(>0)$ and $\Delta f_-(>0)$ by

$$|A(f_n+\Delta f_+)|=|A(f_n-\Delta f_-)|=\frac{1}{\sqrt{2}}|A(f_n)|,$$

then under the condition $\zeta\ll 1$ we have

$$\Delta f_+ + \Delta f_- \doteqdot \frac{\zeta\omega_n}{\pi} .$$

From these two results we get the relation

$$T_\mu \doteqdot \frac{1}{\pi} \frac{1}{\Delta f_+ + \Delta f_-} .$$

For example, when $\Delta f_+ + \Delta f_- = 1$ c.p.s. we have to shift the window $1/\pi$ second to the right of origin. This $\Delta f_+ + \Delta f_-$ is usually called the bandwidth of the filter or of $A(f)$ and our present result shows that the amount of shift T_μ is inversely proportional to the bandwidth of the filter. This knowledge will prevent the danger of using a lag window with $W(T_\mu) \doteqdot 0$. Now we evaluate the bias induced on the present $A(f)$ by the use of windows of trigonometric sum type. It is assumed that every window is shifted by the amount $K = 1/\zeta\omega_n$ and that $T_m > K$. Adopt the approximation

$$A(f_n + f) \doteqdot \frac{-i\omega_n}{2} \frac{1}{\zeta\omega_n + i2\pi f}$$

around the natural frequency f_n. Then we have

$$\frac{1}{2T} \sum_\nu A(f_n - f_\nu) w_\nu \doteqdot \frac{-i\omega_n}{2} \int_0^\infty \exp(-\zeta\omega_n t) W_K(t) dt$$

$$= A(f_n)(\zeta\omega_n) \int_0^\infty \exp(-\zeta\omega_n t) W_K(t) dt .$$

Now

$$\int_0^\infty \exp(-\zeta\omega_n t) W_K(t) \zeta\omega_n dt$$

$$= \sum_{n=-k}^{k} a_n \int_0^{K+T_m} \exp\left(-2\pi i \frac{\nu}{2T_m}(t-K)\right) \exp(-\zeta\omega_n t)\, \zeta\omega_n dt$$

and when we define

$$\rho = \frac{\pi}{\zeta\omega_n} \frac{1}{T_m} \left(\doteqdot \frac{1}{\Delta f_+ + \Delta f_-} \frac{1}{T_m} \right)$$

we get for $k=3$

$$A_D(\zeta) = \int_0^\infty \exp(-\zeta\omega_n t) W_K(t) \zeta\omega_n dt$$

$$= a_0 \left\{ 1 - \exp\left(-\left(1 + \frac{\pi}{\rho}\right)\right)\right\}$$

TABLE 1. Comparison of biases induced on $A(f)$

	$W_1(0, \alpha)$	$W_1(0, \beta/\alpha)$	$W_1(0, \alpha\beta)$	$W_2(0, \alpha)$
a_0	0.3333	0.5272	0.5132	0.2000
$a_1=a_{-1}$	0.3333	0.2364	0.2434	0.2000
$a_2=a_{-2}$	*	*	*	0.2000
$a_3=a_{-3}$	*	*	*	*
$A_D(2)$	0.5050	0.6267	0.6179	0.2146
$A_D(1)$	0.7938	0.8492	0.8452	0.5873
$A_D(1/2)$	0.9293	0.9497	0.9482	0.8338
$A_L(1/4)$	0.9800	0.9859	0.9855	0.9456

	$W_2(2, \alpha)$	$W_2(2, \beta/\alpha)$	$W_2(2, \alpha\beta)$	$W_3(2, \alpha)$
a_0	0.4857	0.4675	0.6398	0.3333
$a_1=a_{-1}$	0.3429	0.2350	0.2401	0.2857
$a_2=a_{-2}$	−0.0587	−0.0588	−0.0600	0.1429
$a_3=a_{-3}$	*	*	*	−0.0952
$A_D(2)$	0.6893	0.7630	0.7594	0.4169
$A_L(1)$	0.9097	0.9330	0.9319	0.8171
$A_L(1/2)$	0.9804	0.9862	0.9860	0.9496
$A_L(1/4)$	0.9978	0.9983	0.9984	0.9916

$$+a_1\frac{2}{1+\rho^2}\left\{\cos\rho+\rho\sin\rho+\exp\left(-\left(1+\frac{\pi}{\rho}\right)\right)\right\}$$

$$+a_2\frac{2}{1+(2\rho)^2}\left\{\cos 2\rho+2\rho\sin 2\rho-\exp\left(-\left(1+\frac{\pi}{\rho}\right)\right)\right\}$$

$$+a_3\frac{2}{1+(3\rho)^2}\left\{\cos 3\rho+3\rho\sin 3\rho+\exp\left(-\left(1+\frac{\pi}{\rho}\right)\right)\right\}.$$

The result of numerical computation of this $A_D(\zeta)$ for the windows treated in the preceding paper [1] is given in table 1, for the values of $\rho=2$, 1, 1/2 and 1/4 which correspond to the values of $T_m=\pi\tilde{T}$, $2\pi\tilde{T}$, $4\pi\tilde{T}$ and $8\pi\tilde{T}$. The result shows that ρ must be kept less than 1 for the ordinary purpose of estimation and that the window $Q(a_0=0.64$ $a_1=a_{-1}=0.24$ $a_2=a_{-2}=-0.06)$ introduced in the preceding paper, which is approximately the same with $W_2(2, \alpha\beta)$, will be quite suited for ordinary use.

4. Sampling variability of the estimate of coherency

As was seen in section 2, the value of coherency is necessary for eʌaluation of the accuracy of our estimate of the frequency response by the use of windows of trigonometric sum type.

$W_2(0, \beta/\alpha)$	$W_2(0, \alpha\beta)$	$W_3(0, \alpha)$	$W_3(0, \beta/\alpha)$	$W_3(0, \alpha\beta)$
0.6202	0.4282	0.1429	0.6129	0.4229
0.2358	0.2433	0.1429	0.2182	0.2124
−0.0459	0.0426	0.1429	−0.0431	0.0434
*	*	0.1429	0.0184	0.0327
0.7326	0.5206	0.1498	0.7135	0.4959
0.9145	0.7851	0.4039	0.8822	0.7335
0.9782	0.9219	0.7336	0.9610	0.8942
0.9957	0.9762	0.9030	0.9885	0.9646

$W_3(2, \beta/\alpha)$	$W_3(2, \alpha\beta)$	$W_3(4, \alpha)$	$W_3(4, \beta/\alpha)$	$W_3(4, \alpha\beta)$
0.6643	0.5571	0.5671	0.7085	0.7029
0.2306	0.2610	0.3247	0.2186	0.2228
−0.0609	−0.0190	−0.1299	−0.0875	−0.0891
0.0041	−0.0205	0.0216	0.0146	0.0149
0.7793	0.6780	0.7725	0.8218	0.8198
0.9385	0.9048	0.9375	0.9526	0.8198
0.9879	0.9777	0.9890	0.9923	0.9923
0.9987	0.9969	0.9992	0.9994	0.9996

function. In practical applications, it is often more necessary for making a decision whether such analysis of linear relation between $x(t)$ and $y(t)$ has a practical meaning or not. We define sample coherency $\hat{\gamma}^2(f)$ at f_μ by

$$\hat{\gamma}^2(f_\mu)=|\hat{A}(f_\mu)|^2\frac{\sum_\nu w_\nu|X(f_\mu-f_\nu)|^2}{\sum_\nu w_\nu|Y(f_\mu-f_\nu)|^2}.$$

This $\hat{\gamma}^2(f_\mu)$ will be an estimate of the true coherency $\gamma^2(f_\mu)$, and the sampling variability of this estimate will be treated in the following. It was shown in the preceding section that the main trouble in the estimation is the bias of the estimate, and that by using the shift operation and lag window, properly combined, we can evade this difficulty. In this section we discuss the sampling variability of the estimate of the coherency under the extremely simplified condition of constant $A(f)$ and constant $p_x(f)$ and $w_\nu=1/n(\nu=1, 2, 3, \cdots, n)$. This condition is identical to that adopted in the paper [8] by Goodman, and we can show that in this case our estimation procedure reduces to that of the classical regression estimate and all the necessary information is directly available from the standard text of mathematical statistics. We adopt the notations

$$U_{X,\nu}=R_eX\left(\frac{\nu}{2T}\right)$$

$$V_{X,\nu}=I_mX\left(\frac{\nu}{2T}\right)$$

$$U_{Y,\nu}=R_eY\left(\frac{\nu}{2T}\right)$$

$$V_{Y,\nu}=I_mY\left(\frac{\nu}{2T}\right)$$

$$U_{N,\nu}=R_eN\left(\frac{\nu}{2T}\right)$$

$$V_{N,\nu}=I_mN\left(\frac{\nu}{2T}\right)$$

$$A=R_eA(f)$$

$$B=I_mA(f)$$

$$\sigma_N^2=\frac{1}{2}p_n(f)$$

$$\sigma_X^2=\frac{1}{2}p_x(f)$$

$$\sigma_Y^2=\frac{1}{2}p_y(f)=|A(f)|^2\sigma_X^2+\sigma_N^2$$

$$\gamma^2=\text{coherency}=\frac{|A(f)|^2\sigma_X^2}{\sigma_Y^2}=(A^2+B^2)\left(\frac{\sigma_X}{\sigma_Y}\right)^2$$

where R_e and I_m denote "the real part of" and "the imaginary part of", respectively. Following the analysis in section 2 we have

$$U_{Y,\nu}=AU_{X,\nu}-BV_{X,\nu}+U_{N,\nu}$$
$$V_{Y,\nu}=BU_{X,\nu}+AV_{X,\nu}+V_{N,\nu}$$

and it is assumed that $U_{X,\nu}$, $V_{X,\nu}$, $U_{N,\nu}$, $V_{N,\nu}$ are mutually independent Gaussion random variables with zero means and variances $D^2(U_{X,\nu})=D^2(V_{X,\nu})=\sigma_X^2$ $D^2(U_{N,\nu})=D^2(V_{N,\nu})=\sigma_N^2=\sigma_Y^2(1-\gamma^2)$. The 4-dimensional probability density function $\phi(u_y, v_y, u_x, v_x)$ of $(U_{Y,\nu}, V_{Y,\nu}, U_{X,\nu}, V_{X,\nu})$ at $(u_y, v\ , u_x, v_x)$ is, therefore, given by

$$\phi(u_y, v_y, u_x, v_x)=\frac{1}{2\pi\sigma_X^2}\exp\left(-\frac{1}{2\sigma_X^2}\{u_x^2+v_x^2\}\right)$$

$$\times\frac{1}{2\pi\sigma_Y^2(1-\gamma^2)}\exp\left(-\frac{1}{2\sigma_Y^2(1-\gamma^2)}\{(u_y-Au_x+Bv_x)^2+(v_y-Bu_x-Av_x)^2\}\right).$$

Now, for a random sample $((U_{Y,\nu}, V_{Y,\nu}, U_{X,\nu}, V_{X,\nu});\ \nu=\nu_0+1, \nu_0+2, \cdots, \nu_0+n)$ the maximum-likelihood estimates $\hat{A}$ and $\hat{B}$ of A and B are given by

$$\hat{A}=\frac{\sum_{\nu=\nu_0+1}^{\nu_0+n} U_{Y,\nu}U_{X,\nu}+\sum_{\nu=\nu_0+1}^{\nu_0+n} V_{Y,\nu}V_{X,\nu}}{\sum_{\nu=\nu_0+1}^{\nu_0+n} U_{X,\nu}^2+\sum_{\nu=\nu_0+1}^{\nu_0+n} V_{X,\nu}^2},\qquad \hat{B}=\frac{\sum_{\nu=\nu_0+1}^{\nu_0+n} V_{Y,\nu}U_{X,\nu}-\sum_{\nu=\nu_0+1}^{\nu_0+n} U_{Y,\nu}V_{X,\nu}}{\sum_{\nu=\nu_0+1}^{\nu_0+n} U_{X,\nu}^2+\sum_{\nu=\nu_0+1}^{\nu_0+n} V_{X,\nu}^2}$$

and the sample coherency $\hat{\gamma}^2$ is given as

$$\hat{\gamma}^2=\frac{(\hat{A}^2+\hat{B}^2)(\sum_{\nu=\nu_0+1}^{\nu_0+n} U_{X,\nu}^2+\sum_{\nu=\nu_0+1}^{\nu_0+n} V_{X,\nu}^2)}{\sum_{\nu=\nu_0+1}^{\nu_0+n} U_{Y,\nu}^2+\sum_{\nu=\nu_0+1}^{\nu_0+n} V_{Y,\nu}^2}.$$

After a moment's consideration, we come to know that we are making here a regression analysis of the form

$$Y=AX_1+BX_2+\varepsilon$$

where

$$\begin{aligned}
Y &=(U_{Y,\nu_0+1}, U_{Y,\nu_0+2}, \cdots, U_{Y,\nu_0+n}, V_{Y,\nu_0+1}, V_{Y,\nu_0+2}, \cdots, V_{Y,\nu_0+n})'\\
X_1 &=(U_{X,\nu_0+1}, U_{X,\nu_0+2}, \cdots, U_{X,\nu_0+n}, V_{X,\nu_0+1}, V_{X,\nu_0+2}, \cdots, V_{X,\nu_0+n})'\\
X_2 &=(-V_{X,\nu_0+1}, -V_{X,\nu_0+2}, \cdots, -V_{X,\nu_0+n}, U_{X,\nu_0+1}, U_{X,\nu_0+2},\cdots, U_{X,\nu_0+n})'\\
\varepsilon &=(U_{N,\nu_0+1}, U_{N,\nu_0+2}, \cdots, U_{N,\nu_0+n}, V_{N,\nu_0+1}, V_{N,\nu_0+2},\cdots, V_{N,\nu_0+n})'
\end{aligned}$$

where $'$ denotes the transpose of the vector. Statistical properties of $\hat{A}$ and $\hat{B}$ are well known and the sample coherency $\hat{\gamma}^2$ is the square of multiple correlation coefficient between (X_1, X_2) and Y. Taking into account of the fact that the inner product $\langle X_1, X_2\rangle$ of vectors X_1 and X_2 is identically equal to zero, we can at once get the representation

$$\langle\varepsilon, \varepsilon\rangle=(\hat{A}-A)^2\langle X_1, X_1\rangle+(\hat{B}-B)^2\langle X_2, X_2\rangle+\langle\hat{\varepsilon}, \hat{\varepsilon}\rangle$$

where $\hat{\varepsilon}=Y-\hat{A}X_1-\hat{B}X_2$ and $\langle,\rangle$ denotes the inner product of vectors. Fisher's lemma can be applied [5, p. 379] to this representation and we know that the right hand side members, when divided by σ_N^2, are mutually independently distributed as χ^2's with d.f. 1, 1 and $2(n-1)$, respectively. The sample coherency $\hat{\gamma}^2$ can be represented as

$$\hat{\gamma}^2=\frac{(\hat{A}^2\langle X_1, X_1\rangle+B\langle X_2, X_2\rangle)(\sigma_N^2)^{-1}}{(\hat{A}^2\langle X_1, X_1\rangle+\hat{B}^2\langle X_2, X_2\rangle+\langle\hat{\varepsilon}, \hat{\varepsilon}\rangle)(\sigma_N^2)^{-1}}$$

and we can see that the numerator is a non-central χ^2-variable with d.f.

2 and non-centrality $\gamma^2\langle X_1, X_2\rangle/(1-\gamma^2)\sigma_X^2$. Taking into account the fact that $\langle X_1, X_1\rangle/\sigma_X^2$ is distributed as a χ^2-variable with d.f. $2n$, we can see that the distribution of the sample coherency $\hat{\gamma}^2$ is that of the square of the sample multiple correlation coefficient when the corresponding population value is equal to γ^2 and a straightforward calculation shows that its probability density $p(\zeta)$ at $\hat{\gamma}^2=\zeta$ is given by

$$p(\zeta)=(1-\eta)^n \sum_{k=0}^{\infty} \frac{1}{\Gamma(n)\Gamma(n-1)} \left(\frac{\Gamma(n+k)}{\Gamma(k+1)}\right)^2 \eta^k \zeta^{k+1-1}(1-\zeta)^{n-1-1}$$

where $\eta=\gamma^2$. This is identical with the formula (4.60) of Goodman's paper [8] if z^2 in the latter is replaced by ζ*.

This $p(\zeta)$ will give an approximation to the true distribution of the sample coherency $\hat{\gamma}^2(f_\mu)$ for general spectral window $\{w_\nu\}$, and for the analysis of the sampling variability of the sample coherency $\hat{\gamma}^2(f_\mu)$, obtained by using a lag window of trigonometric sum type, one should take the nearest integer to $(2T/T_m)(1/2 \sum_{n=k}^{k} |a_n^2|)$ as the n in the above consideration. For the sake of typographical simplicity we hereafter use the letter ζ in place of $\hat{\gamma}^2$. Now, $E(\zeta)$ and $E(\zeta^2)$ are given by

$$E(\zeta)=(1-\eta)^n \sum_{k=0}^{\infty} \frac{\Gamma(n+k)}{\Gamma(n)\Gamma(k+1)} \frac{k+1}{k+n} \eta^k$$

$$E(\zeta^2)=(1-\eta)^n \sum_{k=0}^{\infty} \frac{\Gamma(n+k)}{\Gamma(n)\Gamma(k+1)} \frac{k+1}{k+n} \frac{k+2}{k+n+1} \eta^k .$$

Numerical computations of $E(\zeta)$ and $D^2(\zeta)$ by these formulae for values of n and η which will often be met in practical applications are tabulated in table 2.

Fig. 1 shows the results of a sampling experiment of estimation of the coherency using the windows hamming and Q. Necessary information about the true $A(f)$ of this case will be given in fig's in section 6. It seems that, at least for this example, $E(\zeta)$ and $D^2(\zeta)$ obtained by using the present distribution of ζ provide us with good estimates of the corresponding true values of the distribution of the sample coherency. However, it must be borne in mind that this estimation is valid only when the bias due to smoothing is negligibly small.

For $\rho=\dfrac{\zeta}{1-\zeta}$ we have

$$\rho_0 \equiv E(\rho)=\frac{n}{n-2}\,\bar{\rho}+\frac{1}{n-1}$$

---

* In Goodman's paper, Z and z in (4.8.6.) and (4.16.6) should be read as Z^2 and z^2 respectively.

$$D^2(\rho) = \frac{2n(n-1)}{(n-2)^2(n-3)}\left(\bar{\rho}^2+\bar{\rho}+\frac{1}{2n}\right)$$

where

$$\bar{\rho}=\frac{\eta}{1-\eta};$$

The factor $\bar{\rho}^{-1}=(1-\eta)/\eta$ is necessary for evaluation of the relative accuracy of the estimate of the frequency response function and sometimes we have to substitute its sample value $\rho^{-1}=(1-\zeta)/\zeta$ for this. For evaluation of the error of this last value, the knowledge of $E(\zeta)$ and $D^2(\zeta)$ will conveniently be used. From the relation $\zeta=\rho/(1+\rho)$ we have for $\rho=\rho_0+\Delta\rho$ the local approximation

$$\zeta \doteqdot \frac{\rho_0}{1+\rho_0}+\left(\frac{d\zeta}{d\rho}\right)_{\rho=\rho_0}\Delta\rho.$$

Using this approximation we have

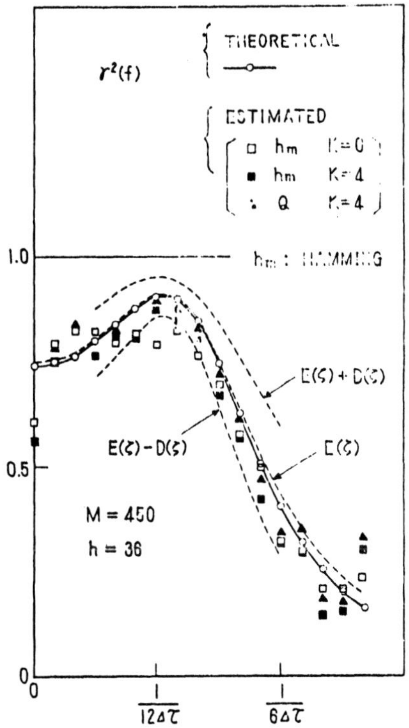

Fig. 1. Sampling experiment of estimation of coherency.

$$E(\zeta)=\frac{\rho_0}{1+\rho_0}$$

$$D^2(\zeta)\doteqdot\frac{2n(n-1)}{(n-2)^2(n-3)}\frac{(\bar{\rho}^2+\bar{\rho}+1/2n)}{(1+\zeta_0)^4}.$$

In table 2 the values obtained by using the present approximation are compared with the corresponding theoretical values. The result is also illustrated in fig. 2. It can be seen that, in this range of parameters, our formulae provide us with estimates which are accurate enough for the purpose of practical applications. Moreover, our approximation shows that $D^2(\zeta)$ is approximately inversely proportional to n, this is quite natural and useful for the design of experiment and analysis. The above result of sampling experiment seems to show that the simplified condition of this section gives a fairly reasonable approximation to the true distribution of the estimate of coherency. From this it can be inferred that it will not be quite useless to mention here of the confidence region

TABLE 2. Comparison of the approximate values with their corresponding theoretical values.

η	$E(\zeta)$					
	$n=5$		$n=10$		$n=20$	
	A*)	T**)	A	T	A	T
0.4	0.59	0.48	0.49	0.44	0.44	0.42
0.5	0.67	0.56	0.58	0.53	0.54	0.51
0.6	0.74	0.64	0.67	0.62	0.63	0.61
0.7	0.81	0.72	0.75	0.71	0.73	0.70
0.8	0.88	0.81	0.84	0.80	0.82	0.80
0.9	0.94	0.90	0.92	0.90	0.91	0.90

η	$D(\zeta)$					
	$n=5$		$n=10$		$n=20$	
	A	T	A	T	A	T
0.4	0.27	0.21	0.17	0.16	0.12	0.12
0.5	0.24	0.20	0.16	0.15	0.11	0.11
0.6	0.20	0.18	0.14	0.14	0.10	0.10
0.7	0.15	0.16	0.11	0.11	0.08	0.08
0.8	0.10	0.12	0.08	0.08	0.06	0.06
0.9	0.05	0.07	0.04	0.05	0.03	0.03

η	$C.V.(\zeta)=E(\zeta)^{-1}D(\zeta)$					
	$n=5$		$n=10$		$n=20$	
	A	T	A	T	A	T
0.4	0.46	0.44	0.36	0.36	0.28	0.28
0.5	0.36	0.36	0.28	0.29	0.21	0.21
0.6	0.27	0.29	0.21	0.22	0.15	0.16
0.7	0.19	0.22	0.14	0.16	0.11	0.11
0.8	0.12	0.15	0.09	0.10	0.07	0.07
0.9	0.06	0.08	0.04	0.05	0.03	0.03

*) A: approximate
**) T: theoretical

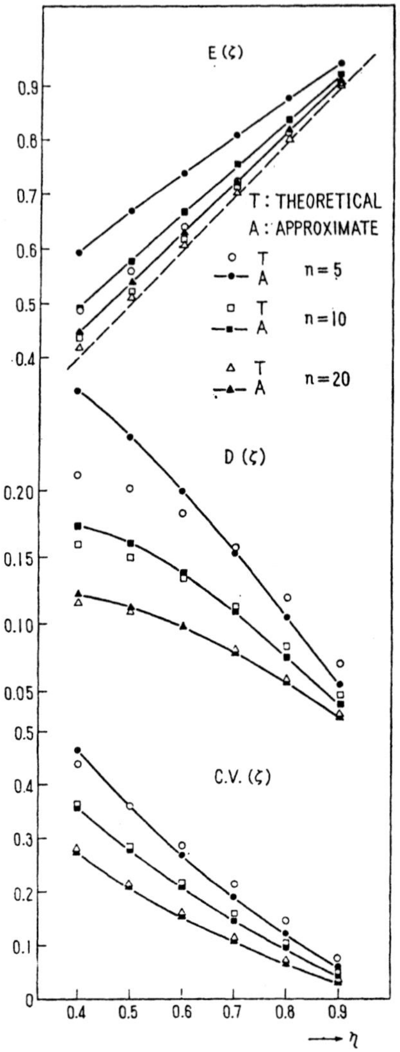

Fig. 2. Comparison of the approximate values with their corresponding theoretical values.

of the estimate of the frequency response function under that condition. We have obtained the relation

$$\langle \varepsilon, \varepsilon \rangle = (\hat{A}-A)^2 \langle X_1, X_1 \rangle + (\hat{B}-B) \langle X_2, X_2 \rangle + \langle \hat{\varepsilon}, \hat{\varepsilon} \rangle$$

where the right-hand side members, when divided by σ_N^2, are mutually independently distributed according to the χ^2-distributions with d.f. 1, 1 and $2(n-1)$, respectively. From this relation we can see that the statistic F which is defined by

$$F=(n-1)\frac{\{(\hat{A}-A)^2+(\hat{B}-B)^2\}\langle X, X\rangle}{\langle\hat{\varepsilon}, \hat{\varepsilon}\rangle},$$

where $\langle X, X\rangle=\langle X_1, X_1\rangle=\langle X_2, X_2\rangle$, is distributed according to the F-distribution with d.f.s 2 and $2(n-1)$.

Therefore the confidence region S of (A, B) with confidence coefficient δ is given as follows.

$$S=\left\{(\alpha, \beta);\ (\hat{A}-\alpha)^2+(\hat{B}-\beta)^2\leq\frac{\langle\hat{\varepsilon}, \hat{\varepsilon}\rangle}{(n-1)\langle X, X\rangle}F(\delta, 2, 2(n-1))\right\}$$

where $F(\delta, 2, 2(n-1))$ is defined by the relation

$$\text{Prob}\ (F\leq F(\delta, 2, 2(n-1)))=\delta.$$

When the lag window of trigonometric sum type is used to get $\hat{A}(f_\mu)$, $\langle\hat{\varepsilon}, \hat{\varepsilon}\rangle/\langle X, X\rangle$ and n in this expression are replaced by

$$(1-\hat{\gamma}^2(f_\mu))\frac{\sum_\nu w_\nu|Y(f_\mu-f_\nu)|^2}{\sum_\nu w_\nu|X(f_\mu-f_\nu)|^2}$$

and the nearest integer to $(2T/T_m)(1/2\sum_{n=-k}^{k}a_n^2)$, respectively, to obtain an approximate region R to the above confidence region for $A(f_\mu)$ with confidence coefficient δ, i.e., the region R in the complex domain is obtained from the following relation.

$$R=\left\{G;\ |\hat{A}(f_\mu)-G|^2\leq\frac{1}{n-1}\left(\frac{\hat{p}_{yy}(f_\mu)}{\hat{p}_{xx}(f_\mu)}-|\hat{A}(f_\mu)|^2\right)F(\delta, 2, 2(n-1))\right\}$$

where

$$n=\text{the nearest integer to }\frac{2T}{T_m}\frac{1}{2\sum_{n=-k}^{k}a_n^2}$$

$$\hat{p}_{yy}(f_\mu)=\sum_\nu w_\nu|Y(f_\mu-f_\nu)|^2$$

$$\hat{p}_{xx}(f_\mu)=\sum_\nu w_\nu|X(f_\mu-f_\nu)|^2.$$

We will here briefly mention about practical meaning of the use of the present confidence region. The results of analysis of sampling variabilities of $\hat{A}(f_\mu)$ and $\hat{\gamma}^2(f_\mu)$ in this and former sections are used mainly for the design of experiment and are taken into consideration *before* the data are obtained. On the other hand *after* the data have been obtained the confidence region will indicate the precision of our estimate $\hat{A}(f_\mu)$ by using the information contained in the present data, where $X(f_\mu)$'s are treated as fixed variables. It can be seen, as was stated in the former paper [2, p. 142], that in these circumstances we may apply, without disturbing the sampling variability of the estimate $\hat{A}(f_\mu)$, any sort of prewhitening operation to the data of input $\{x(t)\}$ after they are obtained, if only the operation is not dependent on the records of output $\{y(t)\}$.

The setup adopted in this section is a quite rough approximation to the real state where the lag window of trigonometric sum type is used and its utility must be checked by some artificial sampling experiments.

5. Computation scheme for the estimation of the frequency response function

In this section a computation scheme will be described to make our estimation procedure of the frequency response function practical. First we shall slightly modify the form of our estimate. Our estimate $\hat{A}(f_\mu)$ of the frequency response function was defined as

$$\hat{A}(f_\mu)=\frac{\sum_\nu w_\nu Y(f_\mu-f_\nu)\overline{X(f_\mu-f_\nu)}}{\sum_\nu w_\nu |X(f_\mu-f_\nu)|^2}.$$

We have

$$\begin{aligned}
&\sum_\nu w_\nu Y(f_\mu-f_\nu)\overline{X(f_\mu-f_\nu)}\\
&=\sum_\nu w_\nu \frac{1}{2T}\int_{-T}^{T}\int_{-T}^{T}\exp\left(-2\pi i\frac{\mu-\nu}{2T}t\right)y(t)\exp\left(2\pi i\frac{\mu-\nu}{2T}s\right)x(s)dtds\\
&=\sum_\nu w_\nu \frac{1}{2T}\int_{-T}^{T}\int_{-T}^{T}\exp\left(-2\pi i\frac{\mu-\nu}{2T}(t-s)\right)y(t)x(s)dtds\\
&=\sum_\nu w_\nu \frac{1}{2T}\int_{-T}^{T}\int_{-T}^{T}\exp\left(-2\pi i\frac{\mu-\nu}{2T}\tau\right)\tilde{y}(\tau+s)x(s)dsd\tau
\end{aligned}$$

where

$$\begin{aligned}
\tilde{y}(t)&=y(t-2T) \quad \text{when} \quad t>T\\
&=y(t+2T) \quad \text{when} \quad t<-T.
\end{aligned}$$

Therefore if we put

$$\tilde{C}_{yx}(\tau)=\frac{1}{2T}\int_{-T}^{T}\tilde{y}(\tau+s)x(s)ds$$

we have

$$\sum_\nu w_\nu Y(f_\mu - f_\nu)\overline{X(f_\mu - f_\nu)} = \sum_\nu w_\nu \int_{-T}^{T} \exp\left(-2\pi i \frac{\mu-\nu}{2T}\tau\right)\tilde{C}_{yx}(\tau)d\tau$$

$$= \int_{-T}^{T} \exp\left(-2\pi i \frac{\mu}{2T}\tau\right) W(\tau)\tilde{C}_{yx}(\tau)d\tau$$

$$= \int_{-T_m}^{T_m} \exp\left(-2\pi i \frac{\mu}{2T}\tau\right) W(\tau)\tilde{C}_{yx}(\tau)d\tau .$$

We define here $C_{yx}(\tau)$ by

$$C_{yx}(\tau) = \frac{1}{2T}\int_{-T}^{T-\tau} y(\tau+s)x(s)ds \qquad \tau \geqq 0$$

$$= \frac{1}{2T}\int_{-T-\tau}^{T} y(\tau+s)x(s)ds \qquad \tau < 0$$

and $\bar{C}_{yx}(\tau)$ by

$$\bar{C}_{yx}(\tau) = \tilde{C}_{yx}(\tau) - C_{yx}(\tau).$$

$$\bar{C}_{yx}(\tau) = \frac{1}{2T}\int_{T-\tau}^{T} y(\tau+s-2T)x(s)ds \qquad \tau \geqq 0$$

$$= \frac{1}{2T}\int_{-T}^{-T-\tau} y(\tau+s+2T)x(s)ds \qquad \tau < 0 .$$

If in the above obtained expression of $\sum_\nu w_\nu Y(f_\mu - f_\nu)\overline{X(f_\mu - f_\nu)}$ we replace $\hat{C}_{yx}(\tau)$ by $C_{xy}(\tau)$, and if T is sufficiently large so that $Ey(s)x(s)$, $Ey(t)y(s)$, and $Ex(t)x(s)$ can all be considered to be negligibly small for $|t-s| > 2(T-T_m)$, we obtain a statistic which has almost the same mean as that of the former one and has a variance smaller than that of the former, i.e., the contribution of $\bar{C}_{yx}(\tau)$ to $\sum_\nu w_\nu Y(f_\mu - f_\nu)\overline{X(f_\mu - f_\nu)}$ can be considered to have a zero-mean and to be orthogonal to the contribution of $C_{yx}(\tau)$. If we correspondingly replace $\sum_\nu w_\nu |X(f_\mu - f_\nu)|^2$ in the definition of $\hat{A}(f_\mu)$ by

$$\int_{-T}^{T} \exp\left(-2\pi i \frac{\mu}{2T}\tau\right) W(\tau)C_{xx}(\tau)d\tau$$

where $C_{xx}(\tau)$ is by definition

$$C_{xx}(\tau) = \frac{1}{2T}\int_{-T}^{T-|\tau|} x(|\tau|+t)x(t)dt,$$

then our new estimate $\hat{A}(f_\mu)$ is given by

$$\hat{A}(f_\mu)$$
$$=\left(\int_{-T}^{T}\exp\left(-2\pi i\frac{\mu}{2T}\tau\right)W(\tau)C_{yx}(\tau)d\tau\right)\left(\int_{-T}^{T}\exp\left(-2\pi i\frac{\mu}{2T}\tau\right)W(\tau)C_{xx}(\tau)d\tau\right)^{-1}$$

This $\hat{A}(f_\mu)$ can be represented in the form

$$\hat{A}(f_\mu)=\frac{\sum_\nu \tilde{w}_\nu Y(\mu/2T-\nu/4T)\,\overline{X(\mu/2T-\nu/4T)}}{\sum_\nu \tilde{w}_\nu |X(\mu/2T-\nu/4T)|^2}$$

where

$$\tilde{w}_\nu=\frac{1}{4T}\int_{-T}^{T}\exp\left(-2\pi i\frac{\nu}{4T}t\right)W(t)dt$$

$$X(f)=\frac{1}{\sqrt{2T}}\int_{-T}^{T}\exp(-2\pi i f t)x(t)dt$$

$$Y(f)=\frac{1}{\sqrt{2T}}\int_{-T}^{T}\exp(-2\pi i f t)y(t)dt\,.$$

When $A(f)$ can be considered to be nearly a constant in the effective range of the smoothing operation, we can get from this expression of $\hat{A}(f_\mu)$ the relation which has formed the basis of the analysis of this paper :

$$\hat{A}(f_\mu)\approx A(f_\mu)+\frac{\sum_\nu \tilde{w}_\nu N(\mu/2T-\nu/4T)\,\overline{X(\mu/2T-\nu/4T)}}{\sum_\nu \tilde{w}_\nu |X(\mu/2T-\nu/4T)|^2}$$

where

$$N(f)=\frac{1}{\sqrt{2T}}\int_{-T}^{T}\exp(-2\pi i f t)n(t)dt.$$

Taking into account the fact that by the present replacement of $\hat{C}_{yx}(\tau)$ and $\hat{C}_{xx}(\tau)$ by $C_{yx}(\tau)$ and $C_{xx}(\tau)$ the mean values of the denominator and numerator of the estimate of $A(f_\mu)$ are affected very little and the variances are reduced, we can see that our analysis of the bias in section 3 maintains its validity for our present new estimate, and further, that sampling variability of the new estimate will not exceed that of the former one. We have adopted $\hat{A}(f_\mu)$ in the stage of analysis of the sampling variability of our estimate because of its simplicity of statistical structure, but by our present argument we can see that $\hat{A}(f_\mu)$ will be more stable than $\hat{A}(f_\mu)$ though the difference will actually be small. For practical applications, therefore we had better use $\hat{A}(f_\mu)$ instead of $\hat{A}(f_\mu)$, and for evaluation of the sampling variability of $\hat{A}(f_\mu)$ and its related quantities the results obtained for $\hat{A}(f_\mu)$ will be applied.

Usually in the practical computation of our estimate only the values of $C_{yx}(\tau)$ and $C_{xx}(\tau)$ at those values of τ which are integral multiples of some fixed constant $\Delta\tau$ are available. If $\Delta\tau$ is choosen to be small enough so that the powers of $x(t)$ and $y(t)$ at frequencies higher than $1/2\Delta\tau$ are negligibly small and

$$p_x(f) \gg \sum_{k\neq 0} p_x\left(f+\frac{k}{\Delta t}\right)$$

$$p_y(f) \gg \sum_{k\neq 0} p_y\left(f+\frac{k}{\Delta t}\right)\ *$$

hold in the range of f of our present concern, we can replace the integrals of $C_{yx}(\tau)$ and $C_{xx}(\tau)$ by the corresponding sums of $C_{xy}(l\Delta\tau)$ and $C_{xx}(l\Delta\tau)$, i.e.,

$$\int_{-T}^{T} \exp\left(2\pi i\frac{\mu}{2T}\tau\right) W(\tau)C_{yx}(\tau)d\tau$$

can be replaced by

$$\Delta\tau\sum_{l} \exp\left(-2\pi i\frac{\mu}{2T}l\Delta\tau\right) W(l\Delta\tau)C_{yx}(l\Delta\tau),**$$

and

$$\int_{-T}^{T} \exp\left(-2\pi i\frac{\mu}{2T}\tau\right) W(\tau)C_{xx}(\tau)d\tau$$

by

$$\Delta\tau\sum_{l} \exp\left(-2\pi i\frac{\mu}{2T}l\Delta\tau\right) W(l\Delta\tau)C_{xx}(l\Delta\tau).$$

For our trigonometric window we have

$$\begin{aligned} W(l\Delta\tau) &= \sum_{n=-k}^{k} a_n \exp\left(2\pi i\frac{n}{2T_m}l\Delta\tau\right) && \text{when } |l\Delta\tau| < T_m \\ &= \frac{1}{2}\left(\sum_{n=-k}^{k} a_n \exp\left(2\pi i\frac{n}{2T_m}l\Delta\tau\right)\right) && \text{when } l\Delta\tau = T_m \\ &= 0 && \text{otherwise,} \end{aligned}$$

* If the mecessary modification of the coherency is admitted, this condition may be replaced by the weaker one $|A(f)|^2 p_x(f) \gg \sum_{k\neq 0} |A(f+k/\Delta\tau)|^2 p_x(f+k/\Delta\tau)$.

** The summation $\sum_l$ is extended all over the l's satisfying $|l\Delta\tau| \leq T$.

and we get our final computation scheme for the estimate $\hat{A}(f)(0 \leqq f \leqq 1/2\Delta\tau)$ of the frequency response function:

1) For $T_m = h\Delta\tau$ we calculate

$$\bar{p}_{yx}(f) = \Delta\tau \sum_{l=-h}^{h} \exp(-2\pi i f l \Delta\tau) C^*_{yx}(l\Delta\tau)$$

$$\bar{p}_{xx}(f) = \Delta\tau \sum_{l=-h}^{k} \exp(-2\pi i f l \Delta\tau) C^*_{xx}(l\Delta\tau)$$

where

$$\begin{aligned} C^*_{yx}(l\Delta\tau) &= C_{yx}(l\Delta\tau) & -h<l<h \\ &= \frac{1}{2} C_{yx}(l\Delta\tau) & l=\pm h \\ C^*_{xx}(l\Delta\tau) &= C_{xx}(l\Delta\tau) & -h<l<h \\ &= \frac{1}{2} C_{xx}(l\Delta\tau) & l=\pm h. \end{aligned}$$

2) We smooth these $\bar{p}_{yx}(f)$ and $\bar{p}_{xx}(f)$ by the smoothing coefficient $\{a_n\}$ to obtain

$$\hat{p}_{yx}(f) = \sum_{n=-k}^{k} a_n \bar{p}_{yx}\left(f - \frac{n}{2T_m}\right)$$

$$\hat{p}_{xx}(f) = \sum_{n=-k}^{k} a_n \bar{p}_{xx}\left(f - \frac{n}{2T_m}\right).$$

3) The estimate $\hat{A}(f)$ is given by

$$\hat{A}(f) = \frac{\hat{p}_{yx}(f)}{\hat{p}_{xx}(f)}.$$

The estimate $\hat{\gamma}^2(f)$ of coherency $\gamma^2(f)$ is given by

$$\hat{\gamma}^2(f) = |\hat{A}(f)|^2 \frac{\hat{p}_{xx}(f)}{\hat{p}_{yy}(f)}$$

where $\hat{p}_{yy}(f)$ is obtained by replacing x by y in the definition of $\hat{p}_{xx}(f)$. The confidence region R for $A(f)$ can be obtained by putting these $\hat{A}(f)$ $\hat{p}_{xx}(f)$ and $\hat{p}_{yy}(f)$ into the formula given in the preceding section. The following approximations which are derived from that of R will also be useful.

$$P_r\left\{\left|\frac{|\hat{A}(f)| - |A(f)|}{|\hat{A}(f)|}\right| \leqq \sqrt{\frac{1}{n-1}\left(\frac{1}{\hat{\gamma}^2(f)} - 1\right) F(\delta, 2, 2(n-1))}\,,\right.$$

$$\left. |\hat{\phi}(f) - \phi(f)| \leqq \sin^{-1}\left(\sqrt{\frac{1}{n-1}\left(\frac{1}{\hat{\gamma}^2(f)} - 1\right) F(\delta, 2, 2(n-1))}\right)\right\} \geqq \delta$$

where

$$\hat{\phi}(f)=\arg \hat{A}(f)$$
$$\phi(f)=\arg A(f)$$
$$n=\text{the nearest integer to } \left(\frac{2T}{T_m}\right)\frac{1}{2\sum_{n=-k}^{k} a_n^2}$$

and $F(\delta, 2, 2(n-1))$ is given by the following relation for $F^2_{2(n-1)}$ which is distributed according to the F-distribution with d.f.s. 2 and $2(n-1)$

$$P_r\{F^2_{2(n-1)} \leqq F(\delta, 2, 2(n-1))\}=\delta,$$

and it is tacitly assumed that $\hat{\gamma}^2(f)$ and the quantity inside the square root are together positive and less than 1. When we use the shifted window $W_K(t)=W(t-K\Delta t)$ to compensate for the variation of phase shift, we usually adopt the following computing scheme

$$\hat{p}_{yx}(f)=\exp(-2\pi ifK\Delta\tau)\hat{p}_{yxK}(f)$$

where $\hat{p}_{yxK}(f)$ is obtained by replacing $C_{yx}(l\Delta\tau)$ by $C_{yx}((l+K)\Delta\tau)$ in the definition of $\hat{p}_{yx}(f)$.

In case where $|K\Delta\tau|$ is not very small compared with T_m, it will be more advisable to recalculate $C_{yx}(l\Delta\tau)$ by using $\{x(t), y(t+K\Delta\tau); -T\leqq t\leqq T\}$ in place of $\{x(t), y(t); -T\leqq t\leqq T\}$ or at least to replace the factor $1/2T$ in the definitions of $C_{xx}(\tau)$ and $C_{yx}(\tau)$ by the factor $1/(2T-|K\Delta\tau|)$. When the original data is given in the form $\{(x(n\Delta t), y(n\Delta t)); n=1, 2, \cdots, M\}$ and $\Delta\tau=m\Delta t$ (m ; positive integer) we use in the above computation formulae those $C_{yx}(l\Delta t)$ and $C_{xx}(l\Delta\tau)$ defined by the following:

$$C_{yx}(l\Delta\tau)=\frac{1}{M}\sum_{n=1}^{M-lm} y((lm+n)\Delta t)x(n\Delta t) \qquad \text{when } l\geqq 0$$

$$=\frac{1}{M}\sum_{n=1-lm}^{M} y((lm+n)\Delta t)x(n\Delta t) \qquad \text{when } l<0$$

$$C_{xx}(l\Delta\tau)=\frac{1}{M}\sum_{n=1}^{M-|l|m} x((|l|m+n)\Delta t)x(n\Delta t).$$

When $\Delta\tau$ is sufficiently small as was assumed in the beginning of this section, the sampling variabilities of these estimates are considered to be almost the same as those of the estimates discussed in the previous sections.

As to the proper choice of the lag window $W(\tau)$ or $\{a_n\}$ the necessary informations are available in the preceding paper and in section 2 of this paper. One thing to be noted here is that, at those frequencies where coherencies are high, one should use $\{a_n\}$ of which the bandwidth

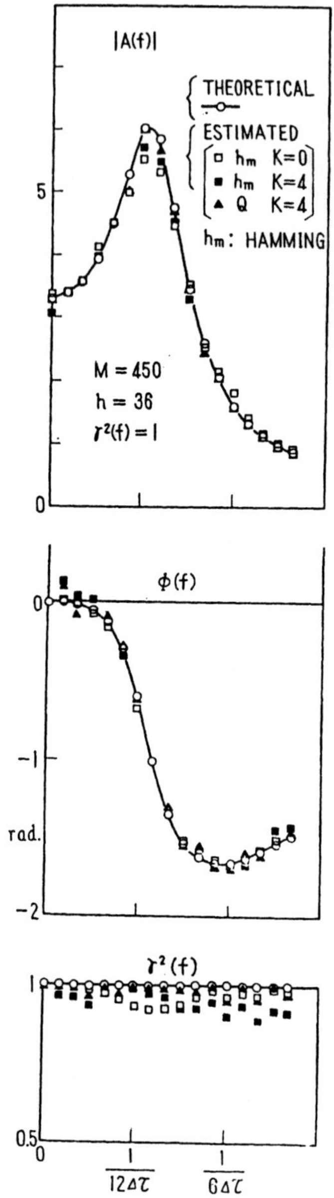

Fig. 3. Effect of the selection of windows. Perfectly coherent case.

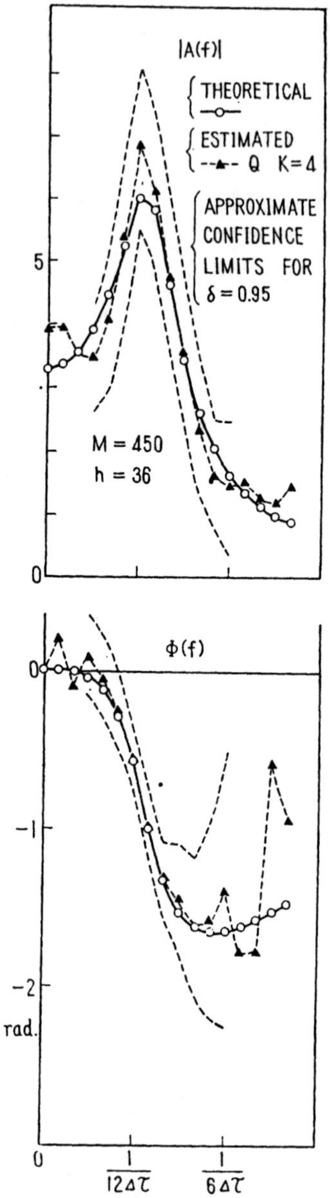

Fig. 4. Estimates of amplitude gain and phase shift with approximate 95% confidence limits. Necessary informations of coherency are given in Fig. 1 of § 4.

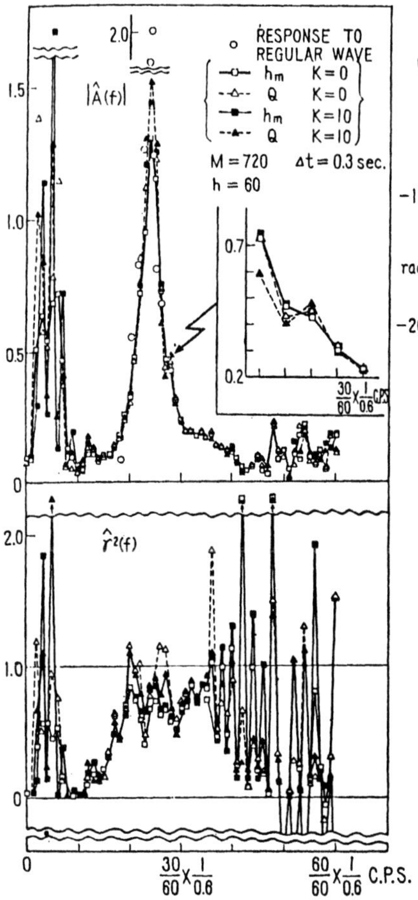

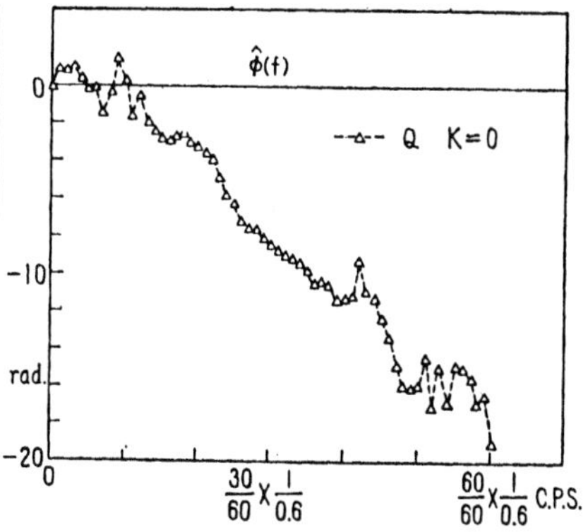

Fig. 5. Estimates of the amplitude gain and phase shift of a ship model, and the corresponding sample coherencies.

h_m stands for hamming.

defined in [1] is rather narrow.

6. Numerical examples

Figs. 3 and 4 give the results of applications of our statistical estimation procedure of the frequency response function, using various $\{a_n\}$ and K, to artificial time series. From these results, we can see that the approximations adopted in this paper do not impair the practical applicability of the results of our discussion. For instance, we can see that our estimates of the phase are fairly free from the bias due to smoothing, however, the gain suffers rather significant bias by the improper selection of the window.

The change of the phase shift in this artificial model is not very rapid, but in practical applications we often meet a system with phase shift varying much more rapidly and, correspondingly, with very sharp peak of the amplitude gain. In these circumstances, use of the properly shifted window with narrow bandwidth becomes more important to obtain a bias-free result. Fig. 5 gives the results of analysis of the response of a ship model. Here the input is the height of wave and the output

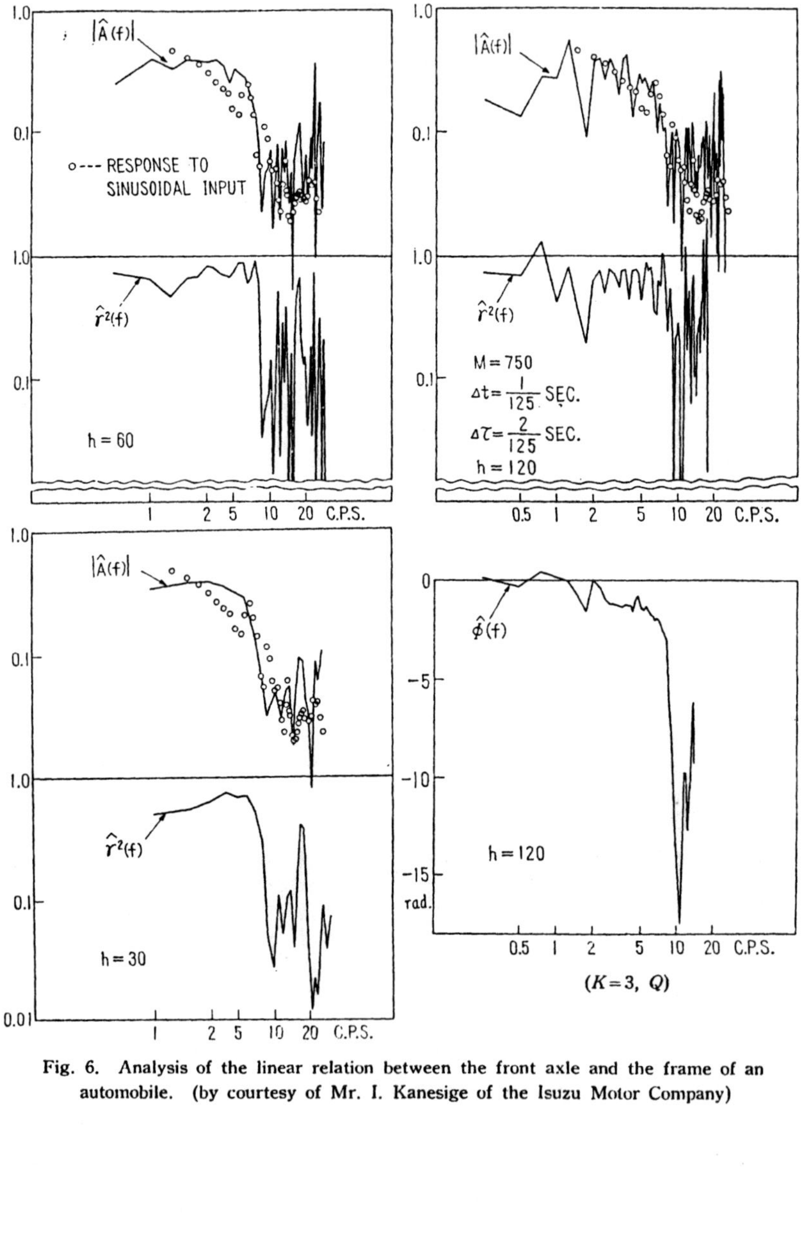

Fig. 6. Analysis of the linear relation between the front axle and the frame of an automobile. (by courtesy of Mr. I. Kanesige of the Isuzu Motor Company)

is the roll of the ship model. In this example we can see that the shifted windows produced more reliable estimates at frequencies ranging from $(20/60)\times(1/0.6)$ c.p.s. to $(30/60)\times(1/0.6)$ c.p.s.. The estimates of the power spectra of the input and output of these examples were given in the preceding paper [1, p. 20, Figs. 2, 3], and taking into account of the shape of the power spectrum of the input we can guess that some type of prewhitening operation is necessary to obtain reliable estimates in the wider range of frequencies. It seems that for this example our new window Q is better suited than that of hamming, and fairly large bias due to smoothing still remains at the peak of the amplitude gain.

Fig. 6 shows the results of the analysis, using Q, of the relation between the oscillation of the front axle and that of the frame of an automobile. We can see that the estimate of the amplitude gain thus obtained is in fairly good agreement with that obtained by the ordinary frequency response test only in the case $T_m=240\Delta t$. This is due to the existence of very sharp peaks around 9 c.p.s. and the estimate of the phase shift explains these circumstances more clearly. The shift $K\Delta\tau$ of the time axis of about 0.8 second will be necessary to get more reliable estimates at this frequency.

We have illustrated these two examples of ship model and automobile for the purpose of showing the complexity of practical problems. For practical applications, much more skillful use of the method will be necessary, but we believe that the use of well designed and properly shifted window will eventually lead to successful results.

Acknowledgment

The present paper is a consequence of the collaboration of many research workers in various research fields who have been interested in the statistical analysis of the time series and have kept continuous contact with the present authors. Here we should like to express our hearty thanks to all of them, especially to Mr. I. Kanesige of the Isuzu Motor Company, Mr. H. Matsumoto of the National Inststute of Animal Health, Dr. S. Takeda of the Transportation Technical Research Institute, and Prof. Y. Kuroda of the Tōkai University for many helpful discussions, and to Dr. K. Matusita of the Institute of Statistical Mathematics for kind support to this project. Thanks are also due to Miss Y. Saigusa and Mrs. T. Isii who have performed all the numerical computations necessary for the present investigations by using a FACOM-128 automatic relay computer and prepared the results for publication.

THE INSTITUTE OF STATISTICAL MATHEMATICS
TRANSPORTATION TECHNICAL RESEARCH INSTITUTE

REFERENCES

[1] H. Akaike, "On the design of lag window for the estimation of spectra," *Ann. Inst. Stat. Math.*, Vol. 14, (1962), pp. 1-21.

[2] H. Akaike, "Undamped oscillation of the sample autocovariance function and the effect of prewhitening operations," *Ann. Inst. Stat. Math.*, Vol. 13 (1960), pp. 127-143.

[3] H. Akaike, "Note on power spectra," *Bull. Japan Statistical Society*, (1960), pp. 59-60 (in Japanese).

[4] R. B. Blackman and J. W. Tukey, "The Measurement of power spectra from the point of view of communications engineering," *Bell System Technical Journal*, Vol. 37 (1958), pp. 185-282, pp. 485-569 (also published separately by Dover (1958)).

[5] H. Cramér, "*Mathematical Methods of Statistics*," Princeton Uuiversity Press, (1946).

[6] J. F. Darzell and Y. Yamanouchi, "The analysis of model test results in irregular head seas to determine the amplitude and phase relations to the waves," *Stevens Institute of Technology Report 708*, (1958).

[7] J. F. Darzell and W. J. Pierson Jr., "The appearent loss of coherency in vector Gaussian processes due to computational procedures with application to ship motions and random sea," *Technical Report of College of Engineering Research Division, Department of Meteorology and Oceanography*, New York University, (1960).

[8] N. R. Goodman, "On the joint estimation of the spectrum cospectrum and quadrature spectrum of two-dimensional stationary Gaussian process," *Scientific Paper No. 10, Engineering Statistical Laboratory*, New York University, (1957).

[9] R. W. Hamming and J. W. Tukey, "Measuring Noise Color" unpublished memorandum.

[10] V. V. Solodovnikov, *Introduction to the Statistical Dynamics of Automatic Control Systems* Dover Publications, Inc. (1960).

[11] Y. Yamanouchi, "On the analysis of the ship oscillations among waves—Part I, II and III," *Journal of the Society of Naval Architect of Japan*, Vol. 109 (1961), pp. 169-183, Vol. 110 (1961), pp. 19-29, Vol. 111 (1962), pp. 103-115, (in Japanese).

Reprinted from *Annals of the Institute of Statistical Mathematics*, Vol. 20, No. 3, 1968, 425-439 by permission of the Institute of Statistical Mathematics

ON THE USE OF A LINEAR MODEL FOR THE IDENTIFICATION OF FEEDBACK SYSTEMS

HIROTUGU AKAIKE

(Received June 12, 1968)

Summary

A basic linear model of stationary stochastic processes is proposed for the analysis of linear feedback systems. The model suggests a simple computational procedure which gives estimates of the response characteristics of the system and the spectra of the noise source. These estimates are obtained through the estimate of the linear predictor of the process, which is obtained by the ordinary least squares method.

The necessary assumption for the validity of the estimation procedure is so general that the procedure can be applied to the analysis of wide variety of practical systems with feedback.

The content of the present paper forms an answer to the problem discussed by the author in a former paper [1].

1. Introduction

The cross-spectral method has been extensively applied to the estimation of frequency response functions [2], [3], [8]. The discrete time parameter model which forms the basis of the cross-spectral method of estimation is of the form

$$x_0(n)=\sum_{j=1}^{K} \sum_m a_{jm}\, x_j(n-m)+u_0(n)\,,$$

where the unobservable noise $u_0(n)$ is assumed to be uncorrelated with, or orthogonal to, the input series $x_j(n)$ $(j=1, 2, \cdots, K)$. This last assumption is essential for the validity of ordinary cross-spectral approach [1].

In many important practical situations, the existence of feedback loops which connect the output $x_0(n)$ to the inputs $x_j(n)$'s is quite common and the present assumption of orthogonality of $u_0(n)$ to $x_j(n)$'s seriously limits the practical applicability of the method. As was briefly touched in the former paper [1] this difficulty may be due to our inability of including the condition of physical realizability, which requires the output of a system to be determined without using the future values of the input, into the cross-spectral method.

In the present paper we treat the problem in the time domain and directly get estimates of the impulse response functions $\{a_{jm}: m=0, 1, 2, \cdots\}$ $(j=1, 2, \cdots, K)$. The estimation procedure can be explained as follows: we first conceptually whiten the spectrum of the additive disturbance $u_0(n)$, or actually a driving input to the system, by a physically realizable linear transformation and then apply the least squares method to get an estimate of the regression coefficients of the linear predictor of the process and finally back-transform the estimated predictor into the original form of the system structure. The only practical condition to assure the validity of the estimation procedure is that, besides the existence of necessary driving input for each $x_j(n)$ such as $u_0(n)$ for $x_0(n)$, there should be some delays in the feedback loops so that we can effectively assume that an instantaneous return from the output to itself through the feedback loops is prohibited. When we are observing a physical process, this condition will be satisfied at least approximately if we limit our attention to some frequency band.

In the next section we shall give a precise description of the basic model and in the following section we propose an estimation procedure. The procedure is directly applicable to practical data and the consistency of the estimates is discussed. Some numerical examples are given to show the practical applicability of the procedure. Possible bias due to the incorrect specification of the model is discussed in the last section.

2. Basic model

Here we consider a set of observation points $\{i; i=0, 1, 2, \cdots, K\}$. The model of the system we are going to treat in this paper is given by the relation

$$x_i(n)=\sum_{j=0}^{K}\sum_{m=0}^{M} a_{ijm}\,x_j(n-m)+u_i(n)$$

$$i=0, 1, 2, \cdots, K, \qquad n=0, 1, 2, \cdots,$$

where $\{u_i(n); n=0, 1, 2, \cdots\}$ is the driving input at i and $\{x_i(n); i=0, 1, \cdots, K, n=0, 1, \cdots\}$ is the response of the system and the initial condition of the system is given by $\{x_i(n); i=0, 1, \cdots, K, n=-1, -2, \cdots, -M\}$. $\{a_{ijm}; m=0, 1, \cdots, M\}$ is the impulse response of the output $x_i(n)$ to the input $x_j(n)$ and we shall assume

$$a_{iim}=0 \qquad i=0, 1, \cdots, K, \quad m=0, 1, 2, \cdots, M.$$

This assumption is a salient feature of our model and if there is a feedback loop from i to i without going through other j's its effect is included into all of the responses $\{a_{ijm}\}$ and also the noise within the loop is included into $u_i(n)$. This fact should be taken into account when we analyze the result of application of our model to a practical problem.

In the following, we shall denote by $|G|$ the determinant of a matrix G and by $(G)_{ij}$ the (i, j) element of G and by I_n the n-dimensional identity matrix. We define

$$A(z) = \sum_{m=0}^{M} A_m z^m ,$$

where z is a complex scalar variable and A_m is a $(K+1)$-dimensional matrix with $(A_m)_{ij} = a_{ijm}$. We shall assume that the absolute values of the roots of characteristic equation $|I_{K+1} - A(z)| = 0$ are all greater than 1. Under this assumption, $B(z) = (I_{K+1} - A(z))^{-1}$ has a Taylor expansion $B(z) = \sum_{m=0}^{\infty} B_m z^m$ with radius of convergence greater than 1. Thus if we put $b_{ijm} = (B_m)_{ij}$ we have $\sum_{m=0}^{\infty} |b_{ijm}| < \infty$ for every (i, j). From the relation $(I_{K+1} - A(z))\, B(z) = I_{K+1}$, we have

$$(I_{K+1} - A_0)\, B(z) - A'(z)\, B(z) = I_{K+1} ,$$

where $A'(z) = \sum_{m=1}^{M} A_m z^m$. Thus we get

$$(I_{K+1} - A_0)\, B_0 = I_{K+1} ,$$

$$(I_{K+1} - A_0)\, B_m = \sum_{k=1}^{m} A_k B_{m-k} ,$$

where it is assumed that $A_k = 0$ (null matrix) for $k > M$. This result shows that b_{ijm} is the response of the system at i at time m when a unit impulse $\{u_j(n)\}$ with $u_j(0) = 1$ and $u_j(n) = 0$ $(n \neq 0)$ is applied at j and other $u_k(n)$'s are all kept equal to zero. We have noticed that under our assumption $\sum_{m=0}^{\infty} |b_{ijm}| < \infty$ holds for every pair (i, j). Thus we can see that our system is absolutely stable in the sense that starting from arbitrary initial condition the response of the system eventually damps out if it is without driving input. It can be shown that our assumption on the absolute values of the characteristic equation $|I_{K+1} - A(z)| = 0$ is just equivalent to this absolute stability of the system. It also should be noticed that the assumption implies $|I_{K+1} - A_0| \neq 0$.

Now we turn our attention to the stochastic situation where $\{u_i(n);\ i = 0, 1, \cdots, K;\ n = 0, \pm 1, \pm 2, \cdots\}$ is a $(K+1)$-dimensional stationary process with zero mean vector and finite second order moments. We assume $\{u_i(n)\}$ $(i=0, 1, \cdots, K)$ to be mutually uncorrelated, or orthogonal, and that each $u_i(n)$ is a regular process in the sense that it admits a one-sided moving average representation $u_i(n)=\sum_{l=0}^{\infty} b_{il}\,\varepsilon_i(n-l)$ with a white noise $\varepsilon_i(n)$ satisfying $E\varepsilon_i(n)=0$, $E\varepsilon_i^2(n)=\sigma_i^2$ (>0) and $E\varepsilon_i(n)\,\varepsilon_i(m)=0$ $(n\neq m)$. Hereafter the convergence and equality of random quantities are to be understood in the sense of mean square. We shall assume $b_{i0}=1$ $(i=0, 1, \cdots, K)$. To assume the regularity of $u_i(n)$ is just equivalent to assuming the influence of its infinitely remote past history to be vanishing in $u_i(n)$ in the sense of mean square. Thus the regularity assumption is a natural one for various practical situations. Now $\sum_{l=1}^{\infty} b_{il}\,\varepsilon_i(n-l)$ is the projection of $u_i(n)$ into the space spanned by its own past history and $\varepsilon_i(n)$ is the innovation. Thus $\sum_{l=1}^{\infty} b_{il}\,\varepsilon_i(n-l)$ can be approximated arbitrarily closely in the sense of mean square by a finite linear combination of $u_i(n-m)$ $(m=1, 2, \cdots)$. Thus, as a simplification, we assume that for some finite L $u_i(n)$ satisfies the relation

$$\text{(A)} \qquad u_i(n)=\sum_{l=1}^{L} c_{il}\,u_i(n-l)+\varepsilon_i(n)\,.$$

As we have assumed the orthogonality between $\{u_i(n)\}$'s, it holds that $E\varepsilon_i(n)\,\varepsilon_j(m)=0$ $(i\neq j)$. Here we define $x_i(n)$ by

$$x_i(n)=\sum_{j=0}^{K}\sum_{m=0}^{\infty} b_{ijm}\,u_j(n-m)\,.$$

Then it can be seen that $\{x_i(n);\ i=0, 1, \cdots, K\}$ satisfies the relation

$$x_i(n)=\sum_{j=0}^{K}\sum_{m=0}^{M} a_{ijm}\,x_j(n-m)+u_i(n)\,.$$

Now let us assume that $\{z_i(n);\ i=0, 1, \cdots, K\}$ satisfies the same relation as $\{x_i(n);\ i=0, 1, \cdots, K\}$ and is stationarily correlated with $\{u_i(n)\}$ or with $\{x_i(n)\}$. Then $y_i(n)=x_i(n)-z_i(n)$ satisfies the relation

$$y_i(n)=\sum_{j=0}^{K}\sum_{m=0}^{M} a_{ijm}\,y_j(n-m) \quad i=0, 1, \cdots, K.$$

From this it follows that

$$\int_{-1/2}^{1/2} \exp(\sqrt{-1}\,2\pi f m)(I_{K+1}-A(\exp(-\sqrt{-1}\,2\pi f)))\,dF_y(f)=O$$
$$m=0,\ \pm 1,\ \pm 2,\ \cdots\ ,$$

where $F_y(f)$ is the matrix spectral function of $\{y_i(n);\ i=0, 1, \cdots, K\}$ and is defined by the relation $Ey_i(n+m)\,y_j(n)=\int_{-1/2}^{1/2}\exp(\sqrt{-1}\,2\pi f m)\,d(F_y(f))_{ij}$

and O is the $(K+1)\times(K+1)$ null matrix. Thus we get, formally,

$$(I_{K+1}-A(\exp(-\sqrt{-1}\,2\pi f)))\,dF_v(f)=O\,,$$

and by multiplying $B(\exp(-\sqrt{-1}\,2\pi f))$ from left we get

$$dF_v(f)=O\,.$$

This shows that under our assumption of the absolute stability of the system, above relation between $\{x_i(n)\,;\,i=0,1,\cdots,K\}$ and $\{u_i(n)\,;\,i=0,1,\cdots,K\}$ uniquely determines $\{x_i(n)\}$ in the sense of mean square. Thus, as our basic model we adopt

$$\text{(B)}\qquad x_i(n)=\sum_{j=0}^{K}\sum_{m=0}^{M}a_{ijm}\,x_j(n-m)+u_i(n)$$

$$i=0,1,\cdots,K,\quad n=0,\pm 1,\pm 2,\cdots.$$

It should be remembered that we are assuming $a_{iim}=0$ and not assuming $u_i(n)$ to be a white noise.

Now by using the relation (A), we can transform the original representation (B) to whiten $u_i(n)$ and get

$$\text{(C)}\qquad x_i(n)=\sum_{j=0}^{K}\sum_{m=0}^{M+L}A_{ijm}\,x_j(n-m)+\varepsilon_i(n)$$

where

$$A_{ii0}=0$$

$$A_{iim}=c_{im}\qquad \text{for}\quad m=1,2,\cdots,L,$$

$$A_{ij0}=a_{ij0}$$

$$A_{ijm}=a_{ijm}-\sum_{l=1}^{m}c_{il}\,a_{ijm-l}\qquad \text{for}\quad m=1,2,\cdots,M+L,$$

and it is assumed that $c_{il}=0$ for $l>L$ and $a_{ijm}=0$ for $m>M$. Conversely $\{a_{ijm}\}$ and $\{c_{im}\}$ can be obtained from $\{A_{ijm}\}$ by the relation

$$c_{im}=A_{iim}\qquad \text{for}\quad m=1,2,\cdots,L,$$

$$a_{ij0}=A_{ij0}$$

$$a_{ijm}=A_{ijm}+\sum_{l=1}^{m}c_{il}\,a_{ijm-l}\qquad \text{for}\quad m=1,2,\cdots,M,$$

where we are assuming $c_{il}=0$ for $l>L$.

Our present representation (C) corresponds to the so called reduced form of mutually related multiple time series when $A_{ij0}=a_{ij0}=0$ for all (i,j) and in this case we can readily apply the method of least squares to get an estimate of $\{A_{ijm}\}$ [5]. In physical processes with continuous time parameter, there are usually lags in the responses and if we limit our attention to some frequency band and select the length of sampling interval between consecutive observations short enough, we shall generally be able to expect this condition to hold for the equi-spaced time-sampled data. But using a too short time interval usually introduces inefficiency of estimation procedure and sometimes it is not quite practical to ask all the A_{ij0}'s to be vanishing. What we need here for the validity of the least squares method is that the values of $x_j(n)$ with $A_{ij0}\neq 0$ should be (practically at least approximately) uncorrelated with $\varepsilon_i(n)$.

From this, we can see that for our requirement to be filled it is necessary and sufficient that

$$(A_0)_{ij}((I_{K+1}-A_0)^{-1})_{ji}=0 \qquad \text{for} \quad i, j=0, 1, \cdots, K.$$

If we are going to approximate a physical process with continuous time parameter by our present model, A_0 will represent the effect of quick responses of the system and A_0^n will have to approximate the effect of quick responses traveling through n observation points during a unit of sampling interval. In this case, we shall have to select the sampling interval short enough so that at least we can expect that every possible quick response of i to i is effectively negligible during the unit of sampling interval. Otherwise, we shall generally never be able to expect the orthogonality of $\varepsilon_i(n)$ to all of $x_j(n)$'s with $a_{ij0}\neq 0$. Thus we assume

$$(A_0^n)_{ii}=0 \qquad \text{for} \quad i=0, 1, \cdots, K \quad \text{and} \quad n=1, 2, \cdots.$$

This condition is satisfied if and only if, after proper rearrangement of the vector components of $\{x_0(n), \cdots, x_K(n)\}$ or relabelling of the observation points, A_0 has zeros on and below the diagonal, i.e., it takes the form

$$A_0=\begin{bmatrix} 0 & * & * \cdots * & * \\ 0 & 0 & * \cdots * & * \\ 0 & 0 & 0 \cdots * & * \\ \vdots & \vdots & \vdots \quad\; \vdots & \vdots \\ 0 & 0 & 0 \cdots 0 & * \\ 0 & 0 & 0 \cdots 0 & 0 \end{bmatrix}.$$

It can be seen that $(A_0)_{ij}((I_{K+1}-A_0)^{-1})_{ji}=0$ holds for this type of A_0. We shall hereafter assume this shape of A_0. Taking into account the relation $A_{ij0}=a_{ij0}$, we can see that in this case (C) takes the form

$$\text{(D)} \qquad x_i(n)=\sum_{j=i+1}^{K} A_{ij0}\,x_j(n)+\sum_{j=0}^{K}\sum_{m=1}^{M+L} A_{ijm}\,x_j(n-m)+\varepsilon_i(n)$$

$$i=0, 1, \cdots, K.$$

This corresponds to the so-called primary form of a linear causal chain system which has been extensively discussed by Wold [11], [12] in the field of econometric model building. It is fairly clear from our observation that our model with present assumption will be useful for the analysis of various physical processes. The assumed shape of A_0 means that, roughly speaking, we are arranging $x_j(n)$'s in the order of the speeds of their responses; starting from $x_0(n)$, the quickest one, to $x_K(n)$, the slowest one.

Obviously our present model is only a crude approximation for a physical process with continuous time parameter and if there exist any ambiguities of the values of estimates of a_{ij0} obtained by applying the model to equi-spaced time-sampled data we should try another analysis using a sample with twice as much frequency of sampling as of the original one and assuming $a_{ij0}=0$. Theoretical discussion of the approximation of continuous time parameter process by one with discrete time parameter would be important but is beyond the scope of the present paper.

For the proof of consistency of the least squares estimate of the coefficients of (D), we have to show the non-singularity of the variance covariance matrix of the regressor $\{x_j(n)\ (j=i+1, \cdots, K),\ x_j(n-m)\ (j=0, 1, \cdots, K;\ m=1, 2, \cdots, H)\}$ for any finite positive integer H. For this we have to notice that

$$(I_{K+1}-A_0)^{-1}=I_{K+1}+A_0+A_0^2+\cdots+A_0^K$$

and thus $(I_{K+1}-A_0)^{-1}$ has all zeros below the diagonal and 1's on the diagonal. The multi-dimensional prediction formula of $x_i(n)$ is then given by

$$x_i(n)=\sum_{j=0}^{K}\sum_{m=1}^{M+L}\sum_{k=0}^{K}((I-A_0)^{-1})_{ik}A_{kjm}x_j(n-m)+\delta_i(n)$$

where

$$\delta_i(n)=\sum_{k=0}^{K}((I-A_0)^{-1})_{ik}\varepsilon_k(n)=\varepsilon_i(n)+\sum_{k=i+1}^{K}((I-A_0)^{-1})_{ik}\varepsilon_k(n)\,.$$

As we have assumed $E\varepsilon_i^2(n)=\sigma_i^2>0$ and $\varepsilon_i(n)$'s are mutually orthogonal, the variance covariance matrix of $\delta_i(n)$ $(i=0, 1, \cdots, K)$ is non-singular. For any set of coefficients $\{B_{jm}\}$ $(m=0, 1, \cdots, H)$ we have

$$E\left|\sum_{j=0}^{K}\sum_{m=0}^{H}B_{jm}x_j(n-m)\right|^2=E\left|\sum_{j=0}^{K}B_{j0}\delta_j(n)\right|^2$$
$$+E\left|\text{linear combination of } x_j(n-m)\ (m=1,2,\cdots,H)\right|^2,$$

and for this quantity to be vanishing we have to ask $B_{j0}=0$ $(j=0, 1, \cdots, K)$ and accordingly $B_{jm}=0$ for all j and m. Thus, we have shown the non-singularity of the variance covariance matrix of $\{x_j(n)\ (j=i+1, \cdots, K),\ x_j(n-m)\ (j=0, 1, \cdots, K;\ m=1, 2, \cdots, H)\}$. Taking into account the fact that $x_i(n)$ admits one-sided moving average representation by $\varepsilon_j(n)$'s, we can also get the present result as a direct consequence of a general theory of regular full rank process ([10], p. 147). We shall use the present result in the next section and see that the assumption $E\varepsilon_i^2(n)=\sigma_i^2>0\ (i=0, 1, \cdots, K)$ is playing a definite role in our estimation procedure.

It is interesting to note that Parzen [7] has stressed the importance of the representation of time series in which the disturbances have the properties that least squares estimates are the efficient estimates. It is stated that "There is no guarantee that such representations exist. What to do in this case represents a development of the theory of statistical inference on stochastic processes which in my opinion would not be merely an extension of classical statistical inference." ([7], p. 45). Our present procedure uses such a representation as a pivot to get the final estimate, and it gives an example that the analysis of multiple time series is not completely reduced to the cross-spectral analysis if the latter is confined to the mere extension of classical regression analysis into the complex domain.

Obviously our transformed representation (D) is a kind of predictor formula for the process $x_i(n)$ and we are going to identify the necessary response characteristics and noise spectra through this representation. Thus it seems to be appropriate to call our present identification procedure predictive identification.

3. Estimation procedure

There is a fundamental paper on the estimation of the coefficients of a multiple autoregressive scheme by Mann and Wald [5]. Relation between the least squares estimation and the causal chain system was discussed by Wold [11], [12]. Also there are papers by Durbin [4] and by Whittle [9] which are closely related with our present subject.

Here we shall only describe a simple, though not necessarily efficient, estimation procedure for direct applications and very briefly discuss the consistency of the estimates. Practical applicability of the method will be illustrated by some numerical examples. An example of application to the analysis of a physical process will be discussed elsewhere [6]. It is quite desirable to have a practical formula for the evaluation of sampling variabilities of our estimates. We shall discuss this in a subsequent paper.

Our estimation procedure is as follows:

I First we arrange the records in the order of the speeds of their responses; the quickest one as $x_0(n)$ and the slowest one as $x_K(n)$.

We properly select the values of L and M. These are guessed values of L and M in the original model (B). For typographical simplicity we shall not distinguish these values from the true values but the difference will be clear from the context. We suggest to do the whole computation for several sets of values of L and M to get informations for the selection of L and M. The values of L and M may change for each set of the following normal equations.

II We solve the normal equation for each i;

$$\sum_{j=i+1}^{K} \hat{A}_{ij0} C(x_j, x_k)(l) + \sum_{j=0}^{K} \sum_{m=1}^{M+L} \hat{A}_{ijm} C(x_j, x_k)(l-m) = C(x_i, x_k)(l)$$

$$\begin{aligned} &l=1, 2, \cdots, M+L \quad \text{for } k=0, 1, 2, \cdots, i, \\ &l=0, 1, 2, \cdots, M+L \quad \text{for } k=i+1, i+2, \cdots, K, \end{aligned}$$

where $C(\xi, \eta)(l)$ denotes the lagged sample covariance $C(\xi, \eta)(l) = \frac{1}{N} \sum_{n=1}^{N-l} \xi(n+l)\, \eta(n)$ of observed sequence $\{\xi(n), \eta(n); n=1, 2, \cdots, N\}$. We are assuming the mean values of $\xi(n)$ and $\eta(n)$ to be vanishing.

III We put

$$\begin{aligned} &\hat{c}_{il} = \hat{A}_{iil} \qquad (l=1, 2, \cdots, L), \\ &\hat{c}_{il} = 0 \qquad (l>L) \end{aligned}$$

and get $\hat{a}_{ijm}$ by the relation

$$\begin{aligned} &\hat{a}_{ij0} = \hat{A}_{ij0}, \\ &\hat{a}_{ijm} = \hat{A}_{ijm} + \sum_{l=1}^{m} \hat{c}_{il}\, \hat{a}_{ijm-l} \quad (m=1, 2, \cdots, M). \end{aligned}$$

These $\{\hat{a}_{ijm}\}$'s are the desired estimates of the impulse response functions $\{a_{ijm}\}$. We assume $\hat{a}_{ij0} = a_{ij0} = 0$ for $j=0, 1, \cdots, i$.

IV We get an estimate $\hat{a}_{ij}(f)$ of the frequency response function $a_{ij}(f) = \sum_{m=0}^{M} a_{ijm} \exp(-2\pi \sqrt{-1}\, fm)$ by

$$\hat{a}_{ij}(f) = \sum_{m=0}^{M} \hat{a}_{ijm} \exp(-2\pi \sqrt{-1}\, fm).$$

V We get an estimate of the power spectral density function $p(u_i)(f)$ of $u_i(n)$ by

$$\hat{p}(u_i)(f) = \frac{S_i^2}{\left|1 - \sum_{l=1}^{L} \hat{c}_{il} \exp(-2\pi \sqrt{-1}\, fl)\right|^2}$$

where

$$S_i^2 = C(x_i, x_i)(0) - \sum_{k=0}^{i} \sum_{l=1}^{M+L} C(x_i, x_k)(l)\hat{A}_{ikl} - \sum_{k=i+1}^{K} \sum_{l=0}^{M+L} C(x_i, x_k)(l)\hat{A}_{ikl}$$

is an estimate of $E\varepsilon_i^2(n)$.

VI We compute an estimate $\hat{b}_{ij}(f)$ of the closed loop frequency response function from j to i by

$$[\hat{b}_{ij}(f)] = [\delta_{ij} - \hat{a}_{ij}(f)]^{-1}$$

where [ ] denotes a $(K+1)\times(K+1)$ matrix and

$$\delta_{ii}=1 \quad \text{and} \quad \delta_{ij}=0 \ (i\neq j)\,.$$

VII The power contribution of the noise source $u_j(n)$ to the output $x_i(n)$ can be estimated by the quantity

$$\hat{q}_{ij}(f) = |\hat{b}_{ij}(f)|^2\, \hat{p}(u_j)(f) \quad j=0, 1, 2, \cdots, K\,.$$

It is expected that at least approximately

$$\hat{p}(x_i)(f) = \sum_{j=0}^{K} \hat{q}_{ij}(f)$$

holds, where $\hat{p}(x_i)(f)$ is an estimate of the power spectral density at f of $x_i(n)$, which is obtained from the record $\{x_i(n);\ n=1, 2, \cdots, N\}$. If this is not the case, some increase of L and/or M would be necessary. The equality is strict when the original model is strict and an infinitely long record and M and L greater than or equal to their true values, respectively, are used for the computation and $\hat{p}(x_i)(f)$ is replaced by its theoretical value.

The quantity $\gamma_{ij}(f)$ given by

$$\gamma_{ij}(f) = \frac{\hat{q}_{ij}(f)}{\sum_{k=0}^{K} \hat{q}_{ik}(f)}$$

will be used to evaluate the relative contribution of $u_j(n)$ to the power of $x_i(n)$ at frequency f, and the quantity $R_{ij}(f)$ defined by

$$R_{ij}(f) = \sum_{k=0}^{j} \gamma_{ik}(f) \quad j=0, 1, 2, \cdots, K-1$$

will conveniently be used for graphical representation.

We shall here very briefly discuss the consistency of our estimates. Under our assumption of A_0, $\{A_{ijm};\ j=0, 1, \cdots, K,\ m=0, 1, \cdots, M\}$ satisfies the normal equation of the step II when $C(x_j, x_k)(l)$'s are replaced by the corresponding $E(x_j(n+l)\,x_k(n))$'s. We have already shown in the preceding section the non-singularity of the variance covariance matrix of $\{x_j(n)\ (j=i+1, \cdots, K),\ x_j(n-m)\ (j=0, 1, \cdots, K,\ m=1, 2, \cdots, H)\}$ (H: any finite positive integer). Thus if we can assume the convergence of $C(x_j, x_k)(l)$ to $E(x_j(n+l)\, x_k(n))$ in probability or with probability one, we can show the convergence of $\hat{A}_{ijm}$, properly defined when the solution of the normal equation does not exist, to A_{ijm} in probability or with probability one, respectively. $\hat{c}_{ii}$, $\hat{a}_{ijm}$, S_i^2 and other related quantities converge to their theoretical values correspondingly. The simplest and sometimes most natural assumption in practical situations would be the assumption of ergodicity of the noise $\{u_i(n);\ i=0, 1, \cdots, K:\ n=0, \pm 1, \pm 2, \cdots\}$. Under this assumption the convergence is with probability one.

To show the practical applicability of our estimation procedure we give here some numerical examples. We have used as a realization of

Table 1

m	a_{01m}	$\hat{a}_{01m}$ ($L=6$, $M=6$)	$\hat{a}_{01m}$ ($L=6$, $M=2$)
0	0	−0.053	−0.052
1	0.12	0.120	0.122
2	0.20	0.178	0.175
3	0.05	−0.033	
4	0	−0.048	
5	0	0.012	
6	0	−0.052	

m	a_{10m}	$\hat{a}_{10m}$ ($L=6$, $M=6$)	$\hat{a}_{10m}$ ($L=6$, $M=2$)
0	0	0	0
1	−0.10	−0.113	−0.111
2	−0.10	−0.069	−0.068
3	−0.10	−0.146	
4	0	−0.067	
5	0	0.010	
6	0	0.050	

Table 2

m	a_{01m}	$\hat{a}_{01m}$ ($L=6$, $M=6$)	a_{10m}	$\hat{a}_{10m}$ ($L=6$, $M=6$)
0	0.25	0.197	0	0
1	0.15	0.148	−0.1	−0.114
2	0.08	0.057	−0.2	−0.169
3	0.03	−0.056	−0.1	−0.151
4	0	−0.050	0	−0.077
5	0	0.005	0	0.007
6	0	−0.055	0	0.064

our noise $\{u_i(n)\ (i=0, 1;\ n=1, 2, \cdots, 500)\}$ the results of two independent observations of a physical process and generated $x_i(n)\ (i=0, 1)$ by the formula (B). We assumed $x_i(n)=0$ for $n \leqq 0$. The results of computation are illustrated in Tables 1 and 2. From the result of Table 1 we can see that our estimate is rather insensitive to the change of M. This tendency has been observed in numerous other practical applications and suggests a kind of robustness of our procedure. The result illustrated in Table 2 is concerned with the numerical example of artificial series reported in [1]. The result shows that our present procedure is quite promising even in the identification of this kind of model where $a_{010} \neq 0$, if only the specification of the model or the ordering of the variables is correct. We shall discuss this last point in the next section. It also should be mentioned that in an example of application to a physical process [6], the present procedure gave a quite reasonable result, where the conventional cross-spectral method of estimation of the frequency response function completely failed.

4. Bias due to incorrect specification

Here we shall analyse the effect of incorrectly assuming $a_{ij0} \neq 0$ for some (i, j). We treat the case where $K=1$, i.e., the 2-dimensional case. Thus as our original form we have

$$x_0(n) = \sum_{m=0}^{M} a_{01m}\, x_1(n-m) + u_0(n)$$

$$x_1(n) = \sum_{m=1}^{M} a_{10m}\, x_0(n-m) + u_1(n)\,.$$

After whitening of $u_0(n)$ and $u_1(n)$ we get

$$x_0(n) = \sum_{m=1}^{L} A_{00m}\, x_0(n-m) + \sum_{m=0}^{M+L} A_{01m}\, x_1(n-m) + \varepsilon_0(n)$$

$$x_1(n) = \sum_{m=1}^{M+L} A_{10m}\, x_0(n-m) + \sum_{m=1}^{L} A_{11m}\, x_1(n-m) + \varepsilon_1(n)\,.$$

We are assuming $A_{100}=0$. If we incorrectly specify the model as

$$x_1(n) = \sum_{m=0}^{M+L} c_{10m}\, x_0(n-m) + \sum_{m=1}^{L} c_{11m}\, x_1(n-m) + \eta(n)$$

and apply the method of least squares we shall have to solve the following normal equation:

$$\sum_{m=0}^{M+L} \hat{c}_{10m}\, C(x_0, x_0)(l-m) + \sum_{m=1}^{L} \hat{c}_{11m}\, C(x_1, x_0)(l-m) = C(x_1, x_0)(l)$$

$$l=0, 1, 2, \cdots, M+L\,,$$

$$\sum_{m=0}^{M+L} \hat{c}_{10m} C(x_0, x_1)(l-m) + \sum_{m=1}^{L} \hat{c}_{11m} C(x_1, x_1)(l-m) = C(x_1, x_1)(l)$$

$$l=1, 2, \cdots, L .$$

Here we assume that our computation is based on an infinitely long record and $C(x_i, x_j)(l) = Ex_i(n+l)\,x_j(n)$. Then, from the definition of the model, we have

$$\sum_{m=1}^{M+L} A_{10m} C(x_0, x_0)(l-m) + \sum_{m=1}^{L} A_{11m} C(x_1, x_0)(l-m) = C(x_1, x_0)(l)$$

$$l=1, 2, \cdots, M+L ,$$

$$\sum_{m=1}^{M+L} A_{10m} C(x_0, x_1)(l-m) + \sum_{m=1}^{L} A_{11m} C(x_1, x_1)(l-m) = C(x_1, x_1)(l)$$

$$l=1, 2, \cdots, L .$$

We also have

$$C(x_1, x_0)(0) = \sum_{m=1}^{L+M} A_{10m} C(x_0, x_0)(-m) + \sum_{m=1}^{L} A_{11m} C(x_1, x_0)(-m) + C(\varepsilon_1, x_0)(0) ,$$

where

$$C(\varepsilon_1, x_0)(0) = A_{010} C(\varepsilon_1, x_1)(0) = A_{010} C(\varepsilon_1, \varepsilon_1)(0) .$$

Thus the bias $B_{1jm} = c_{1jm} - A_{1jm}$ is given as

$$\begin{bmatrix} B_{100} \\ \vdots \\ B_{10M+L} \\ B_{111} \\ \vdots \\ B_{11L} \end{bmatrix} = \begin{bmatrix} C(x_0, x_0)(0) & \cdots C(x_0, x_0)(M+L) & C(x_1, x_0)(-1) & \cdots C(x_1, x_0)(-L) \\ \vdots & \vdots & \vdots & \vdots \\ C(x_0, x_0)(M+L) & \cdots C(x_0, x_0)(0) & C(x_1, x_0)(M+L-1) & \cdots C(x_1, x_0)(M) \\ C(x_0, x_1)(1) & \cdots C(x_0, x_1)(1-M-L) & C(x_1, x_1)(0) & \cdots C(x_1, x_1)(L-1) \\ \vdots & \vdots & \vdots & \vdots \\ C(x_0, x_1)(L) & \cdots C(x_0, x_1)(-M) & C(x_1, x_1)(L-1) & \cdots C(x_1, x_1)(0) \end{bmatrix}^{-1}$$

$$\times \begin{bmatrix} A_{010} C(\varepsilon_1, \varepsilon_1)(0) \\ \vdots \\ 0 \\ 0 \\ \vdots \\ 0 \end{bmatrix} ,$$

where the elements of the last vector are all zeros except the first one. As the inverse matrix is positive definite we can see that the following relation holds:

$$c_{100} = B_{100} = A_{010}\, C(\varepsilon_1, \varepsilon_1)(0) \times (\text{positive number}) .$$

Taking into account the relation $a_{ij0} = A_{ij0}$, this result tells us that by missing to specify $a_{100}=0$ we shall introduce into the estimate of a_{100}, the bias B_{100} which is of the same sign as A_{010} or a_{010}. This effect is clearly seen in our numerical example shown in Table 3. The result was obtained by using the same data as the result of Table 2 but with incorrect specification of the order of variables. The present observation will be of help to understand why it is possible that we sometimes get an estimate of a_{010} with unexpected sign.

Table 3

m	a_{01m}	$\hat{a}_{01m}$ ($L=6$, $M=6$)	a_{10m}	$\hat{a}_{10m}$ ($L=6$, $M=6$)
0	0	0.232	0.25	0
1	−0.1	−0.048	0.15	0.210
2	−0.2	−0.118	0.08	0.072
3	−0.1	−0.119	0.03	−0.067
4	0	−0.010	0	−0.016
5	0	0.045	0	0.011
6	0	0.072	0	−0.051

Acknowledgement

The author expresses his hearty thanks to Dr. T. Nakagawa of Chichibu Cement Company for helpful discussions of the problem. The results of this paper were obtained while Dr. Nakagawa and the author were working on the analysis and control of a cement rotary kiln.

Thanks are also due to Miss E. Arahata for the programming of the necessary computations.

THE INSTITUTE OF STATISTICAL MATHEMATICS

REFERENCES

[1] H. Akaike, "Some problems in the application of the cross-spectral method," *Spectral Analysis of Time Series* (ed. B. Harris), New York, John Wiley (1967), 81-107.

[2] H. Akaike, "On the statistical estimation of the frequency response function of a system having multiple input," *Ann. Inst. Statist. Math.*, 17 (1965), 185-210.

[3] H. Akaike and Y. Yamanouchi, "On the statistical estimation of frequency response function," *Ann. Inst. Statist. Math.*, 14 (1962), 23-56.

[4] J. Durbin, "Estimation of parameters in time-series regression models," *J. R. Statist., Soc.*, Series B, 22 (1960), 139-153.

[5] H. B. Mann and A. Wald, "On the statistical treatment of linear stochastic difference equations," *Econometrica*, 11 (1943), 173-220.
[6] T. Otomo, T. Nakagawa and H. Akaike, "Implementation of computer control of a cement rotary kiln through data analysis," submitted to the 4th congress of IFAC.
[7] E. Parzen, "Analysis and synthesis of linear models for time series," Technical Report No. 4, Department of Statistics, Stanford University, 1966.
[8] E. Parzen, "On empirical multiple time series analysis," *Proc. 5th Berkeley Symposium*, 1 (1967), 305-340.
[9] P. Whittle, "The analysis of multiple stationary time series," *J. R. Statist. Soc.*, Series B, 15 (1953), 125-139.
[10] N. Wiener and P. Masani, "The prediction theory of multivariate stochastic processes, Part I," *Acta Math.*, 98 (1957), 111-150.
[11] H. Wold, "Ends and means in econometric model building," *Probability and Statistics* (ed. V. Grenander), Almqvist & Wiksell, Stockholm, John Wiley & Sons, New York (1959), 355-434.
[12] H. Wold, "Forecasting by the chain principle," *Time Series Analysis* (ed. M.Rosenblatt), John Wiley & Sons, New York (1963), 471-497.

Reprinted from *Annals of the Institute of Statistical Mathematics*, Vol. 21, No. 2, 1969, 243-247 by permission of the Institute of Statistical Mathematics

FITTING AUTOREGRESSIVE MODELS FOR PREDICTION

HIROTUGU AKAIKE

(Received June 17, 1969)

1. Introduction and summary

This is a preliminary report on a newly developed simple and practical procedure of statistical identification of predictors by using autoregressive models. The use of autoregressive representation of a stationary time series (or the innovations approach) in the analysis of time series has recently been attracting attentions of many research workers and it is expected that this time domain approach will give answers to many problems, such as the identification of noisy feedback systems, which could not be solved by the direct application of frequency domain approach [1], [2], [3], [9].

The main difficulty in fitting an autoregressive model

$$X(n)=\sum_{m=1}^{M} a_m X(n-m)+a_0+\varepsilon(n) ,$$

where $X(n)$ is the process being observed and $\varepsilon(n)$ is its innovation which is uncorrelated with $X(l)$ $(l<n)$ and is forming a white noise, lies in the decision of the order M. We assume the mutual independence and strict stationarity of $\{\varepsilon(n)\}$.

There have been extensive investigations of topics which are very closely related with this subject but the definite description of the procedure, which could directly be adopted for practical applications, is quite lacking yet [6], [7]. T. W. Anderson [4] has treated this problem as a multiple decision problem and given a description of a procedure which in some sense has an optimal property. Though the procedure is well described, it contains many constants which are to be determined before its application and the problem of selection of these constants are left open. E. Parzen is advocating the use of autoregressive representation for the estimation of spectra and suggested the use of the maximum likelihood test criterion proposed by P. Whittle for this purpose [8], [9].

The main difficulty in applying this kind of procedures stems from the fact that they are essentially formulated in the form of a successive test of the whiteness of the series against multiple "alternatives." Actually one of the "alternatives" is just the model we are looking for and thus it is very difficult for us to get the feeling of the possible alternatives to set reasonable "significance levels."

To overcome this difficulty we adopt entirely decision theoretic approach where a figure of merit is defined for each model being fitted and the one with the best figure is chosen as our predictor. This figure of merit which we shall call the final prediction error (FPE) is defined as the expected variance of the prediction error when an autoregressive model fitted to the present series of $X(n)$ is applied to another independent realization of $X(n)$, or to the process with one and the same covariance characteristic as that of $X(n)$ and is independent of the present $X(n)$, to make a one step prediction. The notion of FPE can also be utilized for the decision of the constants in Anderson's procedure.

In the practical application of our procedure we compute an estimate of FPE of each autoregressive model within a prescribed sufficiently wide range of possible orders and select the one which gives the minimum of the estimates. This procedure we shall call the FPE scheme.

In this paper we shall only give a description with a brief discussion of the FPE scheme for practical use and the theoretical and numerical details of our investigation of the scheme will be discussed in separate papers.

2. FPE scheme

We consider the situation where a set of data $\{X(n); n=1, 2, \cdots, N\}$ is given.

0) First we replace $X(n)$ by $\tilde{X}(n)=X(n)-\bar{X}$, where $\bar{X}=\frac{1}{N}\sum_{n=1}^{N} X(n)$.

1) We set the upper limit L of the order of autoregressive models to be fitted to the data. L should be chosen large enough not to exclude the efficient model. Also confer the description in the following 2).

2) We calculate the sample autocovariances

$$C_{xx}(l)=\frac{1}{N}\sum_{n=1}^{N-l} \tilde{X}(n+l)\tilde{X}(n) \qquad \text{for} \quad l=0, 1, 2, \cdots, L\,.$$

The value of L here is generally much smaller than the value of L usually considered to be necessary for the estimation of power spectrum by Fourier transforming the windowed sample autocovariance function.

3) Then we try to fit the autoregressive model of order M ($M=1, 2, \cdots, L$) by the least squares method which requires the mean square of

residuals

$$R(a^{(M)})=\frac{1}{N}\sum_{n=1}^{N}\left(\tilde{X}(n)-\sum_{m=1}^{M}a_m^{(M)}\tilde{X}(n-m)\right)^2$$

to be minimized with respect to $\{a_m^{(M)};\ m=1,2,\cdots,M\}$ assuming $\tilde{X}(0)=\tilde{X}(-1)=\cdots=\tilde{X}(-M+1)=0$. The required set of coefficients $\{\hat{a}_m^{(M)};\ m=1,2,\cdots,M\}$ is obtained by solving the normal equation

$$\begin{pmatrix} C_{xx}(0), & C_{xx}(1), & \cdots, & C_{xx}(M-1) \\ C_{xx}(1), & C_{xx}(0), & \cdots, & C_{xx}(M-2) \\ \cdot & \cdot & & \cdot \\ \cdot & \cdot & & \cdot \\ \cdot & \cdot & & \cdot \\ C_{xx}(M-1), & C_{xx}(M-2), & \cdots, & C_{xx}(0) \end{pmatrix} \begin{pmatrix} \hat{a}_1^{(M)} \\ \hat{a}_2^{(M)} \\ \cdot \\ \cdot \\ \cdot \\ \hat{a}_M^{(M)} \end{pmatrix} = \begin{pmatrix} C_{xx}(1) \\ C_{xx}(2) \\ \cdot \\ \cdot \\ \cdot \\ C_{xx}(M) \end{pmatrix}.$$

We denote the value of $R(a^{(M)})$ corresponding to this solution $\{\hat{a}_m^{(M)};\ m=1,2,\cdots,M\}$ by R_M. An estimate $(\mathrm{FPE})_M$ of FPE of the autoregressive model of order M is calculated by the definition

$$(\mathrm{FPE})_M=\left(1+\frac{M+1}{N}\right)S_M\,,$$

where

$$S_M=\frac{N}{N-1-M}R_M\,.$$

We put $R_0=C_{xx}(0)$, $S_0=\frac{N}{N-1}R_0$ and $(\mathrm{FPE})_0=S_0$.

For the purpose of comparison of the magnitudes of $(\mathrm{FPE})_M$ the relative value $(\mathrm{RFPE})_M$ of $(\mathrm{FPE})_M$ defined by

$$(\mathrm{RFPE})_M=\frac{(\mathrm{FPE})_M}{(\mathrm{FPE})_0}$$

can conveniently be used.

The recursive method of solution of the equation is most conveniently applied for the computation of this step.

4) We adopt the value M_0 of M which gives the minimum of $(\mathrm{FPE})_M$ within $M=0,1,2,\cdots,L$ as the order of our model for prediction. $\{\hat{a}_m^{(M)};\ m=0,1,\cdots,M\}$, where $\hat{a}_0^{(M)}=\left(1-\sum_{M=1}^{M}\hat{a}_m^{(M)}\right)\bar{X}$, and $S_M=\frac{N}{N-1-M}R_M$ with $M=M_0$ are adopted as our estimate of the set of coefficients of predictor and that of the innovation variance, respectively.

5) An estimate $\hat{p}_{xx}(f)$ of the power spectrum density $p_{xx}(f)$ of

$\{X(n)-E(X(n))\}$ at frequency f is obtained by the formula

$$\hat{p}_{xx}(f)=\frac{S_M}{\left|1-\sum_{m=1}^{M}\hat{a}_m^{(M)}\exp(-i2\pi fm)\right|^2}\cdot\left(1-\frac{M}{N-1}\right)$$

with $M=M_0$.

3. A brief discussion of FPE scheme

As is suggested by the definition of its estimate, the definition of FPE of the autoregressive model of order M is given by the relation

$$\mathrm{FPE}=\left(1+\frac{M+1}{N}\right)r_M,$$

where r_M is the minimum of $E\left(X(n)-\sum_{m=1}^{M}a_m^{(M)}X(n-m)-a_0^{(M)}\right)^2$ with respect to $\{a_m^{(M)};\ m=0,1,\cdots,M\}$. Obviously r_M is equal to the variance of the innovation $\varepsilon(n)$ when $X(n)$ is generated from $\varepsilon(n)$ by a finite autoregression of order equal to or less than M. In this case FPE gives the asymptotic mean square of the prediction error when the least squares estimate $\{\hat{a}_m^{(M)};\ m=0,1,\cdots,M\}$ is applied to another independent observation of the same process, or to the process with one and the same covariance characteristic and is independent of the present process, to make a one-step prediction. We can see that FPE tends to be large when unnecessarily large value of M is adopted. When M is less than the true order of the process, r_M and its estimate include, beside the contribution of the innovation variance, the contribution of the inevitable bias of the model and thus it tends to be significantly large when a too small value of M is adopted. Thus by seeking the minimum of FPE we shall be able to arrive at an autoregressive model of an order which will not be giving a significant bias and at the same time will not be giving a too big mean square prediction error in the above stated sense.

We have simultaneously applied our procedure, a practical version of Anderson's procedure and a modified version of our original procedure to many artificial and practical time series. At present it seems that our original procedure is the simplest and the most satisfactory one giving good results in wide variety of practical situations.

Also we have computed estimates of some of the power spectra by using the Fourier transforms of the estimates of predictors and estimates of the innovation variances obtained by following our FPE scheme and the results have been compared with the estimates obtained by the classical procedure described by Blackmann and Tukey [5]. The comparison has shown that our new procedure is giving extremely good results suggesting that our estimate is with a good traceability at both high and low power level regions and is well balancing the bias and variance of the estimate, a fact which was scarcely expected by the simple application of the classical procedure.

Experimental applications of FPE scheme with $L=44$ to three real time series with $N=511$, 511 and 524, respectively, have resulted in $M_0=15$, 13 and 5, respectively. Our estimates of the power spectra have shown, though this remains as a subjective judgement at present, a good resolvability which nearly corresponds to that of the classical estimate obtained by applying a hanning type window with truncation point $L=90$ and yet probably with a sampling variability generally smaller than that of the classical estimate obtained with truncation point $L=45$.*) From the series of residuals or estimated innovations obtained by fitting these models we could not find significant trace of deviation from whiteness.

We shall discuss the details of these theoretical and experimental investigations of FPE scheme in subsequent papers.

THE INSTITUTE OF STATISTICAL MATHEMATICS

REFERENCES

[1] H. Akaike, "Some problems in the application of the cross-spectral method," *Spectral Analysis of Time Series* (ed. B. Harris), New York, John Wiley, (1967), 81-107.

[2] H. Akaike, "On the use of a linear model for the identification of feedback systems," *Ann. Inst. Statist. Math.*, 20 (1968), 425-439.

[3] H. Akaike, "A method of statistical identification of discrete time parameter linear system," *Ann. Inst. Statist. Math.*,21 (1969), 225-242.

[4] T. W. Anderson, "Determination of the order of dependence in normally distributed time series," *Time Series Analysis* (ed. M. Rosenblatt), New York, John Wiley, (1963), 425-446.

[5] R. B. Blackman and J. W. Tukey, *The Measurement of Power Spectra*, New York, Dover, 1959.

[6] J. Durbin, "Efficient estimation of parameters in moving-average models," *Biometrika*, 46 (1959), 306-316.

[7] J. Durbin, "The fitting of time series models," *Rev. Int. Statist. Inst.*, 28 (1960), 233-243.

[8] E. Parzen, "Statistical spectral analysis (single channel case) in 1968," Stanford University Statistics Department Technical Report, No. 11, June 10, (1968), 44 pages.

[9] E. Parzen, "Multiple time series modelling," Stanford University Statistics Department Technical Report, No. 12, July 8, (1968), 38 pages.

*) On this point confer the forthcoming paper by the present author entitled "Power spectrum estimation through autoregressive model fitting."

Reprinted from *Annals of the Institute of Statistical Mathematics*, Vol. 22, No. 2, 1970, 203-217 by permission of the Institute of Statistical Mathematics

STATISTICAL PREDICTOR IDENTIFICATION

HIROTUGU AKAIKE

(Received Dec. 26, 1969)

1. Introduction and summary

In a recent paper by the present author [1] a simple practical procedure of predictor identification has been proposed. It is the purpose of this paper to provide a theoretical and empirical basis of the procedure.

Our procedure is based on a figure of merit of a predictor, which is called the final prediction error (FPE) and is defined as the mean square prediction error of the predictor. We consider the application of the least squares method for the identification of the predictor when the stochastic process under observation is an autoregressive process generated from a strictly stationary and mutually independent innovations. The identification is realized by fitting autoregressive models of successive orders within a prescribed range, computing estimates of FPE for the models, and adopting the one with the minimum of the estimates.

The statistical characteristics of these estimates of FPE and the overall procedure are discussed to show the practical utility of the procedure. A modified version of this original procedure is proposed, which shows a consistency, as an estimation procedure of the order of a finite order autoregressive process, which is lacking in the original procedure. The notion of FPE is also applied for the determination of the constants of the decision procedure, which was proposed by T. W. Anderson [2] for the decision of the order of a Gaussian autoregressive process, to provide a third procedure.

Performances of the three types of procedures, the original one, a modified version and that of Anderson's type, are compared by using various realizations of artificial time series. The results show that for practical applications, where the true orders of autoregressive processes would generally be infinite, the original procedure would be the most useful.

Implication of the present identification procedure on the estimation of power spectra will be discussed in a subsequent paper [3].

We shall use the convention of denoting by $(u(l))$ the column vector of $u(l)$ $(l=1, 2, \cdots, M)$ and by v or $(v(l, m))$ the matrix with (l, m) element $v(l, m)$ $(l, m=1, 2, \cdots, M)$. When the dimension M is of special interest we shall add the subscript M and thus u_M and v_M are used for the above u and v. The symbol $'$ will be used to denote the transpose of a matrix or a vector.

2. Definition of FPE of a predictor and the statement of the problem

Here we first introduce a general definition of a figure of merit of a predictor. This is defined simply as the mean square prediction error and is called the FPE (final prediction error) of the predictor, i.e., for a predictor $\hat{X}(n)$ of $X(n)$

$$\text{FPE of } \hat{X}(n)=E(X(n)-\hat{X}(n))^2 . \tag{2.1}$$

In practical situations $\hat{X}(n)$ is given as a function of the recent values of $X(n)$ and the structure or the parameter of the function is determined, or identified, by using the whole past history of $X(n)$. Assuming the dependency of this identified structure on the recent values of $X(n)$ which are to be used to give $\hat{X}(n)$ to be decreasing as the length of the past history used for the identification is increased, we consider the idealized situation where the dependency is completely vanishing. This is equivalent to the situation where the structure of a predictor is identified by using an observation of a process $X(n)$ and, using the structure, the prediction is made with another process $Y(n)$ which is independent of $X(n)$ but with one and the same statistical property as $X(n)$.

When the process $X(n)$ is stationary and the predictor $\hat{Y}(n)$ of $Y(n)$ is linear and given by

$$\hat{Y}(n)=\sum_{m=1}^{M}\hat{a}_M(m)Y(n-m)+\hat{a}_M(0) , \tag{2.2}$$

where $\hat{a}_M(m)$ is a function of $\{X(n)\}$, we have

$$\text{FPE of } \hat{Y}(n)=\sigma^2(M)+\sum_{l=0}^{M}\sum_{m=0}^{M}E(\Delta a_M(l)\Delta a_M(m))V_{M+1}(l, m) , \tag{2.3}$$

where

$$\begin{aligned}\sigma^2(M)&=E(Y(n)-\sum_{m=1}^{M}a_M(m)Y(n-m)-a_M(0))^2\\&=\operatorname*{Min}_{\{a(m)\}}E(Y(n)-\sum_{m=1}^{M}a(m)Y(n-m)-a(0))^2 ,\end{aligned} \tag{2.4}$$

$$V_{M+1}(l, m)=EY(n-l)Y(n-m) \qquad l, m=1, 2, \cdots, M , \tag{2.5}$$

$$V_{M+1}(0, m) = V_{M+1}(m, 0)$$
$$= EY(n) \qquad m=1, 2, \cdots, M ,$$
$$V_{M+1}(0, 0) = 1 ,$$

and

(2.6) $$\Delta a_M(m) = \hat{a}_M(m) - a_M(m) \qquad m=0, 1, \cdots, M ,$$

where $a_M(m)$ is defined by (2.4) and is giving the best (in the sense of mean square) linear predictor. (2.3) shows that the FPE in this case is composed of two components: the first one corresponding to the FPE of the best linear predictor for a given M and the second one due to the statistical deviation of $\hat{a}_M(m)$ from $a_M(m)$. Generally, as the value of M is increased, the first term $\sigma^2(M)$ will decrease but the second term will increase for a finite length of observation of $X(n)$.

Given a set of predictors the definition of FPE naturally suggests the adoption of the predictor with the minimum value of FPE as optimum. The problem we are concerned with in this paper is the realization of a good approximation to the optimum choice of M for the above stated stationary and linear case, using the information obtained by observing $X(n)$ for a finite length of time.

3. FPE of the least squares estimate of an autoregressive model

Hereafter we shall assume that $X(n)$ is a stationary autoregressive process generated by the relation

(3.1) $$X(n) = \sum_{m=1}^{M} a(m)X(n-m) + a(0) + \varepsilon(n) ,$$

where $\varepsilon(n)$'s are mutually independently and identically distributed random variables with $E\varepsilon(n)=0$ and $E\varepsilon^2(n)=\sigma^2$.

Given a set of data $\{X(n);\ n=-M+1, -M+2, \cdots, N\}$, the parameter $\hat{a}_M(m)$ of our predictor is defined as the least squares estimate of $a(m)$, i.e., $\hat{a}_M(m)$ is the solution of

(3.2) $$\sum_{m=1}^{M} C_{xx}(m, l)\hat{a}_M(m) = C_{xx}(0, l) \qquad l=1, 2, \cdots, M ,$$

and

(3.3) $$\hat{a}_M(0) = \bar{X}_0 - \sum \hat{a}_M(m)\bar{X}_m ,$$

where

$$\bar{X}_m = N^{-1} \sum_{n=1}^{N} X(n-m) \qquad (m=0, 1, 2, \cdots, M)$$

and

$$C_{xx}(m, l) = N^{-1} \sum_{n=1}^{M} (X(n-m) - \bar{X}_m)(X(n-l) - \bar{X}_l) .$$

Following the definition of $Y(n)$ given in the preceding section, our predictor $\hat{Y}(n)$ of $Y(n)$ is given in this case by

$$\hat{Y}(n) = \sum_{m=1}^{M} \hat{a}_M(m)(Y(n-m) - \bar{X}_m) + \bar{X}_0 . \tag{3.4}$$

We are assuming that $Y(n)$ is generated by the relation $Y(n) = \sum_{m=1}^{M} a_m Y(n-m) + a_0 + \delta(n)$, where $\delta(n)$ has one and the same statistical property as $\varepsilon(n)$. We have

$$Y(n) - \hat{Y}(n) = \delta(n) - \sum_{m=1}^{M} \Delta a_M(m)\, y(n-m) - (\Delta \bar{X}_0 - \sum_{m=1}^{M} \hat{a}_M(m) \Delta \bar{X}_M) ,$$

where $y(n) = Y(n) - E(Y(n))$ and $\Delta \bar{X}_l = \bar{X}_l - E(X(n))$.

Taking into account the independency of $y(n)$ of Δa_M and $\Delta \bar{X}_l$ we get

$$\begin{aligned} \text{FPE of } \hat{Y}(n) &= E(Y(n) - \hat{Y}(n))^2 \\ &= \sigma^2 + \sum_{m=1}^{M} \sum_{l=1}^{M} E(\Delta a_M(m) \Delta a_M(l)) R_{xx}(l-m) \\ &\quad + E(\Delta \bar{X}_0 - \sum_{m=1}^{M} \hat{a}_M(m) \Delta \bar{X}_m)^2 , \end{aligned} \tag{3.5}$$

where

$$R_{xx}(l-m) = EX(n-l)X(n-m) - (EX(n))^2 .$$

For the asymptotic evaluation of this FPE of $\hat{Y}(n)$ we make use of the following basic theorem.

THEOREM 1. *Under the present assumption of $X(n)$, the limit distribution $\sqrt{N} \Delta \bar{X}_0 = \sqrt{N}(\bar{X}_0 - E(X(n)))$ and $\sqrt{N} \Delta a_M(m) = \sqrt{N}(\hat{a}_M(m) - a(m))$ $(m = 1, 2, \cdots, M)$, when N tends to infinity, is $(M+1)$-dimensional Gaussian with zero mean and the variance matrix*

$$\sigma^2 \begin{pmatrix} \delta^{-2} & 0'_M \\ 0_M & R_M^{-1} \end{pmatrix} , \tag{3.6}$$

where $\delta = 1 - \sum_{m=1}^{M} a(m)$, R_M is the $M \times M$ matrix of $R(l, m) = R_{xx}(l-m)$ and 0 denotes a zero vector.

From the ergodicity of the process $X(n)$ we know that $C_{xx}(l, m)$ converges to $R_{xx}(l-m)$, as N tends to infinity, with probability one. Thus $\hat{a}_M$ is a consistent estimate of a_M, in this case with convergence with probability one. From (3.2) we have, for $l = 1, 2, \cdots, M$,

$$(3.7)\qquad (\hat{a}_M(l))=(C_{xx}(m,l))^{-1}(C_{xx}(0,l)) \\ =(a_M(l))+(C_{xx}(m,l))^{-1}(C_{\epsilon x}(l)) \,,$$

where $C_{\epsilon x}(l)=N^{-1}\sum_{n=1}^{N}\varepsilon(n)(X(n-l)-\bar{X}_l)$. Thus we get

$$(3.8)\qquad (\Delta a_M(l))=(C_{xx}(m,l))^{-1}(C_{\epsilon x}(l)) \,.$$

From the consistency of $C_{xx}(m,l)$ we know that the limit distribution of $\sqrt{N}\,\Delta\bar{X}_0$ and $\sqrt{N}\Delta a_M$ is identical to that of $\sqrt{N}\,\Delta\bar{X}_0$ and $\sqrt{N}\,R_M^{-1}C_{\epsilon x}$. By applying the Diananda's central limit theorem [4] for finitely dependent sequence, as was done by Anderson and Walker [5], we can easily get

LEMMA. *The limit distribution of $\sqrt{N}\,\Delta\bar{X}_0$ and $\sqrt{N}C_{\epsilon x}$, when N tends to infinity, is $(M+1)$-dimensional Gaussian with zero mean and the variance*

$$\sigma^2\begin{pmatrix}\delta^{-2} & 0'_M \\ 0_M & R_M\end{pmatrix}.$$

It should be noted that as the power spectral density of $X(n)-EX(n)$ at zero frequency is $\sigma^2\delta^{-2}$, where $\delta=1-\sum_{m=1}^{M}a_M(m)$, the variance of the limit distribution of $\sqrt{N}\,\Delta\bar{X}_0$ is equal to $\sigma^2\delta^{-2}$. The assertion of Theorem 1 is a direct consequence of this lemma and the observation following (3.8).

Now we return to the evaluation of FPE of $\hat{Y}(n)$. Instead of taking the expectation of $(Y(n)-\hat{Y}(n))^2$ directly as suggested in (3.5) we first take the conditional expectation of $(Y(n)-\hat{Y}(n))^2$ for a given $X(n)$. This we will denote by $E_x(Y(n)-\hat{Y}(n))^2$. From the independency of $Y(n)$ of $X(n)$ we have

$$(3.9)\qquad E_x(Y(n)-\hat{Y}(n))^2=\sigma^2+\sum_{m=1}^{M}\sum_{l=1}^{M}\Delta a_M(m)\Delta a_M(l)R_{xx}(l-m) \\ +(\Delta\bar{X}_0-\sum_{m=1}^{M}\hat{a}_M(m)\Delta\bar{X}_m)^2 \,.$$

Taking into account the fact that in the limit the differences between $\sqrt{N}\,\Delta\bar{X}_0$ and $\sqrt{N}\,\Delta\bar{X}_m$ $(m=1,2,\cdots,M)$ are stochastically vanishing, we can see from the theorem that $N\{E_x(Y(n)-\hat{Y}(n))^2-\sigma^2\}$ has a limit distribution with expectation equal to $(M+1)\sigma^2$. This observation suggests the following definition of $(\text{FPE})_M$ as an asymptotic evaluation of FPE of $\hat{Y}(n)$:

$$\text{(3.10)} \qquad (\text{FPE})_M \text{ of } \hat{Y}(n) = \left(1 + \frac{M+1}{N}\right)\sigma^2 .$$

Our identification procedure of the predictor will be based on some estimate of $(\text{FPE})_M$.

4. An estimate of $(\text{FPE})_M$ and the minimum FPE procedure

From the ergodicity of $X(n)$ we know that

$$\text{(4.1)} \qquad S(M) = C_{xx}(0, 0) - \sum_{l=1}^{M} \hat{a}_M(l) C_{xx}(0, l)$$

is a consistent estimate of σ^2. By (3.2) we have

$$S(M) = C_{xx}(0, 0) - \sum_{l=1}^{M} \sum_{m=1}^{M} \hat{a}_M(l) C_{xx}(m, l) \hat{a}_M(m) ,$$

and by taking into account the relation

$$\sum_{m=1}^{M} \Delta a_M(m) C_{xx}(m, l) = C_{xx}(0, l) - \sum_{m=1}^{M} C_{xx}(m, l) a(m)$$

we get

$$\text{(4.2)} \qquad S(M) = C_{xx}(0, 0) - 2 \sum_{m=1}^{M} a(m) C_{xx}(0, m) + \sum_{l=1}^{M} \sum_{m=1}^{M} a(l) a(m) C_{xx}(m, l) - \sum_{l=1}^{M} \sum_{m=1}^{M} \Delta a_M(l) \Delta a_M(m) C_{xx}(m, l) .$$

From the definition of $C_{xx}(m, l)$ we have

$$\text{(4.3)} \qquad H(M) = C_{xx}(0, 0) - 2 \sum_{m=1}^{M} a(m) C_{xx}(0, m) + \sum_{l=1}^{M} \sum_{m=1}^{M} a(m) a(l) C_{xx}(m, l) = N^{-1} \sum_{n=1}^{N} (\varepsilon(n) - \bar{\varepsilon})^2 ,$$

and we get

$$\text{(4.4)} \qquad E(H(M)) = (1 - N^{-1}) \sigma^2 .$$

From Theorem 1 we know that when we assume the model (3.1) the limit distribution of

$$\text{(4.5)} \qquad Q(M) = N \sum_{l=1}^{M} \sum_{m=1}^{M} \Delta a_M(m) \Delta a_M(l) C_{xx}(m, l)$$

has expectation $M\sigma^2$, i.e.,

$$\text{(4.6)} \qquad E_\infty \{Q(M)\} = M\sigma^2 ,$$

where E_∞ denotes the expectation of the limit distribution of the quantily within the braces, when N tends to infinity. These observations suggest that it would be reasonable to adopt $(1-N^{-1}(M+1))^{-1}S(M)$ as an estimate of σ^2 to define our estimate (FPE)(M) of $(\mathrm{FPE})_M$ by

$$(4.7) \qquad (\mathrm{FPE})(M)=(1+N^{-1}(M+1))(1-N^{-1}(M+1))^{-1}S(M)\,.$$

The discussions in this and the preceding sections naturally lead us to the idea that when there are many predictors obtained by applying the least squares method it would be reasonable for us to pick the one with the minimum value of (FPE)(M). Following this idea, for the identification of the predictor by a single record of $X(n)$, we proceed as follows; we compute (FPE)(M) successively for $M=0, 1, \cdots, L$ (L; preassigned positive integer) and adopt $\hat{a}_M$ with $M=M_0$ to define the predictor, where $(\mathrm{FPE})(M_0)=$the minimum of (FPE)(M) ($M=0, 1, \cdots, L$). This process which was called by the name of FPE scheme in the former paper [1] will hereafter be called the minimum FPE procedure.

5. Statistical properties of (FPE)(M)

To see the practical utility of the minimum FPE procedure we shall have first to analyze the statistical characteristics of (FPE)(M) ($M=0, 1, \cdots, L$) for a fixed model of $X(n)$. We assume that the order of $X(n)$ is K, i.e., $a_K\neq 0$ and $a_m=0$ for $m>K$ in (3.1). We assume $K\geqq 0$ and exclude the case where $K=-1$, with $a_0=0$, from our discussion. We also assume that the set of data is given in a form $\{X(n); n=-L+1, -L+2, \cdots, 1, 2, \cdots, N\}$. From (4.3) we can see that $H(M)$ remains constant for $M\geqq K$ and thus the behavior of $S(M)$ is dependent only on $Q(M)$ of (4.5). From the discussion of Section 3 we know that the limit distribution of $Q(M)$ ($M=K, K+1, \cdots, L$) is identical to that of $NC'_{\epsilon xM}R_M^{-1}C_{\epsilon xM}$, where $C_{\epsilon xM}=(C_{\epsilon x}(l))$ ($l=1, 2, \cdots, M$). As was stated in the lemma of Section 3 the covariance matrix of the limit distribution of $\sqrt{N}\,C_{\epsilon xM}$ is identical to that of $\{X(n-m)-EX(n-m)\colon m=1, 2, \cdots, M\}$ multiplied by σ^2, i.e., $\sigma^2 R_M$. Thus the successive orthonormalization procedure of $X(n-m)-EX(n-m)$ ($m=1, 2, \cdots, M$) can be applied to $\sqrt{N}\,C_{\epsilon xM}$ to give a vector random variable U_M of which limit distribution is the M-dimensional unit normal distribution. The detail of this transformation is already described in [6]. We have $U_M=T_M\sqrt{N}\,C_{\epsilon xM}$ with $\sigma^2 T_M R_M T'_M=I_M$, where I denotes the identity matrix, and the matrix T_M of the transformation has zeros above the diagonal. From the structure of T_M it is readily seen that

$$T_M(l, m)=T_L(l, m) \qquad (l, m=1, 2, \cdots, M) \quad \text{for} \quad M<L\,,$$

i.e., the submatrix of the first $M\times M$ elements of T_L is identical to T_M.

Thus we can see that U_M is the vector of the first M elements of U_L. From this observation we can see that the limit distribution of $Q(M)$ is identical to that of $\sigma^2 \sum_{l=1}^{M} U^2(l)$ $(M=K, K+1, \cdots, L)$. The limit distribution of $U^2(l)$ $(l=1, 2, \cdots, L)$ is then the distribution of mutually independent chi-square variables each with d.f.1 [7]. Thus we get

THEOREM 2. *For $M \geqq K$, $\sigma^{-2}Q(M)$ is asymptotically distributed as the partial sum of the first M terms of a sequence of mutually independently distributed chi-square variables each with d.f.* 1.

Now we proceed to the analysis of the statistical behavior of (FPE)(M) $(M=0, 1, \cdots, L)$. For any positive integer M, we define $a_M(m)$, irrespectively of the order K, as the solution of (3.2) when $C_{xx}(m, l)$ is replaced by $R_{xx}(l-m)$ and define $\sigma^2(M)$ by

$$\sigma^2(M) = R_{xx}(0) - \sum_{m=1}^{M} a_M(m) R_{xx}(m) . \tag{5.1}$$

We shall denote $\Delta a_M(m) = \hat{a}_M(m) - a_M(m)$, where $\hat{a}_M(m)$ is the solution of (3.2). M is not restricted to be equal or larger than the order K. Corresponding to (4.2) it holds that

$$\begin{aligned} S(M) &= C_{xx}(0, 0) - \sum_{l=1}^{M} \hat{a}_M(l) C_{xx}(0, l) \\ &= C_{xx}(0, 0) - 2 \sum_{m=1}^{M} a_M(m) C_{xx}(0, m) + \sum_{l=1}^{M} \sum_{m=1}^{M} a_M(l) a_M(m) C_{xx}(m, l) \\ &\quad - \sum_{l=1}^{M} \sum_{m=1}^{M} \Delta a_M(l) \Delta a_M(m) C_{xx}(m, l) . \end{aligned} \tag{5.2}$$

Ignoring the terms of order N^{-2}, we have approximately

$$\begin{aligned} (\text{FPE})(M_1) - (\text{FPE})(M_2) &= (1 - N^{-1}(M_1+1))^{-1}(1 - N^{-1}(M_2+1))^{-1} \\ &\quad \cdot (S(M_1) - S(M_2) - N^{-1}(M_2 - M_1)(S(M_1) + S(M_2))) , \end{aligned} \tag{5.3}$$

where $0 \leqq M_1, M_2 \leqq L$. If we assume the equality (5.3) to be strict, we have for $M_1 < M_2$

$$\begin{aligned} &\text{Prob}\{(\text{FPE})(M_1) - (\text{FPE})(M_2) > 0\} \\ &\quad \geqq \text{Prob}\{S(M_1) - S(M_2) - 2N^{-1}(M_2 - M_1) S(M_1) > 0\} \\ &\quad \geqq \text{Prob}\{(S(M_1) - S(M_2))((M_2 - M_1) C_{xx}(0))^{-1} > 2N^{-1}\} \end{aligned} \tag{5.4}$$

From (5.2) we have

$$\begin{aligned} S(M_1) - S(M_2) &= \sigma^2(M_1) - \sigma^2(M_2) + \Delta(\sigma^2(M_1) - \sigma^2(M_2)) \\ &\quad + N^{-1}(Q(M_2) - Q(M_1)) , \end{aligned} \tag{5.5}$$

where $\Delta(\sigma^2(M_1) - \sigma^2(M_2))$ is obtained by replacing $R_{xx}(m, l)$ by $\Delta R_{xx}(m) =$

$C_{xx}(m, l)-R_{xx}(m, l)$ in the definition of $\sigma^2(M_1)-\sigma^2(M_2)$ and $Q(M)$ is as defined in (4.5). For M_1 and M_2 which are very small compared with N and for which the differences $a_{M_1}(m)-a_{M_2}(m)$ $(m=1, 2, \cdots, M_2;\ a_{M_1}(m)=0$ for $m>M_1)$ are of the order of $N^{-1/2}$, $R_{xx}^{-1}(0)(\sigma^2(M_1)-\sigma^2(M_2))$ will be of the order of $N^{-1/2}$ from (5.1), while $C_{xx}^{-1}(0, 0)\Delta(\sigma^2(M_1)-\sigma^2(M_2))$ and $C_{xx}^{-1}(0, 0)(Q(M_2)-Q(M_1))N^{-1}$ are stochastically of the order of N^{-1}. Thus the probability of (5.4) will be very nearly equal to 1 in this case. Generally we have, for $M<K$,

$$(5.6)\qquad \lim_{N\to\infty} \text{Prob}\ \{(\text{FPE})(M)-(\text{FPE})(K)>0\}=1\ .$$

By (5.3) and the fact that $S(M)$ is a consistent estimate of σ^2 for $M\geqq K$ we can see that the limit distribution of $N((\text{FPE})(K)-(\text{FPE})(M))$ $(M=K+1, K+2, \cdots, L)$ is identical to that of $N(S(K)-S(M))-2\sigma^2(M-K)$. As it hold that, for $M\geqq K$, $S(K)-S(M)=N^{-1}(Q(M)-Q(K))$, we can see from Theorem 2 that the limit distribution of $N\sigma^{-2}((\text{FPE})(K)-(\text{FPE})(M))+2(M-K)$ is identical to the distribution of the successive sum of $M-K$ chi-square variables $\chi_1^2(i)$ $(i=1, 2, \cdots)$ which are mutually independently distributed each with d.f.1. Thus for this case we have

$$(5.7)\qquad \lim_{N\to\infty} \text{Prob}\ \{(\text{FPE})(K)>(\text{FPE})(M)\} =\text{Prob}\ \{\sum_{i=1}^{M-K} \chi_1^2(i)>2(M-K)\}\ .$$

We can see from (5.6) that by using (FPE)(M) for our minimum FPE procedure the probability of adopting M smaller than K as M_0 will be made arbitrarily small when N is increased indefinitely, while (5.7) shows that for $M>K$ the probability of observing (FPE)(M) small than (FPE)(K) tends to a non-zero constant. This last observation shows that the value M_0 of M adopted by our minimum FPE procedure as the order of the predictor is not a consistent estimate of K. This does not necessarily mean a serious draw back of the procedure for practical applications. The probability itself, as suggested by (5.7), of adopting M_0 larger than K is not necessarily intolerable for practical applications. Further, it will be more common for us to encounter with the situation where the theoretical value of K is considered to be infinity. For this case the result of the foregoing discussion following (5.5) suggests that the probability of M_0 being equal to an M for which $|a_M(m)-a(m)|$ is larger than $N^{-1/2}$ and the corresponding $\sigma^2(M)$ is differing from σ^2 by a quantity greater than $N^{-1/2}R_{xx}(0)$ would be very small. Also (5.5) suggests that in this case (5.7) will hold approximately when the diffference of $\sigma^2(K)$ from σ^2 is made significantly smaller than N^{-1}.

Admittedly our present analysis is quite rough for the range of $M<K$. We will supplement the discussion with numerical examples in Section 8.

The procedure which is obtained by replacing the definition of (FPE)(M) in the minimum FPE procedure by

$$(5.8)\qquad (\mathrm{FPE})^{\alpha}(M)=(1+N^{-\alpha}(M+1))(1-N^{-1}(M+1))^{-1}S(M)\,,$$

where $0<\alpha<1$, will be called the minimum $(\mathrm{FPE})^{\alpha}$ procedure. By this modification we shall certainly obtain the consistency of M_0 as an estimate of K of a finite order autoregressive process. But the modification may add much to the tendency of M_0 taking values too small for the minimization of FPE. In Section 8, the performance of the minimum $(\mathrm{FPE})^{1/4}$ procedure will be compared with that of the original procedure.

6. FPE and Anderson's procedure

T. W. Anderson [2] has given a multiple decision procedure for choosing the order of dependence K in normally distributed time series of the type (3.1). The procedure is such that it is completely specified by, and optimum for, a selection of probabilities p_l $(m<l\leqq q)$ for some preassigned m and q, where p_l is the probability of deciding on the order of dependence to be l when the actual order is less than l. Thus in this procedure we are going to keep small the probabilities $q_l=\sum_{\nu=l}^{q} p_\nu$ $(l=m+1, m+2, \cdots, q)$ of errors of choosing a higher order than necessary. On the other hand, we shall have to keep p_l as large as possible within some allowable limit to maintain the sensitivity of the procedure to non-zero autoregression coefficients. If we evaluate the loss, incurred by adopting a higher order than necessary, by FPE, it would be more natural to control the quantities

$$(6.1)\qquad Q_l=\sum_{\nu=l}^{q}(1+N^{-1}(\nu-l+1))\sigma^2 p_\nu \qquad (l=m+1, m+2, \cdots, q)\,,$$

rather than the probabilities q_l. Obviously Q_{m+1} takes the largest value among Q_l and we decide to pay our attention only to this maximum possible loss. We state the allowable limit of this maximum possible loss relatively to the value of FPE for the order m, i.e., we require Q_{m+1} to be less than or equal to $\rho(1+N^{-1}m)\sigma^2$, where ρ is a small positive quantity such as 0.1 and the like. To keep the sensitivity to a possible non-zero $a(l)$ it is necessary to choose p_l as large as possible, but this also contributes to Q_{m+1} with the corresponding amount of $(1+N^{-1}(l-m))\sigma^2 p_l$. We introduce here the principle of equal harmfulness which states that these losses $(1+N^{-1}(l-m))\sigma^2 p_l$ $(l=m+1, m+2, \cdots, q)$ should all be equal to a positive quantity $\gamma\sigma^2$. By this principle our set of probabilities p_l $(l=m+1, m+2, \cdots, q)$ is determined as follows:

1) Define the allowable relative amount of loss ρ $(<(1+N^{-1}m)^{-1})$.
2) Obtain the value γ by the relation

$$(q-m)\gamma=(1+N^{-1}m)\rho \ . \tag{6.2}$$

3) p_l is given by

$$p_l=(1+N^{-1}(l-m))^{-1}\gamma \qquad (l=m+1,\ m+2,\ \cdots,\ q)\ . \tag{6.3}$$

When we assume that the partial serial correlations, $\hat{a}_M(M)$'s in the formulation of (3.2), are distributed mutually independently and symmetrically around zero when the true order is less than M, the Anderson's procedure is realized by testing the partial serial correlation $\hat{a}_M(M)$ against zero successively for $M=q,\ q-1,\ \cdots,\ m+1$ and taking M_0 equal to the first and the largest M for which $\hat{a}_M(M)$ is decided to be significant. The level of significance β_M and the corresponding critical value δ_M of each test is given by the relations

$$\begin{aligned}&\beta_M=\text{Prob}\ \{|\hat{a}_M(M)|>\delta_M\}\ ,\\ &\beta_q=p_q\ ,\\ &\beta_M=p_M\prod_{l=M+1}^{q}(1-\beta_l)^{-1} \qquad (M=q-1,\ q-2,\ \cdots,\ m+1)\ .\end{aligned} \tag{6.5}$$

If we adopt the approximation that $N|\hat{a}_M(M)|^2$ is distributed as a chi-square variable with d.f.1, δ_M is very simply obtained by using the table of chi-square or Gaussian distribution.

In the following discussion of numerical results we shall exclusively adopt this chi-square approximation along with the constants $\rho=0.1$, $m=0$ and $q=L$.

7. A practical version of the procedures

For practical applications of the three procedures we propose the following modification. Given a set of data $\{X(n);\ n=1,\ 2,\ \cdots,\ N\}$ we replace $\bar{X}_l$ and $C_{xx}(l,\ m)$ in the foregoing description of the procedures by $\bar{X}$ and $C_{xx}(l-m)$, respectively, where by definition

$$\bar{X}=N^{-1}\sum_{n=1}^{N}X(n) \tag{7.1}$$

and

$$C_{xx}(k)=N^{-1}\sum_{n=1}^{N-|k|}(X(n+|k|)-\bar{X})(X(n)-\bar{X})\ . \tag{7.2}$$

By this modification we lose nothing but the relation (4.4) in the preceding discussions. Above all, the result of discussions in Section 5 remains valid and we can expect that the practical usefulness of the original procedures is not affected by this modification.

The modification introduces a great simplification into the computational procedure, especially when we take into account the fact [8, 9] that the computations of (3.2) and (4.1) for $\hat{a}_M(m)$ and $S(M)$ can most easily be carried out by using the recursive relations

$$\begin{aligned}
&\hat{a}_{M+1}(M+1)=(S(M))^{-1}(C_{xx}(M+1)-\sum_{m=1}^{M}\hat{a}_M(m)C_{xx}(M+1-m)) ,\\
(7.3)\quad &\hat{a}_{M+1}(m)=\hat{a}_M(M)-\hat{a}_{M+1}(M+1)\hat{a}_M(M+1-m) \qquad m=1, 2, \cdots, M ,\\
&S(M+1)=S(M)(1-(\hat{a}_{M+1}(M+1))^2) ,
\end{aligned}$$

with the initial values

$$(7.4)\qquad \begin{aligned}&\hat{a}_0(m)=0 ,\\ &S(0)=C_{xx}(0) .\end{aligned}$$

Little difference has been observed between the results obtained by the original and the present versions of the procedures in many applications to artificial time series and the whole numerical results in the following section are obtained by using this practical version.

8. Numerical examples and discussions

Table 1 shows the results of applications of the three procedures, minimum FPE, minimum $(\mathrm{FPE})^{1/4}$ and an Anderson type described in Section 6, with $N=100$ and $L=10$, to ten artificial realizations of the process

$$X(n)=0.3X(n-1)+0.2X(n-2)+0.1X(n-3)+\varepsilon(n) ,$$

where $\varepsilon(n)$'s are mutually independently distributed uniformly over $[-\frac{1}{2}, \frac{1}{2}]$. It can be seen that all the three procedures are showing the tendency of giving M_0 lower than the true order, except the three extreme cases of the Anderson type.

Table 1. Frequency table of adopted order M_0 in ten applications of the three procedures to the process $X(n)=0.3X(n-1)+0.2X(n-2)+0.1X(n-3)+\varepsilon(n)$. $N=100$ and $L=10$.

Adopted order M_0 / Type of procedure	0	1	2	3	10
Anderson with $\rho=0.1$	2	4	1		3
Minimum $(\mathrm{FPE})^{1/4}$	2	3	5		
Minimum (FPE)		2	6	2	

The experiment has exposed the weakness of the Anderson type procedure that, in its present definition, it is not fully protected against adopting extraordinarily large values of M_0. The procedure is also with the difficulty in selecting the value of ρ.

The present results suggest that in spite of its inconsistency, discussed in Section 5, as an estimate of the order of a finite order autoregressive process, M_0 of the minimum FPE procedure will not be giving too large values in practical applications. This point is further backed up by the next example.

We have applied the three procedures with $N=100$ and $L=20$ to the process $X(n)=\varepsilon(n)-0.8\varepsilon(n-1)$, where $\varepsilon(n)$ is as in the former example. In this case $X(n)$ is actually an autoregressive process of infinite order. The orders which gave the estimates $\hat{a}_M$ with the minimum of the one-step prediction error variances in each experiment were identified by numerical computations and are given in the column denoted by " optimum " in Table 2, along with the orderes adopted by the three procedures. The table clearly shows the general tendency of the three procedures giving lower orders than optimum. The differences of the one-step prediction error variances of the optimum predictors and those obtained by the minimum FPE procedure were all relatively small and were at most of the order of 10% of the variance of $\varepsilon(n)$. This shows that for the minimum FPE procedure the present tendency of taking the lower values of orders is not so harmful for prediction.

Table 2. Orders adopted by the three procedures for $X(n)=\varepsilon(n)-0.8\varepsilon(n-1)$ in nine experiments and the corresponding orders which gave the predictors with the minimum one-step prediction error variance in each experiment. $N=100$ and $L=20$.

Type of procedure / Number of the experiment	Anderson $\rho=0.1$	Minimum $(FPE)^{1/4}$	Minimum FPE	Optimum
1	3	3	5	7
2	2	2	6	8
3	3	3	3	6
4	3	3	3	6
5	12	3	3	9
6	1	2	3	6
7	2	2	2	7
8	2	3	4	8
9	1	7	8	6

The results of Tables 1 and 2 both show the wide variability of M_0 of the present Anderson type procedure. Also they show that the tendency of giving lower estimates of orders than optimum is weakest in the minimum FPE procedure. Furthermore, there is no arbitrariness in the definition of the minimum FPE procedure, such as ρ in the Anderson's and α in the minimum $(\text{FPE})^{\alpha}$, except the only one constant L which was common to all the three procedures. These observations suggest that the original minimum FPE procedure would be the most useful for practical applications.

To give a feeling of the behavior of $(\text{FPE})(M)$, one example is depicted in Fig. 1. The figure illustrates the behavior of $(\text{RFPE})(M) = (\text{FPE})(M)((\text{FPE})(0))^{-1}$ for one realization of $X(n)=0.8X(n-1)+\varepsilon(n)$ with $N=100$ and $L=20$. $\varepsilon(n)$ was the same as in the former examples. In practical applications of the minimum FPE procedure to real data, $(\text{RFPE})(M)$ $(M=M_0, M_0+1, \cdots, L)$ has shown a similar behavior to that of Fig. 1 for $M=1, 2, \cdots, L$.

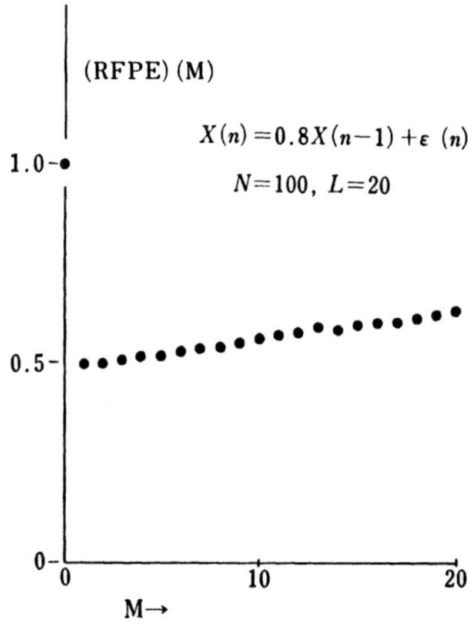

Fig. 1. Behavior of $(\text{RFPE})(M)=(\text{FPE})(M)((\text{FPE})(0))^{-1}$ for a realization of $X(n)=0.8X(n-1)+\varepsilon(n)$.

From the experiences of application to real and artificial data it seems that to set L nearly equal to $0.1N$ would be a reasonable choice for ordinary size of N. Some numerical results of application of the minimum FPE procedure to real data are to be seen elsewhere [3, 6].

THE INSTITUTE OF STATISTICAL MATHEMATICS

REFERENCES

[1] Akaike, H. (1969). Fitting autoregressive models for prediction, *Ann. Inst. Statist. Math.*, **21**, 243-247.

[2] Anderson, T. W. (1963). Determination of the order of dependence in normally distributed time series, *Time Series Analysis* (ed. M. Rosenblatt), New York, John Wiley, 425-446.

[3] Akaike, H. (1970). On a semi-automatic power spectrum estimation procedure, *Proc. 3rd Hawaii International Conference on System Sciences*, 974-977.

[4] Diananda, P. H. (1953). Some probability limit theorems with statistical applications, *Proc. Cambridge Philos. Soc.*, **49**, 239-246.

[5] Anderson, T. W. and Walker, A. M. (1964). On the asymptotic distribution of the autocorrelations of a sample from a linear stochastic process, *Ann. Math. Statist.*, **35**, 1296-1303.

[6] Akaike, H. (1969). Power spectrum estimation through autoregressive model fitting, *Ann. Inst. Statist. Math.*, **21**, 407-419.

[7] Mann, H. B. and Wald, A. (1943). On stochastic limit and order relationships, *Ann. Math. Statist.*, **14**, 217-226.

[8] Durbin, J. (1960). The fitting of time-series models, *Rev. Int. Inst. Stat.*, **28**, 233-244.

[9] Jones, R. H. (1964). Prediction of multivariate time series, *J. of Applied Meteorology*, **3**, 285-289.

Reprinted from *Annals of the Institute of Statistical Mathematics*, Vol. 23, No. 2, 1971, 163-180 by permission of the Institute of Statistical Mathematics

AUTOREGRESSIVE MODEL FITTING FOR CONTROL

HIROTUGU AKAIKE

(Received Dec. 25, 1970)

Summary

The use of a multidimensional extension of the minimum final prediction error (FPE) criterion which was originally developed for the decision of the order of one-dimensional autoregressive process [1] is discussed from the standpoint of controller design. It is shown by numerical examples that the criterion will also be useful for the decision of inclusion or exclusion of a variable into the model. Practical utility of the procedure was verified in the real controller design process of cement rotary kilns.

Introduction

In a practical situation of the autoregressive model fitting the order of the model is not generally known. The order may not be finite and rather will only be an artificial variable for the purpose of developping an approximation to the real world. Thus the decision of the order forms a crucial point in the autoregressive model fitting and any fitting procedure which lacks a description of this point may not be considered to be very efficient for practical applications. The situation is the same for any fitting procedure of finite parameter models and remains at present as a most challenging subject of study which deeply concerns with the practical utility of the whole statistical theories.

The natural approach to this type of problem will be to estimate the possible risk of fitting each model and adopt the one which gives the minimum of the estimates. The main problem in this approach is the choice of the risk function. In recent papers by the present author [1], [2] it has been demonstrated that the use of FPE (final prediction error), which is the mean square one-step prediction error when the set of the fitted coefficients is applied to another independent realization of the process, produces quite reasonable results. This fact is also confirmed by the results of applications of the procedure to many practical data.

Although, from the standpoint of evaluation of the linear transformation of the process under quadratic error criterion, the ultimate purpose of the model fitting may always be considered to be the identification of the spectral characteristics of the process the controller design of the process poses various interesting problems for the model fitting. As will be described briefly in the next section, the autoregressive representation of a multivariate stationary process can directly serve as a starting point of the controller design.

In a real situation of controlling a complex and noisy system another very important problem is the decision on the inclusion or exclusion of a variable into the model. Thus, besides the decision of the order, a procedure must be developed for the decision of the dimensionality of the multivariate vector process.

Extensions of the definition of FPE to multidimensional case have been proposed in [3] and the use of the estimate of generalized variance of the one step prediction error is suggested. Taking into account that this is the only quantity which appears in the maximum likelihood in Gaussian case it seems that this choice is a reasonable one for the purpose of the determination of the spectral characteristics of the process. In the case of controller design we are only interested in the predictability by the model of the system output, since the system input will eventually be under our complete control. This consideration suggests the replacement of the generalized variance by that of the system output variables. This is the procedure to be proposed in the present paper and this definition of the criterion naturally suggests an extension of its use for the purpose of inclusion and exclusion of the input variables. Once the extension of the decision procedure to the selection of input variables is admitted the procedure can also be utilized for the decision of the inclusion and exclusion of the output variables. Naturally the procedure of selecting variables excludes a complete theoretical analysis at present and it must be used with a sound scientific reasoning. The utility of the procedure is illustrated by using real and artificial data.

1. Use of autoregressive models for controller design

As a preliminary for the discussions in the following sections the use of the multivariate autoregressive model for the controller design under the quadratic criterion is briefly described in this section. We assume that the r-dimensional vector of the system output variables and the l-dimensional vector of the system input variables at time n are represented by x_n and y_n, respectively. We will call the components of x_n controlled variables and those of y_n manipulated variables. The $(r+l)$-dimensional vector X_n is defined by

$$(1.1)\qquad X_n = \begin{bmatrix} x_n \\ y_n \end{bmatrix} \begin{matrix} \updownarrow r \\ \updownarrow l \end{matrix} .$$

For the sake of simplicity it is assumed that x_n has a zero mean vector. We assume that X_n admits the following autoregressive representation:

$$(1.2)\qquad X_n = \sum_{m=1}^{M} A_m X_{n-m} + U_n ,$$

where A_m is an $(r+l)\times(r+l)$ matrix and U_n is a random $(r+l)\times 1$ vector satisfying the relations

$$(1.3)\qquad \begin{aligned} &EU_n = 0 \qquad \text{(zero vector)} \\ &EU_n X'_{n-m} = 0 \ \text{(zero matrix)} \qquad \text{for } m \geqq 1 \\ &EU_n U'_m = \delta_{nm} S , \end{aligned}$$

where $\delta_{nm}=1$ $(n=m)$, $=0$ $(n\neq m)$ and S is a positive definite $(r+l)\times(r+l)$ matrix and the symbol $'$ denotes the transpose. From the representation (1.2) we get the following representation of controlled variables, which will be used for the controller design:

$$(1.4)\qquad x_n = \sum_{m=1}^{M} a_m x_{n-m} + \sum_{m=1}^{M} b_m y_{n-m} + w_n ,$$

where a_m, b_m and w_n are given by the relations

$$(1.5)\qquad \begin{aligned} A_m &= \begin{matrix} \updownarrow r \\ \updownarrow l \end{matrix} \overset{\leftarrow r \rightarrow \ \leftarrow l \rightarrow}{\begin{bmatrix} a_m & b_m \\ * & * \end{bmatrix}} \\ U_n &= \begin{matrix} \updownarrow r \\ \updownarrow l \end{matrix} \overset{\leftarrow 1 \rightarrow}{\begin{bmatrix} w_n \\ * \end{bmatrix}} \end{aligned}$$

where $*$ denotes the irrelevant quantities for the present representation.

For the purpose of controller design, (1.4) is transformed into the following state space representation [10]:

$$(1.6)\qquad Z_n = \Phi Z_{n-1} + \Gamma Y_{n-1} + W_n ,$$

where

$$Z_n = \begin{matrix} \updownarrow r \\ \updownarrow r \\ \\ \\ \updownarrow r \end{matrix} \overset{\leftarrow 1 \rightarrow}{\begin{bmatrix} z_n^{(1)} \\ z_n^{(2)} \\ \vdots \\ z_n^{(M)} \end{bmatrix}}$$

$$\Phi=\begin{bmatrix} a_1 & I & 0 & \cdots & 0 \\ a_2 & 0 & I & \cdots & 0 \\ \vdots & \vdots & \vdots & \ddots & \vdots \\ a_{M-1} & 0 & 0 & \cdots & I \\ a_M & 0 & 0 & \cdots & 0 \end{bmatrix}$$

(1.7)
$$\Gamma=\begin{bmatrix} b_1 \\ b_2 \\ \vdots \\ b_M \end{bmatrix}$$

$$Y_{n-1}= y_{n-1}$$

$$W_n=\begin{bmatrix} w_n \\ 0 \\ \vdots \\ 0 \end{bmatrix}$$

and

$$z_n^{(1)}= x_n \ .$$

A simple controller design under quadratic performance criterion proceeds as follows: We assume that the performance of the control is evaluated by the quantity

$$J_T=E\left\{\sum_{n=1}^{T}(Z_n'QZ_n+Y_{n-1}'RY_{n-1})\right\} , \tag{1.8}$$

where Q and R are positive definite matrices of $Mr\times Mr$ and $l\times l$ dimensions and T is a properly chosen large integer and the control input Y_n is chosen so as to make J_T minimum. Obviously J_T admits the representation

$$\begin{aligned} J_T=EW_T'QW_T+E\{&Z_{T-1}'(\Phi'Q\Phi')Z_{T-1}+Y_{T-1}'\Gamma'Q\Phi Z_{T-1} \\ &+Z_{T-1}'\Phi'Q\Gamma Y_{T-1}+Y_{T-1}'(\Gamma'Q\Gamma+R)Y_{T-1}\}+J_{T-1} , \end{aligned} \tag{1.9}$$

and it can be seen that the optimal control Y_{T-1} is given by

$$Y_{T-1}=G_{T-1}Z_{T-1} , \tag{1.10}$$

where

$$G_{T-1}=-(\Gamma'Q\Gamma+R)^{-1}\Gamma'Q\Phi \ .$$

By inserting the result of (1.10) into (1.9) and successively applying Bellman's optimality principle we get

$$Y_n = G_n Z_n \,, \tag{1.11}$$

where $G_n = -(\Gamma' P_{T-n} \Gamma + R)^{-1} \Gamma' P_{T-n} \Phi$ and P_{T-n} is given by the recursive relation

$$P_1 = Q$$

$$\left. \begin{aligned} M_i &= P_{i-1} - P_{i-1}\Gamma(\Gamma' P_{i-1}\Gamma + R)^{-1}\Gamma' P_{i-1} \\ P_i &= \Phi' M_i \Phi + Q \end{aligned} \right\} \quad i = 2, 3, \cdots, T \,. \tag{1.12}$$

When T is sufficiently large G_1 will show little change for the further increase of T and for the control of the stationary process this G_1 is used as the fixed controller gain G and the control is realized by

$$Y_n = G Z_n \,. \tag{1.13}$$

Taking into account the fact that under fairly general condition the stationary process admits an autoregressive representation (1.2) [6], the discussion of this section can be considered to have given a theoretical justification for the controller design based on the model of (1.4) to be a generally useful procedure.

2. Fitting autoregressive models with FPE

In the following discussions we shall adopt the convention to denote by $X(i)$ the ith element of a vector X and by $A(i, j)$ the (i, j)th element of a matrix A.

We assume the model (1.2) of autoregression and consider the case where U_n's are independently identically distributed and X_n is stationary with finite second order moments. Under the assumption of finiteness of all order moments of U_n it was shown by Mann and Wald [5] that distribution of the least squares estimate $\hat{A}_m$ of A_m based on a set of observations $\{X_n;\ n = -M+1, -M+2, \cdots, N\}$ tends to be Gaussian, i.e., the distribution of $\sqrt{N}(\hat{A}_m(i, j) - A_m(i, j))$ $(i, j = 1, 2, \cdots, k,\ m = 1, 2 \cdots, M)$, where $k = r + l$, tends to be Gaussian with a zero mean vector and the covariance corresponding to $EN(\hat{A}_{m_1}(i_1, j_1) - A_{m_1}(i_1, j_1))(\hat{A}_{m_2}(i_2, j_2) - A_{m_2}(i_2, j_2))$ equal to $S(i_1, i_2)R_{xx}^{-1}(m_1, j_1; m_2, j_2)$, where S is the variance matrix of innovations as given by (1.3) and $R_{xx}^{-1}(m_1, j_1; m_2, j_2)$ is the $((m_1-1) \cdot k + j_1, (m_2-1)k + j_2)$th element of the inverse of the matrix R_{xx} which is an $Mk \times Mk$ matrix with the $((m_1-1)k + j_1, (m_2-1)k + j_2)$th element $R_{xx}(m_1, j_1; m_2, j_2)$ equal to $EX_{n-m_1}(j_1)X_{n-m_2}(j_2)$*. Hereafter we denote the ex-

* cf. Proposition in the Appendix.

pectation of a random variable assuming this limiting Gaussian distribution to be exact by E_∞, so that we have

$$E_\infty\{\sqrt{N}(\hat{A}_m(i, j)-A_m(i, j))\}=0$$

and

$$E_\infty\{N(\hat{A}_{m_1}(i_1, j_1)-A_{m_1}(i_1, j_1))(\hat{A}_{m_2}(i_2, j_2)-A_{m_2}(i_2, j_2))\} \\ =S(i_1, i_2)R_{xx}^{-1}(m_1, j_1; m_2, j_2) . \tag{2.1}$$

The one-step prediction error when $\hat{A}_m$ is applied to another independent realization of X_n is given by

$$D_n=\sum_{m=1}^{M}(A_m-\hat{A}_m)X_{n-m}+U_n . \tag{2.2}$$

We take the expectation E_x of $D_n D_n'$ with respect to the realization X_n and get

$$E_x D_n D_n'=S+\sum_{m=1}^{M}\sum_{l=1}^{M}\Delta A_m E X_{n-m}X_{n-l}'(\Delta A_l)' , \tag{2.3}$$

where $\Delta A_m=\hat{A}_m-A_m$. We then consider the expectation of the last term with respect to this limiting distribution. We have

$$E_\infty\left\{N\sum_{m=1}^{M}\sum_{l=1}^{M}\Delta A_m E X_{n-m}X_{n-l}'(\Delta A_l)'\right\}(i, h) \\ =\sum_{m=1}^{M}\sum_{l=1}^{M}\sum_{j=1}^{k}\sum_{g=1}^{k}E_\infty\{N\Delta A_m(i, j)\Delta A_l(h, g)\}R_{xx}(m, j; l, g) ,$$

and by using (2.1)

$$=S(i, h)\sum_{m=1}^{M}\sum_{l=1}^{M}\sum_{j=1}^{k}\sum_{g=1}^{k}R_{xx}^{-1}(m, j; l, g)R_{xx}(m, j; l, g)$$

$$=MkS(i, h) . \tag{2.4}$$

Thus we have

$$E_\infty E_x D_N D_N'=\left(1+\frac{Mk}{N}\right)S , \tag{2.5}$$

and

$$\|E_\infty E_x D_N D_N'\|=\left(1+\frac{Mk}{N}\right)^k\|S\| , \tag{2.6}$$

where $\|A\|$ denotes the determinant of a matrix A. We shall call this last quantity given by (2.6) MFPE (multiple final prediction error) which coincides with our definition of FPE when $k=1$.

The meaning of MFPE is intuitively clear but it can not enjoy the unique status of FPE in one-dimensional case as an index of of the prediction error variance. For the adoption of MFPE as our criterion of the autoregressive model fitting we need a further justification. Since we are interested in the discrimination of the fitted models it will be reasonable to take into account the fact that when the process X_n is Gaussian the only sample function entering into the maximized asymptotic likelihood is $\|\hat{S}_M\|$ [8], where

$$\hat{S}_M=\frac{1}{N}\sum_{n=1}^{N}\left(X_n-\sum_{m=1}^{M}\hat{A}_m X_{n-m}\right)\left(X_n-\sum_{l=1}^{M}\hat{A}_l X_{n-l}\right)' . \tag{2.7}$$

Since $\hat{A}_m$ is obtained by minimizing the trace of $\hat{S}_M$ in (2.7), $\varDelta A_m=\hat{A}_m-A_m$ is minimizing the trace of

$$\hat{S}_M=\frac{1}{N}\sum_{n=1}^{N}\left(U_n-\sum_{m=1}^{M}\varDelta A_m X_{n-m}\right)\left(U_n-\sum_{l=1}^{M}\varDelta A_l X_{n-l}\right)' \tag{2.8}$$

and it must hold that

$$\frac{1}{N}\sum_{n=1}^{N}\left(U_n-\sum_{m=1}^{M}\varDelta A_m X_{n-m}\right)X'_{n-l}=0 . \tag{2.9}$$

From this relation we can get

$$\hat{S}_M=\frac{1}{N}\sum_{n=1}^{N}U_nU'_n-\sum_{l=1}^{M}\sum_{m=1}^{M}\varDelta A_m\frac{1}{N}\sum_{n=1}^{N}X_{n-m}X'_{n-l}(\varDelta A_l)' . \tag{2.10}$$

We have

$$E\left(\frac{1}{N}\sum_{n=1}^{N}U_nU'_n\right)=S , \tag{2.11}$$

and similarly as in (2.5)

$$E_\infty\left\{N\sum_{l=1}^{M}\sum_{m=1}^{M}\varDelta A_m\frac{1}{N}\sum_{n=1}^{N}X_{n-m}X'_{n-l}(\varDelta A_l)'\right\}=MkS . \tag{2.12}$$

where as was defined before E_∞ denotes the expectation of the limit distribution of the quantity within the braces as N tends to infinity. This observation suggests that for the present discussion where the quantity of the order of $1/N$ is playing crucial role it will be reasonable to adopt $(1-Mk/N)^{-1}\hat{S}_M$ as our estimate of $\hat{S}_M$ and accordingly to adopt

$$\left(1-\frac{Mk}{N}\right)^{-k}\|\hat{S}_M\| \tag{2.13}$$

as our estimate of $\|S\|$.

Based on these observations we propose the following procedure for autoregressive model fitting: We fit models with order M $(0 \leqq M \leqq L)$ by using the least square method and adopt the one which gives the minimum of

$$\text{MFPE}(M) = \left(1+\frac{Mk}{N}\right)^k \left(1-\frac{Mk}{N}\right)^{-k} \|\hat{S}_M\| \tag{2.14}$$

as our final choice.

3. System identification for control

The procedure described in the preceding section may suffer some lack of efficiency by assuming one and the same value of M for all the components, though this allows us to fully enjoy the efficient computing procedure developed by Whittle [9]. From the stand point of maximizing the likelihood function in Gaussian case it will be reasonable to focus our attension to the behavior of the estimate of the prediction error variance of any subset of the component variables of X_n when the corresponding components of U_n are independent from the rest. As was discussed in Section 1, in the case of controller design we need only the estimates of the rows of A_m which are giving the system outputs or the controlled variables. In many practical situations, if only sufficient number of variables are taken into account, it is quite possible that the disturbances originating in the controller or the manipulated variables are independent of those originating in the system, in the sense that the set of the components of U_n corresponding to the controlled variables are independent of those corresponding to the manipulated variables. In this case it will be most practical and useful to concentrate our attention to the subset of the controlled variables and replace the definition of MFPE(M) by

$$\text{FPEC}(M) = \left(1+\frac{Mk}{N}\right)^r \left(1-\frac{Mk}{N}\right)^{-r} \|\hat{S}_{r,M}\|, \tag{3.1}$$

where $\hat{S}_{r,M}$ is the $r\times r$ submatrix of $\hat{S}_M$ with the rows and columns corresponding to the controlled varibales. FPEC stands for final prediction error of the controlled variables. By the formulation of Section 1, $\hat{S}_{r,M}$ is the $r\times r$ submatrix in the upper left hand corner of $\hat{S}_M$.

By introducing FPEC(M) our model fitting procedure for the controller design will be realized by replacing MFPE(M) in the case of general autoregressive model fitting by FPEC(M) and by adopting $\hat{a}_m$, $\hat{b}_m$ which are obtained from $\hat{A}_m$ and the definition (1.5) as the coefficients of the model and $\hat{S}_{r,M}$, or $(1-Mr/N)^{-1}\hat{S}_{r,M}$, as an estimate of the variance matrix of the innovations w_n within the system. This procedure will guard us against adopting inadequately large or small values of M for the system due to the effect of the structure of manipulated variables.

It should be mentioned here that by following the line of the discussion of Section 5 of [1] it is possible to show that for $K \leqq M_1 \leqq M_2 \leqq L$, where K is the order of the autoregressive process, i.e., $A_K \neq 0$ and $A_m = 0$ for $m > K$, $N \log_e (\|\hat{S}_{M_1}\| / \|\hat{S}_{M_2}\|)$ is asymptotically distributed as the sum $\sum_{j=M_1+1}^{M_2} \chi_j^2$ of mutually independently distributed chi-square variables χ_j^2 each with k^2 degrees of freedom. Analogously $N \log_e (\|\hat{S}_{r,M_1}\| / \|\hat{S}_{r,M_2}\|)$ is asymptotically distributed as the sum $\sum_{j=M_1+1}^{M_2} \chi_j'^2$ of mutually independently distributed chi-square variables $\chi_j'^2$ each with kr degrees of freedom for $K_r \leqq M_1 < M_2 \leqq L$, where K_r is such that $a_m = 0$ for $m > K_r$ and $\neq 0$ for $m = K_r$. A proof of this fact will be given in Appendix. Statistical properties of MFPE(M) and FPEC(M) can be deduced from these results as in the one dimensional case. It should be noted that as k or r tends to be large the probabilities of adopting higher values of M_0 tend to be small.

The present minimum FPEC procedure will be a reasonable one in practical applications if only the dependence between the subsets of innovations is not quite significant and this will be the case if there exist lags in the responses of the controlled variables to the variations of the manipulated variables and also there does not exist any hidden variable which is influencing on both the controlled and manipulated variables simultaneously. This consideration suggests the utility of evaluation of independence between the innovations of controlled and manipulated variables, since this will give some indication of possible inadequacy of sampling interval for time sampled observations of a continuous process and/or of possible existence of some hidden influencing variables. For this purpose we propose the use of the statistic

$$\lambda = \frac{\|\hat{S}_M\|}{\|\hat{S}_{r,M}\| \, \|\hat{S}_{l,M}\|}, \tag{3.2}$$

where $\hat{S}_{l,M}$ denotes the $l \times l$ matrix in the lower right hand corner of $\hat{S}_M$. When it is desired to test the significance of λ being smaller than 1, the result by Whittle [8] can be used, which tells that asymptotically

$$\xi = -N \log_e \lambda \tag{3.2}$$

is distributed as a chi-square variable with $r \times l$ degrees of freedom.

Here we consider the problem of inclusion and exclusion of variables.

Given a set of controlled variables we adopt the set of manipulated variables which gives the minimum value of FPEC(M) of (3.1) for the modelling by the present data. Thus even if there is a record of another manipulated variable we do not think it profitable to include it into the present model if its inclusion increase the minimum of FPEC. Exclusion of a manipulated variable can be treated analogously. As to the inclusion and exclusion of a controlled variables we have only to place it in the set of manipulated variables and apply the above stated procedure.

It must be stressed here that this kind of simple and general decision may be allowed and useful only when there does not exists any definite structural information of the process under observation. Any structural information available of the process should be paid carefull attention at the time of modelling. FPEC will give an indication of relative merit of each model with respect to the given set of observation data.

4. A practical computation procedure

In practical applications when a set of data $\{X_n;\ n=1, 2, \cdots, N\}$ is given the sample means of each variables are first deleted and the sample covariance functions are computed by the formulae:

$$(4.1)\quad C(X_{n-m}(j), X_{n-l}(h)) \\ =\frac{1}{N}\sum_{n=1}^{N-m+l}(X_n(j)-\overline{X(j)})(X_{n-l+m}(h)-\overline{X(h)}) \qquad \text{for } m\geqq l \\ =\frac{1}{N}\sum_{n=1}^{N-l+m}(X_{n-m+l}(j)-\overline{X(j)})(X_n(h)-\overline{X(h)}) \qquad \text{for } m\geqq l\,,$$

where

$$\overline{X(j)}=\frac{1}{N}\sum_{n=1}^{N} X_n(j)\,.$$

Instead of solving for the least squares estimate $\hat{A}_m$ described in Section 2 we fit the autoregressive model with the second order moments equal to those given by (4.1) up to the order M. This gives the coefficients of autoregression A_m^M by the efficient recursive computational procedure based on a formula given by Whittle [4], [9]. We define $k\times k$ $(k=r+l)$ matrix C_m, for $m=0, 1, \cdots, L$, by

$$(4.2)\qquad C_m(i, j)=C(X_n(i), X_{n-m}(j))\,.$$

The matrices A_m^M $(m=1, 2, \cdots, M,\ M=1, 2, \cdots, L)$ are obtained recursively by the following relation with initials $A_l^0=B_l^0=0$ (zero matrix),

$d_0=f_0=C_0$ and $e_0=C_1$:

$$A_l^{M+1}=\begin{cases} A_l^M-D^M B_{M+1-l}^M & l=1,2,\cdots,M \\ D^M & l=M+1 \end{cases}$$

(4.3)

$$B_l^{M+1}=\begin{cases} B_l^M-E^M A_{M+1-l}^M & l=1,2,\cdots,M \\ E^M & l=M+1\,, \end{cases}$$

where

(4.4)
$$D^M=e_M f_M^{-1}$$
$$E^M=e_M' d_M^{-1}$$

and

(4.5)
$$d_M=C_0-\sum_{l=1}^{M} A_l^M C_l'$$
$$e_M=C_{M+1}-\sum_{l=1}^{M} A_l^M C_{M+1-l}$$
$$f_M=C_0-\sum_{l=1}^{M} B_l^M C_l\,,$$

where $'$ denotes transpose. We replace $\hat{S}_M$ of the preceding section by d_M and taking into account of the effect of adjusting for the mean we modify the definitions of MFPE and FPEC into

$$\text{MFPE}\,(M)=\left(1+\frac{Mk+1}{N}\right)^k\left(1-\frac{Mk+1}{N}\right)^{-k}\|d_M\|$$

and

$$\text{FPEC}\,(M)=\left(1+\frac{Mk+1}{N}\right)^r\left(1-\frac{Mk+1}{N}\right)^{-r}\|d_{r,M}\|\,,$$

respectively, where $d_{r,M}$ is the $r\times r$ submatrix in the upper left hand corner of d_M.

Taking into account the required degrees of freedom of the related statistics, it seems reasonable to keep L within the limit of $N/(10k)$ or $N/(5k)$.

5. Numerical examples

The decision procedure described in the preceding section was applied to the real data of cement rotary kilns of Chichibu Cement Company, Kumagaya, Japan, by Dr. T. Nakagawa and other members of the company. It was found that the procedure is very useful for the design of actual controllers and is eliminating the tedious trial and error process of identification of the basic model. Technical details of the recent results will be published shortly. We shall here content ourselves by presenting some of the results obtained by applying the procedure to a set of old data of which analysis was reported in [7]. The data were obtained while the kiln was under human control. By the result of the former analysis three variables were selected as controlled variables, which will be represented by $x_n(1)$, $x_n(2)$ and $x_n(3)$, and four variables $y_n(1)$, $y_n(2)$, $y_n(3)$ and $y_n(4)$ were identified as manipulated variables. For various combinations of variables the values of the minimum FPEC (M) are given in the Table 1 along with the values M_0 of M which gave the

Table 1. Application to kiln data

Controlled variable	Manipulated variable	M_0	FPEC $(M_0)\times 10^{-4}$
$x_n(1)$, $x_n(2)$, $x_n(3)$		6	2.22927
$x_n(1)$, $x_n(2)$, $x_n(3)$	$y_n(1)$	6	2.23594
$x_n(1)$, $x_n(2)$, $x_n(3)$	$y_n(2)$	7	2.27954
$x_n(1)$, $x_n(2)$, $x_n(3)$	$y_n(3)$	6	2.15254
$x_n(1)$, $x_n(2)$, $x_n(3)$	$y_n(4)$	6	2.17707
$x_n(1)$, $x_n(2)$, $x_n(3)$	$y_n(3)$, $y_n(1)$	6	2.10345
$x_n(1)$, $x_n(2)$, $x_n(3)$	$y_n(3)$, $y_n(2)$	6	2.20113
$x_n(1)$, $x_n(2)$, $x_n(3)$	$y_n(3)$, $y_n(4)$	6	2.08433
$x_n(1)$, $x_n(2)$, $x_n(3)$	$y_n(3)$, $y_n(4)$, $y_n(1)$	6	2.05687
$x_n(1)$, $x_n(2)$, $x_n(3)$	$y_n(3)$, $y_n(4)$, $y_n(2)$	6	2.13427
$x_n(1)$, $x_n(2)$, $x_n(3)$	$y_n(3)$, $y_n(4)$, $y_n(1)$, $y_n(2)$	6	2.09615

minima. The length of observation was $N=511$ and L was set equal to 15. The results in Table 1 show that the variables $y_n(3)$, $y_n(4)$, $y_n(1)$ and $y_n(2)$ are contributing to reduce the FPEC (M_0) in this order and that the inclusion of $y_n(2)$ into the model may not be profitable. Later analysis has shown that this result may be due to the very noisy behavior of $y_n(2)$ at the time of the experiment and the data provided by the later experiments under more natural running conditions have proven the inclusion of $y_n(2)$ necessary. The result is also considered to be due to the fact that the effect of manipulating $y_n(2)$ tends to be significantly non-linear when the system is too noisy or abnormal.

The present use of FPEC may find application in many other situations, such as the case of cross-spectrum estimation by autoregressive model fitting. In this case if uncorrelated observations are included into the fitting procedure this will introduce some bias into the values of M_0, the value of M which gives the minimum of FPEC (M) $(M=0, 1, \cdots, L)$. In the case of the example of Table 1 the value of M_0 which gave the minimum of MFPE for the total seven variables was equal to 4 contrary to 6 giving the minimum of FPEC. Thus a reasonable procedure may be to watch the behaviour or FPEC (M_0) at each inclusion of new variables into the analysis and divide the variables into small number of groups and adopt a different value of M_0 for each group. To show the feasibility of this type of procedure a numerical experiment was performed on the variables $x_n(i)$ $(i=1, 2, 3, 4)$, where $x_n(1)$ and $x_n(2)$ were given by the relations

$$
\begin{aligned}
&x_n(1)=0.5x_{n-1}(1)+0.4x_{n-1}(2)+v_n(1) \\
&x_n(2)=-0.6x_{n-1}(1)+0.7x_{n-1}(2)+0.3x_{n-2}(1)+0.2x_{n-2}(2)+v_n(2)
\end{aligned}
\tag{5.1}
$$

and

$$
\begin{aligned}
&v_n(1)=u_n(1)+0.4u_n(2) \\
&v_n(2)=u_n(2)+0.4u_n(1) ,
\end{aligned}
\tag{5.2}
$$

where $u_n(1)$ and $u_n(2)$ are the sequences of mutually independent random numbers both uniformly distributed over $[-0.5, 0.5]$, and $x_n(3)$ and $x_n(4)$ were realizations of physical processes which were mutually independent and also independent from $x_n(1)$ and $x_n(2)$. The results of application of our procedure are given in Table 2. For the sake of simplicity we denoted the variables for which we have evaluated FPEC (M) as controlled variables and other variables taken into consideration as manipulated variables. FPEC (M) is equal to MFPE (M) when the manipulated variables are absent. The values $L=15$ and $N=500$ were adopted. The

Table 2. Application to artificial data

Controlled variable	Manipulated variable	M_0	FPEC (M_0)	ξ (d.f.)
$x_n(1)$		3	0.558057	
$x_n(1)$	$x_n(2)$	1	0.094309	2.3×10^2 (1)
$x_n(1)$, $x_n(2)$		2	0.004971	
$x_n(1)$, $x_n(2)$	$x_n(3)$	2	0.005031	
$x_n(1)$, $x_n(2)$	$x_n(3)$, $x_n(4)$	2	0.005079	2.3 (4)
$x_n(1)$, $x_n(2)$, $x_n(3)$, $x_n(3)$		4	0.000672	
$x_n(3)$, $x_n(4)$		13	0.003382	

results show that the inclusion of $x_n(2)$ into the analysis along with $x_n(1)$ is profitable but the inclusion of $x_n(3)$ and $x_n(4)$ is unprofitable. Also the inclusion of $x_n(3)$ or $x_n(3)$ and $x_n(4)$ causes an unnecessary increase of M for $x_1(n)$ and $x_2(n)$ and decrease for $x_3(n)$ and $x_4(n)$. Though the almost perfect identifications of orders of the model is partly by coincidence the results give a very clear illustration of the possible use of the procedure discussed in the preceding paragraph. Two of the values of the chi-square variable ξ defined by (3.2) are also illustrated in Table 2 with corresponding degrees of freedom within the parentheses. The values very clearly show the dependence and independence of innovations in the two cases and suggest the utility of this statistic.

6. Concluding remarks

The minimum FPEC procedure described in this paper will be valid even if the manipulate variables $y_n(i)$ contain deterministic components if only it is assured that the sample variance and covariance matrices $C(X_{n-m_1}(j_1), X_{n-m_2}(j_2))$ tend to some definite constant matrices which can replace $EX_{n-m_1}(j_1)X_{n-m_2}(j_2)$ in the definition of R_{xx} of Section 2, which is assumed to be non-singular for any choice of M. In this way the procedure may find an application in the field of econometric model building. It must be born in mind that generally a manipulated variable under completely noise-free linear feedback control is not allowed in the present identification procedure. If it is considered to be desirable to adopt different values of M for each controlled variables we can apply the procedure by assuming each controlled variable in turn as the only output of the system and assuming all the variables except this one as manipulated variables in the definition of FPEC. The use of this procedure has been discussed in [3]. This is equivalent to ignoring the effect of possible correlations within the components of innovations. There are obvious modifications of the procedure to allow finer specifications of the system equations, but except for some special situations the procedure described in the present paper will serve as the most practical and generally useful one for the identification of a system for the purpose of controller design.

Acknowledgements

The result reported in this paper is a direct consequence of the cooperative research of cement rotary kilns by Dr. T. Nakagawa and the present author. Necessary programmings for the computations were performed by Miss E. Arahata.

Appendix

A derivation of the asymptotic distribution of statistics

$$\kappa_M = -N \log_e \| W_r^{-1} \hat{S}_{r,M} \| , \tag{A.1}$$

where

$$W_r = \frac{1}{N} \sum_{n=1}^{N} w_n w_n' \tag{A.2}$$

and w_n is given by (1.5), is presented here. The distribution of these statistics forms a theoretical background of the minimum FPEC procedure proposed in the text. We shall hereafter denote the operation $\frac{1}{N}\sum_{n=1}^{N}$ by $\overline{}$. We sometimes use the notation

$$c_m = \overset{\updownarrow r}{} [\overset{\leftarrow r \rightarrow}{a_m}, \overset{\leftarrow l \rightarrow}{b_m}] \tag{A.3}$$

and also recall the definition $X_n' = [x_n' y_n']$.

We have

$$\begin{aligned} \hat{S}_{r,M} &= \overline{x_n x_n'} - \sum_{m=1}^{M} \hat{c}_m \overline{X_{n-m} x_n'} \\ &= H(M) - N^{-1} Q(M) , \end{aligned} \tag{A.4}$$

where

$$H(M) = \overline{x_n x_n'} - \sum_{m=1}^{M} (c_m \overline{X_{n-m} x_n'} + \overline{x_n X_{n-m}'} c_m') + \sum_{m=1}^{M} \sum_{l=1}^{M} c_m \overline{X_{n-m} X_{n-l}'} c_l' \tag{A.5}$$

and

$$Q(M) = N \sum_{m=1}^{M} \sum_{l=1}^{M} \Delta c_m \overline{X_{n-m} X_{n-l}'} (\Delta c_l)' , \tag{A.6}$$

where $\Delta c_m = \hat{c}_m - c_m$ and $\hat{c}_m$ is the least squares estimate of c_m. We assume $c_m = 0$ for $m \geqq K_r$ and get

$$\begin{aligned} H(M) &= \overline{w_n w_n'} \\ &= W_r \qquad \text{for } M \geqq K_r . \end{aligned} \tag{A.7}$$

We also have for $M \geqq K_r$

$$[\Delta c_1, \Delta c_2, \cdots, \Delta c_M] = [\overline{w_n X_{n-1}'}, \overline{w_n X_{n-2}'}, \cdots, \overline{w_n X_{n-M}'}] C_{xx}^{-1} , \tag{A.8}$$

where C_{xx} is an $Mk \times Mk$ matrix with

$$C_{xx}((l-1)k+i, (m-1)k+j) = \overline{X_{n-l}(i) X_{n-m}(j)} . \tag{A.9}$$

For this case $Q(M)$ can be represented in the form

(A.10) $$Q(M)=N[\overline{wX'}]C_{xx}^{-1}[\overline{wX'}]',$$

where

(A.11) $$X'=[X'_{n-1}, X'_{n-2}, \cdots, X'_{n-M}]$$ and $$[\overline{wX'}]=[\overline{w_nX'_{n-1}}, \cdots, \overline{w_nX'_{n-M}}].$$

The whole discussion will depend on the following

PROPOSITION (Mann and Wald [5]). The limit distribution of $\sqrt{N}[\overline{wX'}]\cdot(i, (l-1)k+j)$ $(i=1, 2, \cdots, r;\ l=1, 2, \cdots, M;\ j=1, 2, \cdots, k)$ is Gaussian with a zero mean vector and with the variance covariance matrix given by the relation

(A.12) $$E_\infty N([\overline{wX'}](i, \cdot)'([\overline{wX'}](j, \cdot))=S(i, j)R_{xx},$$

where $[\ ](i, \cdot)$ denotes the ith row of the matrix $[\ ]$ and $R_{xx}=EXX'$.

Since we have for $M\geqq K_r$

(A.13) $$\|W_r^{-1}\hat{S}_{r,M}\|=\|I-N^{-1}W_r^{-1/2}Q(M)W_r^{-1/2}\|,$$

we have

(A.14) $$-N\log_e\|W_r^{-1}\hat{S}_{r,M}\|=\mathrm{Trace}\,(W_r^{-1/2}Q(M)W_r^{-1/2})+o_p(1),$$

where $o_p(1)$ denotes a term which is stochastically vanishing as N tends to infinity. Thus we have only to concern ourselves with the limiting distribution of $\mathrm{Trace}\,(W_r^{-1/2}Q(M)W_r^{-1/2})$. The limit distribution of this quantity is identical if we replace C_{xx} in the definition of $Q(M)$ by its theoretical value R_{xx} and W_r by S_r which is the $r\times r$ matrix at the upper left corner of S. We know that there is a special orthogonalization procedure of $X_{n-1}, X_{n-2}, \cdots, X_{n-M}$, which is implicitly described in Section 4 and, by using the same notation for the corresponding quantities, is realized in the form

(A.15) $$Z_{n-l}=X_{n-l}-\sum_{m=1}^{l-1}B_m^{l-1}X_{n-l+m} \qquad l=1, 2, \cdots, M.$$

We further orthonormalize the components within Z_{n-l} by a transformation

(A.16) $$V_l=O_lZ_{n-l},$$

where O_l is a $k\times k$ matrix. Combining these transformations we get a transformation T_M

(A.17) $$V = T_M X\,,$$

where $V' = [V_1', V_2', \cdots, V_M']$ and T_M is an $M \times M$ block matrix of $k \times k$ matrices, of which superdiagonal triangular part is filled with $k \times k$ zero matrices. From the structure of T_M we have

(A.18) $$\begin{aligned} EVV' &= I \\ &= T_M R_{xx} T_M'\,, \end{aligned}$$

where I denotes an $Mk \times Mk$ identity matrix, and $Q(M)$ can be replaced by

(A.19) $$Q(M) = N[\overline{wV'}][\overline{wV'}]'\,,$$

where it holds that

(A.20) $$E_\infty N([\overline{wV'}])'(i, \cdot)[\overline{wV'}](j, \cdot) = S_r(i, j)I\,.$$

This shows that in the limit distribution the columns of $\sqrt{N}[\overline{wV'}]$ are distributed independently with covariance matrices equal to S_r and consequently the components of the $r \times Mk$ matrix $\sqrt{N} S_r^{-1/2}[\overline{wV'}]$ are asymptotically distributed mutually independently as Gaussian with zero means and unit variances. Since W_r converges to S_r with probability one as N tends to infinity we can see that the limit distribution of Trace $(W_r^{-1/2} \cdot Q(M) W_r^{-1/2})$ is identical to that of

(A.21) $$\sum_{l=1}^{M} \sum_{j=1}^{k} \sum_{i=1}^{r} \{\sqrt{N}(S_r^{-1/2}[\overline{wV'}])(i, (l-1)k+j)\}^2\,.$$

It is obvious that T_M is identical to the $Mk \times Mk$ matrix at the left upper corner of T_L for $M \leqq L$ and since $V = T_M X$ we can see that (A.21) is identical to the partial sum

(A.22) $$\sum_{l=1}^{M} \chi_{kr}^2(l) \qquad \text{for } M \leqq L\,,$$

where

$$\chi_{kr}^2(l) = \sum_{j=1}^{k} \sum_{i=1}^{r} \{\sqrt{N}(S_r^{-1/2}[\overline{wV'}])(i, (l-1)k+j)\}^2$$

$(l = 1, 2, \cdots, L)$ are defined by putting $M = L$ in (A.21) and are asymptotically mutually independently distributed as chi-square variables with d.f. kr. This gives the asymptotic property of the statistic κ_M defined by (A.1) and completes the proof of the statement made at Section 3 about the asymptotic distribution of $N \log_e(\|\hat{S}_{r,M_1}\| / \|\hat{S}_{r,M_2}\|)$ which is equal to $\kappa_{M_2} - \kappa_{M_1}$ $(k_r \leqq M_1 \leqq M_2 \leqq L)$.

Since we have

$$\text{(A.23)} \quad N\left(\frac{\text{FPEC}(M_2)}{\text{FPEC}(M_1)}-1\right)=2(M_2-M_1)kr-N\log_e\frac{\|\hat{S}_{r,M_1}\|}{\|\hat{S}_{r,M_2}\|}+o_p(1)\ ,$$

we can see from the above result that for $M_2 \geqq M_1 \geqq K_r$ $N(\text{FPEC}(M_2)/\text{FPEC}(M_1)-1)$ will asymptotically behave as a realization of a random walk with variance of each step equal to $2kr$ and with upward drift of amount kr. Thus when kr tends to be significant compared with N the present minimum FPEC procedure may show the tendency to underestimate the order of the process, ignoring the effect of the bias of order kr/N.

THE INSTITUTE OF STATISTICAL MATHEMATICS

REFERENCES

[1] Akaike, H. (1970). Statistical predictor identification, *Ann. Inst. Statist. Math.*, **22**, 203-217.

[2] Akaike, H. (1970). On a semi-automatic power spectrum estimation procedure, *Proc. 3rd Hawaii International Conference on System Sciences*, 974-977.

[3] Akaike, H. (1970). On a decision procedure for system identification, *Preprints, IFAC Kyoto Symposium on System Engineering Approach to Computer Control*, 485-490.

[4] Jones, R. H. (1964). Prediction of multivariate time series, *J. Appl. Meteor.*, **3**, 285-289.

[5] Mann, H. B. and Wald, A. (1943). On the statistical treatment of linear stochastic difference equations, *Econometrica*, **11**, 173-220.

[6] Masani, P. (1966). Recent trends in multivariate prediction theory, *Multivariate Analysis*, (P. R. Krishnaiah ed.), Academic Press, New York, 351-382.

[7] Otomo, T., Nakagawa, T. and Akaike, H. (1969). Implementation of computer control of a cement rotary kiln through data analysis, *Preprints, IFAC 4th Congress, Warszawa*.

[8] Whittle, P. (1953). The analysis of multiple stationary time series, *J. R. Statist. Soc.*, **B 15**, 125-139.

[9] Whittle, P. (1963). On the fitting of multivariate autoregressions, and the approximate canonical factorization of a spectral density matrix, *Biometrika*, **50**, 129-134.

[10] Wong, K. Y., Wiig, K. M. and Allbritton, E. J. (1968). Computer control of the Clarksville cement plant by state space design method, *IEEE Cement Industry Technical Conference, St. Louis, U.S.A., May* 20-24.

Automatica, Vol. 8, pp. 35–48. Pergamon Press, 1972. Printed in Great Britain.

Statistical Approach to Computer Control of Cement Rotary Kilns*

Approche statistique à la commande par calculateur de fours rotatifs à ciment

Statistische Methode der Prozeßregelung von Zementdrehöfen

Статистический подход к управлению, с помощью цифровой вычислительной машины, вращающимися цементными печами

T. OTOMO,† T. NAKAGAWA† and H. AKAIKE‡

Statistical analyses reveal the necessity of the integrated kiln and clinker cooler control of a cement rotary kiln and a practically useful on-line computer control.

Summary—A fully computerized cement rotary kiln process control was tested in a real production line and the results are presented in this paper. The controller design was based on the understanding of the process behavior obtained by careful statistical analyses, and it was realized by using a very efficient statistical identification procedure and the orthodox optimal controller design by the statespace method. All phases of analysis, design and adjustment during the practical application are discussed in detail. Technical impact of the success of the control on the overall kiln installation is also discussed. The computational procedure for the identification is described in an Appendix.

1. INTRODUCTION

THERE are broadly two types of approach to the automation of a cement rotary kiln. The first is the theoretical or analytical approach which bases the control system design on a theoretical model of the kiln process. The paper by PHILLIPS [1] at the Second IFAC Congress describes a typical example of this type of approach. The second is the statistical one which utilizes a statistical model of the process obtained by the data analysis of records of a normally operating process under human control. The paper by WONG *et al.* [2] perhaps provides the most technically advanced description of this type of approach. In a situation like this, it is generally a rule that a final solution is obtained by a proper combination of the two approaches, but, as can be seen from the discussion of the paper by SKULL [3], it seems that the computer control based on a fully theoretical model tends to be unwieldly complex at present. Also the cement rotary kiln process, especially of the long kiln, is known as a notoriously noisy object and, as was discussed by KAISER in a recent review paper [4], it has been considered that the problems in the measurement and identification have precluded complete automation. The purpose of the present paper is to present a case history of analysis, identification, and controller design of a long cement rotary kiln using the statistical approach with successful results in a real production line. The results very clearly show that practically almost complete automation is now possible.

The main difference in the present kiln process controller from those described in the above mentioned papers [1, 2] lies in the fact that it is designed, from the very beginning, for the integrated system of the kiln and the clinker cooler. The necessity of this approach was first made clear by the statistical analysis of the kiln behavior under human control. Based on the understandings of the kiln behavior thus obtained necessary improvements of the kiln control facilities were realized and a set of controlled and manipulated variables were chosen to represent the kiln system. To implement the control a linear model of the kiln process was developed by using a statistical identification procedure. The identification which has hitherto been considered to be a tedious process has become a very simple routine by the introduction of a new statistical decision procedure by one of the present authors [6, 7] and the controller has

* Received 25 July 1969; revised 29 March 1971; revised 26 July 1971. The original version of this paper was presented at the 4th IFAC Congress which was held in Warsaw, Poland during June 1969. It was recommended for publication in revised form by Associate Editor H. A. Spang.

† Chichibu Cement Company, 1-4-6 Marunouchi, Chiyoda-ku, Tokyo, Japan.

‡ The Institute of Statistical Mathematics, 4-6-7 Minami-Azabu, Minato-ku, Tokyo, Japan.

been realized by the orthodox optimal controller design procedure for multivariable linear systems with a quadratic cost criterion.

It must be admitted that the cement rotary kiln process is essentially nonlinear and unstable, but, as will be seen by the numerical results to be presented, it can practically be stablized by a linear controller if sufficient attention is paid only to the very low frequency behavior of the kiln and if abrupt changes of the process condition, such as those caused by the irregularities within the mechanical system or by the dropping of unusally wide areas of coating, do not occur. To compensate for this last type of process irregularity an empirically designed control for the kiln rotation speed has been incorporated into the final form of the controller.

The details of the whole process of realization will successively be described in the following sections and the computational procedure for the identification will be given in Appendix. Since the description is kept as objective as possible, the validity of the analysis and design procedure will easily be checked with a real kiln. The present paper is a completely revised and updated version of a paper presented at the 4th IFAC Congress held in Warszawa from 16–21 June 1969, under the title "Implementation of computer control of a cement rotary kiln through data analysis" [8].

2. DESCRIPTION OF THE KILN PROCESS

Since a very good exposition of cement manufacturing process is already available [4] only a brief description of the kiln process, which is depicted in Fig. 1, will be given in this section, and the symbols and abbreviations shown in the figure are explained in the next section.

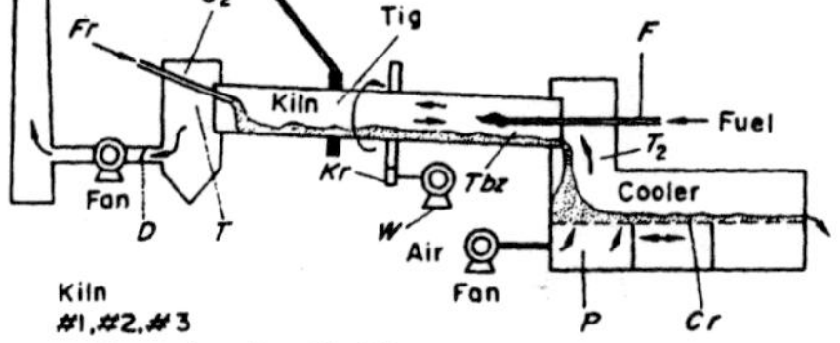

FIG. 1. Rotary kiln and clinker coller system.

The raw materials such as limestone, clay and other mineral products are ground and blended in combination with water, the wet process, to provide the slurry input to the kiln. The blending process is controlled by a computer so that it produces slurry with specified chemical composition. The slurry is fed into the higher end of the kiln and the raw material is moved downwards by the rotation of the kiln. During the travel through the kiln the raw material is first dried, calcined and further heated to reaction temperature to form the clinker after several phases of physical–chemical reactions. The clinker is then quenched and cooled in the clinker cooler and is fed into the grinding mill to form the final product of the cement process. The necessary heat for the reactions within the kiln is supplied by burning fuel at the lower end of the kiln.

Since the reactions are quite complex, theoretical analyses of the kiln process hitherto developed were only useful for the qualitative understanding of the process and were not sufficient to fully describe the dynamic behavior of a real cement kiln. This last point is further stressed by the existence of disturbances within and without the kiln which is so sensitive that even a heavy rain shower can cause a breakdown without a proper control. The speed of the material flow through the kiln must be kept uniform to keep the whole process stationary, but due to the formation of irregular coating and mud rings this can never be expected in a real kiln. The hood draft or the air flow through the kiln should be kept uniform to keep the heat distribution stationary but this can never happen due to the irregular pressure variations within the clinker cooler, which is unavoidable due to the randomness of the travelling speed of material over the cooler grate. Apart from the basic heat and material balance relations, the whole process is definitely stochastic and for the case of a noisy, long kiln, it is necessary to adopt the statistical approach.

3. RESULTS OF SPECTRUM ANALYSIS

The dimensions of the kiln are given in Fig. 1. All the three kilns to be discussed in this paper have the same set of dimensions.

At the very beginning of the data analysis, logging was made every 10 min of fuel rate (*F*), kiln drive power (*W*), exit gas temperature (*T*), intermediate gas temperature (*Tig*), burning zone temperature (*Tbz*), secondary air temperature (*T2*), cooler grate speed (*Cr*), per cent oxygen at the exhaust (0_2), draft damper position (*D*), cooler under great pressure (*P*), kiln speed (*Kr*) and feed rate of material (*Fr*). The measuring points of these quantities are schematically represented in Fig. 1.

Of these quantities *F*, *Cr*, *D* and *Kr* are considered here to be the manipulated variables and others the controlled variables. The kilns were under human control and the records of continuous running for three days or more, which did not contain very abnormal portion of control and were considered to be representing the stationary behavior of the kiln, were used for the analysis. The routine spectrum analysis procedure [5] was applied directly to the data without any filtering for the reduction of the low frequency variation of data.

The power spectra of the controlled variables showed significant peaks around D.C. and the frequency of 1 cycle per 3 hours. A typical example is illustrated in Fig. 2. Cross-spectrum analysis revealed very high coherencies within the variations of controlled variables and rather low coherencies between the variations of manipulated variables and controlled variables. Some of the results are illustrated in Figs. 3 and 4. The multiple coherencies between each one of the controlled variables and the manipulated variables were also computed but they were rather low. Taking into account the fact that the kilns were being operated above their specified levels and the material flow rates were almost at the limits of the cooler capacities, these results of spectrum analysis were considered to be showing the effect of some definite feedback relations within the controlled variables, which were amplifying the above stated significant frequency components within the random disturbances. The results also seemed to suggest the limitation of the performance of human control, but, unfortunately, it was theoretically shown that the clarification of this point was beyond the power of the ordinary cross-spectrum analysis technique [9]. The theoretical analysis of cross-spectral methods by AKAIKE [9] indicated that the results of spectrum analysis of the kiln data were almost useless for the understanding of the feedback system behavior and an entirely new approach was developed [10] and applied to the kiln data. The results are discussed in the next section.

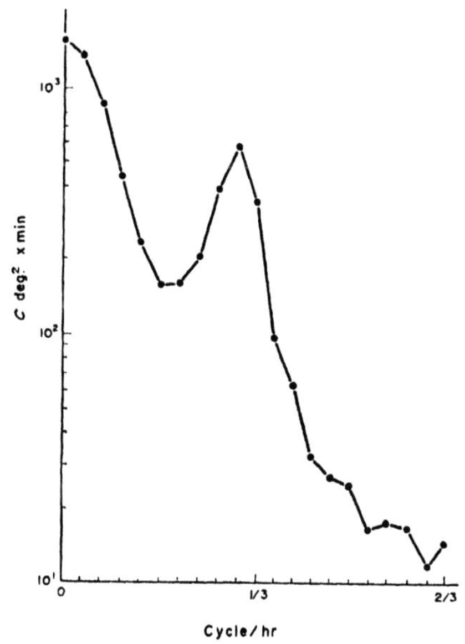

FIG. 2. Power spectrum of exit gas temperature (T).

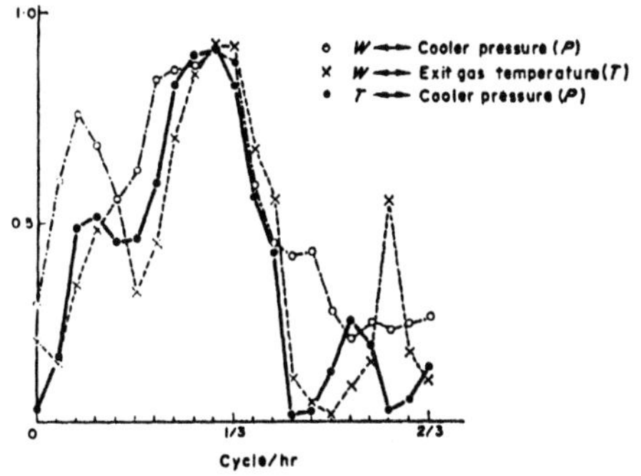

FIG. 3. Coherencies within the controlled variables.

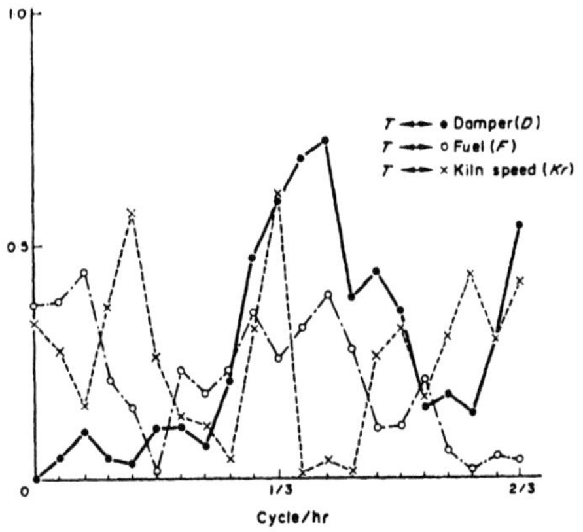

FIG. 4. Coherencies between exit gas temperature (T) and manipulated variables.

4. A PROCEDURE OF STATISTICAL ANALYSIS OF FEEDBACK SYSTEMS

The most serious question was: what factors most contribute to the overall fluctuation of the kiln process? To find an answer to this question, the following structure of the kiln was assumed:

(1) Each variable has its own noise source which is independent of those of other variables.

(2) Each variable is represented as a sum of the effects due to the variations of other variables and its own noise source. By assuming the linearity, these assumptions give the following model of a noisy system under noisy control:

$$x_n(i) = \sum_{j=1}^{K} \sum_{m=1}^{M} a_{ijm} x_{n-m}(j) + u_n(i) \qquad i = 1, 2, \ldots, K \qquad (4.1)$$

where $x_n(i)$ represents the deviation from the mean of the i-th variable at time $n\Delta t$, where Δt is the sampling period, $u_n(i)$ represents the contribution of the noise and a_{ijm} is the impulse response of $x_n(i)$ to $x_{n-m}(j)(j \neq i)$. Following the assumption (2), it is assumed that $a_{iim}=0$ for all m. This means that in the present model each $x_n(i)$ is considered to be the output of a linear system with input $x_n(j)$ $(j \neq i)$ and the whole system is being driven by the noise source $u_n(i)(i=1, 2, \ldots, K)$. If a_{ijm} and the noise characteristics are identified, $x_n(i)$ can be represented as a sum of linear combinations of $u_{n-m}(j)(m=1, 2, \ldots, j=1, 2, \ldots, K)$, i.e. the variation of $x_n(i)$ can be traced back to the original contributions of $u_n(j)$'s. Thus if the assumption of mutual independence, or only of orthogonality, of $u_n(j)$'s is admitted, the mean square variation of $x_n(i)$ can be decomposed into the sum of contributions from each noise source and the frequency domain analysis provides a corresponding power spectrum decomposition.

The identification was made possible through the autoregressive representation of the noise source $u_i(n)$, and the whole procedure of identification and spectrum decomposition is described in detail in [8, 10]. The procedure was first applied to the set of variables $x_n(1)=T$, $x_n(2)=W$ and $x_n(3)=P$. Fig. 5 shows the decomposition of the power spectra in relative scales. The graphs show very clearly that the kiln drive power W is significantly influenced by the disturbances originating in the exit gas temperature T and the cooler under grate pressure P, thus indicating the possibility of reducing the fluctuation of W by properly compensating for the fluctuations of T and P. The importance of the kiln drive power for the purpose of the kiln control has long been advocated by one of the present authors and its characteristics were discussed in detail by NAKAGAWA [11]. The present model was an extremely simplified representation of the kiln process where the possible contribution of the human operator was ignored. But the validity of the result has later been confirmed by comparing it with the result obtained by using an extended model including manipulated variables. This result is reported in [6]. It should be mentioned here that the above stated assumption (1) was fundamental for the validity of the present approach and the later analysis by the autoregressive model fitting of the whole process has shown that this does not differ much from reality.

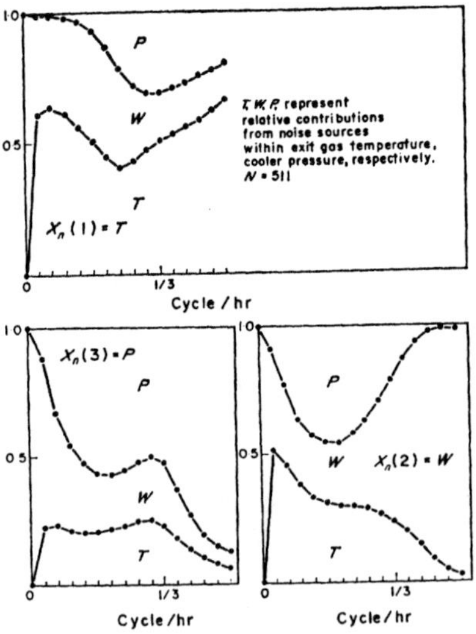

FIG. 5. Relative power contributions of noise sources to variables $x_n(i)$.

Ignoring the possible bias due to the manipulations of some of the other variables, the frequency response function of T to the input variation of fuel rate F was estimated by using the procedure of this section. Based on this estimate a very simple feedback control adjusting the fuel rate proportionally to the difference of the temperature in two successive measurements and with unusually high gain was introduced between T and F. The performance of the controller was just as expected from the estimated frequency response function. The control did not show hunting behavior and did reduce the fluctuation of T significantly but of course was not able to keep the whole process stationary for a long period without the aid of a human operator. This experimental result gave the first partial proof of the feasibility of the controller design by using the result of statistical identification.

Now, as can be seen from the estimates of the impulse response functions $\{a_{ijm}\}$ illustrated in Fig. 6 the response of the exit gas temperature T to the fuel rate F and that of the kiln drive power W to T seem to be of rather long duration.* Compared with this, the response of W to the cooler under grate pressure P seems rather instantaneous. Since P is generally very quickly controlled by changing the cooler grate speed Cr, these observations suggested that for the reduction of the fluctuation of the kiln drive power W an efficient control of the cooler under grate pressure P would be indispensable. This point was also backed by the opinion of one of the experienced operators that the control of P by manipulating the cooler grate speed Cr is vital for the control of the kiln drive power W and thus for the stabilization of the kiln, though at this time the kilns were running above

* At this time of analysis some a_{ij0}'s were included in the model (4.1), see [8, 10].

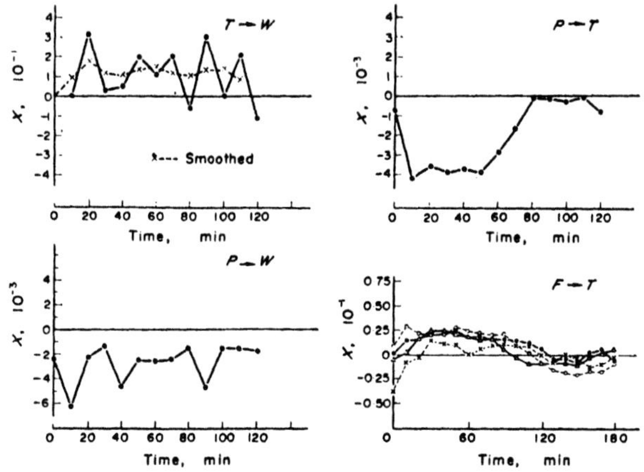

FIG. 6. Estimates of impulse response functions.

their specifications and *Cr*'s were kept at their maximum possible values. This finding strongly indicated that the clinker cooler should be treated as an essential ingredient in the integrated kiln and cooler control. It was decided that the clinker cooler capacity should be increased along with the necessary improvements of instrumentations to realize an overall computer control of the kiln.

5. PROCESS IDENTIFICATION AND CONTROLLER DESIGN

After the completion of improvements of the clinker coolers the implementation of a final controller was contemplated. The autoregressive representation of a stationary stochastic process gives the present value of the process variables as a sum of a linear combination of its past values and a white noise. This representation is generally valid for ordinary stationary random processes in engineering, if only each variable is not in a perfect linear relation with others [12], and the representation allows a very simple procedure of identification and controller design. The use of this type of modeling and controller design in relation to the cement rotary kiln control has already been discussed by WONG *et al.* [2], but it seems that the application was limitted to local controller designs in contrast with the present integrated kiln and cooler controller design and, as will be discussed shortly, the treatment of D.C. components was quite unsatisfactory.

Hereafter, the r-dimensional vector of the controlled variables and the l-dimensional vector of the manipulated variables at time $n\Delta t$ are represented by x_n and y_n, respectively. The $(r+l)$-dimensional vector obtained by augmenting x_n with y_n is denoted by X_n;

$$X_n = \overbrace{\begin{bmatrix} x_n \\ \\ y_n \end{bmatrix}}^{1} \begin{matrix} \updownarrow r \\ \\ \updownarrow l \end{matrix} . \tag{5.1}$$

The autoregressive representation of a stationary X_n is given by

$$X_n = \sum_{m=1}^{M} A_m X_{n-m} + A_0 + U_n, \tag{5.2}$$

where, if we put $k=r+l$, A_m if *a* $k \times k$ matrix, A_0 *a* $k \times 1$ vector and U_n a random $k \times 1$ vector satisfying the relation

$$\begin{aligned} &EU_n = 0 \text{ (zero vector)} \\ &EU_n X'_{n-m} = 0 \text{ (zero matrix) for } m \geqslant 1 \\ &EU_n U'_m = \delta_{nm} S, \end{aligned} \tag{5.3}$$

where $\delta_{nm} = 1(n \neq m)$, $=0(n \neq m)$ and S is a positive definite matrix and the symbol ′ denotes the transpose. The representation (5.2) allows the following representation of the controlled variables, which is directly used for the controller design:

$$x_n = \sum_{m=1}^{M} a_m x_{n-m} + \sum_{m=1}^{M} b_m y_{n-m} + a_0 + u_n, \tag{5.4}$$

where a_m, b_m, a_0 and u_n are given by the relations

$$A_m = \begin{matrix} \\ r \\ l \end{matrix}\overset{\leftarrow r \rightarrow\ \leftarrow l \rightarrow}{\begin{bmatrix} a_m & b_m \\ * & * \end{bmatrix}}, \qquad A_0 = \begin{matrix} \\ r \\ l \end{matrix}\overset{\leftarrow 1 \rightarrow}{\begin{bmatrix} a_0 \\ * \end{bmatrix}},$$

$$U_n = \begin{matrix} \\ r \\ l \end{matrix}\overset{\rightarrow 1 \rightarrow}{\begin{bmatrix} u_n \\ * \end{bmatrix}},$$

where * denotes quantities representing the characteristics of the human operator in the present application and are irrelevant for the controller design. Here if x_n and y_n are measured from their means the constant vector a_0 vanishes and the system behavior can be described by

$$x_n = \sum_{m=1}^{M} a_m x_{n-m} + \sum_{m=1}^{M} b_m y_{n-m} + u_n. \tag{5.5}$$

The identification of (5.5) is realized by using a record of the kiln process under human control. The most difficult problem in the identification of (5.5) is the determination of the order M. This problem has never been studied to the point to yield a practical answer. One of the present authors developed a decision procedure of M for the general autoregressive model (5.2) and the procedure was reported at the IFAC Kyoto Symposium, 1970, with a simple necessary computation scheme for the identification [6]. The procedure has further been modified to be especially useful for the identification of (5.5) and made the hitherto time consuming identification an almost mechanical routine [7]. The necessary computational procedure is described in the Appendix. It must be mentioned here that this type of completely objective decision procedure is entirely new in the statistical literature and by the present authors' knowledge all the hitherto published model fittings were based on test procedures dependent on more or less subjectively chosen levels of significance. At present the complexity of the statistical characteristics of the new procedure precludes fully analytical evaluation of the accuracy of the estimates. When M is assumed to be known the evaluation of statistical characteristics of the estimates is a simple matter, but this assumption almost never holds in practical applications. The statistical stability of the results has only been checked by the Monte Carlo experiments [6, 7]. The fitting procedure described in the Appendix always gives a stable model of the combined system of the kiln and the operator. If the feedback loop gain of the operator in the model is completly suppressed, the model gives the kiln behavior model which may not be stable. For those data now in use for control, the kiln behavior model produced stable responses but with extremely long settling times.

The vector a_0 is generally concerned with the D.C. components of the process, which should reflect the heat and mass balance relations within the kiln. Thus if (5.5) is used carelessly for the implementation of a controller when actually a_0 in (5.4) is significantly different from zero the application of the controller will cause a significant drift of the whole system. It is quite possible that this drift may carry the kiln process system into the region where the linear and stationary approximation no longer holds. Since, in the following, controller design (5.5) is assumed and x_n and y_n are measured from respective set points, this suggests the difficulty inherent in the selection of the set points. The present linear model is valid only around a nominal trajectory which is a constant vector in this case. At present the components of the vector, or the set points, are determined empirically. It seems that the development of a systematic procedure for the determination is quite necessary. For this, further accumulation of running experiences will be needed. The present consideration also suggests that the controller of a cement kiln must be capable of controlling the D.C. components and the identification and controller design procedure described in [2], which is based on the differences of the measurement values at adjacent time points, will definitely be inappropriate for the overall control of a kiln process.

Except for the above stated essential difference in the basic model, the controller design follows the general line described in [2] based on a state space representation of the system, which is derived from (5.5) and is given by

$$Z_n = \Phi Z_{n-1} + \Gamma Y_{n-1} + W_n, \tag{5.6}$$

where

$$Z_n = \begin{matrix} r \\ r \\ \vdots \\ r \end{matrix}\overset{\leftarrow 1 \rightarrow}{\begin{bmatrix} z_{1,n} \\ z_{2,n} \\ \vdots \\ z_{M,n} \end{bmatrix}},$$

$$\Phi=\begin{array}{c} \\ \updownarrow r \\ \updownarrow r \\ \\ \\ \\ \updownarrow r \\ \updownarrow r \end{array}\begin{array}{c} \leftarrow r\rightarrow \quad \leftarrow r\rightarrow \quad \leftarrow r\rightarrow \qquad\qquad \leftarrow r\rightarrow \\ \begin{bmatrix} a_1 & I & 0 & \cdots & 0 \\ a_2 & 0 & I & \cdots & 0 \\ \cdot & \cdot & \cdot & & \cdot \\ \cdot & \cdot & \cdot & & \cdot \\ \cdot & \cdot & \cdot & & \cdot \\ a_{M-1} & 0 & 0 & \cdots & I \\ a_M & 0 & 0 & \cdots & 0 \end{bmatrix}\end{array}$$

$$\Gamma=\begin{array}{c} \updownarrow r \\ \updownarrow r \\ \\ \\ \updownarrow r \end{array}\begin{array}{c} \leftarrow l\rightarrow \\ \begin{bmatrix} b_1 \\ b_2 \\ \cdot \\ \cdot \\ \cdot \\ b_M \end{bmatrix}\end{array},\quad Y_{n-1}=\overset{\leftarrow 1\rightarrow}{\underset{\updownarrow l}{y_{n-1}}},\quad W_n=\begin{array}{c} \updownarrow r \\ \updownarrow r \\ \\ \\ \updownarrow r \end{array}\begin{array}{c} \leftarrow 1\rightarrow \\ \begin{bmatrix} u_n \\ 0 \\ \cdot \\ \cdot \\ \cdot \\ 0 \end{bmatrix}\end{array}$$

and

$$z_{1,n}=\overset{\leftarrow 1\rightarrow}{\underset{\updownarrow r}{x_n}}\;.$$

$z_{i,n}$'s ($i=2, 3, \ldots, M$) are computed from the past history of x_n and y_n by (5.6) and $z_{1,n}$ is given as the present observation of x_n. If $z_{1,n}$ is computed as defined by the right hand side of (5.6) and assuming $u_n=0$, this gives the one step prediction of x_n. The controller is designed so as to give the minimum of the quadratic criterion with positive semidefinite Q and R

$$J_H=E\left\{\sum_{n=1}^{H}(Z'_nQZ_n+Y'_{n-1}RY_{n-1})\right\},$$

where H is a predetermined positive integer and E denotes expectation. H is generally chosen large enough so that any further increase of H would not introduce significant change of the controller gain. The optimal controller in this case is given in a form

$$Y_n=GZ_n, \tag{5.7}$$

where as is well known G is obtained by the following iterative computation [13]:

$$\begin{aligned} &P_0=Q, \\ &\left.\begin{aligned} M_i&=P_{i-1}-P_{i-1}\Gamma(\Gamma'P_{i-1}\Gamma+R)^{-1}\Gamma'P_{i-1} \\ P_i&=\Phi'M_i\Phi+Q \end{aligned}\right\}, \\ &\qquad\qquad i=1, 2, \ldots, H-1 \\ &G=-(\Gamma'P_{H-1}\Gamma+R)^{-1}\Gamma'P_{H-}\Phi_1. \end{aligned} \tag{5.8}$$

In the computation of G for a large system the accuracy of the digits is quickly lost by the accumulation of rounding errors as H is increased and it is necessary to avoid unnecessarily large value of H. The performance of the realized control can numerically be checked by introducing various test inputs such as unit impulse or white noise as u_n in (5.5) with $y_{n-1}=GZ_{n-1}$. For the initial design of the controller it will be practical to use the white noise u_n which possesses the statistical characteristics of u_n obtained by the identification and for a given Q to adjust R so as to keep the variances of the manipulated variables nearly equal to those under human control or to those values which are technically allowable. The resulting variances of the controlled variables may then be used to adjust Q. This procedure can easily be realized by using the Monte Carlo method.

In the realization of the multivariable control of the kiln only the first three components of the state vector, the actual outputs of the kiln, were weighted positively, so that the matrix Q was limited to the following simple form:

$$Q=\underset{\updownarrow}{3M}\overset{\leftarrow 3M\rightarrow}{\begin{bmatrix} \overset{\leftarrow 3\rightarrow}{\underset{\updownarrow 3}{Q_1}} & 0 \\ 0 & 0 \end{bmatrix}}. \tag{5.9}$$

Further, Q_1 and R were both limited to be diagonal. Experience suggests that it is reasonable to start the design with the weights $Q_1(i, i)$ and $R(j, j)$ for the process output $x_n(i)$ and the manipulated variable $y_n(j)$ put equal to the inverses of the estimates of the variances of the white noise $u_n(i)$ and that of $y_n(j)$ under human control, respectively.

6. ON-LINE CONTROL IN PRODUCTION LINE

The following variables were adopted for the controller design:

$$x_n(1)=W,\quad x_n(2)=P,\quad x_n(3)=T,$$
$$y_n(1)=F,\quad y_n(2)=Kr,\quad y_n(3)=D,\quad y_n(4)=Cr,$$

where $x_n(i)$ and $y_n(j)$ represent the i-th and j-th components of x_n and y_n, respectively. In realizing a control these quantities are respectively measured from their set points. The observation and control period Δt was 4 min. For the identification running records $\{x_n;\ n=1, 2, \ldots, N\}$, about 2 days of kiln data under human control were used. The value of the order M of the model depends on the spectral characteristics of the process, the sampling interval Δt and the value of N. When N is small M tends to be small, giving a rough estimate of the process characteristics. It must be very clearly understood that our knowledge of a process is essentially limited by the length of available data and for the present case with $\Delta t=4$ min and $N=720$ the decision procedure, described in Appendix, suggested the adoption of M of about 5 or 6, i.e. the model which takes into account the past history of the kiln and control for about a ½ hour. For the controller design using an IBM 1800 computer, taking into account the effect of rounding errors, $H=10$ was used in (5.8). Since a precise manipulation of the kiln speed Kr was impractical due to the mechanical limitation of the controller implementation, the controller output $y_n(2)$ was suppressed, at the initial phase of experiment, by introducing an extremely large weight as the second diagonal element of R, and Kr was essentially kept constant. Before implementing the overall control, a local control loop for the single-input single-output system of the cooler grate speed Cr and the cooler under grate pressure P was tentatively designed by the present procedure and tested on line. The control was extremely effective in stabilizing the overall behavior of the kiln. This experimentally confirmed the conclusion of Section 4 and provided the second partial proof of the feasibility of the desired control.

At the initial implementation of the control, Kr was kept constant and the operator was completely excluded from the control system. One of the successful records of application at this phase is illustrated in Table 1. In this case the result of identification of #2 kiln which had the same dimensions and structure as #1 kiln was applied to #1 kiln and the result demonstrated the desirable insensitiveness of the controller design. It can be seen that the mechanical breakdowns are now more serious than the process disorders. It has become quickly clear that to compensate for some significant variations of the kiln drive power W, which are generally considered to be due to the irregularities of the material flow and the burning zone activity, manipulation of the kiln speed Kr would be the most useful and necessary. Due to the mechanical limitations which precluded the desired fine control of Kr, manual intervention by the operator was tried first. Since a change of Kr directly influences W through an electric connection rather than through the process, an empirically designed logical filter was applied to W to provide an output W_2 to be used in place of the original W. This W_2 is supposed to be free of the temporally effect of manipulation of Kr but to gradually follow the change of the base line of W. The function of the filter is given as follows:

TABLE 1. PART OF RUNNING RECORDS OF #1 KILN

Date	Content water of slurry (%)	Fuel consumption (Kl)	Product (ton)‡	Fuel product (l/ton)	Remarks
1970 July 6	35·3	159·8	1200	133·2	12.00 ON-LINE START
7	35·1	159·1	1213	131·2	ON-LINE
8	35·1	160·1	1222	131·0	ON-LINE
9	35·1	160·7	1229	130·8	ON-LINE
10	35·0	160·5	1242	129·1	ON-LINE
11	—	—	—	—	9.40 ON-LINE OFF*
15	34·8	156·5	1237	126·6	MANUAL
16	35·0	158·1	1282	128·2	MANUAL
17	34·2	156·7	1263	124·0	15.00 ON-LINE START
18	34·5	155·1	1252	123·9	ON-LINE
19	34·5	156·5	1252	125·0	ON-LINE
20	34·4	156·9	1262	124·0	ON-LINE
21	34·3	156·8	1256	124·8	ON-LINE
22	33·8	155·3	1274	121·9	ON-LINE
23	—	—	—	—	13.00 ON-LINE OFF†

* Due to abnormal slurry input.
† Due to a trouble in the cooling fan motor.
‡ Kr kept approximately equal to 60 r.p.h.

$$W_{2,n}=W_n-d_n$$
$$d_n=cd_{n-1}+e_n \qquad (6.1)$$

$e_n=0$ when Kr was not manipulated during $[(n-2)\Delta t,\ (n-1)\Delta t)$
$=W(t+\Delta t)-W(t)$ when Kr was manipulated at t in the interval $[(n-2)\Delta t,\ (n-1)\Delta t)$,

where c is less than but nearly equal to 1 and $W(t)$ denotes the value of W at time t and other variables are measured at $n\Delta t$'s. This type of filtering was necessary when Kr was manipulated significantly during the data gathering phase of identification, even when the manipulation of Kr was not intended. Also it should be mentioned that the human operator usually manipulates Kr significantly when the process condition is extremely bad and this sometimes introduces quite large and probably non-linear effects into the process behavior. This difficulty which is inherent in including Kr into the present linear model was also recognized by a statistical analysis as described elsewhere [7,

Section 5]. A result of applying this controller design procedure to #3 kiln is illustrated in Table 2 which provides a comparison of the performances of a human operator and the computer control with intermittent manual *Kr* modification. The reductions of the variances of controlled variables are quite significant. It can be seen through the values of the variances that the modificantion of the kiln speed *Kr* during the computer control was only slight. #3 kiln is analogous in size and structure to the #1 and #2 kilns.

TABLE 2. COMPARISON OF VARIANCE OF #3 KILN

	Human control* Data length $N=680$†		Computer control Data length $N=660$‡		
	Mean	Variance	Mean	Variance	Set point
$W(kw)$	180·010	164·519	204·662	26·265	*
$W2(kw)$	168·188	50·848	204·795	29·204	203·000
$P(mmH_2O)$	110·188	228·596	122·345	59·746	124·100
T(°C)	172·398	23·277	173·612	4·843	173·740
$F(m^3)$	7·024	0·015	7·686	0·0045	7·653
Kr(r.p.h.)	55·244	2·301	57·429	0·147	58·405
D(°)	45·319	4·181	52·140	0·262	50·728
Cr(r.p.m.)	21·911	3·034	22·586	8·370	23·798

* Data used for identification.
† $\Delta t=4$ min.

A final form of the controller was realized by including empirically designed control loops of *Kr* and *F*. The control loop of the kiln speed *Kr* is depicted in Fig. 7 and is based on the observation that extraordinary reduction of the kiln drive power *W* can most simply be recovered by the reduction of *Kr*.

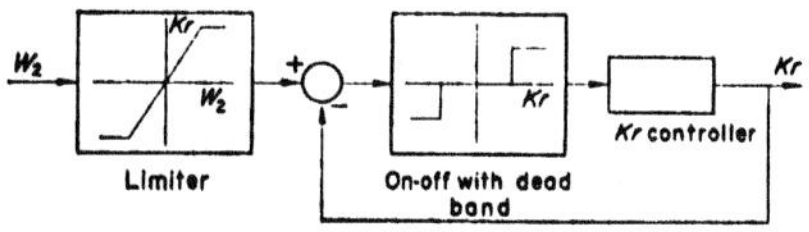

FIG. 7. Empirically designed kiln speed control.

Since the manipulation of *Kr* significantly affects the model, its effect is monitored by the one-step prediction error of *W*, which is the deviation of the true observed value from the predicted value obtained by the model (5.6). This error is smoothed slightly and when the smoothed error is above a prescribed level the information is fed back to reduce the gain of the fuel rate *F*. Though the design and adjustment of these control loops were performed only empirically, the addition of these loops highly improved the robustness of the control without significant deterioration of the performance of the original control loop under normal operating conditions. The whole program can now control the kiln except during the very initial phase of start up, and the full automation of start up is now being contemplated.

Some of the power spectra of the controlled variables under human and computer control, respectively designated by manual and computer for #1 and #3 kiln, are illustrated in Figs. 8 and 9. The significant reduction of power at the lower frequency band very clearly shows the superiority of the present controller over human control.

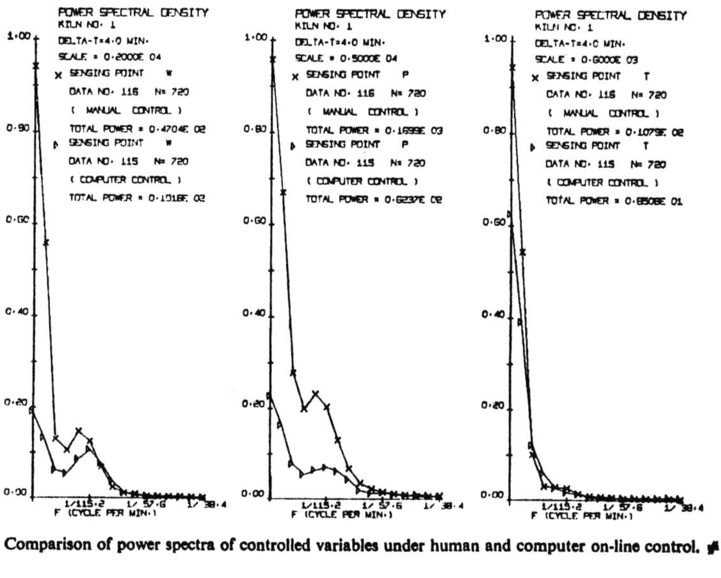

FIG. 8. Comparison of power spectra of controlled variables under human and computer on-line control. #1 kiln.

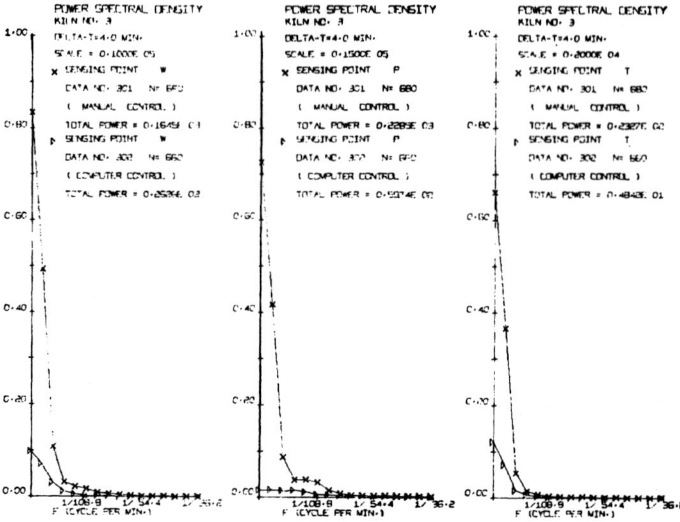

FIG. 9. Comparison of power spectra of controlled variables under human and computer on-line control. #3 kiln.

7. OBSERVATIONS DURING THE RUNNING EXPERIENCE

First of all it should be remembered here that the present design procedure was developed for the stabilization of the exit gas temperature T, the cooler under grate pressure P, and the kiln drive power W. The running experience shows that this generally leads to a controller design which can keep the burning zone in a quite satisfactory condition. Validity of this fact can be seen from the discussion of the long running records at the end of this section. The present controller has made the observation of the kiln process behavior possible when it is completely free from the human intervention. This greatly improved the S/N ratio in observing the effects of some of the extraneous disturbances. The effect of the voltage fluctuation of 5 per cent of the electric power supply, definitely affecting the kiln condition, can now very clearly be seen. For the desirable compensation of this effect a very fine control of the kiln speed Kr, which is beyond the accuracy of the present instrument, is required. Some of the cause of the significant fluctuations of the kiln drive power W was also traced back to the inhomogeneity of the dust flow which is being returned into the kiln as the secondary raw material. These findings are really important contributions of the present controller since they disclosed the origins of significant disorders, which had not been recognizable due to the high noise level under the human control. The company has already started the introduction of necessary modifications of the system installation. Also all the experiences with the present controller are now being used to determine the specifications of newly planned kilns.

In one application it was experienced that the response of the kiln was quite contrary to that expected from the result of identification. The response strongly suggested the lack of oxygen and it was found that the cooler damper which controls the air flow through the cooler was set at an extraordinary position. By resetting the damper the kiln quickly regained the expected response characteristics. This experience very clearly shows the utility of the kiln model as the basic reference in understanding the process abnormality. It suggests the inclusion of cooler damper position into the controller as a manipulated variable, though it has been generally kept constant during the human control, and also of air rate as another controlled variable. The one-step prediction error or the difference between the real observation of process variables and the one-step prediction obtained by the kiln model (5.6) seems to be very useful in detecting the abnormalities of the process. The full expolitation of this information will be a subject of future study. Also the fact that the fitted kiln models gave very slow settling responses, which preclude the practicability of the kiln process without control, suggests that the present statistical procedure is giving fairly reasonable approximations to the process behavior.

The experience of running the kilns by the present controller design confirmed the authors' expectation that the present combination of the linear controller and the empirically designed and properly adjusted kiln rotation control would be sufficient for use in production line even before the above stated modification of the kiln system installation. The controller kept #1 kiln running for more than three weeks with only two short interruptions of 4 or 5 hours which were caused by the cooler screw breakdown. The running had to be ended by the demand of management. The standard deviations, or root mean squares, of the variables and other informations related with this period are given in Table 3. The power spectra of the controlled variables are illustrated in Fig. 10.

TABLE 3. STANDARD DEVIATIONS OF LONG RUN ON-LINE RECORDS OF #1 KILN

	1st week* 4–11 April 1971 Data length $N=2500$		2nd week† 11–18 April 1971 Data length $N=2500$		3rd week* 18–25 April 1971 Data length $N=2480$	
	Mean	Standard deviation	Mean	Standard deviation	Mean	Standard deviation
$W(kw)$	215·867	6·013	215·526	7·715	215·705	9·497
$W_2(kw)$	216·399	4·750	215·975	6·327	215·594	9·425
$P(mmH_20)$	98·616	4·990	99·096	6·629	105·141	9·243
T(°C)	144·218	3·952	140·919	3·241	145·547	2·671
$F(m^3)$	7·194	0·102	7·172	0·121	7·215	0·108
Kr(r.p.h.)	59·916	0·397	59·572	0·503	58·342	0·626
D(°)	59·884	1·379	58·292	1·763	1·763	0·641
Cr(r.p.m.)	28·336	1·822	28·060	1·529	28·348	2·329

* The kind of cement was changed from normal to high early strength cement during the on-line control.

† The kind of cement was changed from high early strength to normal cement during the on-line control.

The data length was adjusted to be nearly equal for the three sets of records.

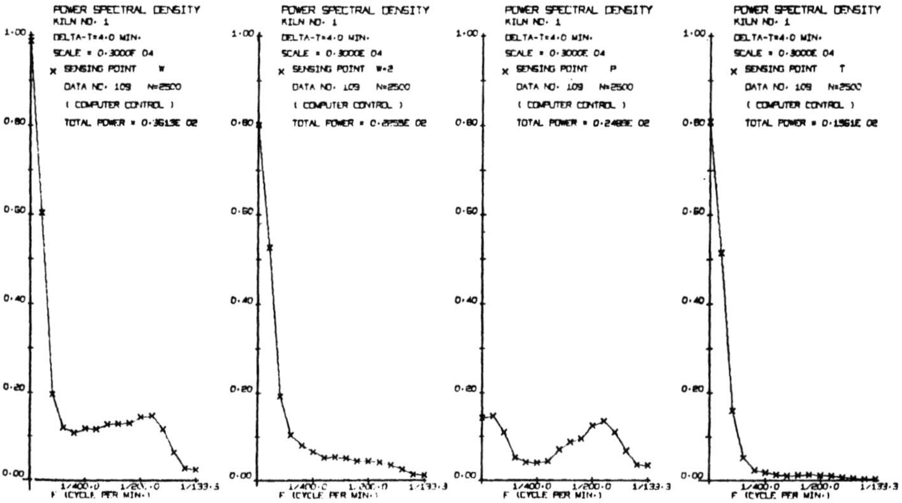

FIG. 10. Power spectra of 1 week history of controlled variables.

The difference between the power spectra of W and its modification W_2 shows that the kiln speed was manipulated significantly. It seems that the action of this kiln rotation control is not fully compensated by the linear controller and causing the humps of the power spectrum of the cooler pressure P and the kiln drive power W. In spite of this, the standard deviations of Table 3 are sufficiently low for practical applications. Taking into account the length of the period of continuous operation, these values will assure the practical utility of the present controller. It would be quite desirable to present numerical results of longer records of operation, but with one IBM 1800, doing many on-line jobs and controlling two or three kilns, this was much too demanding the programmer's effort, though it might not be impossible.

The formation and falling of mud rings are not so serious under the computer control as they were under human control, and the fuel consumption rate seems to be lower under the computer control. Since there are fluctuations of the conditions of raw materials and changes of operating condition due to the demand of management, very precise numerical evaluation of these effects is not possible at present.

8. CONCLUSIONS

The results reported in this paper clearly demonstrates that an almost complete automation of the cement kiln process can be realized by a statistical approach. The simplicity of the realized control scheme is remarkable and it seems that once the feasibility of this type of approach is confirmed it will be a simple matter to repeat it on other kilns.

One of the most significant technical contribution of the present paper is its proof of necessity and feasibility of the controller design for the integrated kiln and clinker cooler system. This has been realized through the statistical data analysis and controller design. Another contribution is the realization of a useful kiln rotation speed control through the empirical observation of the process behavior. The final form of the controller has been used in a real production line and the results show that the controller performance is quite satisfactory for practical use.

The main limitation of the present controller is with its setpoints, or the D.C. values of the related variables, which have to be provided from outside. This point will be a subject of further study.

The present controller has kept kilns running for an extended period of time without intervention of human operator and with low values of variances of controlled variables, and has revealed various causes of extraneous disturbance. This must be considered to be an indirect but fundamental contribution of the controller.

Since the kiln process changes parameter values very slowly and also because the condition of operation is changed by the demand of management, the accumulation of a very long records of operation is necessary for the precise evaluation of the average performance. In spite of this, it is already completely clear that the present computer control is by far the superior to human control and it did open up the almost unlimited possibility of further improvements of the kiln process control. The success is founded on the detailed statistical data analyses at the initial phase of the study and this will have to be repeated when the present approach is tried on objects other than the cement rotary kiln.

It should be noted that all the results have been obtained with very close cooperation between the people of management, instrumentation and control, and statistical analysis and design, respectively represented by the present authors. It is the authors' hope that the present paper will serve to fill the gap between the theory and practice of automatic control.

Acknowledgements—The authors wish to thank Messrs. S. Tsukada Y. Yagihara and T. Kominami who worked on the implementationf of the on-line control, and Miss E, Arahata who worked on the programming of the statistical methods.

REFERENCES

[1] R. A. PHILLIPS: Automation of a Portland cement plant using a digital control computer. *Automatic and Remote Control, Proc. 2nd IFAC Congress, Applications and Components*, pp. 347–357, Butterworths, London (1964).

[2] K. Y. WONG, K. M. WIIG and E. J. ALLBRITTON: Computer control of the Clarksville cement plant by state space design method. *IEEE Cement Industry Technical Conference, St. Louis, U.S.A.*, 20–24 May (1968).

[3] A. SKULL: The process control/operational research interface. *Operational Research Quarterly* 21, Special Conference Issue, 21–36 (1970).

[4] V. A. KAISER: Computer control in the cement industry. *Proc. IEEE* **58**, 70–77 (1970).

[5] G. M. JENKINS and D. G. WATTS: *Spectral Analysis and Its Applications*, Holden Day, San Francisco (1968).

[6] H. AKAIKE: On a decision procedure for system identification. *Preprints of Papers, IFAC Kyoto Symposium on System Engineering Approach to Computer Control*, pp. 485–490 (1970).

[7] H. AKAIKE: Autoregressive model fitting for control. *Ann. Inst. Statist. Math.* **23** 163–180 (1971).

[8] T. OTOMO, T. NAKAGAWA and H. AKAIKE: Implementation of computer control of a cement rotary kiln through data analysis. *Preprints, Tech. Session 66, IFAC 4th Congress, Warszawa*, 115–140 (1969).

[9] H. AKAIKE: Some problems in the application of the cross-spectral method. *Advanced Seminar on Spectral Analysis of Time Series* (Ed. B. HARRIS), John Wiley, New York, 81–107 (1967).

[10] H. AKAIKE: On the use of a linear model for the identification of feedback systems. *Ann. Inst. Statist. Math.* **20**, 425–439 (1968).

[11] T. NAKAGAWA: Study on the control of cement rotary kiln. Ph.D. Dissertation, University of Tokyo (1964). (In Japanese.)

[12] P. MASANI: Recent trends in multivariate prediction theory. *Multivariate Analysis* (Ed. P. R. KRISHNAIAH), Academic Press, New York, 351–382 (1966).
[13] J. T. TOU: *Optimum Design of Digital Control Systems.* Academic Press, New York (1963).
[14] H. AKAIKE: Information theory and an extension of the maximum likelihood principle. *Research Memo. No.* 46. *The Institute of Statistical Mathematics*, 25 pages, April (1971). Presented at the Second International Symposium on Information Theory, Tsahkadsor, Armenian SSR, 2–8 September (1971).

APPENDIX

Identification procedure for an autoregressive model

For an identification of the model (5.5) a very efficient procedure is available [7]. Given a set of observed data $\{X_n;\ n=1, 2, \ldots, N\}$, the identification proceeds as described in the following where, for the notations, reference should be made to section 5 of the text. Also for typographical convenience, the i-th component of X_n is denoted by $X_n(i)$. The first r of $X_n(i)$'s $(i=1, 2, \ldots, k$, $k=r+l)$ form the vector of controlled variables x_n and the last l the vector of manipulated variables y_n.

(1) Define $\tilde{X}_n(i)$ $(=1, 2, \ldots, k, n=1, 2, \ldots, N)$ by

$$\tilde{X}_n(i)=X_n(i)-\overline{X(i)},$$

where

$$\overline{X(i)}=\frac{1}{N}\sum_{n=1}^{N} X_n(i).$$

(2) For $m=0, 1, 2, \ldots, L$, where L is the maximally allowable order of the model and should generally be kept below $N/5k$, define the (i, j)th element of a $k\times k$ matrix C_m by

$$C_m(i, j)=\frac{1}{N}\sum_{n=1}^{N-m} \tilde{X}_{n+m}(i)\tilde{X}_n(j).$$

(3) Successively compute $k\times k$ matrices $A_m(M)$ $(m=1, 2, \ldots, M)$ and $S(M)$ for $M=1, 2, \ldots, L$ by the following recursive formulae:

$$S(M)=C_0-\sum_{m=1}^{M} A_m(M)C_m'$$

$$R(M)=C_{M+1}-\sum_{m=1}^{M} A_m(M)C_{M+1-m}$$

$$Q(M)=C_0-\sum_{m=1}^{M} B_m(M)C_m$$

$$D(M)=R(M)(Q(M))^{-1}$$

$$E(M)=(R(M))'(S(M))^{-1}$$

$$A_m(M+1)=A_m(M)-D(M)B_{M+1-m}(M) \quad m=1, 2, \ldots, M$$
$$=D(M) \quad m=M+1$$

$$B_m(M+1)=B_m(M)-E(M)A_{M+1-m}(M) \quad m=1, 2, \ldots, M$$
$$=E(M) \quad m=M+1,$$

where $'$ denotes transpose and

$$S(0)=Q(0)=C_0, \quad R(0)=C_1.$$

(4) Also compute FPEC(M) $(M=1, 2, \ldots, L)$ by

$$\mathrm{FPEC}(M)=\left(1+\frac{Mk+1}{N}\right)^r\left(1-\frac{Mk+1}{N}\right)^{-r} \times \|S_{rxr}(M)\|,$$

where $S_{rxr}(M)$ denotes the rxr submatrix in the upper left hand corner of $S(M)$ and $\|\ \ \|$ denotes the determinant.

(5) Adopt the value of M which gives the minimum of FPEC(M) $(M=1, 2, \ldots, L)$ as the order of the model and determine the corresponding necessary matrices of coefficients a_m and b_m $(m=1, 2, \ldots, M)$ of (5.5) by the relation

$$A_m(M)=\begin{array}{c} \\ \updownarrow r \\ \updownarrow l \end{array}\begin{array}{cc} \leftarrow r\rightarrow & \leftarrow l\rightarrow \\ \left[\begin{array}{c} a_m \\ * \end{array}\right. & \left.\begin{array}{c} b_m \\ * \end{array}\right] \end{array}.$$

For the details of statistical characteristics of this procedure and a further extended use of FPEC(M) for the variable selection reference should be made to [7]. Information theoretic implication of this type of procedure was first discussed recently by AKAIKE [14].

Résumé—Une commande automatique, entièrement réalisée par calculateur, de fours rotatifs à ciment a été essayée, en production réélle, et les résultats sont exposés dans le présent article. L'étude du régulateur a été basée sur la compréhension du comportement du procédé obtenue au moyen d'analyses statistiques approfondies; il a été réalisé en utilisant une méthode très efficace d'identification statistique et une étitude classique du régulateur optimal au moyen de la méthode de l'espace des états. Toutes les phases de l'analyse, de l'étude et des règlages pendant l'application pratique sont discutées en detail. L'influence technique du succès de la commande sur l'ensemble de l'installation des fours est également discutée. La méthode de calcul pour l'identification est décrite en annexe.

Zusammenfassung—Ein voll rechnergesteurter Zementdrehofenprozeß wurde in einem realen Produktionsgang getestet und hier die Resultate dargestellt. Der Reglerentwurf wurde auf Grund der Einsicht über das Prozeßverhalten vorgenommen und zwar durch sorgfältige statistische Analyse und unter Benutzung einer sehr wirksamen statistischen Indentifikationsprozedur realisiert und der Entwurf des orthodoxen optimalen Reglers mit der Zustandsmethode durchgeführt. Alle Phasen der Analyse, des Entwurfs und der Justierung während der praktischen Anwendung werden im Einzelnen diskutiert. Der technische Einfluß des Erfolges der Regelung auf die gesamte Ofeninstallation wird ebenfalls diskutiert. Die rechnerische Prozedur für die Identifikation wird in einem Anhang beschrieben.

Резюме—Автоматическое управление вращающимися цементными печами, полностью осуществнленое с помощью цифровой вычислительной машины, было испытано, в реальной продукции, и результаты приведены в настоящей статье. Исследование регулятора основывалось на понятии поведения процесса полученном с помощью углубленных статистических анализов; Он был осуществлен используя весьма эффективный метод статистического опознавания и классический рассчёт оптимального регулятора с помощью метода пространства состояний. Детально обсуждаются все фазы анализа, рассчёта и настроек в течении практического применения. Также обсуждается техническое влияние успеха управления на общность установки печей. В приложении, описывается метод рассчёта для опознавания.

Automatica, Vol. 17, No. 1, pp. 143–155, 1981
Printed in Great Britain

0005-1098/81/010143-13 $02.00/0
Pergamon Press Ltd.

Statistical Identification for Optimal Control of Supercritical Thermal Power Plants*

H. NAKAMURA† and H. AKAIKE‡

A digital optimal controller based on the autoregressive model of a thermal power plant significantly improves the system performance with respect to that obtained with a conventional PID controller.

Key Words—Identification; autoregressive model; computer control; PID control; optimal control; power plant control; APC; ADC; AIC.

Abstract—The use of a multivariate autoregressive model for the implementation of a new practical optimal control of a supercritical thermal power plant is discussed. The control is realized by identifying the system characteristics of the plant under the conventional PID control by the autoregressive model fitting and then implementing the digital control to correct the defect of the analog control. The procedure of identification and the controller implementation is described in detail by using the experimental results of a real plant. The results clearly demonstrate the advantage of the new controller over the conventional PID controller. The experience of the commercial operation of the plant confirms that the new controller is extremely robust against the gradual change of the plant characteristics, and this shows the practical utility of the identification procedure on which the design of the controller is based.

INTRODUCTION

IN A THERMAL power plant, which is a typical example of multivariable system, control loops within the boiler process show significant mutual interactions. Under the conventional PID controller, it is not easy to fully compensate for these interactions and adjust the plant to fast and large load changes with the controlled variables kept within the prescribed ranges. This difficulty of controlling a mutually interacting multivariable system has been one of the principal factors that set the limit to the response of a thermal power plant to the load changes required for the load-frequency control (LFC) of an electric power system. The use of a quadratic optimal controller was considered by the Kyushu Electric Power Company to solve this difficulty.

The basic problem for the realization of an optimal controller is how to obtain a practically useful state space representation of the object system. There are mainly two types of approaches for this purpose—the theoretical and statistical.

In the theoretical approach, a set of simultaneous partial differential equations that describe the dynamic balance of the energy and mass within the process are established and linearized under appropriate assumptions. The state equation is derived from these linearized equations. However, as can be seen from the recent discussion in the ASME Transactions (Smoak, 1976), it is not easy to obtain a model which is useful for practical applications. This is due to the fact that the order of the model generally tends to be very high. Also another and more important source of difficulty is the existence of the noise within the process which precludes the possibility of confirming the adequacy of the model by a simple one-shot testing.

In the statistical approach, a statistical model is fitted to the plant data obtained under normal, or approximately normal, operating conditions. By using an appropriate statistical model this procedure provides a model which takes into account the characteristics of both the process and the noise.

As the basic model for the statistical approach we adopt the autoregressive (AR) model. This model has been applied to the cement kiln process with success and a detailed report of the experience is described in Otomo, Nakagawa and Akaike (1972). The AR model of the process leads to a simple determination of the state space representation, and once the state equation is derived, the implementation of an optimal

*Received 20 March 1980; revised 20 August 1980. The original version of this paper was presented at the 5th IFAC Symposium on Identification and System Parameter Estimation which was held in Darmstadt, Federal Republic of Germany during September, 1979. The published Proceedings of this IFAC Meeting may be ordered from: Pergamon Press Limited, Headington Hill Hall, Oxford OX3 0BW, England. This paper was recommended for publication in revised form by associate editor A. Van Cauwenberghe.

†Research Laboratory, Kyushu Electric Power Co., Inc., 497-1 Aizo, Shiobara, Minami-ku, Fukuoka 815, Japan.
‡The Institute of Statistical Mathematics, 4-6-7 Minami-Azabu, Minato-ku, Tokyo 106, Japan.

control is realized with the orthodox design method of a linear quadratic controller. Our decision to use AR model instead of more general autoregressive moving average (ARMA) model is due to the fact that we could not get a significantly better fit to our data by using an ARMA model which required extensive amount of computing.

The effectiveness of this approach in the control of a thermal electric power plant was first checked by a digital simulation based on the model identified by an existing record of a plant operation (Akaike, 1978b) and then by applying the technique to an analog power plant model set up on an analog–digital hybrid simulator (Nakamura, Hirano and Akaike, 1978). Only after these careful checkings of feasibility the procedure was applied to a 500 MW fossil-fueled power plant equipped with a supercritical boiler.

The implementation of the optimal control of the actual plant proceeded in several steps. These were the analysis of the plant dynamics to decide on the proper choice of variables to be used in the final model, the check of the appropriateness of the state equation and the feedback gain matrix by digital simulation, the analysis of the effect of process nonlinearity, and the final determination of the feedback gain by field tests.

The first optimal control system thus completed was put into operation in February 1978. The result was a remarkable reduction of fluctuations of temperatures at various parts of the boiler system. With its improved response to the load change the plant has been contributing quite effectively to the load-frequency control of the power system, thus demonstrating the importance of an appropriate identification. This first result was obtained by Buzen No. 1 unit (500 MW supercritical) of the Kyushu Electric Power Company. Similar results were obtained by Shinkokura No. 3 and No. 4 units (both 600 MW supercritical) which were put into commercial operation by the same company in February and November 1979, respectively. Since June 1980, Buzen No. 2 unit (500 MW supercritical) has also been in operation under the same type of control.

In this paper, we will first describe rather in detail our experience of the system identification and optimal controller design and then discuss briefly the operating experience of these plants by the new optimal control. The performance of this new controller is compared with that of a conventional controller to illustrate the necessity of the system identification for a boiler process control.

Lecrique and his co-workers (1978) reported a very successful result of application of a statistical approach to identification and control developed by Richalet and his colleagues (1978) to thermal power plant control. The fundamental idea of this approach is very close to ours. However their basic model is based on a direct expression of the impulse responses which requires a large memory of past histories. In our present approach this difficulty is avoided by the use of an AR model which leads to an efficient identification of the necessary state space representation.

DIFFICULTY OF POWER PLANT CONTROL

In a large capacity high-pressure high-temperature boiler for power generation, deviations of steam temperatures at the boiler outlet must be kept within one or two per cent of their rated values to maintain the nominal operating efficiency and insure the safety and maximum equipment life of the plant.

The main purpose of the boiler control is to allow the increase or reduction of steam generation as fast as possible in response to the load command issued from the system's dispatch center, while satisfying the above-mentioned operating conditions. The conventional control so far used for this purpose is the analog PID control, commonly called APC (Automatic Plant Control). In this section, we will explain the problems of APC when the plant is subjected to load changes.

In order to give a general idea of the boiler control, we will briefly discuss the steam temperature control which has been considered to be most difficult in the boiler control. Figure 1 is a conceptual view of the temperature control of an FW (Foster Wheeler) type supercritical boiler. For convenience sake, we will hereafter refer to the plant variables by the following abbreviated symbols:

MWD: Megawatt demand or load command issued from the system's dispatch center to the plant
SHT: Superheater outlet steam temperature
RHT: Reheater outlet steam temperature
WWT: Waterwall outlet fluid temperature
FR: Fuel to feedwater ratio
SP: Flow rate of the spray water to the superheater tube
GD: Opening of the gasdamper in the rear path of the boiler shell.

Certainly there are many other important variables but only the variables which were selected by a preliminary analysis to be described in a later section are retained here.

Plant variables SHT and RHT are regulated by the three manipulated variables FR, SP, and

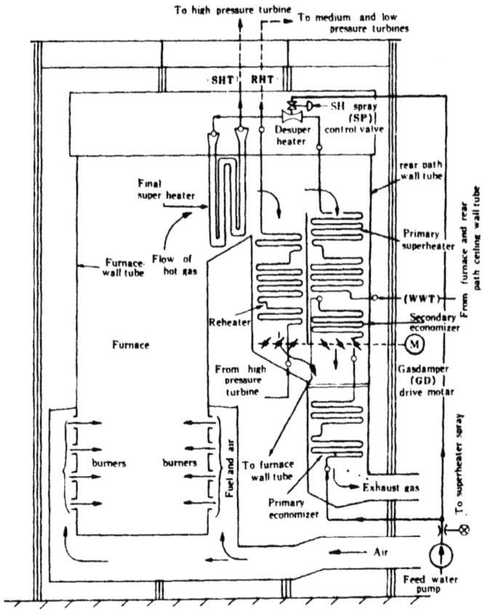

FIG. 1. Conceptual view of boiler system.

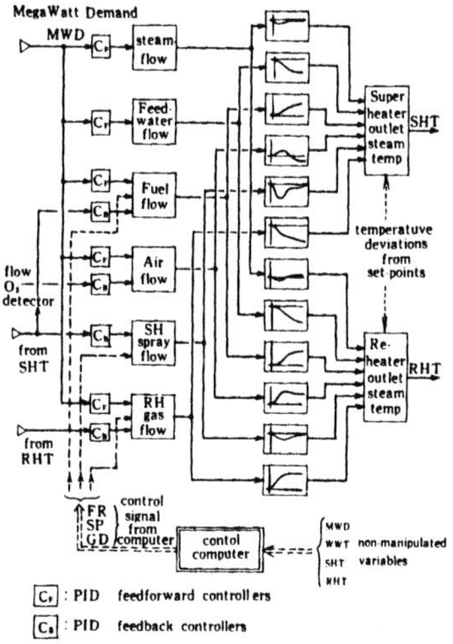

FIG. 2. Response of steam temperatures to step change of process inputs.

GD. The manipulation of FR has influence on both SHT and RHT in the same direction. When FR increases both SHT and RHT are raised. SP is the portion (3 or 4 per cent) of the feedwater injected into the midway of the primary and the secondary superheaters and SHT control is partially realized by manipulating this quantity. Its effect on SHT is quick but temporary, and GD control aims to regulate RHT. The manipulation of GD causes the change in the distribution of hot gas that flows along the heat conducting surfaces of the reheater and superheater. The increase of GD increases the hot gas flow along the reheater and raises RHT, but it lowers SHT by reducing the gas flow along the superheater. Consequently, the GD control acts on SHT and RHT in an opposite manner.

Figure 2 is a graphical representation of the dynamic behavior of the SHT and RHT. Each of the frames in the center of Fig. 2 relates the manipulated variables on the left to SHT and RHT on the right, and the dynamic response of the temperature to the stepwise increase of the corresponding manipulated variables is illustrated. As shown in Fig. 2 manipulation of each manipulated variable has effects on SHT and RHT differently in direction, phase, and amplitude.

At first, the turbine controller regulates the steam supply to the turbine by manipulating the turbine governor to meet the MWD change. At the same time, the C_F's for the feedwater, fuel and air adjust the feedwater, fuel and air flow rates to the values required by the load demand. If thermal–hydraulic balance is kept appropriately during the load change, the fluctuations of the steam temperature will remain small. Due to the difference of the temperature responses to the manipulation of each manipulated variable it is not practical to expect such an ideal performance of the feedforward control, especially for large and rapid load changes. The reduction of the deviations of the steam temperatures from their set-points is then realized by the feedback control from SHT to FR and SP, and that from RHT to GD. However, as the manipulation of each variable affects more than one controlled variable in different ways, the feedback control loops interact with each other through the boiler process and form a typical mutually interacting multivariable system. By APC, the conventional PID control, it is not easy to realize a satisfactory multiple loop feedback system. This has been the principal factor that limits the load changing rate of a thermal power plant and suggests the need of a more sophisticated control system.

OPTIMAL CONTROL SYSTEM

Today, DDC (Direct Digital Control) prevails in the computer control of industrial processes. In DDC, a digital computer completely replaces the analog controller. In contrast with this, we adopted a system named ADC (Analog Digital Control) in which a digital computer cooperates with the conventional PID controller.

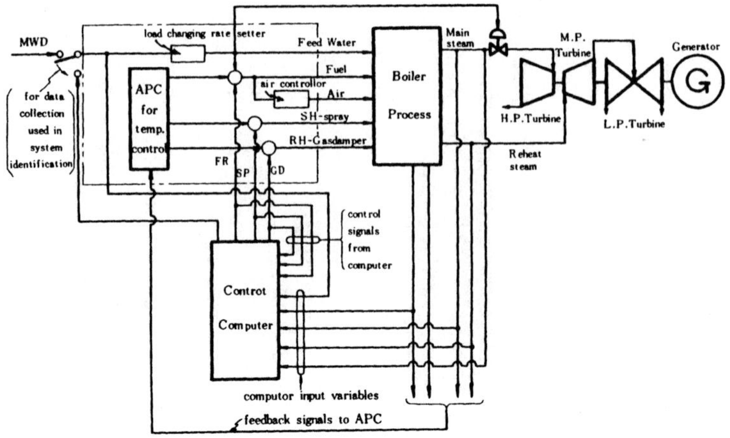

FIG. 3. Conceptual diagram of ADC system.

Figure 3 shows a schematic diagram of the concept of ADC system. In ADC, a plant control computer is added on top of the conventional PID controller APC. As can be seen by Fig. 3, the control by ADC is performed by the sum of the signals issued from both the PID controller and the computer (also see Fig. 2).

ADC has the following advantages over DDC:

(1) With DDC, the control of the off-set, or stationary deviation of controlled variables from their set-points, is rather difficult. With ADC, such an off-set can easily be compensated by the PID controller. Thus, the development of troublesome computer programs to cancel the off-set is unnecessary with ADC.

(2) ADC has high reliability compared with DDC. In case of the computer outage, the analog PID controller (APC) which is always in operation in the system immediately takes over the plant control. This also allows easy and safe adjustment of the controller. From the point of view of plant instrumentation, the ADC system may be considered as a dual instrumentation of controllers. However, when we take into account the large losses that would be caused by a plant trip due to malfunction of the controller or a miss operation during the phase of controller tuning, the advantage of ADC as a practical optimal control system of a large power plant is quite significant.

A SCHEME OF SYSTEM IDENTIFICATION AND CONTROLLER IMPLEMENTATION

The analysis and optimal controller implementation are performed according to the scheme shown in Fig. 4.

In step 1, since the control is realized as the digital control of the system operating under the analog PID controller (APC), the plant is kept under APC and the plant data to be used for the identification are collected. The usual difficulty of feedback system identification appears in this case and to avoid this random test signals are injected into the system from the computer. The data are collected at equi-spaced time intervals.

In step 2 (MULCOR), auto- and cross-covariance matrices of the system variables are computed.

In step 2 (FPE), a multi-variate AR model of the system variables, including the process and manipulated variables, is identified. The information criterion AIC, or FPE (Akaike, 1976, 1978b), is used for the determination of the order of the model.

Step 4 through step 6 perform the analysis of system characteristics. In step 4, statistical properties of the system noises and their relative contributions to the process variables are analyzed. In step 5, power spectral densities of the system variables are analyzed, and the impulse response functions between the system variables are computed in step 6. These analyses produce useful information for the selection of system variables to be used in the final modeling.

In step 7 (FPEC), AR model fitting is performed for the purpose of controller design. This is based on the system variables which were selected by the preliminary investigation. From the AR model thus obtained, a state equation of the system is derived.

In step 8 (OPTDES), a gain matrix for state feedback control is obtained by using the state equation and a quadratic criterion function.

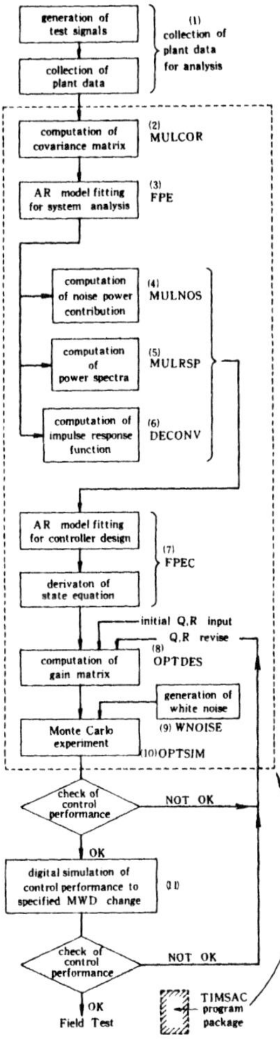

FIG. 4. Procedure of system identification and controller implementation.

Dynamic programming (DP) method is used for this purpose.

In step 9 (WNOISE), the white noise with the covariance matrix equal to that of the innovation of the non-manipulated variables is generated.

In step 10 (OPTSIM), Monte Carlo experiments are performed to examine the appropriateness of the feedback gain matrix by using the white noise generated in step 9.

The abbreviations given in the above parentheses correspond to the names of the programs included in a complete package for the analysis and control of time series, named TIMSAC (*Time Series Analysis and Control*) and included in the book by Akaike and Nakagawa (1972). A copy of TIMSAC on computer tape is available from Division of Mathematical Sciences, University of Tulsa, Tulsa, Oklahoma 74104, U.S.A.

The response characteristics of a power plant show a strong dependence on the load condition. To cope with this non-linearity a load adaptive adjustment of control parameters, to be described in the section on practical considerations, is introduced. In step 11, simulation is performed to check the performance of this adjustment against specific load changes.

Gain matrix computation is repeated until appropriate ones are found by the simulation. Several candidates of the gain matrices are retained for the final field test.

The procedure shown in Fig. 4 was applied to Buzen No. 1 unit which is a supercritical thermal power plant with the specifications given in Table 1. The results are discussed in the following sections.

TABLE 1. SPECIFICATIONS OF BOILER AND TURBINE

Boiler
type : IHI-FW oncethrough, reheat type steam generator,
maximum allowable working pressure : 274kg/cm^{2}G,
superheater outlet steam temperature : 541℃,
reheater outlet steam temperature : 540℃,
steam flow at maximum continuous rating : 1750t/h,
combustion system : opposed firing parallel gasdamper,
fuel : heavy oil,
draft system : forced draft
manufacturer : ISHIKAWAZIMA-HARIMA HEAVY, INDUSTRY
Turbine
type : tandem, 3cylinders, 4flows, reheat, regenerating type,
steam conditions :
246kg/cm^2, 538℃ at main stop valve,
45.3kg/cm^2, 538℃ at reheat stop valve,
output : 500MW,
r.p.m. : 3600,
manufacturer : TOKYO SHIBAURA ELECTRIC CO.

POWER PLANT IDENTIFICATION AND CONTROL

In this section the process of power plant identification and optimal controller implementation is explained following the steps described in the preceding section.

System identification

For the identification, four test signals were generated by the computer and injected into the signal injection points of MWD, FR, SP, and GD through D/A converters. MWD forms the largest disturbance to the plant and FR, SP, GD are the principal manipulated variables for the steam temperature control.

The values of the process variables which seemed relevant for the analysis of the dynamic behavior of the controlled variables were recorded by the computer together with those of the test signals at every 40 s for about 9 h. The time interval of 40 s may at first sight look too long for the control of a boiler system. However, as will be seen by the records illustrated in Figs. 9 and 11, most of the power of the temperature fluctuations are limited to the frequency range below one cycle per 10 min. This observation and the fact that the difficulty of identification increases as the sampling interval is reduced led us to the choice of this interval length. The data points recorded in the experiment were 832. The four test signals were generated from four different quasi-random sequences and fed into first-order lag digital filters to generate the test signals with appropriate frequency characteristics as shown in the top frame of Fig. 6. The amplitude of the test signals was adjusted in the preliminary test so that each of them produced approximately the same amount of steam temperature fluctuations. All the four test signals were then applied to the system simultaneously. Figure 5 shows a portion of the record of the experiment. As can be seen by the figure, the fluctuations of the steam temperatures during the experiment were kept within $\pm 3°C$.

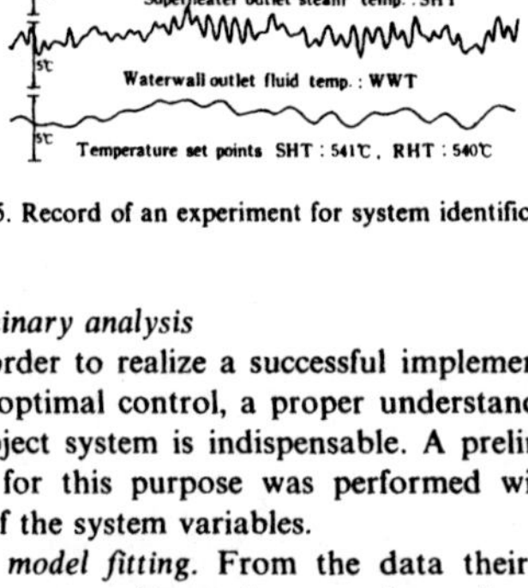

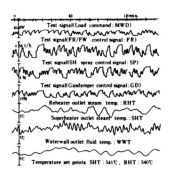

FIG. 5. Record of an experiment for system identification.

Preliminary analysis

In order to realize a successful implementation of an optimal control, a proper understanding of the object system is indispensable. A preliminary study for this purpose was performed with the data of the system variables.

AR model fitting. From the data their mean values were subtracted to produce the vector time series $\mathbf{X}(n)$, $n=1, 2, \ldots, N$. To $\mathbf{X}(n)$ we fitted AR model defined by

$$\mathbf{X}(n)=\sum_{m=1}^{M} \mathbf{A}(m)\mathbf{x}(n-m)+\mathbf{U}(n), \qquad (1)$$

where $\mathbf{A}(m)$, $m=1, 2, \ldots, M$, are the coefficient matrices, M is the order of the model, and $\mathbf{U}(n)$ the innovation vector. The maximum likelihood estimates of $\mathbf{A}(m)$'s are obtained under the assumption of Gaussian process by solving the Yule–Waker equation. A Levinson type efficient computational procedure is available to solve the equation. M is determined as the order that minimizes the information criterion AIC.

Analysis of innovation. Analysis of the correlation coefficients between the elements of innovation $U(n)$ in (1) is useful for the selection of system variables for final use. If some of the process variables (or manipulated variables) show strong correlations among their innovations, their simultaneous inclusion into the model must be avoided. A typical example which explains the application of this idea to the thermal power plant data is given in Akaike (1978a).

Power contribution analysis. By assuming the orthogonality between the elements of the innovation $\mathbf{U}(n)$, the relative power contribution of the elements of $\mathbf{U}(n)$ to an element of $\mathbf{X}(n)$ can be computed; i.e. the mean square variation of $x_i(n)$, the ith component of $\mathbf{X}(n)$, is decomposed into the sum of the contributions from $u_j(n)$, the jth component of $\mathbf{U}(n)$. The corresponding decomposition of the power spectrum is also possible. This procedure was applied to the data of the system variables, including MWD, steam pressure and fluid temperatures along the boiler tube. By the analysis to be explained below, the variables which were found less relevant to the behaviors of SHT and RHT were eliminated and the four variables, MWD, WWT, SHT and RHT, were picked up in addition to the three manipulated variables, FR, SP and GD.

The lower three frames of Fig. 6 show the decompositions of the power spectra of WWT, SHT and RHT in relative scales. It can be seen that to SHT the contributions from MWD, WWT, FR and SP are significant in the low frequency band, while in the frequency band higher than about 0.15 cycles per min the contribution of SP is dominant. As to RHT, the contribution of GD in the low frequency band and that of MWD in the high frequency band are significant. It is also seen that WWT, which seems to be an index of thermal-hydraulic balance in the evaporator, will effectively be controlled in the low frequency band by the manipulation of FR and GD.

These observations led to the conclusion that inclusion of WWT and MWD into the non-

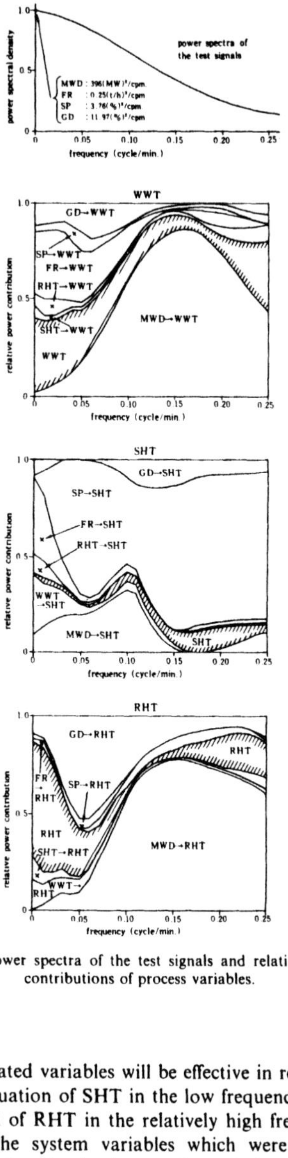

FIG. 6. Power spectra of the test signals and relative power contributions of process variables.

manipulated variables will be effective in reducing the fluctuation of SHT in the low frequency band and that of RHT in the relatively high frequency band. The system variables which were finally adopted for the implementation of the optimal controller are as follows:

Non-manipulated variables

MWD:	megawatt demand (rate of change)
WWT:	water wall outlet fluid temperature (rate of change)
SHT:	superheater outlet steam temperature (deviation from set-point)
RHT:	reheater outlet steam temperature (deviation from set-point)

Manipulated variables

FR:	fuel rate control signal from computer
SP:	super heater spray control signal from computer
GD:	gas damper opening control signal from computer.

Optimal controller design

State space representation of the system dynamics. To obtain a state equation, the k-dimensional system variable vector $\mathbf{X}(n)$ is divided into two subvectors $\mathbf{x}(n)$ and $\mathbf{y}(n)$, where $\mathbf{x}(n)$ is an r-dimensional vector composed of non-manipulated variables, and $\mathbf{y}(n)$ is an l-dimensional vector composed of manipulated variables. Then we fit the model

$$\mathbf{x}(n)=\sum_{m=1} a_m \mathbf{x}(n-m) + \sum_{m=1}^{M} b_m \mathbf{y}(n-m)+\mathbf{u}(n) \qquad (2)$$

to $\mathbf{x}(n)$ and $\mathbf{y}(n)$ $(n=1, 2,\ldots, N)$, where $\mathbf{u}(n)$ is an r vector composed of the innovations of $\mathbf{x}(n)$, and a_m, b_m are obtained from the coefficient matrices $\mathbf{A}(m)$ of (1) by the relation

$$\mathbf{A}(m)= \overset{\longleftarrow r \longrightarrow \quad \longleftarrow l \longrightarrow}{\begin{bmatrix} \mathbf{a}_m & \mathbf{b}_m \\ * & * \end{bmatrix}} \begin{matrix} \updownarrow r \\ \updownarrow l \end{matrix} \; \Big\updownarrow k,$$

where $*$ denotes the quantities irrelevant for the controller design.

The order of the model, M in (2), is determined by the FPEC criterion (Otomo, Nakagawa and Akaike, 1972, p. 47), or equivalently by the AIC defined by assuming the Gaussian process and the independence between the innovations of $\mathbf{x}(n)$ and $\mathbf{y}(n)$. We define the r-dimensional vectors $\mathbf{x}_0(n), \mathbf{x}_1(n),\ldots, \mathbf{x}_{M-1}(n)$ by

$$\mathbf{x}_0(n)=\mathbf{x}(n)$$

$$\mathbf{x}_k(n)= \sum_{m=k+1}^{M} \mathbf{a}_m \mathbf{x}(n+k-m) + \mathbf{b}_m \mathbf{y}(n+k-m) \qquad k=1,2,\ldots,M-1.$$

Then we get

$$\mathbf{x}_0(n)=\mathbf{a}_1\mathbf{x}_0(n-1)+\mathbf{b}_1\mathbf{y}(n-1) + \mathbf{x}_1(n-1)+\mathbf{u}(n)$$

$$\mathbf{x}_k(n)=\mathbf{a}_{k+1}\mathbf{x}_0(n-1)+\mathbf{b}_{k+1}\mathbf{y}(n-1) + \mathbf{x}_{k+1}(n-1), \quad k=1,2,\ldots,M-2$$

$$\mathbf{x}_{M-1}(n)=\mathbf{a}_M\mathbf{x}_0(n-1)+\mathbf{b}_M\mathbf{y}(n-1).$$

In the above equations, $\mathbf{x}_k(n)$ is the vector expressing the effect of $\mathbf{x}$ and $\mathbf{y}$ at $n-1, n-2, \ldots, n-M$ on $\mathbf{x}$ at $n+k$ $(k=1, 2, \ldots, M-1)$.

If we define $\mathbf{Z}(n)$ by

$$\mathbf{Z}(n)=\begin{bmatrix} \mathbf{x}_0(n) \\ \mathbf{x}_1(n) \\ \vdots \\ \mathbf{x}_{M-1}(n) \end{bmatrix} \begin{matrix} \updownarrow r \\ \\ Mr, \\ \\ \end{matrix}$$

we get the desired state equation

$$\left.\begin{aligned} \mathbf{Z}(n)&=\mathbf{\Phi}\cdot\mathbf{Z}(n-1)+\mathbf{\Gamma}\cdot\mathbf{y}(n-1)+\mathbf{u}(n) \\ \mathbf{X}(n)&=\mathbf{H}\cdot\mathbf{Z}(n) \end{aligned}\right\}, \quad (3)$$

where

$$\mathbf{\Phi}=\begin{bmatrix} \mathbf{a}_1 & \mathbf{I} & \mathbf{0} & \ldots & \mathbf{0} \\ \mathbf{a}_2 & \mathbf{0} & \mathbf{I} & \ldots & \mathbf{0} \\ \vdots & & & & \vdots \\ \mathbf{a}_{M-1} & \mathbf{0} & \mathbf{0} & \ldots & \mathbf{I} \\ \mathbf{a}_M & \mathbf{0} & \mathbf{0} & \ldots & \mathbf{0} \end{bmatrix} \quad (\text{blocks } r \times r;\ \text{overall } Mr \times Mr),$$

$$\mathbf{\Gamma}=\begin{bmatrix} \mathbf{b}_1 \\ \mathbf{b}_2 \\ \vdots \\ \mathbf{b}_{M-1} \\ \mathbf{b}_M \end{bmatrix} \quad (\text{blocks } r \times l;\ Mr \text{ rows}),$$

$$\mathbf{U}(n)=\begin{bmatrix} \mathbf{u}(n) \\ \mathbf{0} \\ \vdots \\ \mathbf{0} \end{bmatrix} \begin{matrix} \updownarrow r \\ \\ Mr, \\ \\ \end{matrix}$$

$$\mathbf{H}=[\mathbf{I} \quad \mathbf{0} \quad \ldots \quad \mathbf{0}] \quad (r \text{ rows}, \ Mr \text{ columns; } \mathbf{I} \text{ is } r\times r)$$

Computation of state-feedback gain matrix

The criterion function for the determination of the feedback gain matrix is defined by

$$J=E\sum_{i=1}^{K}(\mathbf{Z}'(i)\cdot\mathbf{Q}\cdot\mathbf{Z}(i) + \mathbf{y}'(i-1)\cdot\mathbf{R}\cdot\mathbf{y}(i-1)) \quad (4)$$

where denotes the transpose, $\mathbf{Z}(i)$ and $\mathbf{y}(i-1)$ are the vectors defined in (3), and E denotes the statistical expectation. $\mathbf{Q}$ and $\mathbf{R}$ in (4) are non-negative definite matrices whose elements specify the weights to adjust the variances of $\mathbf{Z}(i)$ and $\mathbf{y}(i)$.

By minimizing (4) with respect to $\mathbf{y}(i-1)$ $(i=1, 2, \ldots, K)$ by the DP procedure we obtain $\mathbf{y}(1)=\mathbf{G}(1)\mathbf{Z}(1)$. In the optimal control, we set the gain matrix $\mathbf{G}$ equal to $\mathbf{G}(1)$ and the optimal input at each time point n is given by $\mathbf{y}(n)=\mathbf{G}\mathbf{Z}(n)$. In (4), K, or the span of DP, is an integer which is chosen so that a further increase of K would not cause a significant change of G. To avoid the numerical difficulty the use of unnecessarily large values of K must be avoided.

Practical considerations in optimal controller implementation

Adjustment of PID controllers. Since ADC consists of APC and computer control, proper adjustment of APC is very important to obtain a desirable control performance. For this purpose, the thermo-hydraulic balance of the boiler was adjusted for both steady and transient states by the adjustment of PID controllers within the feedforward loops that control feedwater and fuel rates. The analog function generators and controllers that determine the water to fuel ratio and the overfueling or underfueling rate were carefully tuned to realize a good adjustment. As the result of this, even under the APC, steam temperature fluctuations for both step and ramp changes in MWD were remarkably reduced.

Further, the superheater spray valve and the gasdamper were adjusted to take the middle positions of their control ranges at steady state so that their linearities will be kept for large load disturbances. Such preparations are quite important to obtain the expected control performance.

Load adaptive adjustment of control parameters. It is well known that the dynamic properties of a thermal power plant vary depending upon the plant load. This is the most significant non-linearity in the case of a thermal power plant. To compensate for this non-linear effect, two pairs of state equation and feedback gain matrix were obtained, one for the load level of 485 MW and the other for 265 MW, and the state equation and gain matrix were defined at each control time by a linear interpolation of the two pairs, defined as a linear function of MWD.

Checking by a power plant simulator. The effectiveness of the parameter adjustment for the compensation of the non-linearity was first checked by a series of experiments on a hybrid power plant simulator (Nakamura, Hirano and Akaike, 1978). Figure 6 shows the result of one particular experiment by a hybrid simulator developed by the Central Research Institute of Electric Power Industry of Japan. A full-scale model of a

500 MW supercritical power plant with APC was set up on the analog-digital power plant simulator to simulate the dynamics of the actual plant. The solid lines in Fig. 7 show the response of steam temperatures and the control signals to a 200 MW rampwise load change without the load adaptive adjustment of control parameters, while the dotted lines show the response when the parameters were adjusted by the interpolation procedure.

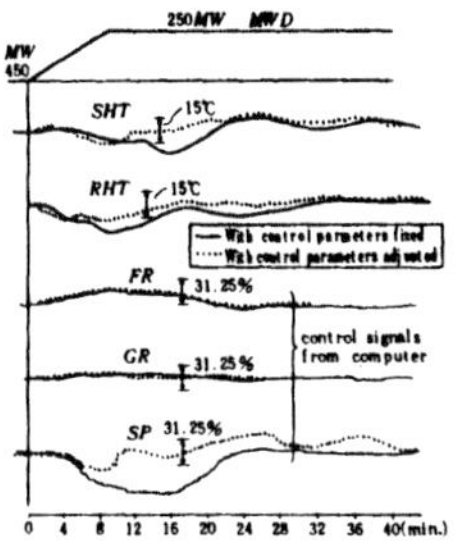

FIG. 7. Effect of load-adaptive parameter adjustment.

The results shown in Fig. 7 demonstrate the effectiveness of the load adaptive parameter adjustment. Both the fluctuations of steam temperatures and the amplitudes of the control signals are remarkably reduced by the adoption of this parameter adjustment.

Checking by digital simulation

The appropriateness of the state equation and the gain matrix is also checked by digital simulations. It is often considered that by the use of the optimal controller of the present type we lose the physical insight of the controller. In the multivariable feedback system the high dimensionality of the variables precludes the straightforward physical interpretation of related parameters.

Monte Carlo experiment. In the Monte Carlo experiment, we realize the transition of the state vector $\mathbf{Z}(n)$ in (3) by defining $\mathbf{u}(n)$ at each n by the white noise whose statistical properties are identical to those of the innovation of the AR model. The moves of WWT, SHT and RHT in $\mathbf{x}_0(n)$ thus obtained represent the temperature fluctuations of the system driven by the process noise. In this experiment, if we replace $\mathbf{y}(n-1)$ in (3) by the vector $\mathbf{y}^0(n-1)=\mathbf{GZ}(n-1)$, the performance of the optimal control can be estimated. If a zero vector is used instead of $\mathbf{y}^0(n-1)$, the behavior of the system without computer control, namely the performance of APC, is simulated. An example of the Monte Carlo experiment is illustrated in Akaike (1978b).

Response evaluation by a specific load change. As described in Table 2, $\mathbf{x}_0(n)$ of the state vector $\mathbf{Z}(n)$ in (3) consists of the four components, MWD, WWT, SHT and RHT. To evaluate the response of the system to a significant change of the load command the state transition is performed by replacing the MWD element of $\mathbf{x}_0(n-1)$ at each step by a specific value of the load command at that time, while keeping the elements of $\mathbf{u}(n)$ in (3) equal to zero. The control performance with or without computer control is obtained in the same manner as in the case of Monte Carlo experiment, i.e. by replacing $\mathbf{y}(n-1)$ with $\mathbf{y}^0(n-1)$ or by a zero vector.

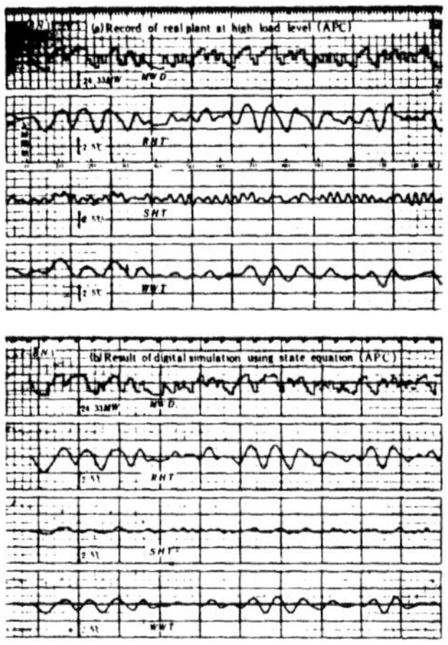

FIG. 8. Comparison of real plant behavior and the digital simulation of APC.

Figure 8 shows the result of a simulation at the high load level. In Fig. 8, (a) represents the record of the actual plant obtained under the APC and (b) is the result of the simulation of the same operating condition. These records show quite a good agreement with each other. From this result it was concluded that the state equation derived by the procedure described in the preceding section provided an adequate representation of the power plant. A similar experiment was also carried out for the low load level and the appropriateness of the state equation was confirmed.

Another examples of the digital simulation is illustrated in Fig. 9 which shows a result of the simulation to estimate control performance of ADC. In Fig. 9, (a) shows the response of APC and (b) that of ADC. In this simulation, the load adaptive adjustment of the parameters of the state equation and the feedback gain matrix was performed. The responses of the real and

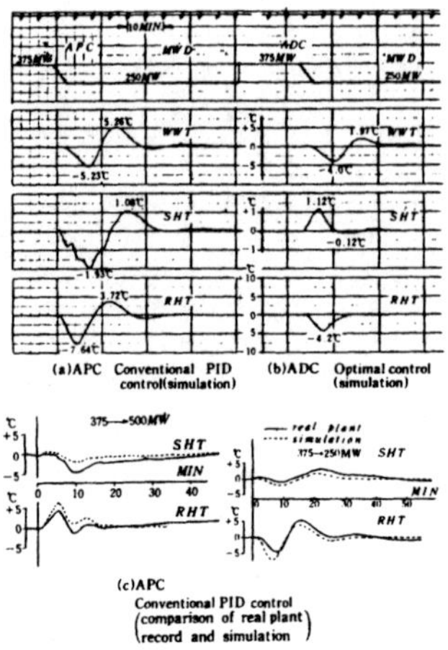

FIG. 9. Estimation of control performance of ADC by the digital simulation.

identified plant under APC are illustrated in (c) to confirm the adequateness of the identified model. The comparison of (a) and (b) suggests a remarkable improvement of the temperature control by the adoption of ADC.

Final field test

Adjustment of the feedback gain. The most elaborate job in the implementation of the quadratic optimal regulator is the determination of the feedback gain matrix **G**. Since **G** is determined by the weighting matrix **Q** and **R** in the DP computation, the problem is how to choose proper **Q** and **R**. The matrix that was used in the actual computation was $Mr \times Mr$ matrix ($r=4$, $M=11$, for the high load level) with zeros except for the diagonal elements which correspond to $\mathbf{x}_0(n)$ in (3). As for the matrix **R**, an $l \times l$ ($l=3$) diagonal matrix was used. The element of **Q** corresponding to MWD of $\mathbf{x}_0(n)$ was always put equal to zero to allow free movement of MWD and the rest were each put equal to the reciprocal of the variance of the element of $\mathbf{u}(n)$ determined by the identification. The diagonal elements of **R** were put equal to the reciprocals of the variances of the manipulated variables during the identification experiment. The design of the gain matrix **G** was started with these **Q** and **R** and was then revised by taking into account the results of simulations.

Final determination of the gain matrix was made in a trial-and-error manner by the experiment with the actual plant. In the experiment the appropriateness of **G** was first checked by observing the control performance under a step change in MWD. The feedback gain matrix of the optimal controller was tuned at each load level by a stepwise load change. Figure 10 shows the process of controller tuning at the low load level. The figure shows the responses of the system variables under the optimal control designed by various choices of the weighting matrices **Q** and **R**. These are the diagonal matrices defined with the diagonal elements diag. **Q** = [ ] and diag. **R** = [ ] given in the figure. The tuning was started with diag. **Q** = [0, 23, 70. 220] and diag. **R** = [320, 4.5, 3], and through the process shown in Fig. 10 the feedback gain matrix coresponding to diag. **Q** = [0, 10, 70, 220] and diag. **R** = [800, 24, 16] was selected as the final choice. By a similar process the feedback gain matrix corresponding to diag. **Q** = [0, 10, 80, 240] and diag. **R** = [600, 30, 12] was chosen for the high load level. The appropriateness of the gain matrix was also checked by observing the amplitude and frequency characteristics of the control signals issued from the computer to the process.

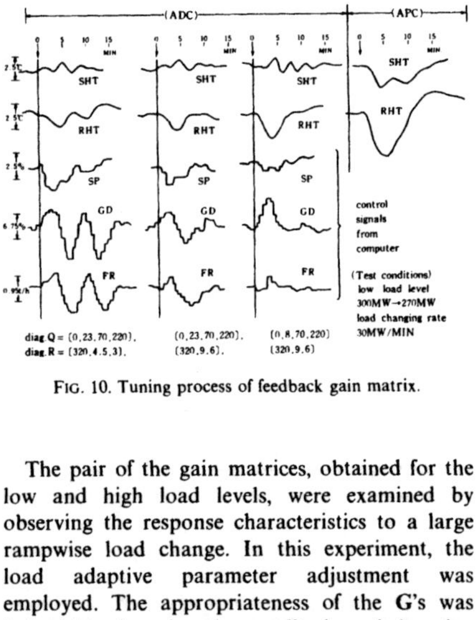

FIG. 10. Tuning process of feedback gain matrix.

The pair of the gain matrices, obtained for the low and high load levels, were examined by observing the response characteristics to a large rampwise load change. In this experiment, the load adaptive parameter adjustment was employed. The appropriateness of the **G**'s was judged by observing the amplitude and damping characteristics of the process and manipulated variables.

Control performance. The optimal control of the plant by ADC is realized as follows:

when $\mathbf{x}(n)$, the vector consisting of the actual values of MWD, WWT, SHT, and RHT at the control time n, is taken into the computer, $\mathbf{x}_0(n)$ of $\mathbf{Z}(n)$, which is calculated at the previous control time, is replaced by $\mathbf{x}(n)$. The gain matrix is then adjusted to the present load. The optimal manipulation vector is calculated as the product of G and $\mathbf{Z}(n)$ and sent to the process via the process input/output devices. The parameters of the state equation are then adjusted to the present load, and the computer computes $\mathbf{Z}(n)$ to be used at the next control time. The control interval is the same as the sampling interval of the data used for the identification and controller design, namely 40 s. The whole process of identification and control of ADC can be realized by a mini-computer. The detail of the computer control algorithm is discussed by Kato and co-workers (1978).

Figure 11 shows the record of temperature responses to 250 MW rampwise load changes. On the left and right hand side of Fig. 11 are shown the responses under ADC and APC, respectively. As can be seen by this figure the improvement of the temperature control by the adoption of the optimal control is remarkable, while the signals manipulating the plant remain well within the allowable ranges. Taking into account the fact that the APC was already very finely adjusted, this improvement obtained by ADC must be considered as a demonstration of the importance of an appropriate identification in control.

OPERATING EXPERIENCES

The comparison of control performance of ADC and APC under routine operation

The ADC system of Buzen No. 1 unit was put into routine operation in February 1978. Figure 12 shows the records of steam temperatures of the plant under ADC and APC operating under the load command from the power system's dispatch center. At the middle of Fig. 12 the control is switched from ADC to APC. By comparing the records on both sides of Fig. 12, the increase of low frequency fluctuations of steam temperatures by switching to APC can be seen quite clearly. This demonstrates the superiority of ADC to APC. At the time of switching the outputs from the digital computer to the control equipments are gradually reduced to zero with the aid of some analog memories, and there are no significant transients due to the switching. Similar results are also reported by Nakamura and co-workers (1979).

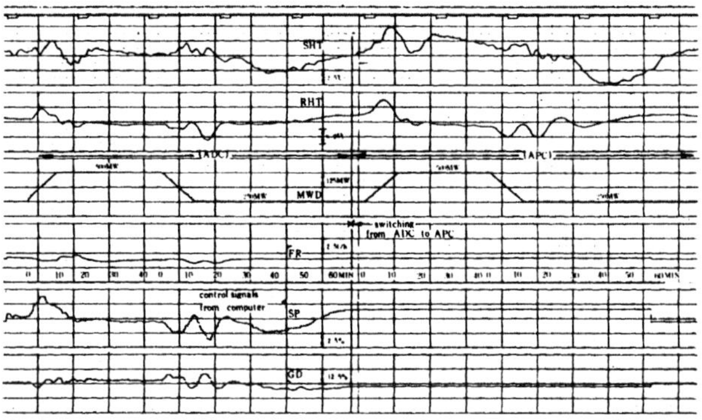

FIG. 11. Comparison of control performances of ADC and APC (Field test results).

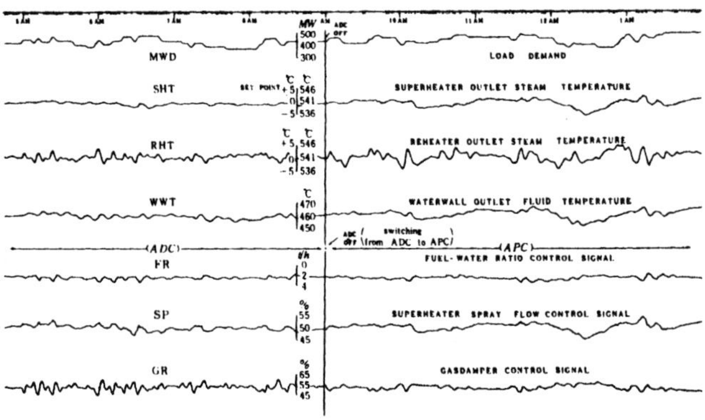

FIG. 12. Comparison of ADC and APC in commercial operation.

Stability of ADC

It was observed that ADC is quite useful in stabilizing the system against extraordinarily large disturbances. Figure 13 shows an example in which the control was switched from APC to ADC during the disturbance caused by the starting of one of the two forced fans. It can be seen that by putting ADC into service the plant disturbance is quickly suppressed and the system is swiftly restored to its normal operation.

After about one year's continuous operation the plant was shut down for regular inspection. When the plant was restarted, in spite of some change of the performance of the plant due to the cleaning of the boiler tubes, the ADC produced completely satisfactory control without any modifications of the parameters of the state equation and the feedback gain. This suggests a significant robustness of the present ADC.

Coordination between analog and digital control

In view of the sensitivity of the performance of the analog control to the change of the system characteristics, very critical tuning of the analog control under a particular condition of the plant is not quite recommendable. Our experience suggests that it is best to leave the work of prediction to the digital part and let the analog part concentrate on the control of the low frequency behavior of the plant. Also this makes the tuning of both the analog and digital control easier.

Performance against non-linearity

In the case of Shinkokura No. 3 unit, it was observed that, in spite of the existence of

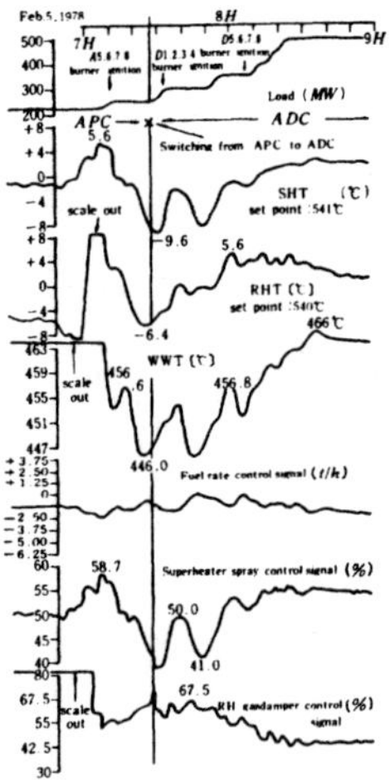

FIG. 13. Control performance of ADC during extraordinary plant disturbance.

considerable hysteresis of the gasdamper action, the optimal controller designed on the basis of the linear AR model again produced far better control, particularly of RHT, than the conventional PID controller.

Long run performance

By the survey of the Federation of Electric Power Companies, Buzen No. 1 unit was rated as No. 1 and 3 among all the Japanese thermal power plants in terms of nominal efficiency (total output MW hour/total calories of input fuel) during the early and late six months of 1979, respectively. Taking into account that the unit was under very frequent and significant load changes and that the sulphur content of fuel was rather high this result is quite remarkable.

Both Buzen No. 1 and Shinkokura No. 3 units established records of running over 390 days without interruption. Although these results are not entirely due to the contribution of ADC, the fact that these units are both equipped with ADC cannot be ignored. It is the performance of these units that lead the Kyushu Electric Power Company to the decision to equip Buzen No. 2 unit, another 500 MW supercritical unit, with ADC.

CONCLUSION

The optimal control system ADC based on the multivariate AR model fitting has been found quite practical and useful for the control of thermal electric power plants. The ADC system produces excellent control under normal operating condition and also acts as a very effective stabilizer when the plant is exposed to an extraordinary disturbance. This usefulness and the simplicity of its implementation makes the ADC a very practical new control system of thermal electric power plants.

The satisfactory performance of ADC in commercial operation confirms the fact that a successful identification is the basis of a successful control.

Acknowledgements—The authors are very grateful to Mr M. Uchida of Kyushu Electric Power Company, Messrs T. Kitami and H. Mizutani of Central Research Institute of Electric Power Industry, Messrs N. Kato and K. Kawai of Tokyo Shibaura Electric Company, and many others who participated in the implementation of ADC. Thanks are also due to Professors M. Takada, K. Sekoguchi and A. Mohri of Kyushu University for their helpful advice on the boiler control. The authors are also grateful to the referees for helpful comments.

REFERENCES

Akaike, H. and T. Nakagawa (1972). *Statistical Analysis and Control of Dynamic Systems*, Saiensu-sha, Tokyo (1972). (In Japanese, with a computer program package TIMSAC written in FORTRAN IV with English comments.)

Akaike, H. (1976). Canonical correlation analysis of time series and the use of an information criterion. In R. K. Mehra and D. G. Lainiotis (Eds.), *System Identification: Advances and Case Studies*. Academic Press, New York, pp. 27–96.

Akaike, H. (1978a). On the identification of state space models and their use in control. In D. R. Brillinger and G. C. Tiao (Ed.), *Direction in Time Series*. The Institute of Mathematical Statistics, Hyward, California, pp. 175–187.

Akaike, H. (1978b). On newer statistical approaches to parameter estimation and structure identification. In A. Niemi (Ed.), *A Link Between Science and Applications of Automatic Control*, Vol. 3, Pergamon Press, Oxford, pp. 1877–1887.

Kato, N., K. Kawai, J. Miyake, M. Uchida and H. Nakamura (1978). Application of computer control system COPOS to fossil fired power generation plant. *3rd USA-JAPAN Computer Conference, 1978*, pp. 505–509.

Lecrique, M., M. Tessier, A. Rault and J. L. Testud (1978). *Multivariable Control of A Steam Generator. Characteristics and Results*. In A. Niemi (Ed.), *A Link Between Science and Applications of Automatic Control*, Vol. 1, Pergamon Press, Oxford, pp. 73–80.

Nakamura, H., T. Hirano and H. Akaike (1978). Statistical approach to the optimal control of a thermal power plant. *Trans. IEE Japan*, **98-B**, 601 (in Japanese).

Nakamura, H., M. Uchida, T. Kitami and H. Akaike (1979). Application of optimal control system to a supercritical thermal power plant. *1979 Control of Power System Conference and Exposition Conference Record*. IEEE, New York, pp. 10–14.

Otomo, T., T. Nakagawa, and H. Akaike (1972). Statistical approach to computer control of cement rotary kilns. *Automatica*, **8**, 35.

Richalet, J., A. Rault, J. L. Testud and J. Papon (1978). Model predictive heuristic control: application to industrial processes. *Automatica* **14**, 413–428.

Smoak, R. A. (1976). On the application of modern control to power plants. *ASME Trans., J. Dyn. Syst., Measurement Control*, **98**, 320.

Reprinted from *Breakthroughs in Statistics, Vol.I, Foundations and Basic Theory*, S. Kotz and N.L.Johnson, eds., Springer-Verlag, New York, 1992, 610-624 by permission of Akademia Kiado and Springer-Verlag, New York. (Originally published in Proceeding of the Second International Symposium on Information Theory, B.N. Petrov and F. Caski, eds., Akademiai Kiado, Budapest, 1973, 267-281)

Information Theory and an Extension of the Maximum Likelihood Principle

Hirotogu Akaike
Institute of Statistical Mathematics

Abstract

In this paper it is shown that the classical maximum likelihood principle can be considered to be a method of asymptotic realization of an optimum estimate with respect to a very general information theoretic criterion. This observation shows an extension of the principle to provide answers to many practical problems of statistical model fitting.

1. Introduction

The extension of the maximum likelihood principle which we are proposing in this paper was first announced by the author in a recent paper [6] in the following form:

Given a set of estimates $\hat{\theta}$ of the vector of parameters θ of a probability distribution with density function $f(x|\theta)$ we adopt as our final estimate the one which will give the maximum of the expected log-likelihood, which is by definition

$$E \log f(X|\hat{\theta}) = E \int f(x|\theta) \log f(x|\hat{\theta})\, dx, \tag{1.1}$$

where X is a random variable following the distribution with the density function $f(x|\theta)$ and is independent of $\hat{\theta}$.

This seems to be a formal extension of the classical maximum likelihood principle but a simple reflection shows that this is equivalent to maximizing an information theoretic quantity which is given by the definition

$$E \log\left(\frac{f(X|\hat{\theta})}{f(X|\theta)}\right) = E \int f(x|\theta) \log\left(\frac{f(x|\hat{\theta})}{f(x|\theta)}\right) dx. \tag{1.2}$$

The integral in the right-hand side of the above equation gives the Kullback-Leibler's mean information for discrimination between $f(x|\hat{\theta})$ and $f(x|\theta)$ and is known to give a measure of separation or distance between the two distributions [15]. This observation makes it clear that what we are proposing here is the adoption of an information theoretic quantity of the discrepancy between the estimated and the true probability distributions to define the loss function of an estimate $\hat{\theta}$ of θ. It is well recognized that the statistical estimation theory should and can be organized within the framework of the theory of statistical decision functions [25]. The only difficulty in realizing this is the choice of a proper loss function, a point which is discussed in details in a paper by Le Cam [17].

In the following sections it will be shown that our present choice of the information theoretic loss function is a very natural and reasonable one to develop a unified asymptotic theory of estimation. We will first discuss the definition of the amount of information and make clear the relative merit, in relation to the asymptotic estimation theory, of the Kullback–Leibler type information within the infinitely many possible alternatives. The discussion will reveal that the log-likelihood is essentially a more natural quantity than the simple likelihood to be used for the definition of the maximum likelihood principle.

Our extended maximum likelihood principle can most effectively be applied for the decision of the final estimate of a finite parameter model when many alternative maximum likelihood estimates are obtained corresponding to the various restrictions of the model. The log-likelihood ratio statistics developed for the test of composite hypotheses can most conveniently be used for this purpose and it reveals the truly statistical nature of the information theoretic quantities which have often been considered to be probabilistic rather than statistical [21].

With the aid of this log-likelihood ratio statistics our extended maximum likelihood principle can provide solutions for various important practical problems which have hitherto been treated as problems of statistical hypothesis testing rather than of statistical decision or estimation. Among the possible applications there are the decisions of the number of factors in the factor analysis, of the significant factors in the analysis of variance, of the number of independent variables to be included into multiple regression and of the order of autoregressive and other finite parameter models of stationary time series.

Numerical examples are given to illustrate the difference of our present approach from the conventional procedure of successive applications of statistical tests for the determination of the order of autoregressive models. The results will convincingly suggest that our new approach will eventually be replacing many of the hitherto developed conventional statistical procedures.

2. Information and Discrimination

It can be shown [9] that for the purpose of discrimination between the two probability distributions with density functions $f_i(x)$ $(i = 0, 1)$ all the necessary information are contained in the likelihood ratio $T(x) = f_1(x)/f_0(x)$ in the sense that any decision procedure with a prescribed loss of discriminating the two distributions based on a realization of a sample point x can, if it is realizable at all, equivalently be realized through the use of $T(x)$. If we consider that the information supplied by observing a realization of a (set of) random variable(s) is essentially summarized in its effect of leading us to the discrimination of various hypotheses, it will be reasonable to assume that the amount of information obtained by observing a realization x must be a function of $T(x) = f_1(x)/f_0(x)$.

Following the above observation, the natural definition of the mean amount of information for discrimination per observation when the actual distribution is $f_0(x)$ will be given by

$$I(f_1, f_0; \Phi) = \int \Phi\left(\frac{f_1(x)}{f_0(x)}\right) f_0(x)\, \mathrm{d}x, \tag{2.1}$$

where $\Phi(r)$ is a properly chosen function of r and $\mathrm{d}x$ denotes the measure with respect to which $f_i(x)$ are defined. We shall hereafter be concerned with the parametric situation where the densities are specified by a set of parameters θ in the form

$$f(x) = f(x|\theta), \tag{2.2}$$

where it is assumed that θ is an L-dimensional vector, $\theta = (\theta_1, \theta_2, \ldots, \theta_L)'$, where $'$ denotes the transpose. We assume that the true distribution under observation is specified by $\theta = \uptheta = (\uptheta_1, \uptheta_2, \ldots, \uptheta_L)'$. We Will denote by $I(\theta, \uptheta; \Phi)$ the quantity defined by (2.1) with $f_1(x) = f(x|\theta)$ and $f_0(x) = f(x|\uptheta)$ and analyze the sensitivity of $I(\theta, \uptheta; \Phi)$ to the deviation of θ from $\uptheta$. Assuming the regularity conditions of $f(x|\theta)$ and $\Phi(r)$ which assure the following analytical treatment we get

$$\frac{\partial}{\partial\theta_1} I(\theta, \uptheta; \Phi)|_{\theta=\uptheta} = \int \left(\frac{\mathrm{d}}{\mathrm{d}r}\Phi(r)\frac{\partial r}{\partial\theta_l}\right)_{\theta=\uptheta} f_\uptheta\, \mathrm{d}x = \Phi(1)\int\left(\frac{\partial f_\theta}{\partial\theta_l}\right)_{\theta=\uptheta} \mathrm{d}x \tag{2.3}$$

$$\begin{aligned}\frac{\partial^2}{\partial\theta_l\partial\theta_m} I(\theta, \uptheta; \Phi)|_{\theta=\uptheta} &= \int\left[\left(\frac{\mathrm{d}^2}{\mathrm{d}r^2}\Phi(r)\right)\left(\frac{\partial r}{\partial\theta_l}\right)\left(\frac{\partial r}{\partial\theta_m}\right)\right]_{\theta=\uptheta} f_\uptheta\, \mathrm{d}x \\ &\quad + \int\left[\left(\frac{\mathrm{d}}{\mathrm{d}r}\Phi(r)\right)\left(\frac{\partial^2 r}{\partial\theta_l\partial\theta_m}\right)\right]_{\theta=\uptheta} f_\uptheta\, \mathrm{d}x \\ &= \ddot{\Phi}(1)\int\left[\left(\frac{\partial f_\theta}{\partial\theta_l}\frac{1}{f_\theta}\right)\left(\frac{\partial f_\theta}{\partial\theta_m}\frac{1}{f_\theta}\right)\right]_{\theta=\uptheta} f_\uptheta\, \mathrm{d}x \\ &\quad + \dot{\Phi}(1)\int\left(\frac{\partial^2 f_\theta}{\partial\theta_l\partial\theta_m}\right)_{\theta=\uptheta} \mathrm{d}x,\end{aligned} \tag{2.4}$$

where r, $\dot{\Phi}(1)$, $\ddot{\Phi}(1)$ and f_θ denote $\frac{f(x|\theta)}{f(x|\theta)}$, $\left.\frac{d\Phi(r)}{dr}\right|_{r=1}$, $\left.\frac{d^2\Phi(r)}{dr^2}\right|_{r=1}$ and $f(x|\theta)$, respectively, and the meaning of the other quantities will be clear from the context. Taking into account that we are assuming the validity of differentiation under integral sign and that $\int f(x|\theta)\,dx = 1$, we have

$$\int \left(\frac{\partial f}{\partial \theta_l}\right) dx = \int \left(\frac{\partial^2 f}{\partial \theta_l \partial \theta_m}\right) dx = 0. \tag{2.5}$$

Thus we get

$$I(\theta, \theta; \Phi) = \Phi(1) \tag{2.6}$$

$$\frac{\partial}{\partial \theta_l} I(\theta, \theta; \Phi)|_{\theta=\theta} = 0 \tag{2.7}$$

$$\frac{\partial^2}{\partial \theta_l \partial \theta_m} I(\theta, \theta; \Phi)|_{\theta=\theta} = \ddot{\Phi}(1) \int \left[\left(\frac{\partial f_\theta}{\partial \theta_l}\frac{1}{f_\theta}\right)\left(\frac{\partial f_\theta}{\partial \theta_m}\frac{1}{f_\theta}\right)\right]_{\theta=\theta} f_\theta\, dx. \tag{2.8}$$

These relations show that $\ddot{\Phi}(1)$ must be different from zero if $I(\theta, \theta; \Phi)$ ought to be sensitive to the small variations of θ. Also it is clear that the relative sensitivity of $I(\theta, \theta; \Phi)$ is high when $\left|\frac{\ddot{\Phi}(1)}{\Phi(1)}\right|$ is large. This will be the case when $\Phi(1) = 0$. The integral on the right-hand side of (2.8) defines the (l, m)th element of Fisher's information matrix [16] and the above results show that this matrix is playing a central role in determining the behaviour of our mean information $I(\theta, \theta; \Phi)$ for small variations of θ around θ. The possible forms of $\Phi(r)$ are e.g. $\log r$, $(r-1)^2$ and $r^{1/2}$ and we cannot decide uniquely at this stage.

To restrict further the form of $\Phi(r)$ we consider the effect of the increase of information by N independent observations of X. For this case we have to consider the quantity

$$I_N(\theta, \theta; \Phi) = \int \Phi \frac{\prod_{i=1}^{N} f(x_i|\theta)}{\prod_{i=1}^{N} f(x_i|\theta)} \prod_{i=1}^{N} f(x_i|\theta)\, dx_1 \ldots dx_N. \tag{2.9}$$

Corresponding to (2.5), (2.6) and (2.7) we have

$$I_N(\theta, \theta; \Phi) = I(\theta, \theta; \Phi) \tag{2.10}$$

$$\frac{\partial}{\partial \theta_l} I_N(\theta, \theta; \Phi)|_{\theta=\theta} = 0 \tag{2.11}$$

$$\frac{\partial^2}{\partial \theta_l \partial \theta_m} I_N(\theta, \theta; \Phi)|_{\theta=\theta} = N \frac{\partial^2}{\partial \theta_l \partial \theta_m} I(\theta, \theta; \Phi)|_{\theta=\theta}. \tag{2.12}$$

These equations show that $I_N(\theta, \theta; \Phi)$ is not responsive to the increase of information and that $\frac{\partial^2}{\partial\theta_l\partial\theta_m} I_N(\theta, \theta; \Phi)|_{\theta=\theta}$ is in a linear relation with N. It can be seen that only the quantity defined by

$$\frac{\partial \prod_{i=1}^{N} f(x_i|\theta)}{\partial\theta_l} \frac{1}{\prod_{i=1}^{N} f(x_i|\theta)}\Bigg|_{\theta=\theta} = \sum_{i=1}^{N} \left(\frac{\partial f(x_i|\theta)}{\partial\theta_l}\frac{1}{f_\theta}\right)_{\theta=\theta} \tag{2.13}$$

is concerned with the derivation of this last relation. This shows very clearly that taking into account the relation

$$\frac{\partial f(x|\theta)}{\partial\theta_l}\frac{1}{f_\theta} = \frac{\partial \log f(x|\theta)}{\partial\theta_l}, \tag{2.14}$$

the functions $\frac{\partial}{\partial\theta_l} \log f(x|\theta)$ are playing the central role in the present definition of information. This observation suggests the adoption of $\Phi(r) = \log r$ for the definition of our amount of information and we are very naturally led to the use of Kullback-Leibler's definition of information for the purpose of our present study.

It should be noted here that at least asymptotically any other definition of $\Phi(r)$ will be useful if only $\Phi(1)$ is not vanishing. The main point of our present observation will rather be the recognition of the essential role being played by the functions $\frac{\partial}{\partial\theta_l} \log f(x|\theta)$ for the definition of the mean information for the discrimination of the distributions corresponding to the small deviations of θ from θ.

3. Information and the Maximum Likelihood Principle

Since the purpose of estimating the parameters of $f(x|\theta)$ is to base our decision on $f(x|\hat{\theta})$, where $\hat{\theta}$ is an estimate of θ, the discussion in the preceding section suggests the adoption of the following loss and risk functions:

$$W(\theta, \hat{\theta}) = (-2) \int f(x|\theta) \log\left(\frac{f(x|\hat{\theta})}{f(x|\theta)}\right) dx \tag{3.1}$$

$$R(\theta, \hat{\theta}) = EW(\theta, \hat{\theta}), \tag{3.2}$$

where the expectation in the right-hand side of (3.2) is taken with respect to the distribution of $\hat{\theta}$. As $W(\theta, \hat{\theta})$ is equal to 2 times the Kullback-Leibler's information for discrimination in favour of $f(x|\theta)$ for $f(x|\hat{\theta})$ it is known that $W(\theta, \hat{\theta})$ is a non-negative quantity and is equal to zero if and only if $f(x|\theta) = f(x|\hat{\theta})$ almost everywhere [16]. This property is forming a basis of the proof of consistency of the maximum likelihood estimate of θ [24] and indicates the close relationship between the maximum likelihood principle and the information theoretic observations.

When N independent realizations x_i ($i = 1, 2, \ldots, N$) of X are available, (-2) times the sample mean of the log-likelihood ratio

$$\frac{1}{N}\sum_{i=1}^{N} \log\left(\frac{f(x_i|\hat{\theta})}{f(x_i|\theta)}\right) \tag{3.3}$$

will be a consistent estimate of $W(\theta, \hat{\theta})$. Thus it is quite natural to expect that, at least for large N, the value of $\hat{\theta}$ which will give the maximum of (3.3) will nearly minimize $W(\theta, \hat{\theta})$. Fortunately the maximization of (3.3) can be realized without knowing the true value of θ, giving the well-known maximum likelihood estimate $\hat{\theta}$. Though it has been said that the maximum likelihood principle is not based on any clearly defined optimum consideration [18; p. 15] our present observation has made it clear that it is essentially designed to keep minimum the estimated loss function which is very naturally defined as the mean information for discrimination between the estimated and the true distributions.

4. Extension of the Maximum Likelihood Principle

The maximum likelihood principle has mainly been utilized in two different branches of statistical theories. The first is the estimation theory where the method of maximum likelihood has been used extensively and the second is the test theory where the log-likelihood ratio statistic is playing a very important role. Our present definitions of $W(\theta, \hat{\theta})$ and $R(\theta, \hat{\theta})$ suggest that these two problems should be combined into a single problem of statistical decision. Thus instead of considering a single estimate of θ we consider estimates corresponding to various possible restrictions of the distribution and instead of treating the problem as a multiple decision or a test between hypotheses we treat it as a problem of general estimation procedure based on the decision theoretic consideration. This whole idea can be very simply realized by comparing $R(\theta, \hat{\theta})$, or $W(\theta, \hat{\theta})$ if possible, for various $\hat{\theta}$'s and taking the one with the minimum of $R(\theta, \hat{\theta})$ or $W(\theta, \hat{\theta})$ as our final choice. As it was discussed in the introduction this approach may be viewed as a natural extension of the classical maximum likelihood principle. The only problem in applying this extended principle in a practical situation is how to get the reliable estimates of $R(\theta, \hat{\theta})$ or $W(\theta, \hat{\theta})$. As it was noticed in [6] and will be seen shortly, this can be done for a very interesting and practically important situation of composite hypotheses through the use of the maximum likelihood estimates and the corresponding log-likelihood ratio statistics.

The problem of statistical model identification is often formulated as the problem of the selection of $f(x|{}_k\theta)$ ($k = 0, 1, 2, \ldots, L$) based on the observations of X, where ${}_k\theta$ is restricted to the space with ${}_k\theta_{k+1} = {}_k\theta_{k+2} = \cdots = {}_k\theta_L =$

0. k, or some of its equivalents, is often called the order of the model. Its decision is usually the most difficult problem in practical statistical model identification. The problem has often been treated as a subject of composite hypothesis testing and the use of the log-likelihood ratio criterion is well established for this purpose [23]. We consider the situation where the results x_i ($i = 1, 2, \ldots, N$) of N independent observations of X have been obtained. We denote by ${}_k\hat{\theta}$ the maximum likelihood estimate in the space of ${}_k\theta$, i.e., ${}_k\hat{\theta}$ is the value of ${}_k\theta$ which gives the maximum of the likelihood function $\prod_{i=1}^{N} f(x_i|{}_k\theta)$. The observation at the end of the preceding section strongly suggests the use of

$$ {}_k\omega_L = -\frac{2}{N}\sum_{i=1}^{N}\log\left(\frac{f(x_i|{}_k\hat{\theta})}{f(x_i|{}_L\hat{\theta})}\right) \tag{4.1}$$

as an estimate of $W(\theta, {}_k\hat{\theta})$. The statistics

$$ {}_k\eta_L = N \times {}_k\omega_L \tag{4.2}$$

is the familiar log-likelihood ratio test statistics which will asymptotically be distributed as a chi-square variable with the degrees of freedom equal to $L - k$ when the true parameter θ is in the space of ${}_k\theta$. If we define

$$ W(\theta, {}_k\theta) = \inf_{{}_k\theta} W(\theta, {}_k\theta), \tag{4.3}$$

then it is expected that

$$ {}_k\omega_L \to W(\theta, {}_k\theta) \text{ w.p.l.} $$

Thus when $NW(\theta, {}_k\theta)$ is significantly larger than L the value of ${}_k\eta_L$ will be very much larger than would be expected from the chi-square approximation. The only situation where a precise analysis of the behaviour of ${}_k\eta_L$ is necessary would be the case where $NW(\theta, {}_k\theta)$ is of comparable order of magnitude with L. When N is very large compared with L this means that $W(\theta, {}_k\theta)$ is very nearly equal to $W(\theta, \theta) = 0$. We shall hereafter assume that $W(\theta, \theta)$ is sufficiently smooth at $\theta = \theta$ and

$$ W(\theta, \theta) > 0 \qquad \text{for} \qquad \theta \neq \theta. \tag{4.4}$$

Also we assume that $W(\theta, {}_k\theta)$ has a unique minimum at ${}_k\theta = {}_k\theta$ and that ${}_L\theta = \theta$. Under these assumptions the maximum likelihood estimates $\hat{\theta}$ and ${}_k\hat{\theta}$ will be consistent estimates of θ and ${}_k\theta$, respectively, and since we are concerned with the situation where θ and ${}_k\theta$ are situated very near to each other, we limit our observation only up to the second-order variation of $W(\theta, {}_k\hat{\theta})$. Thus hereafter we adopt, in place of $W(\theta, {}_k\hat{\theta})$, the loss function

$$ W_2(\theta, {}_k\hat{\theta}) = \sum_{l=1}^{L}\sum_{m=1}^{L} ({}_k\hat{\theta}_l - \theta_l)({}_k\hat{\theta}_m - \theta_m)C(l, m)(\theta), \tag{4.5}$$

where $C(l, m)(\theta)$ is the (l, m)th element of Fisher's information matrix and is given by

$$C(l, m)(\theta) = \int \left(\frac{\partial f_\theta}{\partial \theta_l}\frac{1}{f_\theta}\right)\left(\frac{\partial f_\theta}{\partial \theta_m}\frac{1}{f_\theta}\right) f_\theta \, dx = -\int \left(\frac{\partial^2 \log f}{\partial \theta_l \partial \theta_m}\right) f_\theta \, dx. \tag{4.6}$$

We shall simply denote by $C(l, m)$ the value of $C(l, m)(\theta)$ at $\theta = \theta$. We denote by $\|\theta\|_c$ the norm in the space of θ defined by

$$\|\theta\|_c^2 = \sum_{l=1}^{L} \sum_{m=1}^{L} \theta_l \theta_m C(l, m). \tag{4.7}$$

We have

$$W_2(\theta, {}_k\hat{\theta}) = \|{}_k\hat{\theta} - \theta\|_c^2. \tag{4.8}$$

Also we redefine ${}_k\theta$ by the relation

$$\|{}_k\theta - \theta\|_c^2 = \operatorname*{Min}_{{}_k\theta} \|{}_k\theta - \theta\|_c^2. \tag{4.9}$$

Thus ${}_k\theta$ is the projection of θ in the space of ${}_k\theta$'s with respect to the metrics defined by $C(l, m)$ and is given by the relations

$$\sum_{m=1}^{k} C(l, m)_k\theta_m = \sum_{m=1}^{L} C(l, m)\theta_m \quad l = 1, 2, \ldots, k. \tag{4.10}$$

We get from (4.8) and (4.9)

$$W_2(\theta, {}_k\hat{\theta}) = \|{}_k\theta - \theta\|_c^2 + \|{}_k\hat{\theta} - {}_k\theta\|_c^2. \tag{4.11}$$

Since the definition of $W(\theta, \hat{\theta})$ strongly suggests, and is actually motivated by, the use of the log-likelihood ratio statistics we will study the possible use of this statistics for the estimation of $W_2(\theta, {}_k\hat{\theta})$. Taking into account the relations

$$\begin{aligned} \sum_i \frac{\partial \log f(x_i|\hat{\theta})}{\partial \theta_m} &= 0, \qquad m = 1, 2, \ldots, L, \\ \sum_i \frac{\partial \log f(x_i|{}_k\hat{\theta})}{\partial \theta_m} &= 0, \qquad m = 1, 2, \ldots, k, \end{aligned} \tag{4.12}$$

we get the Taylor expansions

$$\begin{aligned} \sum_{i=1}^{N} \log f(x_i|{}_k\theta) &= \sum_{i=1}^{N} \log f(x_i|\hat{\theta}) + \frac{1}{2} \sum_{m=1}^{L} \sum_{l=1}^{L} N({}_k\theta_m - \hat{\theta}_m)({}_k\theta_l - \hat{\theta}_l) \\ &\quad \times \frac{1}{N} \sum_{i=1}^{N} \frac{\partial^2 \log f(x_i|\hat{\theta} + \varrho({}_k\theta - \hat{\theta}))}{\partial \theta_m \partial \theta_l} \\ &= \sum_{i=1}^{N} \log f(x_i|{}_k\hat{\theta}) + \frac{1}{2} \sum_{m=1}^{k} \sum_{l=1}^{k} N({}_k\theta_m - {}_k\hat{\theta}_m)({}_k\theta_l - {}_k\hat{\theta}_l) \\ &\quad \times \frac{1}{N} \sum_{i=1}^{N} \frac{\partial^2 \log f(x_1|{}_k\hat{\theta} + \varrho_k({}_k\theta - {}_k\hat{\theta}))}{\partial \theta_m \partial \theta_l}, \end{aligned}$$

where the parameter values within the functions under the differential sign denote the points where the derivatives are taken and $0 \leq \varrho_k, \varrho \leq 1$, a convention which we use in the rest of this paper. We consider that, in increasing the value of N, N and k are chosen in such a way that $\sqrt{N}({}_k\theta_m - \theta_m)$ ($m = 1, 2, \ldots, L$) are bounded, or rather tending to a set of constants for the ease of explanation. Under this circumstance, assuming the tendency towards a Gaussian distribution of $\sqrt{N}(\hat{\theta} - \theta)$ and the consistency of ${}_k\hat{\theta}$ and $\hat{\theta}$ as the estimates of ${}_k\theta$ and θ we get, from (4.6) and (4.13), an asymptotic equality in distribution for the log-likelihood ratio statistic ${}_k\eta_L$ of (4.2)

$$ {}_k\eta_L = N\|\hat{\theta} - {}_k\theta\|_c^2 - N\|{}_k\hat{\theta} - {}_k\theta\|_c^2. \tag{4.14}$$

By simple manipulation

$$ {}_k\eta_L = N\|{}_k\theta - \theta\|_c^2 + N\|\hat{\theta} - \theta\|_c^2 - N\|{}_k\hat{\theta} - {}_k\theta\|_c^2 - 2N(\hat{\theta} - \theta, \theta - \theta)_c, \tag{4.15}$$

where $(,)_c$ denotes the inner product defined by $C(l, m)$. Assuming the validity of the Taylor expansion up to the second order and taking into account the relations (4.12) we get for $l = 1, 2, \ldots, k$

$$\begin{aligned} &\frac{1}{\sqrt{N}} \sum_{i=1}^{N} \frac{\partial}{\partial \theta_l} \log f(x_i|{}_k\theta) \\ &= \sum_{m=1}^{k} \sqrt{N}({}_k\theta_m - {}_k\hat{\theta}_m)\frac{1}{N}\sum_{i=1}^{N} \frac{\partial^2 \log f(x_i|{}_k\hat{\theta} + \varrho_k({}_k\theta - {}_k\hat{\theta}))}{\partial\theta_m \partial\theta_l} \\ &= \sum_{m=1}^{L} \sqrt{N}({}_k\theta_m - \hat{\theta}_m)\frac{1}{N}\sum_{i=1}^{N} \frac{\partial^2 \log f(x_i|\hat{\theta} + \varrho({}_k\theta - \hat{\theta}))}{\partial\theta_m \partial\theta_l}. \end{aligned} \tag{4.16}$$

Let C^{-1} be the inverse of Fisher's information matrix. Assuming the tendency to the Gaussian distribution $N(0, C^{-1})$ of the distribution of $\sqrt{N}(\hat{\theta} - \theta)$ which can be derived by using the Taylor expansion of the type of (4.16) at $\theta = \theta$, we can see that for N and k with bounded $\sqrt{N}({}_k\theta_m - \theta_m)$ ($m = 1, 2, \ldots, L$) (4.16) yields, under the smoothness assumption of $C(l, m)(\theta)$ at $\theta = \theta$, the approximate equations

$$\sum_{m=1}^{k} \sqrt{N}({}_k\theta_m - {}_k\hat{\theta}_m)C(l, m) = \sum_{m=1}^{L} \sqrt{N}({}_k\theta_m - \hat{\theta}_m)C(l, m) \quad l = 1, 2, \ldots, k. \tag{4.17}$$

Taking (4.10) into account we get from (4.17), for $l = 1, 2, \ldots, k$,

$$\sum_{m=1}^{k} \sqrt{N}({}_k\theta_m - {}_k\hat{\theta}_m)C(l, m) = \sum_{m=1}^{L} \sqrt{N}(\theta_m - \hat{\theta}_m)C(l, m). \tag{4.18}$$

This shows that geometrically ${}_k\hat{\theta} - {}_k\theta$ is (approximately) the projection of $\hat{\theta} - \theta$ into the space of ${}_k\theta$'s. From this result it can be shown that $N\|\hat{\theta} - \theta\|_c^2 - N\|{}_k\hat{\theta} - {}_k\theta\|_c^2$ and $N\|{}_k\hat{\theta} - {}_k\theta\|_c^2$ are asymptotically independently distributed as chi-square variables with the degrees of freedom $L - k$ and k, respectively. It can also be shown that the standard deviation of the asymptotic distribution of $N(\hat{\theta} - \theta, {}_k\theta - \theta)_c$ is equal to $\sqrt{N}\|{}_k\theta - \theta\|_c$. Thus if $N\|_k\theta - \theta\|_c^2$ is of comparable magnitude with $L - k$ or k and these are large integers then the contribution of the last term in the right hand side of (4.15) remains relatively insignificant. If $N\|_k\theta - \theta\|_c^2$ is significantly larger than L the contribution of $N(\hat{\theta} - \theta, {}_k\theta - \theta)_c$ to ${}_k\eta_L$ will also relatively be insignificant. If $N\|_k\theta - \theta\|_c^2$ is significantly smaller than L and k again the contribution of $N(\hat{\theta} - \theta, {}_k\theta - \theta)_c$ will remain insignificant compared with those of other variables of chi-square type. These observations suggest that from (4.11), though $N^{-1}{}_k\eta_L$ may not be a good estimate of $W_2(\theta, {}_k\hat{\theta})$,

$$r(\hat{\theta}, {}_k\hat{\theta}) = N^{-1}({}_k\eta_L + 2k - L) \tag{4.19}$$

will serve as a useful estimate of $EW_2(\theta, {}_k\hat{\theta})$, at least for the case where N is sufficiently large and L and k are relatively large integers.

It is interesting to note that in practical applications it may sometimes happen that L is a very large, or conceptually infinite, integer and may not be defined clearly. Even under such circumstances we can realize our selection procedure of ${}_k\hat{\theta}$'s for some limited number of k's, assuming L to be equal to the largest value of k. Since we are only concerned with finding out the ${}_k\hat{\theta}$ which will give the minimum of $r(\hat{\theta}, {}_k\hat{\theta})$ we have only to compute either

$${}_k\nu_L = {}_k\eta_L + 2k \tag{4.20}$$

or

$${}_k\lambda_L = -2\sum_{i=1}^{N} \log f(x_i|{}_k\hat{\theta}) + 2k. \tag{4.21}$$

and adopt the ${}_k\hat{\theta}$ which gives the minimum of ${}_k\nu_L$ or ${}_k\lambda_L$ $(0 \le k \le L)$. The statistical behaviour of ${}_k\lambda_L$ is well understood by taking into consideration the successive decomposition of the chi-square variables into mutually independent components. In using ${}_k\lambda_L$ care should be taken not to lose significant digits during the computation.

5. Applications

Some of the possible applications will be mentioned here.

1. Factor Analysis

In the factor analysis we try to find the best estimate of the variance covariance matrix Σ from the sample variance covariance matrix using the model $\Sigma = AA' + D$, where Σ is a $p \times p$ dimensional matrix, A is a $p \times m$ dimensional $(m < p)$ matrix and D is a non-negative $p \times p$ diagonal matrix. The method of the maximum likelihood estimate under the assumption of normality has been extensively applied and the use of the log-likelihood ratio criterion is quite common. Thus our present procedure can readily be incorporated to help the decision of m. Some numerical examples are already given in [6] and the results are quite promising.

2. Principal Component Analysis

By assuming $D = \delta I (\delta \geq 0, I$; unit matrix) in the above model, we can get the necessary decision procedure for the principal component analysis.

3. Analysis of Variance

If in the analysis of variance model we can preassign the order in decomposing the total variance into chi-square components corresponding to some factors and interactions then we can easily apply our present procedure to decide where to stop the decomposition.

4. Multiple Regression

The situation is the same as in the case of the analysis of variance. We can make a decision where to stop including the independent variables when the order of variables for inclusion is predetermined. It can be shown that under the assumption of normality of the residual variable we have only to compare the values $s^2(k)\left(1 + \frac{2k}{N}\right)$, where $s^2(k)$ is the sample mean square of the residual after fitting the regression coefficients by the method of least squares where k is the number of fitted regression coefficients and N the sample size. k should be kept small compared with N. It is interesting to note that the use of a statistics proposed by Mallows [13] is essentially equivalent to our present approach.

5. Autoregressive Model Fitting in Time Series

Though the discussion in the present paper has been limited to the realizations of independent and identically distributed random variables, by following the approach of Billingsley [8], we can see that the same line of discussion can be extended to cover the case of finite parameter Markov processes. Thus in the case of the fitting of one-dimensional autoregressive model $X_n = \sum_{m=1}^{k} a_m X_{n-m} + \varepsilon_n$ we have, assuming the normality of the process X_n, only to adopt k which gives the minimum of $s^2(k)\left(1 + \frac{2k}{N}\right)$ or equivalently $s^2(k)\left(1 + \frac{k}{N}\right)\left(1 - \frac{k}{N}\right)^{-1}$, where $s^2(k)$ is the sample mean square of the residual after fitting the kth order model by the method of least squares or some of its equivalents. This last quantity for the decision has been first introduced by the present author and was considered to be an estimate of the quantity called the final prediction error (FPE) [1, 2]. The use of this approach for the estimation of power spectra has been discussed and recognized to be very useful [3]. For the case of the multi-dimensional process we have to replace $s^2(k)$ by the sample generalized variance or the determinant of the sample variance-covariance matrix of residuals. The procedure has been extensively used for the identification of a cement rotary kiln model [4, 5, 19].

These procedures have been originally derived under the assumption of linear process, which is slightly weaker than the assumption of normality, and with the intuitive criterion of the expected variance of the final one step prediction (FPE). Our present observation shows that these procedures are just in accordance with our extended maximum likelihood principle at least under the Gaussian assumption.

6. Numerical Examples

To illustrate the difference between the conventional test procedure and our present procedure, two numerical examples are given using published data.

The first example is taken from the book by Jenkins and Watts [14]. The original data are described as observations of yield from 70 consecutive batches of an industrial process [14, p. 142]. Our estimates of FPE are given in Table 1 in a relative scale. The results very simply suggest, without the help of statistical tables, the adoption of $k = 2$ for this case. The same conclusion has been reached by the authors of the book after a detailed analysis of significance of partial autocorrelation coefficients and by relying on a somewhat subjective judgement [14, pp. 199–200]. The fitted model produced an estimate of the power spectrum which is very much like their final choice obtained by using Blackman-Tukey type window [14, p. 292].

The next example is taken from a paper by Whittle on the analysis of a seiche record (oscillation of water level in a rock channel) [26; 27, pp. 37–38]. For this example Whittle has used the log-likelihood ratio test statistics in successively deciding the significance of increasing the order by one and adopted $k = 4$. He reports that the fitting of the power spectrum is very poor. Our procedure applied to the reported sample autocorrelation coefficients obtained from data with $N = 660$ produced a result showing that $k = 65$ should be adopted within the k's in the range $0 \leq k \leq 66$. The estimates of

Table 1. Autoregressive Model Fitting.

k	0	1	2	3	4	5	6	7
FPE_k^*	1.029	0.899	0.895	0.921	0.946	0.097	0.983	1.012

* $FPE_k = s^2(k)\left(1 + \frac{k+1}{N}\right)\left(1 - \frac{k+1}{N}\right)^{-1} \Big/ s^2(0)$

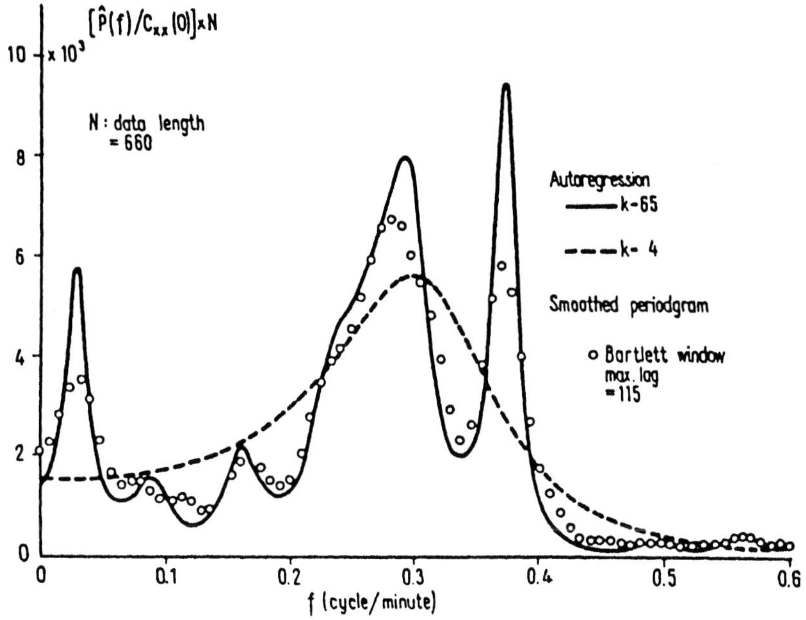

Figure 1. Estimates of the seiche spectrum. The smoothed periodgram of $x(n\,\Delta t)$ $(n = 1, 2, \ldots, N)$ is defined by

$$\Delta t \cdot \sum_{-l}^{l} \left(1 - \frac{|s|}{l}\right) C_{xx}(s) \cos(2\pi f s\,\Delta t),$$

$$\text{where } l = \text{max. lag}, \; C_{xx}(s) = \frac{1}{N} \sum_{n=1}^{N-|s|} \tilde{x}(|s| + n)\tilde{x}(n),$$

$$\text{where } \tilde{x}(n) = x(n\,\Delta t) - \bar{x} \quad \text{and} \quad \bar{x} = \frac{1}{N} \sum_{n=1}^{N} x(n\,\Delta t).$$

the power spectrum are illustrated in Fig. 1. Our procedure suggests that $L = 66$ is not large enough, yet it produced very sharp line-like spectra at various frequencies as was expected from the physical consideration, while the fourth order model did not give any indication of them. This example dramatically illustrates the impracticality of the conventional successive test procedure depending on a subjectively chosen set of levels of significance.

7. Concluding Remarks

In spite of the early statement by Wiener [28; p. 76] that entropy, the Shannon-Wiener type definition of the amount of information, could replace Fisher's definition [11] the use of the information theoretic concepts in the statistical circle has been quite limited [10, 12, 20]; The distinction between Shannon-Wiener's entropy and Fisher's information was discussed as early as in 1950 by Bartlett [7], where the use of the Kullback-Leibler type definition of information was implicit. Since then in the theory of statistics Kullback-Leibler's or Fisher's information could not enjoy the prominent status of Shannon's entropy in communication theory, which proved its essential meaning through the source coding theorem [22, p. 28].

The analysis in the present paper shows that the information theoretic consideration can provide a foundation of the classical maximum likelihood principle and extremely widen its practical applicability. This shows that the notion of informations, which is more closely related to the mutual information in communication theory than to the entropy, will play the most fundamental role in the future developments of statistical theories and techniques.

By our present principle, the extensions of applications 3) ~ 5) of Section 5 to include the comparisons of every possible kth order models are straightforward. The analysis of the overall statistical characteristics of such extensions will be a subject of further study.

Acknowledgement

The author would like to express his thanks to Prof. T. Sugiyama of Kawasaki Medical University for helpful discussions of the possible applications

References

1. Akaike, H., Fitting autoregressive models for prediction. *Ann. Inst. Statist. Math.* **21** (1969) 243–217.
2. Akaike., H., Statistical predictor identification. *Ann. Inst. Statist. Math.* **22** (1970) 203–217.
3. Akaike, H., On a semi-automatic power spectrum estimation procedure. *Proc. 3rd Hawaii International Conference on System Sciences*, 1970, 974–977.
4. Akaike, H., On a decision procedure for system identification, Preprints, *IFAC Kyoto Symposium on System Engineering Approach to Computer Control.* 1970, 486–490.
5. Akaike, H., Autoregressive model fitting for control. *Ann. Inst. Statist. Math.* 23 (1971) 163–180.
6. Akaike, H., Determination of the number of factors by an extended maximum likelihood principle. Research Memo. 44, Inst. Statist. Math. March, 1971.
7. Bartlett, M. S., The statistical approach to the analysis of time-series. *Symposium on Information Theory* (mimeographed Proceedings), Ministry of Supply, London, 1950, 81–101.
8. Billingsley, P., *Statistical Inference for Markov Processes.* Univ. Chicago Press, Chicago 1961.
9. Blackwell, D., Equivalent comparisons of experiments. *Ann. Math. Statist.* 24 (1953) 265–272.
10. Campbell, L.L., Equivalence of Gauss's principle and minimum discrimination information estimation of probabilities. *Ann. Math. Statist.* 41 (1970) 1011–1015.

11. Fisher, R.A., Theory of statistical estimation. *Proc. Camb. Phil. Soc.* **22** (1925) 700–725, *Contributions to Mathematical Statistics*. John Wiley & Sons, New York, 1950, paper 11.
12. Good, I.J. Maximum entropy for hypothesis formulation, especially for multidimensional contingency tables. *Ann. Math. Statist.* **34** (1963) 911–934.
13. Gorman, J.W. and Toman, R.J., Selection of variables for fitting equations to data. *Technometrics* **8** (1966) 27–51.
14. Jenkins, G.M. and Watts, D.G., *Spectral Analysis and Its Applications*. Holden Day, San Francisco, 1968.
15. Kullback, S. and Leibler, R.A., On information and sufficiency. *Ann. Math Statist.* **22** (1951) 79–86.
16. Kullback, S., *Information Theory and Statistics*. John Wiley & Sons, New York 1959.
17. Le Cam, L., On some asymptotic properties of maximum likelihood estimates and related Bayes estimates. *Univ. Calif. Publ. in Stat.* 1 (1953) 277–330.
18. Lehmann, E.L., Testing Statistical Hypotheses. John Wiley & Sons, New York 1969.
19. Otomo, T., Nakagawa, T. and Akaike, H. Statistical approach to computer control of cement rotary kilns. 1971. *Automatica* **8** (1972) 35–48.
20. Rènyi, A., Statistics and information theory. *Studia Sci. Math. Hung.* **2** (1967) 249–256.
21. Savage, L.J., The Foundations of Statistics. John Wiley & Sons, New York 1954.
22. Shannon, C.E. and Weaver, W., *The Mathematical Theory of Communication*. Univ. of Illinois Press, Urbana 1949.
23. Wald, A., Tests of statistical hypotheses concerning several parameters when the number of observations is large. *Trans. Am. Math. Soc.* **54** (1943) 426–482.
24. Wald, A., Note on the consistency of the maximum likelihood estimate. *Ann Math. Statist.* 20 (1949) 595–601.
25. Wald, A., Statistical Decision Functions. John Wiley & Sons, New York 1950.
26. Whittle, P., The statistical analysis of seiche record. *J. Marine Res.* **13** (1954) 76–100.
27. Whittle, P., *Prediction and Regulation*. English Univ. Press, London 1963.
28. Wiener, N., *Cybernetics*. John Wiley & Sons, New York, 1948.

A New Look at the Statistical Model Identification

HIROTUGU AKAIKE, MEMBER, IEEE

***Abstract*—The history of the development of statistical hypothesis testing in time series analysis is reviewed briefly and it is pointed out that the hypothesis testing procedure is not adequately defined as the procedure for statistical model identification. The classical maximum likelihood estimation procedure is reviewed and a new estimate minimum information theoretical criterion (AIC) estimate (MAICE) which is designed for the purpose of statistical identification is introduced. When there are several competing models the MAICE is defined by the model and the maximum likelihood estimates of the parameters which give the minimum of AIC defined by**

$$\text{AIC} = (-2)\log(\text{maximum likelihood}) + 2(\text{number of independently adjusted parameters within the model}).$$

MAICE provides a versatile procedure for statistical model identification which is free from the ambiguities inherent in the application of conventional hypothesis testing procedure. The practical utility of MAICE in time series analysis is demonstrated with some numerical examples.

I. Introduction

In spite of the recent development of the use of statistical concepts and models in almost every field of engineering and science it seems as if the difficulty of constructing an adequate model based on the information provided by a finite number of observations is not fully recognized. Undoubtedly the subject of statistical model construction or identification is heavily dependent on the results of theoretical analyses of the object under observation. Yet it must be realized that there is usually a big gap between the theoretical results and the practical procedures of identification. A typical example is the gap between the results of the theory of minimal realizations of a linear system and the identification of a Markovian representation of a stochastic process based on a record of finite duration. A minimal realization of a linear system is usually defined through the analysis of the rank or the dependence relation of the rows or columns of some Hankel matrix [1]. In a practical situation, even if the Hankel matrix is theoretically given, the rounding errors will always make the matrix of full rank. If the matrix is obtained from a record of observations of a real object the sampling variabilities of the elements of the matrix will be by far the greater than the rounding errors and also the system will always be infinite dimensional. Thus it can be seen that the subject of statistical identification is essentially concerned with the art of approximation which is a basic element of human intellectual activity.

As was noticed by Lehman [2, p. viii], hypothesis testing procedures are traditionally applied to the situations where actually multiple decision procedures are required. If the statistical identification procedure is considered as a decision procedure the very basic problem is the appropriate choice of the loss function. In the Neyman–Pearson theory of statistical hypothesis testing only the probabilities of rejecting and accepting the correct and incorrect hypotheses, respectively, are considered to define the loss caused by the decision. In practical situations the assumed null hypotheses are only approximations and they are almost always different from the reality. Thus the choice of the loss function in the test theory makes its practical application logically contradictory. The recognition of this point that the hypothesis testing procedure is not adequately formulated as a procedure of approximation is very important for the development of practically useful identification procedures.

A new perspective of the problem of identification is obtained by the analysis of the very practical and successful method of maximum likelihood. The fact that the maximum likelihood estimates are, under certain regularity conditions, asymptotically efficient shows that the likelihood function tends to be a quantity which is most sensitive to the small variations of the parameters around the true values. This observation suggests the use of

$$S(g;f(\cdot|\theta)) = \int g(x)\log f(x|\theta)\,dx$$

as a criterion of "fit" of a model with the probabilistic structure defined by the probability density function $f(x|\theta)$ to the structure defined by the density function $g(x)$. Contrary to the assumption of a single family of density $f(x|\theta)$ in the classical maximum likelihood estimation procedure, several alternative models or families defined by the densities with different forms and/or with one and the same form but with different restrictions on the parameter vector θ are contemplated in the usual situation of identification. A detailed analysis of the maximum likelihood estimate (MLE) leads naturally to a definition of a new estimate which is useful for this type of multiple model situation. The new estimate is called the minimum information theoretic criterion (AIC) estimate (MAICE), where AIC stands for an information theoretic criterion recently introduced by the present author [3] and is an estimate of a measure of fit of the model. MAICE is defined by the model and its parameter values which give the minimum of AIC. By the introduction of MAICE the problem of statistical identification is explicitly formulated as a problem of estimation and the need of the subjective judgement required in the hypothesis testing procedure for the decision on the levels of significance is completely eliminated. To give an explicit definition of MAICE and to discuss its characteristics by comparison with the conventional identification procedure based on estimation

Manuscript received February 12, 1974; revised March 2, 1974.
The author is with the Institute of Statistical Mathematics, Minato-ku, Tokyo, Japan.

and hypothesis testing form the main objectives of the present paper.

Although MAICE provides a versatile method of identification which can be used in every field of statistical model building, its practical utility in time series analysis is quite significant. Some numerical examples are given to show how MAICE can give objectively defined answers to the problems of time series analysis in contrast with the conventional approach by hypothesis testing which can only give subjective and often inconclusive answers.

II. Hypothesis Testing in Time Series Analysis

The study of the testing procedure of time series started with the investigation of the test of a simple hypothesis that a single serial correlation coefficient is equal to 0. The utility of this type of test is certainly too limited to make it a generally useful procedure for model identification. In 1947 Quenouille [4] introduced a test for the goodness of fit of autoregressive (AR) models. The idea of the Quenouille's test was extended by Wold [5] to a test of goodness of fit of moving average (MA) models. Several refinements and generalizations of these test procedures followed [6]–[9] but a most significant contribution to the subject of hypothesis testing in time series analysis was made by Whittle [10], [11] by a systematic application of the Neyman–Pearson likelihood ratio test procedure to the time series situation.

A very basic test of time series is the test of whiteness. In many situations of model identification the whiteness of the residual series after fitting a model is required as a proof of adequacy of the model and the test of whiteness is widely used in practical applications [12]–[15]. For the test of whiteness the analysis of the periodgram provides a general solution.

A good exposition of the classical hypothesis testing procedures including the tests based on the periodgrams is given in Hannan [16].

The fitting of AR or MA models is essentially a subject of multiple decision procedure rather than that of hypothesis testing. Anderson [17] discussed the determination of the order of a Gaussian AR process explicitly as a multiple decision procedure. The procedure takes a form of a sequence of tests of the models starting at the highest order and successively down to the lowest order. To apply the procedure to a real problem one has to specify the level of significance of the test for each order of the model. Although the procedure is designed to satisfy certain clearly defined condition of optimality, the essential difficulty of the problem of order determination remains as the difficulty in choosing the levels of significance. Also the loss function of the decision procedure is defined by the probability of making incorrect decisions and thus the procedure is not free from the logical contradiction that in practical applications the order of the true structure will always be infinite. This difficulty can only be avoided by reformulating the problem explicitly as a problem of approximation of the true structure by the model.

III. Direct Approach to Model Error Control

In the field of nontime series regression analysis Mallows introduced a statistic C_p for the selection of variables for regression [18]. C_p is defined by

$$C_p = (\hat{\sigma}^2)^{-1}\,(\text{residual sum of squares}) - N + 2p,$$

where $\hat{\sigma}^2$ is a properly chosen estimate of σ^2, the unknown variance of the true residual, N is the number of observations, and p is the number of variables in regression. The expected value of C_p is roughly p if the fitted model is exact and greater otherwise. C_p is an estimate of the expected sum of squares of the prediction, scaled by σ^2, when the estimated regression coefficients are used for prediction and has a clearly defined meaning as a measure of adequacy of the adopted model. Defined with this clearly defined criterion of fit, C_p attracted serious attention of the people who were concerned with the regression analyses of practical data. See the references of [18]. Unfortunately some subjective judgement is required for the choice of $\hat{\sigma}^2$ in the definition of C_p.

At almost the same time when C_p was introduced, Davisson [19] analyzed the mean-square prediction error of stationary Gaussian process when the estimated coefficients of the predictor were used for prediction and discussed the mean-square error of an adaptive smoothing filter [20]. The observed time series x_i is the sum of signal s_i and additive white noise n_i. The filtered output $\hat{s}_i$ is given by

$$\hat{s}_i = \sum_{j=-M}^{L} \beta_j x_{i+j}, \qquad (i = 1,2,\cdots,N)$$

where β_j is determined from the sample x_i $(i = 1,2,\cdots,N)$. The problem is how to define L and M so that the mean-square smoothing error over the N samples $E[(1/N)\sum_{i=1}^{N}(s_i - \hat{s}_i)^2]$ is minimized. Under appropriate assumptions of s_i and n_i Davisson [20] arrived at an estimate of this error which is defined by

$$\hat{\sigma}_N^2[M,L] = s^2 + 2\hat{c}(M + L + 1)/N,$$

where s^2 is an estimate of the error variance and $\hat{c}$ is the slope of the curve of s^2 as a function of $(M + L)/N$ at "larger" values of $(L + M)/N$. This result is in close correspondence with Mallows' C_p, and suggests the importance of this type of statistics in the field of model identification for prediction. Like the choice of $\hat{\sigma}^2$ in Mallows' C_p the choice of $\hat{c}$ in the present statistic $\hat{\sigma}_N^2[M,L]$ becomes a difficult problem in practical application.

In 1969, without knowing the close relationship with the above two procedures, the present author introduced a fitting procedure of the univariate AR model defined by $y_i = a_1 y_{i-1} + \cdots + a_p y_{i-p} + x_i$, where x_i is a white noise [21]. In this procedure the mean-square error of the one-step-ahead prediction obtained by using the least squares estimates of the coefficients is controlled. The mean-square error is called the final prediction error (FPE) and when the data y_i $(i = 1,2,\cdots,N)$ are given its estimate is defined by

$$\mathrm{FPE}(p) = \{(N+p)/(N-p)\} \cdot (\hat{C}_0 - \hat{a}_{p1}\hat{C}_1 - \cdots - \hat{a}_{pp}\hat{C}_p),$$

where the mean of y_t is assumed to be 0, $\hat{C}_l = (1/N)\sum_{t=1}^{N-l} y_{t+l}y_t$ and $\hat{a}_{pi}$'s are obtained by solving the Yule–Walker equation defined by $\hat{C}_l$'s. By scanning p successively from 0 to some upper limit L the identified model is given by the p and the corresponding $\hat{a}_{pi}$'s which give the minimum of FPE(p) ($p = 0,1,\cdots,L$). In this procedure no subjective element is left in the definition of FPE(p). Only the determination of the upper limit L requires judgement. The characteristics of the procedure was further analyzed [22] and the procedure worked remarkably well with practical data [23], [24]. Gersch and Sharp [25] discussed their experience of the use of the procedure. Bhansali [26] reports very disappointing results, claiming that they were obtained by Akaike's method. Actually the disappointing results are due to his incorrect definition of the related statistic and have nothing to do with the present minimum FPE procedure. The procedure was extended to the case of multivariate AR model fitting [27]. A successful result of implementation of a computer control of cement kiln processes based on the results obtained by this identification procedure was reported by Otomo and others [28].

One common characteristic of the three procedures discussed in this section is that the analysis of the statistics has to be extended to the order of $1/N$ of the main term.

IV. Mean Log-Likelihood as a Measure of Fit

The well known fact that the MLE is, under regularity conditions, asymptotically efficient [29] shows that the likelihood function tends to be a most sensitive criterion of the deviation of the model parameters from the true values. Consider the situation where $x_1,x_2,\cdots,x_N$ are obtained as the results of N independent observations of a random variable with probability density function $g(x)$. If a parametric family of density function is given by $f(x|\theta)$ with a vector parameter θ, the average log-likelihood, or the log-likelihood divided by N, is given by

$$(1/N)\sum_{i=1}^{N} \log f(x_i|\theta), \tag{1}$$

where, as in the sequel of the present paper, log denotes the natural logarithms. As N is increased indefinitely, this average tends, with probability 1, to

$$S(g;f(\cdot|\theta)) = \int g(x) \log f(x|\theta)\, dx,$$

where the existence of the integral is assumed. From the efficiency of MLE it can be seen that the (average) mean log-likelihood $S(g;f(\cdot|\theta))$ must be a most sensitive criterion to the small deviation of $f(x|\theta)$ from $g(x)$. The difference

$$I(g;f(\cdot|\theta)) = S(g;g) - S(g;f(\cdot|\theta))$$

is known as the Kullback–Leibler mean information for discrimination between $g(x)$ and $f(x|\theta)$ and takes positive value, unless $f(x|\theta) = g(x)$ holds almost everywhere [30]. These observations show that $S(g;f(\cdot|\theta))$ will be a reasonable criterion for defining a best fitting model by its maximization or, from the analogy to the concept of entropy, by minimizing $-S(g;f(\cdot|\theta))$. It should be mentioned here that in 1950 this last quantity was adopted as a definition of information function by Bartlett [31]. One of the most important characteristics of $S(g;f(\cdot|\theta))$ is that its natural estimate, the average log-likelihood (1), can be obtained without the knowledge of $g(x)$. When only one family $f(x|\theta)$ is given, maximizing the estimate (1) of $S(g;f(\cdot|\theta))$ with respect to θ leads to the MLE $\hat{\theta}$.

In the case of statistical identification, usually several families of $f(x|\theta)$, with different forms of $f(x|\theta)$ and/or with one and the same form of $f(x|\theta)$ but with different restrictions on the parameter vector θ, are given and it is required to decide on the best choice of $f(x|\theta)$. The classical maximum likelihood principle can not provide useful solution to this type of problems. A solution can be obtained by incorporating the basic idea underlying the statistics discussed in the preceding section with the maximum likelihood principle.

Consider the situation where $g(x) = f(x|\theta_0)$. For this case $I(g;f(\cdot|\theta))$ and $S(g;f(\cdot|\theta))$ will simply be denoted by $I(\theta_0;\theta)$ and $S(\theta_0;\theta)$, respectively. When θ is sufficiently close to θ_0, $I(\theta_0;\theta)$ admits an approximation [30]

$$I(\theta_0;\theta_0 + \Delta\theta) = (\tfrac{1}{2})\|\Delta\theta\|_J^2,$$

where $\|\Delta\theta\|_J^2 = \Delta\theta' J \Delta\theta$ and J is the Fisher information matrix which is positive definite and defined by

$$J_{ij} = E\left\{\frac{\partial \log f(X|\theta)}{\partial\theta_i}\frac{\partial \log f(X|\theta)}{\partial\theta_j}\right\},$$

where J_{ij} denotes the (i,j)th element of J and θ_i the ith component of θ. Thus when the MLE $\hat{\theta}$ of θ_0 lies very close to θ_0 the deviation of the distribution defined by $f(x|\hat{\theta})$ from the true distribution $f(x|\theta_0)$ in terms of the variation of $S(g;f(\cdot|\theta))$ will be measured by $(\tfrac{1}{2})\|\hat{\theta} - \theta_0\|_J^2$. Consider the situation where the variation of θ for maximizing the likelihood is restricted to a lower dimensional subspace Θ of θ which does not include θ_0. For the MLE $\hat{\theta}$ of θ_0 restricted in Θ, if $\boldsymbol{\theta}$ which is in Θ and gives the maximum of $S(\theta_0;\theta)$ is sufficiently close to θ_0, it can be shown that the distribution of $N\|\hat{\theta} - \boldsymbol{\theta}\|_J^2$ for sufficiently large N is approximated under certain regularity conditions by a chi-square distribution the degree of freedom equal to the dimension of the restricted parameter space. See, for example, [32]. Thus it holds that

$$E_\infty 2NI(\theta_0;\hat{\theta}) = N\|\boldsymbol{\theta} - \theta_0\|_J^2 + k, \tag{2}$$

where E_∞ denotes the mean of the approximate distribution and k is the dimension of Θ or the number of parameters independently adjusted for the maximization of the likelihood. Relation (2) is a generalization of the expected prediction error underlying the statistics discussed in the preceding section. When there are several models it will be natural to adopt the one which will give the minimum of $EI(\theta_0;\hat{\theta})$. For this purpose, considering the situation where these models have their θ's very close to θ_0, it becomes necessary to develop some estimate of $N\|\theta - \theta_0\|_J^2$ of (2). The relation (2) is based on the fact that the asymptotic distribution of $\sqrt{N}(\hat{\theta} - \theta)$ is approximated by a Gaussian distribution with mean zero and variance matrix J^{-1}. From this fact if

$$2\left(\sum_{i=1}^{N} \log f(x_i|\theta_0) - \sum_{i=1}^{N} \log f(x_i|\hat{\theta})\right) \tag{3}$$

is used as an estimate of $N\|\theta - \theta_0\|_J^2$ it needs a correction for the downward bias introduced by replacing θ by $\hat{\theta}$. This correction is simply realized by adding k to (3). For the purpose of identification only the comparison of the values of the estimates of $EI(\theta_0;\hat{\theta})$ for various models is necessary and thus the common term in (3) which includes θ_0 can be discarded.

V. Definition of an Information Criterion

Based on the observations of the preceding section an information criterion AIC of θ is defined by

$$\mathrm{AIC}(\hat{\theta}) = (-2)\log(\text{maximum likelihood}) + 2k,$$

where, as is defined before, k is the number of independently adjusted parameters to get $\hat{\theta}$. $(1/N)\mathrm{AIC}(\hat{\theta})$ may be considered as an estimate of $-2ES(\theta_0;\hat{\theta})$. IC stands for information criterion and A is added so that similar statistics, BIC, DIC etc., may follow. When there are several specifications of $f(x|\theta)$ corresponding to several models, the MAICE is defined by the $f(x|\hat{\theta})$ which gives the minimum of $\mathrm{AIC}(\hat{\theta})$. When there is only one unrestricted family of $f(x|\theta)$, the MAICE is defined by $f(x|\hat{\theta})$ with $\hat{\theta}$ identical to the classical MLE. It should be noticed that an arbitrary additive constant can be introduced into the definition of $\mathrm{AIC}(\hat{\theta})$ when the comparison of the results for different sets of observations is not intended. The present definition of MAICE gives a mathematical formulation of the principle of parsimony in model building. When the maximum likelihood is identical for two models the MAICE is the one defined with the smaller number of parameters.

In time series analysis, even under the Gaussian assumption, the exact definition of likelihood is usually too complicated for practical use and some approximation is necessary. For the application of MAICE there is a subtle problem in defining the approximation to the likelihood function. This is due to the fact that for the definition of AIC the log-likelihoods must be defined consistently to the order of magnitude of 1. For the fitting of a stationary Gaussian process model a measure of the deviation of a model from a true structure can be defined as the limit of the average mean log-likelihood when the number of observations N is increased indefinitely. This quantity is identical to the mean log-likelihood of innovation defined by the fitted model. Thus a natural procedure for the fitting of a stationary zero-mean Gaussian process model to the sequence of observations $y_1, y_2, \cdots, y_N$ is to define a primitive stationary Gaussian model with the l-lag covariance matrices $R(l)$, which are defined by

$$R(l) = (1/N)\sum_{n=1}^{N-l} y_{n+l}y_n', \qquad l = 0,1,2,\cdots,N-1$$
$$= 0, \qquad l = N, N+1, \cdots$$

and fit a model by maximizing the mean log-likelihood of innovation or equivalently, if the elements of the covariance matrix of innovation are within the parameter set, by minimizing the log-determinant of the variance matrix of innovation, N times of which is to be used in place of the log-likelihood in the definition of AIC. The adoption of the divisor N in the definition of $R(l)$ is important to keep the sequence of the covariance matrices positive definite. The present procedure of fitting a Gaussian model through the primitive model is discussed in detail in [33]. It leads naturally to the concept of Gaussian estimate developed by Whittle [34]. When the asymptotic distribution of the normalized correlation coefficients of y_n is identical to that of a Gaussian process the asymptotic distribution of the statistics defined as functions of these coefficients will also be independent of the assumption of Gaussian process. This point and the asymptotic behavior of the related statistics which is required for the justification of the present definition of AIC is discussed in detail in the above paper by Whittle. For the fitting of a univariate Gaussian AR model the MAICE defined with the present definition of AIC is asymptotically identical to the estimate obtained by the minimum FPE procedure.

AIC and a primitive definition of MAICE were first introduced by the present author in 1971 [3]. Some early successful results of applications are reported in [3], [35], [36].

VI. Numerical Examples

Before going into the discussion of the characteristics of MAICE its practical utility is demonstrated in this section.

For the convenience of the readers who might wish to check the results by themselves Gaussian AR models were fitted to the data given in Anderson's book on time series analysis [37]. To the Wold's three series artificially generated by the second-order AR schemes models up to the 50th order were fitted. In two cases the MAICE's were the second-order models. In the case where the MAICE was the first-order model, the second-order coefficient of the generating equation had a very small absolute value compared with its sampling variability and the one-step-ahead prediction error variance was smaller for the MAICE than for the second-order model defined with the MLE's of the coefficients. To the classical series of Wolfer's sunspot numbers with $N = 176$ AR models up to the 35th order were fitted and the MAICE was the eighth-order model. AIC attained a local minimum at the second order. In the case of the series of Beveridge's wheat price index with $N = 370$ the MAICE among the AR model up to the 50th order was again of the eighth order. AIC attained a local minimum at the second order which was adopted by Sargan [38]. In the light of the discussions of these series by Anderson, the choice of eight-order models for these two series looks reasonable.

Two examples of application of the minimum FPE procedure, which produces estimates asymptotically equivalent to MAICE's, are reported in [3]. In the example taken from the book by Jenkins and Watts [39, section 5.4.3] the estimate was identical to the one chosen by the authors of the book after a careful analysis. In the case of the seiche record treated by Whittle [40] the minimum FPE procedure clearly suggested the need of a very high-order AR model. The difficulty of fitting AR models to this set of data was discussed by Whittle [41, p. 38].

The procedure was also applied to the series E and F given in the book by Box and Jenkins [12]. Second- or third-order AR model was suggested by the authors for the series E which is a part of the Wolfer's sunspot number series with $N = 100$. The MAICE among the AR models up to the 20th order was the second-order model. Among the AR models up to the 10th order fitted to the series F with $N = 70$ the MAICE was the second-order model, which agrees with the suggestion made by the authors of the book.

To test the ability of discriminating between AR and MA models ten series of y_n ($n = 1,\cdots,1600$) were generated by the relation $y_n = x_n + 0.6x_{n-1} - 0.1x_{n-2}$, where x_n was generated from a physical noise source and was supposed to be a Gaussian white noise. AR models were fitted to the first N points of each series for $N = 50$, 100, 200, 400, 800, 1600. The sample averages of the MAICE AR order were 3.1, 4.1, 6.5, 6.8, 8.2, and 9.3 for the successively increasing values of N. An approximate MAICE procedure which is designed to get an initial estimate of MAICE for the fitting of Markovian models, described in [33], was applied to the data. With only a few exceptions the approximate MAICE's were of the second order. This corresponds to the AR–MA model with a second-order AR and a first-order MA. The second- and third-order MA models were then fitted to the data with $N = 1600$. Among the AR and MA models fitted to the data the second-order MA model was chosen nine times as the MAICE and the third-order MA was chosen once. The average difference of the minimum of AIC between AR and MA models was 7.7, which roughly means that the expected likelihood ratio of a pair of two fitted models will be about 47 for a set of data with $N = 1600$ in favor of MA model.

Another test was made with the example discussed by Gersch and Sharp [25]. Eight series of length $N = 800$ were generated by an AR–MA scheme described in the paper. The average of the MAICE AR orders was 17.9 which is in good agreement with the value reported by Gersch and Sharp. The approximate MAICE procedure was applied to determine the order or the dimension of the Markovian representation of the process. For the eight cases the procedure identically picked the correct order four. AR–MA models of various orders were fitted to one set of data and the corresponding values of AIC(p,q) were computed, where AIC(p,q) is the value of AIC for the model with AR order p and MA order q and was defined by

$$\text{AIC}(p,q) = N \log (\text{MLE of innovation variance}) + 2(p+q).$$

The results are AIC(3,2) = 192.72, AIC(4,3) = 66.54, AIC(4,4) = 67.44, AIC(5,3) = 67.48 AIC(6,3) = 67.65, and AIC(5,4) = 69.43. The minimum is attained at $p = 4$ and $q = 3$ which correspond to the true structure. Fig. 1 illustrates the estimates of the power spectral density obtained by applying various procedures to this set of data. It should be mentioned that in this example the Hessian of the mean log-likelihood function becomes singular at the true values of the parameters for the models with p and q simultaneously greater than 4 and 3, respectively. The detailed discussion of the difficulty connected with this singularity is beyond the scope of the present paper. Fig. 2 shows the results of application of the same type of procedure to a record of brain wave with $N = 1420$. In this case only one AR–MA model with AR order 4 and MA order 3 was fitted. The value of AIC of this model is 1145.6 and that of the MAICE AR model is 1120.9. This suggests that the 13th order MAICE AR model is a better choice, a conclusion which seems in good agreement with the impression obtained from the inspection of Fig. 2.

VII. Discussions

When $f(x|\theta)$ is very far from $g(x)$, $S(g;f(\cdot|\theta))$ is only a subjective measure of deviation of $f(x|\theta)$ from $g(x)$. Thus the general discussion of the characteristics of MAICE will only be possible under the assumption that for at least one family $f(x|\theta)$ is sufficiently closed to $g(x)$ compared with the expected deviation of $f(x|\hat{\theta})$ from $f(x|\theta)$. The detailed analysis of the statistical characteristics of MAICE is only necessary when there are several families which satisfy this condition. As a single estimate of $-2NES(g;f(\cdot|\hat{\theta}))$, -2 times the log-maximum likelihood will be sufficient but for the present purpose of "estimating the difference" of $-2NES(g;f(\cdot|\hat{\theta}))$ the introduction of the term $+2k$ into the definition of AIC is crucial. The disappointing results reported by Bhansali [26] were due to his incorrect use of the statistic, equivalent to using $+k$ in place of $+2k$ in AIC.

When the models are specified by a successive increase of restrictions on the parameter θ of $f(x|\theta)$ the MAICE procedure takes a form of repeated applications of conventional log-likelihood ratio test of goodness of fit with automatically adjusted levels of significance defined by the terms $+2k$. When there are different families approximating the true likelihood equally well the situation will at least locally be approximated by the different parametrizations of one and the same family. For these cases the significance of the difference of AIC's between two models will be evaluated by comparing it with the variability of a chi-square variable with the degree of freedom

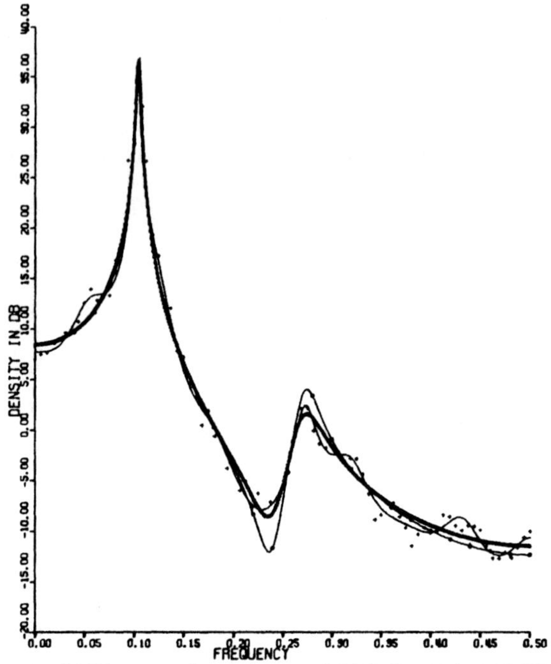

Fig. 1. Estimates of an AR–MA spectrum: theoretical spectrum (solid thin line with dots), AR–MA estimate (thick line), AR estimate (solid thin line), and Hanning windowed estimate with maximum lag 80 (crosses).

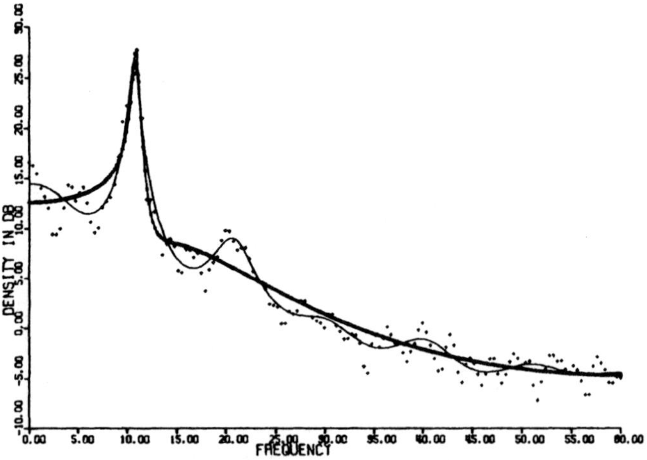

Fig. 2. Estimates of brain wave spectrum: AR–MA estimate (thick line), AR estimate (solid thin line), and Hanning windowed estimate with maximum lag 150 (crosses).

equal to the difference of the k's of the two models. When the two models form separate families in the sense of Cox [42], [43] the procedure developed by Cox and extended by Walker [44] to time series situation may be useful for the detailed evaluation of the difference of AIC.

It must be clearly recognized that MAICE can not be compared with a hypothesis testing procedure unless the latter is defined as a decision procedure with required levels of significance. The use of a fixed level of significance for the comparison of models with various number of parameters is wrong since this does not take into account the increase of the variability of the estimates when the number of parameters is increased. As will be seen by the work of Kennedy and Bancroft [45] the theory of model building based on a sequence of significance tests is not sufficiently developed to provide a practically useful procedure.

Although the present author has no proof of optimality of MAICE it is at present the only procedure applicable to every situation where the likelihood can be properly defined and it is actually producing very reasonable results without very much amount of help of subjective judgement. The successful results of numerical experiments suggest almost unlimited applicability of MAICE in the fields of modeling, prediction, signal detection, pattern recognition, and adaptation. Further improvements of definition and use of AIC and numerical comparisons of MAICE with other procedures in various specific applications will be the subjects of further study.

VIII. Conclusion

The practical utility of the hypothesis testing procedure as a method of statistical model building or identification must be considered quite limited. To develop useful procedures of identification more direct approach to the control of the error or loss caused by the use of the identified model is necessary. From the success of the classical maximum likelihood procedures the mean log-likelihood seems to be a natural choice as the criterion of fit of a statistical model. The MAICE procedure based on AIC which is an estimate of the mean log-likelihood provides a versatile procedure for the statistical model identification. It also provides a mathematical formulation of the principle of parsimony in the field of model construction. Since a procedure based on MAICE can be implemented without the aid of subjective judgement, the successful numerical results of applications suggest that the implementations of many statistical identification procedures for prediction, signal detection, pattern recognition, and adaptation will be made practical with MAICE.

Acknowledgment

The author is grateful to Prof. T. Kailath, Stanford University, for encouraging him to write the present paper. Thanks are also due to Prof. K. Sato, Nagasaki University, for providing the brain wave data treated in Section V.

References

[1] H. Akaike, "Stochastic theory of minimal realization," this issue, pp. 667–674.
[2] E. L. Lehman, *Testing Statistical Hypothesis*. New York: Wiley, 1959.
[3] H. Akaike, "Information theory and an extension of the maximum likelihood principle," in *Proc. 2nd Int. Symp. Information Theory, Supp. to Problems of Control and Information Theory*, 1972, pp. 267–281.
[4] M. H. Quenouille, "A large-sample test for the goodness of fit of autoregressive schemes," *J. Roy. Statist. Soc.*, vol. 110, pp. 123–129, 1947.
[5] H. Wold, "A large-sample test for moving averages," *J. Roy. Statist. Soc., B*, vol. 11, pp. 297–305, 1949.
[6] M. S. Barlett and P. H. Diananda, "Extensions of Quenouille's test for autoregressive scheme," *J. Roy. Statist. Soc., B*, vol. 12, pp. 108–115, 1950.
[7] M. S. Bartlett and D. V. Rajalakshman, "Goodness of fit test for simultaneous autoregressive series," *J. Roy. Statist. Soc., B*, vol. 15, pp. 107–124, 1953.
[8] A. M. Walker, "Note on a generalization of the large sample goodness of fit test for linear autoregressive schemes," *J. Roy. Statist. Soc., B*, vol. 12, pp. 102–107, 1950.
[9] ——, "The existence of Bartlett-Rajalakshman goodness of fit G-tests for multivariate autoregressive processes with finitely dependent residuals," *Proc. Cambridge Phil. Soc.*, vol. 54, pp. 225–232, 1957.
[10] P. Whittle, *Hypothesis Testing in Time-Series Analysis*. Uppsala, Sweden: Almqvist and Wiksell, 1951.
[11] ——, "Some recent contributions to the theory of stationary processes," *A Study in the Analysis of Stationary Time Series*. Uppsala, Sweden: Almqvist and Wiksell, 1954, appendix 2.
[12] G. E. P. Box and G. M. Jenkins, *Time Series, Forecasting and Control*. San Francisco, Calif.: Holden-Day, 1970.
[13] I. Gustavsson, "Comparison of different methods for identification of industrial processes," *Automatica*, vol. 8, pp. 127–142, 1972.
[14] R. K. Mehra, "On the identification of variances and adaptive Kalman filtering," *IEEE Trans. Automat. Contr.*, vol. AC-15, pp. 175–184, Apr. 1970.
[15] R. K. Mehra, "On-line identification of linear dynamic systems with applications to Kalman filtering," *IEEE Trans. Automat. Contr.*, vol. AC-16, pp. 12–21, Feb. 1971.
[16] E. J. Hannan, *Time Series Analysis*. London, England: Methuen, 1960.
[17] T. W. Anderson, "Determination of the order of dependence in normally distributed time series," in *Time Series Analysis*, M. Rosenblatt, Ed. New York: Wiley, 1963, pp. 425–446.
[18] C. L. Mallows, "Some comments on C_p," *Technometrics*, vol. 15, pp. 661–675, 1973.
[19] L. D. Davisson, "The prediction error of stationary Gaussian time series of unknown covariance," *IEEE Trans. Inform. Theory*, vol. IT-11, pp. 527–532, Oct. 1965.
[20] ——, "A theory of adaptive filtering," *IEEE Trans. Inform. Theory*, vol. IT-12, pp. 97–102, Apr. 1966.
[21] H. Akaike, "Fitting autoregressive models for prediction," *Ann. Inst. Statist. Math.*, vol. 21, pp. 243–247, 1969.
[22] ——, "Statistical predictor identification," *Ann. Inst. Statist. Math.*, vol. 22, pp. 203–217, 1970.
[23] ——, "On a semiautomatic power spectrum estimation procedure," in *Proc. 3rd Hawaii Int. Conf. System Sciences*, 1970, pp. 974–977.
[24] R. H. Jones, "Autoregressive spectrum estimation," in *3rd Conf. Probability and Statistics in Atmospheric Sciences, Preprints*, Boulder, Colo., June 19–22, 1973.
[25] W. Gersch and D. R. Sharpe, "Estimation of power spectra with finite-order autoregressive models," *IEEE Trans. Automat. Contr.*, vol. AC-18, pp. 367–379, Aug. 1973.
[26] R. J. Bhansali, "A Monte Carlo comparison of the regression method and the spectral methods of prediction," *J. Amer. Statist. Ass.*, vol. 68, pp. 621–625, 1973.
[27] H. Akaike, "Autoregressive model fitting for control," *Ann. Inst. Statist. Math.*, vol. 23, pp. 163–180, 1971.
[28] T. Otomo, T. Nakagawa, and H. Akaike, "Statistical approach to computer control of cement rotary kilns," *Automatica*, vol. 8, pp. 35–48, 1972.
[29] H. Cramer, *Mathematical Methods of Statistics*. Princeton, N. J.: Princeton Univ. Press, 1946.
[30] S. Kullback, *Information Theory and Statistics*. New York: Wiley, 1959.
[31] M. S. Bartlett, "The statistical approach to the analysis of time series," in *Proc. Symp. Information Theory*, London, England, Ministry of Supply, 1950, pp. 81–101.
[32] P. J. Huber, "The behavior of maximum likelihood estimates under nonstandard conditions," in *Proc. 5th Berkeley Symp. Mathematical Statistics and Probability*, vol. 1, pp. 221–233, 1967.

[33] H. Akaike, "Markovian representation of stochastic processes and its application to the analysis of autoregressive moving average processes," *Ann. Inst. Statist. Math.*, to be published.
[34] P. Whittle, "Gaussian estimation in stationary time series," *Bull. Int. Statist. Inst.*, vol. 39, pp. 105–129, 1962.
[35] H. Akaike, "Use of an information theoretic quantity for statistical model identification," in *Proc. 5th Hawaii Int. Conf. System Sciences*, pp. 249–250, 1972.
[36] H. Akaike, "Automatic data structure search by the maximum likelihood," in *Computer in Biomedicine Suppl. to Proc. 5th Hawaii Int. Conf. on System Sciences*, pp. 99–101, 1972.
[37] T. W. Anderson, *The Statistical Analysis of Time Series.* New York: Wiley, 1971.
[38] J. D. Sargan, "An approximate treatment of the properties of the correlogram and periodgram," *J. Roy. Statist. Soc. B*, vol. 15, pp. 140–152, 1953.
[39] G. M. Jenkins and D. G. Watts, *Spectral Analysis and its Applications.* San Francisco, Calif.: Holden-Day, 1968.
[40] P. Whittle, "The statistical analysis of a seiche record," *J. Marine Res.*, vol. 13, pp. 76–100, 1954.
[41] P. Whittle, *Prediction and Regulation.* London, England: English Univ. Press, 1963.
[42] D. R. Cox, "Tests of separate families of hypotheses," in *Proc. 4th Berkeley Symp. Mathematical Statistics and Probability*, vol. 1, 1961, pp. 105–123.
[43] D. R. Cox, "Further results on tests of separate families of hypotheses," *J. Roy. Statist. Soc., B*, vol. 24, pp. 406–425, 1962.
[44] A. M. Walker, "Some tests of separate families of hypotheses in time series analysis," *Biometrika*, vol. 54, pp. 39–68, 1967.
[45] W. J. Kennedy and T. A. Bancroft, "Model building for prediction in regression based upon repeated significance tests," *Ann. Math. Statist.*, vol. 42, pp. 1273–1284, 1971.

Hiortugu Akaike (M'72), for a photograph and biography see page 674 of this issue.

Reprinted from *Annals of the Institute of Statistical Mathematics*, Vol. 26, No. 3, 1974, 363-387 by permission of the Institute of Statistical Mathematics

MARKOVIAN REPRESENTATION OF STOCHASTIC PROCESSES AND ITS APPLICATION TO THE ANALYSIS OF AUTOREGRESSIVE MOVING AVERAGE PROCESSES

HIROTUGU AKAIKE

(Received Sept. 18, 1973; revised March 28, 1974)

Summary

The problem of identifiability of a multivariate autoregressive moving average process is considered and a complete solution is obtained by using the Markovian representation of the process. The maximum likelihood procedure for the fitting of the Markovian representation is discussed. A practical procedure for finding an initial guess of the representation is introduced and its feasibility is demonstrated with numerical examples.

1. Introduction

The state space representation of a system is a fundamental concept in modern control theory. When a discrete-time system is time-invariant and linear the state space representation of the system is given in the form

$$\begin{aligned} v_{n+1} &= Av_n + Bu_{n+1} \\ y_n &= Cv_n , \end{aligned} \tag{1.1}$$

where n denotes the time and u_n is a $q\times 1$ vector of the input to the system, y_n is an $r\times 1$ vector of the output and v_n is a $p\times 1$ vector of the state. A, B and C are respectively $p\times p$, $p\times q$ and $r\times p$ matrices. The use of the state space representation for the design of optimal control under a quadratic cost function is well-known in the engineering literature. The state space representation of a system has been discussed in some statistical literature (for example, Whittle [19] and Akaike [1]), however, it has not yet been fully exploited by statisticians. This may partly be due to the somewhat abstract definition of the concept of state, as is described by Kalman and others ([10], Chap. 10). The state is sometimes vaguely understood as a condensed representation of information from the present and past, such that the future behaviour of the system can completely be described by the knowledge of the present state and the future input. This idea finds a precise mathematical formulation when the system is stochastic, i.e. when the input u_n and the output y_n are stochastic processes. It was shown (Akaike [6]) that by the analysis of canonical correlations between the set of the present and future output and the set of the present and past input a Markovian representation of a stochastic system can be obtained. The Markovian representation is a stochastic analogue of (1.1) and is given by

$$
\begin{aligned}
v_{n+1} &= Av_n + Bz_{n+1} \\
y_n &= Cv_n + w_n ,
\end{aligned} \tag{1.2}
$$

where w_n is uncorrelated with v_n. In (1.2) z_{n+1} is the innovation of the input u_n at time $n+1$ and is defined by $u_{n+1} - u_{n+1|n}$, where $u_{n+1|n}$ is the projection of u_{n+1} on its past which is defined as the mean square closure of the space of finite linear combinations of the components of $u_n, u_{n-1}, \cdots$. When $u_n = y_n$, w_n vanishes from (1.2) and a Markovian representation of a stationary stochastic process y_n is given in the form

$$
\begin{aligned}
v_{n+1} &= Av_n + Bz_{n+1} \\
y_n &= Cv_n .
\end{aligned} \tag{1.3}
$$

This representation gives y_n as the output of a stochastic system which is time invariant and linear and driven by a white noise input z_n. v_n is called the state of the process.

The purpose of the present paper is to discuss the relation of this Markovian representation of a stationary stochastic process y_n with the familiar autoregressive moving average (AR-MA) representation

$$
y_n + B_1 y_{n-1} + \cdots + B_M y_{n-M} = z_n + A_1 z_{n-1} + \cdots + A_L z_{n-L} . \tag{1.4}
$$

First a proof of the equivalence of the Markovian and AR-MA representations, obtained by showing the existence of direct transformations from one to the other, is given. Hannan [9] discussed the problem of identifiability or the uniqueness of the AR-MA representation that is especially difficult when the process is multivariate. The simplest type of identifiability is the one which is called the block-identifiability in the present paper. The determination procedure of the AR-MA coefficient matrices A_i and B_j from the covariance matrices of y_n, under the block-identifiability condition, leads to a natural stochastic interpretation of Rissanen's [15] block triangularization procedure of block Hankel matrices. This result illustrates the inherent relationship between the problem of identification, or the determination of the Markovian or AR-MA representation, of the process y_n and the Hankel type matrices.

By a further analysis of the Markovian representation it becomes clear that the identifiability problem of the AR-MA representation can be solved completely without any restriction such as block-identifiability. This result demonstrates the merit of the Markovian representation for the purpose of analysis of stochastic systems. Generally there is not a unique structure, or a set of special forms of the matrices A, B and C in the Markovian representation (1.3), with a minimum number of undetermined parameters and that can represent every y_n with the same minimal possible dimension of v_n in the representation (1.3). Thus the identification, or the determination of the Markovian or AR-MA representation, must proceed in two steps, the first step is the selection of a special structure and the second is the determination of the parameters in the structure. Once an exhaustive set of special structures is specified the statistical identification or the determination of the structure and the parameters based on the observations of a Gaussian process can be realized through the maximum likelihood procedure with the aid of an information theoretic criterion (Akaike [2], [3]).

The statistical identification is realized by using the Markovian representation and the results may be used directly for the purpose of analysis and implementation of control of a multivariate stochastic system without recourse to the AR-MA representation. In particular in the last section of the paper it is shown that under a very mild assumption consistent estimates of the structure and the parameters within the matrix A of a special Markovian representation can be obtained by a simple procedure. This is a fundamental contribution to the subject of statistical identification of multivariate stochastic systems. The procedure is based on the canonical correlation analysis between the present and future and the present and past observations of the process and provides an initial guess of the structure and the parameters to be used for the maximum likelihood procedure. This is a significant example of the use of the canonical correlation concept in relation to the time series analysis. Numerical examples are given to show the feasibility of the procedure. Towards the end of the paper it is suggested that a special Markovian representation may be useful to give an answer to the problem of reduction of the number of measurements of a complex stochastic system. This problem was raised by Priestley and others [13].

Throughout the present paper the closure in the sense of mean square of the linear space of finite linear combinations of the components of the random vectors $x_1, x_2, \cdots$ will be denoted by $R(x_1, x_2, \cdots)$

and called the space spanned by the components of, or, simply, the space spanned by, $x_1, x_2, \cdots$. The i-step ahead predictor at time n of a stochastic process y_n is defined as the projection of y_{n+i} on $R(y_n, y_{n-1}, \cdots)$ and denoted by $y_{n+i|n}$. If $S=R(x_1, x_2, \cdots)$ holds for a linear space S, the set of the components of $x_1, x_2, \cdots$ is called a system of generators of S. In the present paper the qualities of random variables are understood in the sense of mean square.

2. Autoregressive-moving average processes and Markovian representations

The autoregressive moving average process (AR-MA process) y_n is defined by

$$y_n+B_1y_{n-1}+\cdots+B_My_{n-M}=z_n+A_1z_{n-1}+\cdots+A_Lz_{n-L}\ , \tag{2.1}$$

where B_i and A_j are the matrices of coefficients, y_n and z_n are $r\times 1$ vectors and z_n is a white noise with $\mathrm{E}\,z_n=0$, zero vector, and $\mathrm{E}\,(z_nz_n')=G$ and for $i\neq 0$ $\mathrm{E}\,(z_nz_{n-i}')=0$, zero matrix, and $\mathrm{E}\,y_nz_{n+i}'=0$ for $i=1, 2,\cdots$. It is assumed that the characteristic equations $\left|\lambda^MI+\sum_{i=1}^{M}\lambda^{M-i}B_i\right|=0$ and $\left|\lambda^LI+\sum_{j=1}^{L}\lambda^{L-j}A_j\right|=0$ have zeros outside the unit circle. This assumption assures that y_n can be expressed in a form

$$y_n=\sum_{m=0}^{\infty}W_mz_{n-m} \tag{2.2}$$

with $W_0=I$, identity matrix, and z_n is the innovation of y_n at time n, i.e. $z_n=y_n-y_{n|n-1}$ and $y_{n|n-1}$ is the one-step ahead predictor of y_n at time $n-1$. To get a Markovian representation of y_n it is only necessary to analyze the structure of the predictors $y_{n|n}, y_{n+1|n},\cdots$. Let $x_{|n}$ denote the projection of x on $R(y_n, y_{n-1},\cdots)$, then $y_{n+i|n}$ satisfies the relation

$$\begin{aligned} y_{n+i|n}&+B_1y_{n+i-1|n}+\cdots+B_My_{n+i-M|n}\\ &=z_{n+i|n}+A_1z_{n+i-1|n}+\cdots+A_{M-1}z_{n+i-L|n}\ , \end{aligned} \tag{2.3}$$

where $y_{n+h|n}=y_{n+h}$ for $h=0, -1,\cdots$, and $z_{n+h|n}=0$ for $h=0, 1,\cdots$. For $i\geqq L+1$ the right-hand side of (2.3) vanishes. Thus $y_{n+i|n}$ $(i=0, 1,\cdots)$ can be expressed as linear transforms of $y_{n|n}, y_{n+1|n},\cdots, y_{n+K-1|n}$, where $K=\max(M, L+1)$, and the components of these vectors form a system of generators of the linear space spanned by the components of $y_{n+i|n}$ $(i=0, 1,\cdots)$. Especially it holds that

$$y_{n+K|n}=-B_1y_{n+K-1|n}-B_2y_{n+K-2|n}\cdots-B_Ky_{n|n}\ , \tag{2.4}$$

where by definition $B_m=0$ for $m=M+1, M+2,\cdots, K$. From (2.2) one can get

$$y_{n+i+1|n+1}=y_{n+i+1|n}+W_i z_{n+1} . \tag{2.5}$$

From (2.4) and (2.5) it can be seen that the vector $v_n=(y'_{n|n}, y'_{n+1|n},\cdots, y'_{n+K-1|n})'$ provides a Markovian representation

$$v_{n+1}=\begin{bmatrix} 0 & I & 0 & \cdots & 0 \\ 0 & 0 & I & \cdots & 0 \\ \vdots & \vdots & \vdots & \ddots & \vdots \\ 0 & 0 & 0 & \cdots & I \\ -B_K & -B_{K-1} & -B_{K-2} & \cdots & -B_1 \end{bmatrix} v_n+\begin{bmatrix} W_0 \\ W_1 \\ \vdots \\ W_{K-2} \\ W_{K-1} \end{bmatrix} z_{n+1}$$

$$y_n=[\,I \;\; 0 \;\; 0 \cdots 0\,]v_n . \tag{2.6}$$

This result shows that an AR-MA process always has a Markovian representation.

It should be noted that W_m's are the impulse response matrices of the time-invariant linear system defined by (2.1) with the input z_n and the output y_n. The jth column $W_m(\cdot j)$ of W_m is obtained by the relation

$$\begin{aligned} &W_m(\cdot j)+B_1W_{m-1}(\cdot j)+\cdots+B_MW_{m-M}(\cdot j) \\ &\quad =D_n(\cdot j)+A_1D_{n-1}(\cdot j)+\cdots+A_LD_{n-L}(\cdot j) , \end{aligned}$$

where $W_m(\cdot j)=0$, zero vector, for $m<0$ and $D_n(\cdot j)$ denotes the jth column of a matrix D_n which is by definition equal to I, identity matrix, for $n=0$ and 0, zero matrix, for $n\neq 0$. Thus numerically it is a simple matter to derive the Markovian representation (2.6) from the AR-MA representation (2.1).

Now suppose a process y_n has a Markovian representation

$$\begin{aligned} v_{n+1}&=Av_n+Bz_{n+1} \\ y_n&=Cv_n , \end{aligned} \tag{2.7}$$

where it is assumed that v_n is a $p\times 1$ vector of the state and z_n is the innovation of y_n. If the characteristic polynomial of A is given by $|\lambda I-A|=\lambda^p+\sum_{m=1}^{p} a_m\lambda^{p-m}$ then by the Cayley-Hamilton theorem $A^p+\sum_{m=1}^{p} a_mA^{p-m}=0$. From (2.7), $v_{n+i}=A^iv_n+A^{i-1}Bz_{n+1}+\cdots+Bz_{n+i}$ and it follows that y_n has an AR-MA representation

$$y_{n+p}+a_1y_{n+p-1}+\cdots+a_py_n=z_{n+p}+C_1z_{n+p-1}+\cdots+C_{p-1}z_{n+1} \tag{2.8}$$

where

$$C_i=C(A^i+a_1A^{i-1}+\cdots+a_iI)B .$$

In (2.8) the autoregressive coefficients are scalars, or, equivalently, constant diagonal matrices. This result shows that any stationary stochastic process with the Markovian representation (2.7) also has an AR-MA representation (2.8).

Thus at least theoretically there is no distinction between the Markovian and the AR-MA representations of a stationary stochastic process. The results obtained by the analysis of the Markovian representation can be used for the analysis of the AR-MA representation. This fact is fully utilized in the following discussion of the identifiability problem of the AR-MA representation. In (2.7), predictors are given simply by

$$y_{n+i|n}=CA^i v_n \qquad i=0, 1, \cdots . \tag{2.9}$$

Thus the components of v_n form a system of generators of the space $R(y_{n|n}, y_{n+1|n}, \cdots)$ which is spanned by the predictors $y_{n|n}, y_{n+1|n}, \cdots$. This space $R(y_{n|n}, y_{n+1|n}, \cdots)$ will hereafter be called the predictor space. When the dimension of the predictor space of an arbitrary stationary process y_n is finite there is a finite K which satisfies the relation (2.4) and y_n has a Markovian representation (2.6). Thus the finiteness of the dimension of the predictor space is the fundamental characterization of a process with Markovian or AR-MA representation. And one of the two Markovian representations of a process y_n, the states of which are defined by the elements of the predictor space, can be obtained from the other by a linear transformation of the state of the latter. It is now obvious that the dimension, as a vector, of the state of a Markovian representation which is defined by using a basis of the predictor space as its state is minimal. The dimension of this basis, or the dimension of the predictor space, is a characteristic of the stochastic system which generates the process from its innovations and will be called the dimension of the system. Also the process is called a process with p-dimensional dynamics.

3. Block-identifiability of autoregressive moving average processes

Although the AR-MA representation of a stationary stochastic process has been used as one of the basic models of time series analysis there is a serious conceptual difficulty inherent in this model. This is the non-uniqueness of the representation. When (2.1) holds there are infinitely many other representations of the same process, for example those which are obtained from (2.1) by the transformations which replace n of (2.1) by some $n-k$ ($k>1$), premultiply it with an $r\times r$ matrix D_k and add to the original (2.1). When y_n is a univariate process, the representation can be made unique by requiring the orders

L and M of (2.1) to be minimal. When y_n is a multivariate process this requirement of minimal order is not necessarily sufficient to make the representation unique. Under the assumption of non-singularity of the covariance matrix $G=(z_n z_n')$ Hannan [9] gave a necessary and sufficient condition for an AR-MA process y_n to have a unique representation of the form (2.1).

It is trivially true that the uniqueness of a representation is a prerequisite for the development of consistent estimation procedures. But this fact should not be considered as meaning the impracticability of developing a general estimation procedure of AR-MA models without identifiability conditions. If the purpose of fitting an AR-MA model is only to get an estimate of the covariance structure of the process under observation any one of the possible equivalent representations can serve for the purpose, if only it can be specified properly. The practicability of this specification procedure is the main subject of the present paper.

Although a general estimation procedure without any assumption of identifiability of the model is developed in the later sections it will be useful for the understanding of the subject to discuss the relation between the AR-MA representation (2.1) and the Markovian representation (2.6) under the assumption of a simplest type of identifiability. Under the assumption of non-singularity of $G=\mathrm{E}(z_n z_n')$, W_m is uniquely determined by (2.2) and satisfies the relation

$$A_i = W_i + B_1 W_{i-1} + \cdots + B_M W_{i-M}, \tag{3.1}$$

where $W_m=0$ for $m<0$. Thus the representation (2.1) is uniquely determined from (2.6) if the variance matrix of v_n is non-singular. This is a simple sufficient condition for the identifiability of the AR-MA representation. Consider a general situation where a stationary Markovian process v_n is defined by

$$v_{n+1} = Av_n + Bz_{n+1}, \tag{3.2}$$

where z_n is a zero mean $q\times 1$ white noise with $\mathrm{E}(z_n z_n')=G$ and $\mathrm{E}(z_n z_{n-i}')=0$ $(i\neq 0)$ and v_n is $s\times 1$ and is an element of the space spanned by $z_n, z_{n-1}, \cdots$. It is assumed that G is non-singular. The dimension of $R(v_n)$, the space spanned by the components of v_n, is determined by the following matrix which is often called the controllability matrix in control engineering literature:

$$C_s = [B, AB, A^2B, \cdots, A^{s-1}B]. \tag{3.3}$$

When the rank of C_s is less than s there is a non-zero $s\times 1$ vector g such that $g'C_s=0$. This means $\mathrm{E}(g'v_n z_{n-i}')=0$ for $i=0, 1, \cdots, s-1$. As was discussed below (2.7) there is a set of coefficients a_m $(m=1, 2, \cdots, s)$

which satisfy the relation $A^s+\sum_{m=1}^{s} a_m A^{s-m}=0$. By using this relation, $\mathrm{E}(v_n z'_{n-i})$ $(i\geqq 0)$ can always be expressed as a linear combination of $\mathrm{E}(v_n z'_n), \mathrm{E}(v_n z'_{n-1}),\cdots, \mathrm{E}(v_n z'_{n-s+1})$. Since $\mathrm{E}(g'v_n z'_{n-i})=0$ holds for $i=0, 1, \cdots, s-1$, this means that $\mathrm{E}(g'v_n z'_{n-i})=0$ for all non-negative values of i. As v_n is an element of the space spanned by $z_n, z_{n-1},\cdots$, this means $g'v_n=0$. Thus the distribution of v_n is degenerate. When the rank C_s is equal to s there is no non-zero g for which $\mathrm{E}(g'v_n z'_{n-i})=0$ $(i=0, 1,\cdots)$ and the distribution of v_n is non-degenerate. Thus the present identifiability condition can readily be checked by analyzing the rank of the matrix C_s of (3.3) with A and B defined by (3.2) and (2.6), where W_i's are obtained by (3.1) from A_i's and B_j's. When an AR-MA process satisfies the present identifiability condition the dimension of the stochastic system or the dimension of the predictor space is, from (2.6), equal to $Kr=\max(Mr, (L+1)r)$, where r is the dimension of y_n. When y_n is multivariate $(r>1)$ the dimension of the stochastic system defined by y_n need not always be an integral multiple of r. Thus it is clear that the class of the AR-MA processes which satisfy the present condition is a rather limited one and will not be wide enough as a basis to develop a fully efficient statistical identification procedure of AR-MA processes on it.

Hereafter the identifiability described in the preceding paragraph will symbolically be called the block-identifiability and the process which satisfies the condition of block-identifiability will be called block-identifiable.

4. Determination of AR-MA coefficients from covariance sequence under the block-identifiability assumption

For an r-dimensional AR-MA process (2.1) the AR coefficient matrices $B_1, B_2,\cdots, B_M$ satisfy the Yule-Walker equation

$$C(L+i)+B_1C(L+i-1)+\cdots+B_MC(L+i-M)=0 \qquad i=1, 2,\cdots, M \tag{4.1}$$

where $C(j)=\mathrm{E}\, y_n y'_{n-j}$. (4.1) can be expressed in a matrix form

$$\begin{aligned}(4.2)\quad [-C(L+1), -C(L+2),\cdots, -C(L+M)]\\ =[B_1, B_2,\cdots, B_M]\end{aligned}$$

$$\cdot\begin{bmatrix} C(L) & C(L+1) & \cdots & C(L+M-1) \\ C(L-1) & C(L) & \cdots & C(L+M-2) \\ \vdots & \vdots & & \vdots \\ C(L-M+1) & C(L-M+2) & \cdots & C(L) \end{bmatrix}.$$

The last $Mr \times Mr$ matrix of (4.2) is a block Toeplitz matrix and a recursive numerical procedure for the solution of (4.2) was discussed by Whittle [18] and Akaike [4]. When the process satisfies the block-identifiability condition the numerical procedure produces a unique solution to (4.2). When y_n is known to be block-identifiable but the values of L and M are unknown the solution must be tried for various combinations of L and M. The block Toeplitz matrix type formulation of the Yule-Walker equation can be seen to be unsuitable for this case. Instead the block Hankel matrix type formulation, which is to be discussed shortly, is much more natural and numerically efficient.

The block-identifiability condition is equivalent to the assumption of linear independence of the components of the predictors $y_{n|n}, y_{n+1|n}, \cdots, y_{n+K-1|n}$, where $K = \max(M, L+1)$. Since $\mathrm{E}(y_{n+i|n}y'_{n-j}) = \mathrm{E}(y_{n+i}y'_{n-j})$ $(i, j=0, 1, \cdots)$ the analysis of dependence of the components of the predictors $y_{n|n}, y_{n+1|n}, \cdots$ is reduced to the analysis of dependence of the elementary rows of the block Hankel matrix of the covariance between $(y_n, y_{n+1}, \cdots)$ and $(y_n, y_{n-1}, \cdots)$ defined by

$$\begin{bmatrix} C(0) & C(1) & C(2) & \cdots \\ C(1) & C(2) & C(3) & \cdots \\ C(2) & C(3) & C(4) & \cdots \\ \vdots & \vdots & \vdots & \end{bmatrix}. \tag{4.3}$$

Under the block-identifiability condition the analysis can be made by a blockwise procedure. This is realized by successively finding a sequence of linear transforms $y^{(i)}_{n+i} = y_{n+i} + B_1(i)y_{n+i-1} + \cdots + B_i(i)y_n$ of $y_n, y_{n+1}, \cdots$ such that $y^{(i)}_{n+i}$ is orthogonal to $y_n, y_{n-1}, \cdots, y_{n-i+1}$. The first transformation is obtained by replacing y_{n+i+1} by $y^{(1)}_{n+i+1} = y_{n+i+1} + B_1(1)y_{n+i}$ $(i=0, 1, \cdots)$ with $B_1(1)$ defined by the relation $\mathrm{E}(y^{(1)}_{n+1}y'_n) = 0$, or $B_1(1) = -C(1)\cdot C(0)^{-1}$. The covariance matrix $\mathrm{E}\{(y'_n, y^{(1)\prime}_{n+1}, y^{(1)\prime}_{n+2}, \cdots)'(y'_n, y'_{n-1}, y'_{n-2}, \cdots)\}$ has its (2, 1)th block element equal to a zero matrix and the matrix below the first block-row is a block Hankel matrix. In the second step it is desired to define $y^{(2)}_{n+i}$ in such a way that $y^{(2)}_{n+2}$ is orthogonal to y_n and y_{n-1}. To realize this, first $y^{(1)}_{n+2+i}$ is transformed into $z^{(1)}_{n+2+i} = y^{(1)}_{n+2+i} + B_{22}y_{n+i}$ $(i=0, 1, \cdots)$ with B_{22} satisfying the relation $\mathrm{E}(z^{(1)}_{n+2}y'_n) = 0$. $z^{(1)}_{n+2+i}$ is further transformed into $y^{(2)}_{n+2+i} = z^{(1)}_{n+2+i} + B_{21}y^{(1)}_{n+1+i}$ with B_{21} satisfying $\mathrm{E}(y^{(2)}_{n+2}y'_{n-1}) = 0$. By the first of these two transformations the (3, 1)th block element of the covariance matrix is turned into a zero matrix and by the second the (3, 2)th block element is also turned into a zero matrix. By the Hankel property the (4, 1)th block element $\mathrm{E}(y^{(2)}_{n+3}y'_n)$ of $\mathrm{E}\{(y'_n, y^{(1)\prime}_{n+1}, y^{(2)\prime}_{n+2}, y^{(2)}_{n+3}, \cdots)'(y'_n, y'_{n-1}, y'_{n-2}, \cdots)\}$ is a zero matrix. The kth step of transformation is defined by $z^{(k-1)}_{n+k+i} = y^{(k-1)}_{n+k+i} + B_{k2}y^{(k-2)}_{n+k-2+i}$ and $y^{(k)}_{n+k+i} = z^{(k-1)}_{n+k+i} + B_{k1}y^{(k-1)}_{n+k-1+i}$ with B_{k2} and B_{k1} defined by the relations $\mathrm{E}(z^{(k-1)}_{n+k+i}\cdot$

$y'_{n-k+2})=0$ and $\mathrm{E}\,(y^{(k)}_{n+k+1}y'_{n-k+1})=0$, respectively. Now $\mathrm{E}\,(y^{(k)}_{n+k}y'_{n-j})=0$ for $j=0, 1, \cdots, k-1$. When k is equal to $K\,(=\max(M, L+1))$ then $\mathrm{E}\,(y^{(k)}_{n+k}\cdot y'_{n-j})=0$ for $j=0, 1, \cdots$, and the representation

$$y^{(K)}_{n+K}=y_{n+K}+B_1y_{n+K-1}+\cdots+B_Ky_n \tag{4.4}$$

gives the desired set of AR coefficient matrices $B_1, B_2, \cdots, B_K$. Incidentally, at the kth step of the above stated procedure the $(k+1)\times(k+1)$ block matrix $\mathrm{E}\,\{(y'_n, y^{(1)\prime}_{n+1}, \cdots, y^{(k)\prime}_{n+k})'(y'_n, y'_{n-1}, \cdots, y'_{n+k})\}$ takes the form of upper triangular block matrix. If the initial covariance matrix (4.3) is replaced by $\mathrm{E}\,\{(y'_n, y'_{n+1}, \cdots)'(u'_n, u'_{n-1}, \cdots)\}$, where u_n is a stochastic process stationarily correlated with y_n, the above procedure gives a stochastic interpretation of a block triangularization procedure of block Hankel matrices developed by Rissanen [15].

Once the AR coefficient matrices $B_1, B_2, \cdots, B_K$ are obtained $y^{(K)}_{n+K}$ defined by (4.4) satisfies the representation

$$y^{(K)}_{n+K}=z_{n+K}+A_1z_{n+K-1}+\cdots+A_{K-1}z_{n+1}\,.$$

Thus the problem of determination of the MA coefficient matrices $A_1, A_2, \cdots, A_{K-1}$ of the AR-MA process y_n reduces to the problem of the determination of the MA coefficient matrices of a simple MA process $y^{(K)}_{n+K}$. A numerical solution to this problem is given also by Rissanen [15] using a blockwise recursive procedure.

5. Special Markovian representations and their use for identification

It is tempting to think that once the dimension p of the system is given the minimum number of necessary parameters to define the Markovian representation of a stationary stochastic process y_n is determined. In fact this is not true for the multivariate case. To see this a special Markovian representation is considered here. The representation is obtained by defining its state v_n as the vector of the first p linearly independent components of the pr-dimensional vector $(y'_{n|n}, y'_{n+1|n}, \cdots, y'_{n+p-1|n})'$. Now define $Y_n(k)$ $(k=1, 2, \cdots)$ by the relation

$$Y_n(jr+i)=y_{n+j}(i)_{|n}\;,$$

where $y_n(k)$ denotes the kth component of y_n. Denote by H the set of the integers $k_1, k_2, \cdots, k_p$ such that $v_n=(Y_n(k_1), Y_n(k_2), \cdots, Y_n(k_p))'$. The set H has a special characteristic:

$$k+r\notin H\,, \qquad \text{when } k\notin H\,. \tag{5.1}$$

From the definition of $Y_n(k)$ it holds that

$$Y_{n+1}(k)_{|n}=Y_n(k+r)\,.$$

Thus the transition matrix A which satisfies the relation $v_{n+1|n}=Av_n$ is determined by the relations

(5.2a) $$Y_{n+1}(k_i)_{|n}=Y_n(k_j) \qquad \text{for } k_i+r=k_j$$

(5.2b) $$= \sum_{j\,:\,k_j<k_i+r} A_{ij}Y_n(k_j) \qquad \text{otherwise},$$

where the last summation extends over the j's such that $k_j<k_i+r$. The observation matrix C which satisfies the relation $y_n=Cv_n$ is determined by the relations $(i=1, 2, \cdots, r)$

(5.3a) $$y_n(i)=Y_n(k_j) \qquad \text{for } i=k_j$$

(5.3b) $$= \sum_{j\,:\,k_j<i} C_{ij}Y_n(k_j) \qquad \text{otherwise},$$

where the last summation extends over j's such that $k_j<i$. Denote the vector of the innovations $\{Y_{n+1}(k_j)-Y_{n+1}(k_j)_{|n};\ j=1, 2, \cdots, t\}$ by z_{n+1}, where t is the maximum of j such that $k_j \leqq r$. The Markovian representation is given by $v_{n+1}=Av_n+Bz_{n+1}$ and $y_n=Cv_n$, where A and C are determined by (5.2) and (5.3) and B is the matrix of regression coefficients of v_{n+1} on z_{n+1}. The matrix B can be obtained from the elements of the impulse response matrices of the system to the input z_n. From the definitions of v_n and z_n it is clear that the present Markovian representation is uniquely determined from the covariance structure of y_n. Thus the set of the r-dimensional stationary processes y_n with p-dimensional dynamics is decomposed into mutually exclusive subsets, each of which is characterized by a set H of the p integers $k_1, k_2, \cdots, k_p$ $(\leqq pr)$ with the characteristic (5.1) and admits a unique Markovian representation of its elements with the transition and the observation matrices, A and C, of the forms respectively described by (5.2) and (5.3). Now for each subset specified by H consider an $r\times(p+1)$ matrix S of which (i, j)th element $S(i, j)$ is 1 when for each element y_n of the subset $y_{n+j-1}(i)_{|n}$ is retained in v_n, i.e. when $i+(j-1)r \in H$, and 0 otherwise. The $p+1$st column of S is always a zero vector. The matrix S takes a special form where each row has first several, or no, elements equal to 1 and others equal to 0. The total number of 1's within S is equal to p. Shift the columns of S one step to the right and fill in 1's in the empty fisrt column to define another $r\times(p+1)$ matrix S_+. Define $T=S+S_+$. The 1's in the first column of T correspond to the i's of (5.3b) and the 1's in other columns correspond to k_i's of (5.2b). Starting at the (1, 1)th element of T, calculate the number of 2's columnwise until the ith 1 $(i=1, 2, \cdots, r)$. The sum of these numbers is equal to the number of parameters within A_{ij} and C_{ij} of (5.2) and (5.3). The number of parameters within B is $p\times t$, where t is the dimension of z_n. From the definition of the subset it is obvious that no further reduc-

tion of the number of parameters is possible. Now it is easy to see by some examples that for a given p there may be different patterns of the distribution of 1's within S which require different number of parameters within A and C.

The special representation discussed above can be applied to produce an ultimate answer to the identifiability problem of the AR-MA representation of a stationary process with a finite dimensional dynamics, under the assumption of non-singularity of the variance matrix of its innovations. For this case C takes the form $C=[I\ 0]$. Denote by p_i the number of 1's in the ith row of the above defined matrix S. By rewriting (5.2b) the following representation is obtained:

$$y_{n+p_i}(i)_{|n}=\sum_{m=0}^{p_i} C_m(i\cdot)y_{n+p_i-m|n}\,, \tag{5.4}$$

where $C_m(i\cdot)$ is a $1\times r$ vector determined by (5.2b) and $C_0(i,j)$, the $(1,j)$th element of $C_0(i\cdot)$, is always equal to zero for $j\geqq i$. Define $q=\max(p_1, p_2, \cdots, p_r)$. From (5.4) one can get a relation

$$y_{n+q|n}=\sum_{m=0}^{q} C_m y_{n+q-m|n}\,,$$

which gives an AR-MA type representation

$$\begin{aligned}(I-C_0)y_{n+q}-C_1y_{n+q-1}-\cdots-C_qy_n\\ =D_0z_{n+q}+D_1z_{n+q-1}+\cdots+D_{q-1}z_{n+1}\,,\end{aligned} \tag{5.5}$$

where $C_m(i,j)$, the (i,j)th element of C_m, is put equal to the $(1,j)$th element of $C_m(i\cdot)$ defined by (5.4) or equal to zero, if undefined by (5.4). From the definition of C_0, $I-C_0$ is non-singular and an AR-MA representation of y_n is given by

$$y_n+B_1y_{n-1}+\cdots+B_qy_{n-q}=z_n+A_1z_{n-1}+\cdots+A_{q-1}z_{n-q+1}\,, \tag{5.6}$$

where $B_i=-(I-C_0)^{-1}C_i$ and $A_i=(I-C_0)^{-1}D_i$. Thus it has become clear that without any assumptions such as the block-identifiability, a stationary process with a finite dimensional dynamics and a non-singular innovation variance matrix always has a uniquely identifiable AR-MA representation (5.6). This result has been obtained with the aid of the special Markovian representation introduced in this section and very clearly shows the advantage of the Markovian representation over the AR-MA representation for the purpose of multivariate stochastic system analysis.

Since the Markovian representation of a stationary stochastic process y_n is a state space representation of a time-invariant linear system which generates y_n from the input z_n, the innovation of y_n, any special Markovian representation can be defined by using a special state space representation of the corresponding system. The subject of the special state space representations of time-invariant discrete-time linear systems has been discussed extensively and several alternatives of the special Markovian representation introduced in this section can easily be introduced (Akaike [7]). Especially by replacing the definition of $Y_n(k)$ by

$$Y_n((i-1)p+j+1)=y_{n+j}(i)_{|n} \qquad i=1,2,\cdots,r;\ j=0,1,\cdots,p-1,$$

and searching for the first p linearly independent components of $Y_{n,p}=(Y_n(1), Y_n(2),\cdots, Y_n(rp))$ one can get another basis v_n of the predictor space. By this choice of the basis, the definition of z_n in the Markovian representation can be replaced by the vector of innovations of those components of y_n which are retained in v_n to make the matrix B unique. The assumption of non-singularity of the innovation matrix is now unnecessary and the unique Markovian representation thus obtained gives a corresponding unique AR-MA representation.

Once a special representation is specified, at least conceptually there is no difficulty in developing a statistical identification procedure based on the maximum likelihood method to be described in the next section. The only difficulty which prevents the practical application of the procedure is caused by the existence of the vast number of possible choices of the basis of the predictor space. It is almost prohibitive to perform the maximum likelihood computation for every possible choice of the basis. Thus the feasibility of the procedure is almost entirely depenent on how to get good initial guesses of the dimension of the system and the structure of the desired basis. A solution to this problem is given in the last section.

It should be mentioned here that in an unpublished paper by Rissanen [14] the special representation discussed at the beginning of this section was used implicitly to develop a consistent estimation procedure of the parameters of a multivariate autoregressive process.

6. Maximum likelihood procedure and information theoretic criterion

When a stationary Gaussian process y_n has a Markovian representation

(6.1a) $$v_{n+1}=Av_n+Bz_{n+1}$$

(6.1b) $$y_n=Cv_n,$$

the representation can conveniently be used to define an approximation to the likelihood function. Under the assumption of non-singularity of the innovation variance matrix, $\mathrm{E}(z_n z_n')$, y_n can be expressed in the form

(6.2) $$y_n=CAv_{n-1}+z_n.$$

When a record of observations (y_n; $n=1, 2,\cdots, N$) is given, by assuming $v_0=0$ a set of realization of z_n ($n=1, 2,\cdots, N$) can be obtained by (6.2) and (6.1a). The logarithm of the approximate likelihood function is given by using the realization of z_n in the form

$$(6.3) \qquad -\frac{N}{2}\{r \log 2\pi + \log|G| + \mathrm{tr}\,(G^{-1}C_0)\} ,$$

where $C_0=\frac{1}{N}\sum_{n=1}^{N} z_n z_n'$ and G is the assumed covariance matrix of the innovations. This result corresponds to the asymptotic evaluation of the Gaussian likelihood given by Whittle ([16], (5.1)). If the Fourier transform $Y(f)$ of (y_n; $n=1, 2,\cdots, N$) is defined by

$$Y(f)=\frac{1}{\sqrt{N}}\sum_{n=1}^{N}\exp(-i2\pi f n)y_n$$

and $V(f)$ and $Z(f)$ by

$$V(f)=\frac{1}{\sqrt{N}}\sum_{n=1}^{N}\exp(-i2\pi f n)v_n$$

$$Z(f)=\frac{1}{\sqrt{N}}\sum_{n=1}^{N}\exp(-i2\pi f n)z_n ,$$

then by neglecting the effect of end conditions it holds that

$$Z(f)=[C\{I-\exp(-i2\pi f)A\}^{-1}B]^{-1}Y(f) .$$

By using the relation

$$C_0=\int_{-1/2}^{1/2} Z(f)Z^*(f)df ,$$

where * denotes the conjugate transpose, one can get an explicit representation of (6.3) in terms of the matrices A, B, C and G and the Fourier transform of y_n. Thus if only a non-redundant parametrization of A, B, C is available the maximum likelihood procedure can be realized numerically by following the line of approach developed for the AR-MA representation (Akaike [5]). The special Markovian representation introduced in the preceding section can directly be used for this purpose. The importance of this non-redundant parametrization of a Markovian representation in statistical identification of a stochastic system can be understood more clearly by persuing its analogy to the factor analysis model. If v_n in (6.1) is replaced by $u_n=Tv_n$, with T non-singular, another Markovian representation of y_n is given by $u_{n+1}=TAT^{-1}u_n+TBz_{n+1}$ and $y_n=CT^{-1}u_n$. Thus it is obvious that there is no meaning in trying to find a unique Markovian representation without any further restrictions. The situation is quite similar to the case of the indeterminacy of the factor analysis model (see, for example, Lawley and Maxwell [12]). The dimension of the system plays a role similar to that of the number of factors in factor analysis. As was already seen in the preceding section, the dimension is not sufficient to determine the minimum number of parameters necessary to define the Markovian representation and the parametrization is much more complicated than that of the factor analysis. As will be made clear in the following discussion of the use of an information theoretic criterion for the decision of the models, the ultimate use of these models, the Markovian or the AR-MA representation of a stationary stochastic process and the factor analysis model, is to provide a reliable estimate of the related covariance structure by controlling the number of parameters within the model when there is not sufficient prior information to limit to a unique model. The decision on the dimension and the structure of the basis or on the number of factors is obviously the crucial point in applying these models to real data.

By using (6.3) or directly from the result given by Whittle [16], the logarithm of the approximate likelihood function of the observations (y_n; $n=1, 2, \cdots, N$) can be given by

$$\text{(6.4)} \qquad -\frac{N}{2}\left[r \log 2\pi + \log |G| + \int_{-1/2}^{1/2} \operatorname{tr}\{Y(f)Y^*(f)P^{-1}(f)\}df\right],$$

where $P(f)=[C\{I-\exp(-i2\pi f)A\}^{-1}B]G[C\{I-\exp(-2\pi f)A\}^{-1}B]^*$ is the spectral density matrix of the assumed model. By taking the expectation of (6.4) $Y(f)Y^*(f)$ is replaced by $P_N(f)=\mathrm{E}\{Y(f)Y^*(f)\}$ which converges to the true spectral density matrix $P_\infty(f)$ as N tends to infinity. The negative of the expectation of a log-likelihood is, ignoring an additive constant, identical to the mean amount of information for discrimination between the assumed model and the true distribution per observation from true distribution as defined by Kullback and Leibler ([11], (2.4)). Thus if the negative of the expectation of the likelihood (6.4) is divided by N this gives the average of the mean information for discrimination. This quantity will be called the average information of the assumed model from the true model, or simply the average information, obtained from ($y_1, y_2, \cdots, y_N$). When N is made infinite the average information obtained from ($y_1, y_2, \cdots, y_N$) tends to a quantity given by

$$\text{(6.5)} \qquad \frac{1}{2}\left[r \log 2\pi + \log |G| + \int_{-1/2}^{1/2} \operatorname{tr}\{P_\infty(f)P^{-1}(f)\}df\right].$$

From the definition it would be reasonable to call this quantity the average information for discrimination of the assumed model from the true model. The maximum likelihood estimates of the parameters are obtained by maximizing (6.4), which is equivalent to minimizing (6.5) with $P_\infty(f)$ replaced by $Y(f)Y^*(f)$. Thus the maximum likelihood estimation procedure is equivalent to first constructing a stationary Gaussian process with the spectral density matrix $Y(f)Y^*(f)$, or the covariance matrix defined by $C(j)=\frac{1}{N}\sum_{n=1}^{N-j} y_{n+j}y'_n$ $(j=0,1,2,\cdots,N-1)$, $C(-j)=C(j)'$ and $C(j)=0$ for $|j|\geqq N$, and then minimizing the average information for discrimination of the model being fitted from this constructed Gaussian process model. Since it is well known that $Y(f)Y^*(f)$ can not even be a consistent estimate of $P_\infty(f)$ the above interpretation of the maximum likelihood estimates gives a clear indication of the difficulty inherent in the model fitting by the maximum likelihood procedure. When the assumed model is too flexible with too many number of parameters, then the estimated covariance structure will come very close to the one given by $Y(f)Y^*(f)$. In that case the estimate would be unreliable. When the assumed model is too inflexible with too small number of parameters the estimated covariance structure may not be able to sufficiently approximate the true structure. In the case of the Markovian model fitting the flexibility is controlled by the dimension of the system and the decision on the dimension becomes crucial for the success of the fitting procedure.

It has been found that a statistic defined by

$$(-2)\log_e(\text{maximum likelihood}) + 2(\text{number of adjusted parameters}) \tag{6.6}$$

is useful for the purpose of the above stated decision on the model flexibility (Akaike [2], [3]). In the present situation this statistic is meant to be an estimate of $2N$ times the average information for discrimination of the assumed model from the true model. The first term of (6.6) stands for the penalty of the badness of fit and the second term for the penalty of increased unreliability. (6.6) tells us that if the badness of fit is identical for two models the one with less number of parameters should be preferred. The definition of the above penalty is based on the chi-square approximation of the limiting distribution of the difference of $2\log_e$(maximum likelihood) from its expectation when the model is exact. For the justification of this chi-square approximation in the time series situation, see Whittle [17]. For the Markovian representations the number of adjusted parameters should be defined as the minimum number of parameters required to define the model and if the special representation introduced in the preceding section is used the model is specified by the dimension of the system and the set of indices H which specify the choice of a basis of the predictor space. For univariate case only the dimension of the system is sufficient to specify a model. From the standpoint of statistical model fitting or identification by using (6.6) the Markovian representation of a stationary stochastic process is merely one choice of the model which will give a good approximation to an arbitrary covariance structure with rather small number of parameters.

7. Determination of dimension and basis of predictor space

Under the assumption of non-singularity of the innovation variance matrix, the search for a basis of the predictor space of y_n, which gives a special Markovian representation discussed in Section 5, can be realized by the search of the first p linearly independent elementary rows of the block Hankel matrix (4.3). If p is not known but its upper bound q is known then the search can be completed by the analysis of dependence of the qr elementary rows within the first q block rows. From the Markovian representation (2.7) the covariance matrices $C(i)=\mathrm{E}\, y_{n+i}y'_n$ $(i=0, 1, \cdots)$ can be expressed in the form

$$C(i)=CA^iPC', \tag{7.1}$$

where $P=\mathrm{E}\, v_nv'_n$ and A is a $p\times p$ matrix. As was discussed below (2.7), there is a set of constants a_m $(m=1, 2, \cdots, p)$ such that $A^p+a_1A^{p-1}+\cdots+a_pI=0$. From (7.1) it also holds that

$$C(i+p)=-\sum_{m=1}^{p} a_mC(i+p-m) \qquad i=0, 1, \cdots . \tag{7.2}$$

This shows that in the block Hankel matrix (4.3) any block column can be expressed as a linear combination of the first p block columns. Thus, when only q is known, the analysis of dependence of the elementary rows within the first q block rows of (4.3) can be realized by the analysis of dependence within the elementary rows of the $qr\times qr$ matrix

$$C_q=\begin{bmatrix} C(0) & C(1) & \cdots & C(q-1) \\ C(1) & C(2) & \cdots & C(q) \\ \vdots & \vdots & & \vdots \\ C(q-1) & C(q) & \cdots & C(2q-2) \end{bmatrix}. \tag{7.3}$$

C_q can be expressed in the form $C_q=\mathrm{E}\, y_+y'_-$, where $y_+=(y'_n, y'_{n+1}, \cdots, y'_{n+q-1})'$, and $y_-=(y'_n, y'_{n-1}, \cdots, y'_{n-q+1})'$, and the matrix of the regression coefficients of y_+ on y_- is given as $C_qD_q^{-1}$, where $D_q=\mathrm{E}\,(y_-y'_-)$ and it is assumed that D_q is non-singular. The dimension p of the system, or the rank of C_q, is equal to the dimension of the linear space spanned by the components of the projection of y_+ on the linear space spanned by the components of y_- and thus is equal to the number of non-zero canonical correlation coefficients between y_+ and y_-. When the process y_n is ergodic, $\tilde{C}(j)$ defined by

$$\tilde{C}(j)=\frac{1}{N}\sum_{n=1}^{N-j}(y_{n+j}-\bar{y})(y_n-\bar{y})'$$

where $\bar{y}=\left(\frac{1}{N}\right)\sum_{n=1}^{N}y_n$, is a consistent estimate of $C(j)$ for $j=0,1,\cdots,2q-2$, and the sample canonical correlation coefficients obtained by replacing $C(j)$ by $\tilde{C}(j)$ in the definition of the canonical correlation coefficients give consistent estimates of the theoretical canonical correlation coefficients, which might be useful for the decision on the dimension of the system.

The practical utility of the above stated estimates of the canonical correlation coefficients was checked with an artificially generated process. The original process y_n is defined by

$$y_n-0.9y_{n-1}+0.4y_{n-2}=z_n+0.8z_{n-1},$$

where y_n is a scalar and z_n is a Gaussian white noise with $\mathrm{E}\,z_n=0$ and $\mathrm{E}\,z_n^2=1$. For this case the dimension p of the system is equal to 2 and the Markovian representation (2.6) with $K=2$ gives a minimal representation

$$v_{n+1}=\begin{bmatrix}0 & 1\\ -0.4 & 0.9\end{bmatrix}v_n+\begin{bmatrix}1\\ 1.7\end{bmatrix}z_{n+1}$$

$$y_n=[1 \quad 0]v_n.$$

Assuming $y_0=y_{-1}=z_0=0$ two sets of records $\{y_n;\ n=1,2,\cdots,550\}$ were obtained. The first fifty points of each record were discarded to suppress the effect of the initial transients and the remaining two sets of records each with the data length $N=500$ were used for the analysis. These data are designated as data #1 and 2. The results of the canonical correlation analysis are given in Table 1. The values of r_i^2, the squared sample canonical correlation coefficient, strongly suggest the choice $p=2$, the correct order in this example. The coefficients which determine the canonical variables showed consistent behaviour within the two sets of data for the canonical variables corresponding to the first two largest canonical correlation coefficients but were quite inconsistent for the rest of the variables. Analogous results were obtained for other two sets of data, data #3 and 4, with $N=100$ and are also illustrated in Table 1. In the ordinary canonical correlation analysis with independent multivariate observations the variable $\chi^2_{(i)}$ will asymptotically be distributed as a chi-square variable with the corresponding

Table 1

i	Data #1				Data #2			
	r_i^2	$\chi^2_{(i)}$	d.f. (i)	I.C. (i)	r_i^2	$\chi^2_{(i)}$	d.f. (i)	I.C. (i)
0		∞		∞		∞		∞
1	1.000	612.78	16	580.78	1.000	647.89	16	615.89
2	0.701	9.79	9	−8.10*	0.717	17.27	9	−0.73
3	0.011	4.19	4	−3.76	0.020	7.09	4	−0.91
4	0.008	0.05	1	−1.95	0.013	0.76	1	−1.23*
5	0.000	0	0	0	0.002	0	0	0

i	Data #3				Data #4			
	r_i^2	$\chi^2_{(i)}$	d.f. (i)	I.C. (i)	r_i^2	$\chi^2_{(i)}$	d.f. (i)	I.C. (i)
0		∞		∞		∞		∞
1	1.000	125.59	16	93.59	1.000	87.09	16	55.09
2	0.688	10.28	9	−7.72*	0.538	10.78	9	−7.22*
3	0.069	3.22	4	−4.78	0.090	1.50	4	−6.50
4	0.022	0.99	1	−1.01	0.014	0.14	1	−1.86
5	0.010	0	0	0	0.001	0	0	0

r_i^2 = square of the ith largest sample canonical correlation coefficient between $(y_n, y_{n+1}, \cdots, y_{n+4})'$ and $(y_n, y_{n-1}, \cdots, y_{n-4})'$

$\chi^2_{(i)} = -N \log_e \prod_{j=i+1}^{5} (1-r_j^2)$ $N=500$ for data #1 and 2

100 for data #3 and 4

d.f. (i) = Difference of the number of independent parameters between the full rank model and the rank i model

I.C. (i) $=\chi^2_{(i)} - 2$(d.f. (i))

* denotes the minimum of I.C. (i)

number of degrees of freedom indicated by d.f. (i) when the theoretical values of r_j^2 $(j>i)$ are equal to zero (Anderson [8], p. 327). The values of $\chi^2_{(i)}$ and its d.f. (i) for the maximum possible value of i are set equal to zero. Also when $r_i^2=1$, $\chi^2_{(i-1)}$ and I.C. $(i-1)$ are put equal to infinity. It is not clear to what degree the asymptotic chi-square distribution can approximate the distribution of $\chi^2_{(i)}$ in the present time series situation, but the statistic I.C. $(i)=\chi^2_{(i)}-2$(d.f. (i)) will be useful as a variant of the information theoretic criterion discussed in Section 6. In this statistic the first term $\chi^2_{(i)}$ stands for the increase of badness of fit of the model by the introduction of the assumption $r_j^2=0$ $(j>i)$ and the second term -2(d.f. (i)) stands for the decrease of unreliability by the restriction of the model. If the chi-square approximation of the distribution of $\chi^2_{(i)}$ is not valid the performance of the decision procedure may depend on the structure of the process under observation. The best choice of the dimension p of the system is given as the value of i for which I.C. (i) is the minimum. In Table 1 the values of I.C. (i) show that the best choice of the dimension is given by $p=2$, the true value of the dimension, for data #1, 3 and 4. For the data #2 $p=4$ is chosen as optimal but the difference between I.C. (i)'s $(i=2, 3, 4)$ are almost meaningless compared with the expected sampling variabilities of the related statistics under the assumption of the validity of the chi-square approximations.

If the sample canonical correlation coefficients are obtained by treating the data $(y'_{ns}, y'_{ns+1}, \cdots, y'_{(n+1)s-1})'$ and $(y'_{ns}, y'_{ns-1}, \cdots, y'_{ns-t+1})'$, with s equal to a positive integer and $n=1, 2, \cdots, N$, as if they were N independent observations, then if t is sufficiently large the asymptotic distribution of the corresponding $\chi^2_{(i)}$ will be approximated by a chi-square distribution with the number of degrees of freedom equal to $(sr-ir)\cdot(tr-ir)$, where r is the dimension of the vector y_n. This is due to the fact that the residual of y_{ns+t} after subtracting the projection on the space spanned by the components of $(y'_{ns}, y'_{ns-1}, \cdots, y'_{ns-t+1})'$ will approximately be independent for different values of n. For this case the statistic I.C. $(i)=\chi^2_{(i)}-2(\text{d.f.}\ (i))$ will behave as the difference between the information theoretic criterion corresponding to a restricted model for which only the $i-1$ largest canonical correlations between $(y'_{ns}, y'_{ns+1}, \cdots, y'_{(n+1)s-1})'$ and $(y'_{ns}, y'_{ns-1}, \cdots, y'_{ns-t+1})'$ are not assumed to be equal to zero and that corresponding to the unrestricted model. In spite of its structural simplicity of this modified definition of the sample canonical correlation coefficients their sampling variabilities will generally be larger compared with those of the sample canonical correlation coefficients obtained by simply replacing $C(j)$ by the sample covariance matrix $\tilde{C}(j)$ in the definition of the canonical correlation coefficients.

The decision on the value of p is sufficient for the fitting of a univariate $(r=1)$ AR-MA or Markovian model. For the multivariate $(r>1)$ case, the above stated criterion I.C. (i) can also be used to decide on the first p linearly independent elementary rows of (7.3), where the decision is also made on the value of p itself. First calculate the sample canonical correlation coefficients between $u=(y_+(1), y_+(2), \cdots, y_+(r+1))'$ and y_-, where $y_+(k)$ denotes the kth component of y_+, $y_+=(y'_n, y'_{n+1}, \cdots, y'_{n+q-1})'$, $y_-=(y'_n, y'_{n-1}, \cdots, y'_{n-q+1})'$ and it is assumed that the dimension p of the system is less than q. If the criterion attains its minimum at $i=r+1$, i.e. I.C. $(r)>$I.C. $(r+1)$ $(=0)$, then retain $y_+(r+1)$ in u, otherwise drop $y_+(r+1)$ from u. When $y_+(r+1)$ is dropped from u discard $y_+(jr+1)$ $(j=2, 3, \cdots, q-1)$ from y_+ and pick up the canonical variable $a'_{r+1}u$ which corresponds to the minimum canonical correlation coefficient where a_{r+1} is an $(r+1)$-dimensional vector $(a_{r+1}(1), a_{r+1}(2), \cdots, a_{r+1}(r+1))'$. An estimate of A_{ij} of (5.2b) with $k_i=1$ is given by $-(a_{r+1}(j))\cdot(a_{r+1}(r+1))^{-1}$. Now include $y_+(r+2)$ into u to define the new u and repeat the analysis to decide on the rejection of $y_+(r+2)$ from u. When $y_+(r+2)$ is rejected, $y_+(jr+2)$ $(j=2, 3, \cdots, q-1)$ are discarded from y_+

and an estimate of A_{ij} of (5.2b) with $k_i=2$ is obtained analogously as in the case with $y_+(r+1)$. Repeat the canonical correlation analysis and decision with the remaining components of y_+ until the last one. The u at this final stage defines a choice on the minimal basis and the corresponding estimates of A_{ij}'s of (5.2b) provide an estimate of the transition matrix A. Many other variants of the selection procedures are conceivable.

The feasibility of the above stated type procedure on the decision of the basis was checked with an artificially generated two-dimensional process y_n which is defined by

$$(7.4)\qquad y_n+\begin{bmatrix} -0.9 & 0 \\ 0 & -1.5 \end{bmatrix} y_{n-1}+\begin{bmatrix} 0.4 & 0 \\ 0 & 1.2 \end{bmatrix} y_{n-2}+\begin{bmatrix} 0 & 0 \\ 0 & -0.448 \end{bmatrix} y_{n-3}$$
$$=z_n+\begin{bmatrix} 0.8 & 0 \\ 0 & 0 \end{bmatrix} z_{n-1}\,,$$

where z_n is a Gaussian white noise with zero mean and unit variance matrix. Assuming the zero initial condition, i.e. $y_0=y_{-1}=y_{-2}=z_0=0$, two realizations of z_n ($n=1, 2,\cdots, 550$) were used to generate two realizations of y_n ($n=1, 2,\cdots, 550$) and the first fifty points were discarded from both realizations. The following numerical results are based on these two sets of data, # 5 and # 6 each with $N=500$. The dimension p of the system defined by (7.4) is 5 and the first 5 independent components of $(y'_{n|n}, y'_{n+1|n}, \cdots)$ are $y_n(1)_{|n}, y_n(2)_{|n}, y_{n+1}(1)_{|n}, y_{n+1}(2)_{|n}, y_{n+2}(2)_{|n}$ and the process is not block-identifiable. By following the discussion of Section 5 and using $v_n=(y_n(1)_{|n}, y_n(2)_{|n}, y_{n+1}(1)_{|n}, y_{n+1}(2)_{|n}, y_{n+2}(2)_{|n})'$ as the state of the representation one can get a minimal Markovian representation of the process in the following form:

$$v_{n+1}=\begin{bmatrix} 0 & 0 & 1 & 0 & 0 \\ 0 & 0 & 0 & 1 & 0 \\ -0.4 & 0 & 0.9 & 0 & 0 \\ 0 & 0 & 0 & 0 & 1 \\ 0 & 0.448 & 0 & -1.2 & 1.5 \end{bmatrix} v_n+\begin{bmatrix} 1 & 0 \\ 0 & 1 \\ 1.7 & 0 \\ 0 & 1.5 \\ 0 & 1.05 \end{bmatrix} z_{n+1}$$

$$y_n=\begin{bmatrix} 1 & 0 & 0 & 0 & 0 \\ 0 & 1 & 0 & 0 & 0 \end{bmatrix} v_n\,.$$

When autoregressive models were fitted to the data using a criterion which is equivalent to the present information criterion (Akaike [1]), the orders chosen for the simple bivariate autoregressive representations were 6 and 5, respectively, for data # 5 and data # 6. Based on this observation $q=5$ was used as a tentative choice of q in the definition of $y_-=(y'_n, y'_{n-1},\cdots, y'_{n-q+1})'$. Canonical correlation analysis between $u_m=$

$(y_+(1), y_+(2), \cdots, y_+(m))$ and y_- were performed, where $y_+(i)$ is the ith component of $y_+=(y'_n, y'_{n+1}, \cdots, y'_{n+q-1})'$. The results are illustrated in Table 2. From the behaviour of the information criterion, which is analogously defined as in Table 1, $y_+(5)_{|n}$, or $y_{n+2}(1)_{|n}$, is judged to be linearly dependent on the preceding components of $u_{m|n}$ for the both sets of data at $m=5$. At $m=6$, $y_+(6)_{|n}$ is judged to be linearly independent of the preceding components of $u_{m|n}$. At $m=7$ and 8 the selected number of linearly independent components remains at 5 for the both sets of data. This shows that $y_+(7)_{|n}$ and $y_+(8)_{|n}$ are judged to be linearly dependent on the preceding components of $u_{r|n}$ and the

Table 2

m	i	Data #5			Data #6		
		$\chi^2_{(i)}$	d.f. (i)	I.C. (i)	$\chi^2_{(i)}$	d.f. (i)	I.C. (i)
3	2	405.57	8	389.57	375.03	8	359.03
	3	0	0	0 *	0	0	0 *
4	2	824.73	16	792.73	722.09	16	690.09
	3	395.32	7	381.32	315.49	7	301.49
	4	0	0	0 *	0	0	0 *
5	2	951.76	24	903.76	855.29	24	807.29
	3	419.12	14	391.12	341.06	14	313.06
	4	2.84	6	−9.16*	8.75	6	−3.25*
	5	0	0	0	0	0	0
6	2	1059.98	32	995.98	958.95	32	894.95
	3	517.25	21	475.25	427.50	21	385.50
	4	29.30	12	5.30	36.84	12	12.84
	5	1.65	5	−8.35*	8.47	5	−1.53*
	6	0	0	0	0	0	0
7	2	1113.74	40	1033.74	1042.71	40	962.71
	3	526.16	28	470.16	449.31	28	393.31
	4	35.61	18	−0.39	45.04	18	9.04
	5	7.57	10	−12.43*	15.62	10	−4.38*
	6	0.72	4	−7.28	6.83	4	−1.17
	7	0	0	0	0	0	0
8	2	1115.40	48	1019.40	1044.81	48	948.81
	3	527.45	35	457.45	451.79	35	381.79
	4	37.35	24	−10.66	47.68	24	−0.32
	5	9.26	15	−20.74*	18.19	15	−11.81*
	6	2.06	8	−13.94	8.93	8	−7.07
	7	0.52	3	−5.48	2.08	3	−3.92
	8	0	0	0	0	0	0

* denotes the minimum of I.C. (i)

search for the basis is terminated. The selected basis is $(y_+(1)_{|n}, y_+(2)_{|n}, y_+(3)_{|n}, y_+(4)_{|n}, y_+(6)_{|n})$ for the both sets of data, which is identical to the theoretical result. The present identity between the experimental and theoretical result is obtained only by chance, but the behaviour of the statistics in Table 2 gives a clear feeling of the utility of this type of procedure for general practical applications. In this experiment, for the simplicity of the computer program, the discarding of the components judged to be dependent was not implemented. The definition of $\chi^2_{(i)}$ was analogous to that of Table 1 but with N replaced by $N-0.5(2m+2q+1)$. This difference of the definition of $\chi^2_{(i)}$ has little effect on the present application.

It is almost certain that if the $(i+1)$st largest canonical correlation coefficient r_{i+1} is equal to zero then $N^{-a}\chi^2_{(i)}$ $(0<a<1)$ will converge to zero in probability as N tends to infinity. If this convergence is assumed and I.C. (i) is replaced by i.c. $(i)=N^{-1}\chi^2_{(i)}-2N^{a-1}(\text{d.f.}\,(i))$ the present decision procedure provides consistent estimates of the desired basis and the corresponding matrix A. This can easily be seen from the fact that $N^{-1}\chi^2_{(i)}$ converges in probability to a positive constant and dominates the behaviour of i.c. (i) when r_{i+1} is not zero while $-2N^{a-1}\cdot$d.f. (i) dominates when r_{i+1} is zero. This result may be considered to be a fundamental one in the subject of statistical identification of multivariate stochastic systems with finite dimensional dynamics. In practical applications where the dimension of the system is infinite the replacement of I.C. (i) by i.c. (i) will stress the tendency to pick up a rather low dimensional model.

Once the estimate of the basis is determined, an estimate of the corresponding matrix B can be obtained from any estimate of W_m $(m=1, 2,\cdots)$ of (2.2). An estimate of W_m is obtained by fitting an autoregressive model and computing the response of the system to an impulsive input. This will be useful to produce an initial estimate of B to start the maximum likelihood computation.

It should be remembered here that by a modification of the search procedure for a basis of the predictor space, which was described in Section 5, the assumption of non-singularity of the innovation variance matrix can be eliminated. With this modification the procedure described in this section will give a minimal set of components of y_n which are judged to be useful for the description of the stochastic system and retained for the further analysis. The application of this type of approach to the problem raised by Priestley and others [13] concerning the reduction of the number of measurements in the control of a complex system will be a subject of further study. Also the procedures described in this section are only for the initial determination of the dimension or the basis of the predictor space. The final decision will be made by comparing the values of the information criterion defined by the maximized likelihood for several possible choices of the dimension or the basis. Theoretical analysis of the asymptotic distribution of $\chi^2_{(i)}$ is also a subject of further study.

The procedure described in this paper provides only a starting point for the development of practical procedures of fitting AR-MA models. Much remains to be done for the development of a computationally and statistically efficient fitting procedure.

Acknowledgements

The author wishes to thank Professor M. B. Priestley, Dr. T. Subba Rao, Dr. H. Tong and Mrs. V. Haggan for many stimulating discussions and to Miss E. Arahata for conducting the numerical experiment. The author is also grateful to Professor W. Gersch for the helpful comments. Thanks are especially due to Professor E. J. Hannan for kindly pointing out a serious error in the original version of Section 3. The research was supported by a Science Research Council grant at the University of Manchester Institute of Science and Technology.

INSTITUTE OF STATISTICAL MATHEMATICS

REFERENCES

[1] Akaike, H. (1971). Autoregressive model fitting for control, *Ann. Inst. Statist. Math.*, **23**, 163-180.

[2] Akaike, H. (1972). Use of an information theoretic quantity for statistical model identification, *Proc. 5th Hawaii International Conference on System Sciences*, 249-250.

[3] Akaike, H. (1973). Information theory and an extension of the maximum likelihood principle, *Proc. 2nd International Symposium on Information Theory*, (B. N. Petrov and F. Csaki eds.), *Akademiai Kiado*, Budapest, 267-281.

[4] Akaike, H. (1973). Block Toeplitz matrix inversion, *SIAM J. Appl. Math.*, **24**, 234-241.

[5] Akaike, H. (1973). Maximum likelihood identification of Gaussian autoregressive moving average models, *Biometrika*, **60**, 255-265.

[6] Akaike, H. (1973). Markovian representation of stochastic processes by canonical variables, to be published in *SIAM J. Control.*

[7] Akaike, H. (1973). Stochastic theory of minimal realizations, to be published in *IEEE Trans. Automat. Contrl.*

[8] Anderson, T. W. (1958). *An Introduction to Multivariate Statistical Analysis*, Wiley, New York.

[9] Hannan, E. J. (1969). The identification of vector mixed autoregressive-moving average systems, *Biometrika*, **56**, 223-225.

[10] Kalman, R. E., Falb, P. L. and Arbib, M. A. (1969). *Topics in Mathematical System Theory*, McGraw-Hill, New York.

[11] Kullback, S. and Leibler, R. A. (1951). On information and sufficiency, *Ann. Math. Statist.*, **22**, 79-86.

[12] Lawley, D. N. and Maxwell, A. E. (1971). *Factor Analysis as a Statistical Method*, Butterworths, London.

[13] Priestley, M. B., Subba Rao, T. and Tong, H. (1972). Identification of the structure of multivariable stochastic systems, to appear in *Multivariate Analysis* III, Ed. P. R. Krishnaiah, Academic Press.
[14] Rissanen, J. (1972). Estimation of parameters in multi-variate random processes, unpublished.
[15] Rissanen, J. (1973). Algorithms for triangular decomposition of block Hankel and Toeplitz matrices with application to factorising positive matrix polynomials, *Mathematics of Computation*, **27**, 147-154.
[16] Whittle, P. (1953). The analysis of multiple stationary time series, *J. R. Statist. Soc.*, B, **15**, 125-139.
[17] Whittle, P. (1962). Gaussian estimation in stationary time series, *Bulletin L'Institut International de Statistique*, **39**, 2e Livraison, 105-129.
[18] Whittle, P. (1963). On the fitting of multivariate autoregressions, and the approximate factorization of a spectral density matrix, *Biometrika*, **50**, 129-134.
[19] Whittle, P. (1969). A view of stochastic control theory, *J. R. Statist. Soc.*, A, **132**, 320-334.

Ann. Inst. Statist. Math.
30 (1978), Part B, 499-504

COVARIANCE MATRIX COMPUTATION OF THE STATE VARIABLE OF A STATIONARY GAUSSIAN PROCESS

HIROTUGU AKAIKE

(Received Oct. 16, 1978; revised Dec. 7, 1978)

1. Summary

A recursive procedure for the computation of one-step ahead predictions for a finite span of time series data by a Gaussian autoregressive moving average model can be realized by using the Markovian representation of the model. The covariance matrix of the stationary state variable of the Markovian representation is required to implement a computational procedure of the predictions. A simple computational procedure of the covariance matrix which does not need an iterative method is obtained by using a canonical representation of the autoregressive moving average process. The recursive computation of the predictions realized by using this procedure provides a computationary efficient method of exact likelihood evaluation of a Gaussian autoregressive moving average model.

2. Recursive computation of one-step ahead predictions

Assume that a stationary scalar zero-mean Gaussian process $y(n)$ is defined by the relation

$$z(n+1)=Fz(n)+Gx(n+1)\,, \qquad y(n)=Hz(n) \tag{2.1}$$

where $x(n)$ is a scalar white noise independent of $z(n-1), z(n-2),\cdots$, and $z(n)$ is the state variable which is a stationary p-vector process and the matrices F, G, H are $p\times p$, $p\times 1$ and $1\times p$, respectively. When $y(0), y(1),\cdots, y(n-1)$ are given, the one-step ahead prediction of $z(n)$ is defined by

$$z(n\,|\,0, n-1)=\text{projection of } z(n) \text{ onto the linear space spanned by the components of } y(0), y(1),\cdots, y(n-1)\,,$$

and the one-step ahead prediction of $y(n)$ is given by $y(n\,|\,0, n-1)=Hz(n\,|\,0, n-1)$.

The computation of $z(n\,|\,0, n-1)$ can be done recursively by using the relations

$$e(n)=y(n)-Hz(n|0, n-1) ,$$

$$z(n+1|0, n)=Fz(n|0, n-1)+K_n r_n^{-1}e(n) , \tag{2.2}$$

$$z(0|0, -1)=0 ,$$

where $r_n = \mathrm{E}\, e(n)^2$ and $K_n = \mathrm{E}\, z(n+1)e(n)$, the Kalman gain vector. By the Kalman filtering procedure, which is a standard procedure of recursive computation of the predictions, K_n and R_n are computed by the relations

$$K_n = FP(n|n-1)H' , \qquad r_n = HP(n|n-1)H' , \tag{2.3}$$

where $'$ denotes transpose and $P(n|n-1)=\mathrm{E}\,(z(n)-z(n|0, n-1))(z(n)-z(n|0, n-1))'$ and is obtained by the relations

$$\begin{aligned} &P(n+1|n)=FP(n|n-1)F'+GqG'-K_n r_n^{-1}K_n' , \\ &P(0|-1)=P_0 , \end{aligned} \tag{2.4}$$

where $q=\mathrm{E}\, x(n)^2$ and $P_0=\mathrm{E}\, z(0)z(0)'$, the covariance matrix of the stationary state vector. Thus the numerical evaluation of P_0 forms the starting point of the Kalman filtering procedure.

3. Prediction of ARMA process

When $y(n)$ is a stationary Gaussian autoregressive moving average (ARMA) process defined by

$$\begin{aligned} y(n)+b_1 y(n-1)+\cdots+b_M y(n-M) \\ =x(n)+a_1 x(n-1)+\cdots+a_L x(n-L) \end{aligned} \tag{3.1}$$

there is a Markovian representation (2.1) defined by

$$F=\begin{bmatrix} 0 & 1 & 0 & \cdots & 0 \\ 0 & 0 & 1 & \cdots & 0 \\ \vdots & \vdots & \vdots & \ddots & \vdots \\ 0 & 0 & 0 & \cdots & 1 \\ -b_K & -b_{K-1} & -b_{K-2} & \cdots & -b_1 \end{bmatrix}, \qquad G=\begin{bmatrix} w_0 \\ w_1 \\ \vdots \\ w_{K-2} \\ w_{K-1} \end{bmatrix}, \tag{3.2}$$

$$H=[1 \ \ 0 \ \ 0 \cdots 0] ,$$

where $a_L \neq 0$, $b_M \neq 0$ and $K=\max(M, L+1)$ and the w_i's are the impulse responses of the system (3.1), i.e., $(w_0, w_1, \cdots, w_{K-1})=(y(0), y(1), \cdots, y(K-1))$ under the assumption that $y(-1)=y(-2)=\cdots=y(-M)=0$ and $x(n)=1$, for $n=0$, 0, otherwise. The corresponding state variable $z(n)$

is defined by

$$z_i(n) = y(n+i\,|\,n) \qquad i=0, 1, \cdots, K-1 , \tag{3.3}$$

where $z_i(n)$ denotes the $i+1$st component of $z(n)$ and $y(n+i\,|\,n)$ the projection of $y(n+i)$ onto the space spanned by the components of $y(n)$, $y(n-1), \cdots$; see, for example, Akaike ([1], [2]).

Thus the recursive procedure for the computation of one-step ahead predictions for a finite span of data can be applied to the autoregressive moving average process (3.1) via the representation (2.1) with the matrices defined by (3.2), if only the covariance matrix P_0 of the state variable $z(n)$ is obtained.

4. Covariance matrix of the state variable

Under the assumption that the characteristic polynomial

$$B(z) = 1 + b_1 z + \cdots + b_M z^M \tag{4.1}$$

has the zero's outside the unit circle $y(n)$ can be represented in the form

$$y(n) = \sum_{m=0}^{\infty} w_m x(n-m) , \tag{4.2}$$

where the summation denotes the limit in the mean square and w_m's are the impulse responses of the system defined by (3.1). By using (4.2) we get for $i \geqq 0$ the representation

$$y(n+i\,|\,n) = \sum_{m=0}^{\infty} w_{i+m} x(n-m)$$

or the relation

$$y(n+i) = \sum_{m=-i}^{-1} w_{i+m} x(n-m) + y(n+i\,|\,n) .$$

From this last relation we get for $j \geqq i$ the relation

$$\mathrm{E}\, y(n+i) y(n+j) = \sigma^2 \sum_{m=0}^{i-1} w_m w_{m+j-i} + \mathrm{E}\, y(n+i\,|\,n) y(n+j\,|\,n) , \tag{4.3}$$

where σ^2 is the variance of $x(n)$. When the ARMA model (3.1) is given the computation of w_n's is straightforward. Thus it is only the computational procedure of $\mathrm{E}\, y(n+i) y(n+j)$ that is required for the evaluation of the covariance matrix of the state variable defined by (3.3).

By multiplying the both sides of (3.1) by $y(n-k)$ and taking the expectations we get the relations ($k=0, 1, \cdots, K$)

$$R(-k)+b_1R(-k+1)+\cdots+b_KR(-k+K)$$
$$=a_kw_0+a_{k+1}w_1+\cdots+a_{K-1}w_{K-1-k}\,,$$

where $b_m=0$ for $m>M$, $a_0=1$ and $a_l=0$ for $l>L$, and $R(k)=\mathrm{E}\,y(n+k)\cdot y(n)$. In the matrix form these relations are given by

$$\begin{bmatrix} 1 & b_1 & b_2 & \cdots b_{K-1} & b_K \\ b_1 & 1+b_2 & b_3 & \cdots b_K & 0 \\ \vdots & \vdots & \vdots & \vdots & \vdots \\ b_{K-1} & b_{K-2}+b_K & b_{K-3}\cdots 1 & & 0 \\ b_K & b_{K-1} & b_{K-2}\cdots b_1 & & 1 \end{bmatrix} \begin{bmatrix} S(1) \\ S(2) \\ \vdots \\ S(K) \\ S(K+1) \end{bmatrix} = \begin{bmatrix} C(1) \\ C(2) \\ \vdots \\ C(K) \\ C(K+1) \end{bmatrix}, \tag{4.4}$$

where $C(m)=a_{m-1}w_0+a_mw_1+\cdots+a_{K-1}w_{K-m}$ and $S(m)=R(m-1)$.

To solve the equation (4.4) for $S(m)$'s we want to transform the matrix on the left-hand side into lower triangular form with 1's on the diagonal. For notational convenience we put $b_i^K=b_i$ $(i=1,2,\cdots,K)$. We subtract from the ith row b_K^K times the $(K+2-i)$th row and divide the resulting ith row by $1-(b_K^K)^2$ for $i=1,2,\cdots,[(K+2)/2]$, where $[x]$ denotes the integer part of x. Hereafter the same operation as that applied to the elements of the matrix on the left-hand side is applied to the elements of the vector $[C(1),C(2),\cdots,C(K+1)]'$. By the present operation the b_i^K's in the first $[(K+2)/2]$ rows of the matrix are transformed into b_i^{K-1}'s which are defined by

$$b_i^{K-1}=\frac{b_i^K-b_K^Kb_{K-i}^K}{1-(b_K^K)^2}\,, \qquad i=1,2,\cdots,K-1\,, \tag{4.5}$$

and

$$b_K^{K-1}=0\,.$$

From (4.5) we get the relation

$$b_i^{K-1}=b_i^K-b_K^Kb_{K-i}^{K-1}\,, \qquad i=1,2,\cdots,K-1\,.$$

Thus by subtracting b_K^K times the $(K+2-i)$th row from the ith row for $i=[(K+2)/2]+1,[(K+2)/2]+2,\cdots,K+1$, the b_i^K's within the matrix are all replaced by b_i^{K-1}'s. When $b_K^K=0$ the above operation is non-effective and can be skipped. At the next stage, since we have $b_K^{K-1}=0$, we can restrict our attention to the first K rows and apply the same type of operation to transform b_i^{K-1}'s into b_i^{K-2}'s with $b_{K-1}^{K-2}=0$. The operation can also be skipped when $b_{K-1}^{K-1}=0$. By repeating the same type of operations K times we get

$$
\begin{bmatrix} 1 & 0 & 0 & \cdots 0 & 0 \\ b_1^1 & 1 & 0 & \cdots 0 & 0 \\ b_2^2 & b_1^2 & 1 & \cdots 0 & 0 \\ \vdots & \vdots & \vdots & \vdots & \vdots \\ b_{K-1}^{K-1} & b_{K-2}^{K-1} & b_{K-3}^{K-1} \cdots & 1 & 0 \\ b_K^K & b_{K-1}^K & b_{K-2}^K \cdots & b_1^K & 1 \end{bmatrix} \begin{bmatrix} S(1) \\ S(2) \\ S(3) \\ \vdots \\ S(K) \\ S(K+1) \end{bmatrix} = \begin{bmatrix} D(1) \\ D(2) \\ D(3) \\ \vdots \\ D(K) \\ D(K+1) \end{bmatrix} . \tag{4.6}
$$

The desired solution is then obtained by the recursive relations

$$
\begin{aligned} & S(m) = D(m) - \sum_{i=1}^{m-1} b_i^{m-1} S(m-i) , \qquad m = 2, 3, \cdots, K+1 , \\ & S(1) = D(1) . \end{aligned} \tag{4.7}
$$

The description of the above procedure shows that the equation (4.4) has a unique solution if only $|b_M^M| = 1$ does not hold for $M = 1, 2, \cdots, K$. It is already well known that $|b_M^M| < 1$ $(M = 1, 2, \cdots, K)$ is necessary and sufficient for the characteristic equation (4.1) to have zero's outside the unit circle; see, for example, Szaraniec [6]. The non-singularity of the matrix of (4.4) has been used by Kitagawa [4]. Thus under the present assumption of stationarity the equation (4.4) uniquely determines $R(m)$ $(m = 0, 1, \cdots, K)$. From these $R(m)$'s we can get, via (4.3),

$$
P_0(i, j) = R(j-i) - \sigma^2 \sum_{m=0}^{i-1} w_m w_{m+j-i} , \qquad j \geqq i ,
$$

where $P_0(i, j)$ denotes the (i, j)-element of the covariance matrix P_0 of the state variable and $P_0(j, i) = P_0(i, j)$.

5. Application to exact likelihood computation

For a set of data $(y(1), y(2), \cdots, y(N))$ the exact likelihood of a zero-mean stationary Gaussian ARMA model (3.1) is given by

$$
\prod_{n=1}^{N} \left(\frac{1}{2\pi\sigma^2 r_n} \right)^{1/2} \exp \left[-\frac{1}{2\sigma^2 r_n} e(n)^2 \right] ,
$$

where $e(n)$ and r_n are obtained by (3.2), (2.2), (2.3) and (2.4) under the assumption that $q=1$ and σ^2 denotes the assumed value of the variance of $x(n)$.

Once P_0 is given a computational procedure of $e(n)$ and r_n given by Morf, Sidhu and Kailath [5], which is more efficient than that by (2.4), can be used for the computation of the exact likelihood. A computer program for the maximum likelihood computation of a Gaussian ARMA model by this procedure is already developed by Akaike et al. [3]. The computational efficiency of the likelihood computation procedure is confirmed through the comparison with other procedures.

Acknowledgements

The present author is indebted to T. Ozaki and G. Kitagawa who invited his attention to the subject of exact likelihood computation of an ARMA model.

THE INSTITUTE OF STATISTICAL MATHEMATICS

REFERENCES

[1] Akaike, H. (1974). Markovian representation of stochastic processes and its application to the analysis of autoregressive moving-average processes, *Ann. Inst. Statist. Math.*, **26**, 363-387.

[2] Akaike, H. (1976). Canonical correlation analysis of time series and the use of an information criterion, In *System Identification: Advances and Case Studies*, R. K. Mehra and D. G. Lainiotis, eds., Academic Press, New York, 27-96.

[3] Akaike, H., Kitagawa, G., Arahata, E. and Tada, F. (1979). TIMSAC-78, *Computer Science Monographs*, No. 11, The Institute of Statistical Mathematics, Tokyo.

[4] Kitagawa, G. (1977). On a search procedure for the optimal AR-MA order, *Ann. Inst. Statist. Math.*, **29**, B, 319-332.

[5] Morf, M., Sidhu, G. S. and Kailath, T. (1974). Some new algorithms for recursive estimation in constant, linear, discrete-time systems, *IEEE Trans. Automat. Contr.*, AC-**19**, 315-323.

[6] Szaraniec, E. (1973). Stability, instability and aperiodicity tests for linear discrete systems, *Automatica*, **9**, 513-516.

Ann. Inst. Statist. Math.
30 (1978), Part B, 185-197

ANALYSIS OF CROSS CLASSIFIED DATA BY AIC

YOSIYUKI SAKAMOTO AND HIROTUGU AKAIKE

(Received Mar. 19, 1977; revised Dec. 13, 1977)

Abstract

The purpose of the present paper is to propose a simple but practically useful procedure for the analysis of multidimensional contingency tables of survey data. By the procedure we can determine the predictor on which a specific variable has the strongest dependence and also the optimal combination of predictors. The procedure is very simply realized by the search for the minimum of the statistic AIC within a set of models proposed in this paper. The practical utility of the procedure is demonstrated by the results of some successful applications to the analysis of the survey data of the Japanese national character. The difference between the present procedure and the conventional test procedure is briefly discussed.

1. Introduction

Tables 1.1, 1.2 and 1.3 are a part of the survey results obtained by the 1973 nation wide survey of the Japanese national character [6], [7]. The question asked was: "On the whole in Japan, which sex do you think has the more difficult life, men or women?" These tables evidently show that the answer to this question depends most significantly on sex, among the three demographic factors.

How can we form such a judgement? How can we evaluate the strength of the dependence of the answer on those three factors?

Table 1.1

		Sex		Total
		Male (S_1)	Female (S_2)	
Which sex has more difficult life?	Men (W_1)	904	790	1694
	Women (W_2)	491	870	1361
Total		1395	1660	3055

Table 1.2

	Age				Total
	20-29 (A_1)	30-39 (A_2)	40-49 (A_3)	50 yrs & over (A_4)	
Men	454	408	363	469	1694
Women	324	319	303	415	1361
Total	778	727	666	884	3055

Table 1.3

	Rural vs. urban breakdown				Total
	6 Metropolitan cities (R_1)	Other cities		Rural (R_4)	
		Pop.: 200,000 & over (R_2)	Pop.: Under 200,000 (R_3)		
Men	322	388	525	459	1694
Women	217	304	445	395	1361
Total	539	692	970	854	3055

Perhaps we are tacitly assuming that the dependences can be evaluated by some standard which is common to all these situations.

In this paper we first propose a search procedure for the predictor on which a specific variable has the strongest dependence and then propose a procedure to search for the optimal combination of predictors. Here 'optimal' means that the combination demonstrates the most significant dependence between the variable predicted and the predictors. To solve these problems we propose the use of some models which describe the dependence relations among the variables. The discrepancy of a model fitted to a set of observed data by the method of maximum likelihood is evaluated by the statistic AIC defined by the following [1], [2]:

$$\text{AIC}=(-2)\log(\text{maximized likelihood})+2k\,, \tag{1.1}$$

where log denotes the natural logarithm and k is the number of parameters within the model which are adjusted to attain the maximum of the likelihood.

The introduction of AIC is based on the entropy maximization principle: formulate the object of statistical inference as the estimation of the true distribution from the data and try to find the estimate which will maximize the expected entropy. The entropy is a natural measure of discrimination between the true and the estimated probability distribution, $f(x)$ and $g(x:\theta)$, and is defined by

$$B(f:g(\cdot\,;\theta))=-\int f(x)\log\{f(x)/g(x\,;\theta)\}\,dx \tag{1.2}$$

$$=\mathrm{E}\log g(x;\theta)-\mathrm{E}\log f(x)\ .$$

A large value of the entropy $B(f:g(\cdot;\theta))$ means that the distribution $g(x;\theta)$ is a good approximation to the true distribution $f(x)$.

Consider the situation where the family of models $\bigcup_{k=1}^{L}\{g(x;{}_k\theta)\}$ is given, where $g(x;{}_k\theta)$ is specified by the vector of parameters ${}_k\theta=(\theta_1,\theta_2,\cdots,\theta_k,0_{k+1},\cdots,0_L)$ $(k=1,\cdots,L)$ and it is assumed that $f(x)=g(x;{}_p\theta)=g(x;\theta_0)$ for some p $(1\leqq p\leqq L)$. Here 0_k denotes a prescribed value of θ_k. Denote by ${}_k\hat{\theta}$ the maximum likelihood estimate of the parameter ${}_k\theta$, then the familiar log likelihood ratio statistic is given by ${}_k\eta_L=(-2)\{\log g(x;{}_k\hat{\theta})-\log g(x;{}_L\hat{\theta})\}$ and the statistic $({}_k\eta_L+2k-L)/n$ is an asymptotically unbiased estimate of $-\mathrm{E}\,B(g(\cdot;\theta_0),g(\cdot;{}_k\hat{\theta}))$ [1]. For the purpose of comparison of $g(x;{}_k\hat{\theta})$, the common constant $2\log g(x;{}_L\hat{\theta})-L$ is ignored and we get an information criterion (AIC) of (1.1). We regard a model with a smaller AIC as a better one, as it is expected to have a larger entropy. The model with the minimum AIC will be called the minimum AIC estimate or MAICE. Detailed discussions of these concepts are found in [1], [2].

To illustrate the use of AIC we consider the classical test of independence. From the point of view of the statistic AIC the conventional test of independence of a two-way contingency table $\{n(i,j): i=1,\cdots,r,\ j=1,\cdots,c\}$ is regarded as a comparison of the unrestricted model and the independence model defined by $p(i,j)=p(i,\cdot)p(\cdot,j)$, where $p(i,j)$ denotes the probability of observing a combination (i,j), $p(i,\cdot)=\sum_j p(i,j)$ and $p(\cdot,j)=\sum_i p(i,j)$. The corresponding maximum likelihood estimates of $p(i,j)$ are given by $n(i,j)/n$ and $\{n(i,\cdot)n(\cdot,j)\}/n^2$, respectively. Due to the constraint $\sum_i\sum_j p(i,j)=1$, the number of free parameters in the first model is $(rc-1)$ and, due to the constraints $\sum_j p(\cdot,j)=1$ and $\sum_i p(i,\cdot)=1$, that in the second is $(r-1)+(c-1)$. The AIC's for these models are respectively given by

$$\text{(1.3)}\qquad \mathrm{AIC}_1=(-2)\sum_i\sum_j n(i,j)\log\{n(i,j)/n\}+2(rc-1)$$

$$\text{(1.4)}\qquad \mathrm{AIC}_0=(-2)\sum_i\sum_j n(i,j)\log\{n(i,\cdot)n(\cdot,j)/n^2\}+2(r+c-2)\ .$$

The definition of AIC suggests that the independence model should be adopted if AIC_0 is smaller than AIC_1, otherwise the dependence model should be adopted. If we follow this suggestion we take the MAICE as our choice. This defines the MAICE procedure.

In the case of the analysis of the Tables 1.1, 1.2 and 1.3, it would be reasonable to assume that in evaluating the dependence between the opinion and a factor we are neglecting the effects of the remaining two factors. This idea leads us to a set of models to be defined in the next section.

2. The simplest models and their AIC's

Assume that a k-way contingency table consists of a variable to be predicted (denoted by i_1) and the $k-1$ predictors (denoted by $i_2, i_3, \cdots, i_k$). We denote the joint probability by $p(i_1, i_2, \cdots, i_k)$ and the cell frequency by $n(i_1, i_2, \cdots, i_k)$ $(\sum_{i_1,\cdots,i_k} n(i_1, i_2, \cdots, i_k)=n)$, where i_j is used to represent one of the values $1, 2, \cdots, C_{i_j}$ which are taken by the variable i_j $(j=1, 2, \cdots, k)$. In these representations we will simply discard a variable when a sum is taken with respct to its values. For example, we put

$$(2.1)\qquad \begin{aligned} &p(i_1, i_2, \cdots, i_{k-1})=\sum_{i_k} p(i_1, i_2, \cdots, i_k) \\ \text{and}\quad & \\ &n(i_1, i_2, \cdots, i_{k-2})=\sum_{i_{k-1}} n(i_1, i_2, \cdots, i_{k-1}) . \end{aligned}$$

First consider the search for a single predictor on which the variable to be predicted has the strongest dependence. The simplest model which is in accordance with the observation at the end of the preceding section can be obtained by assuming the simplest possible structure which completely ignores the dependence between the variables left out of our consideration. This is given by

$$(2.2)\qquad p(i_1, \cdots, i_k)=p(i_1, i_l) \prod_{j=2, j\neq l}^{k} p(i_j) \qquad l=2, 3, \cdots, k .$$

The log likelihood of a model belonging to (2.2) for a sample with cell frequencies $n(i_1, \cdots, i_k)$ is given by

$$L=\sum_{i_1,\cdots,i_k} n(i_1, \cdots, i_k) \log \left\{p(i_1, i_l) \prod_{j=2, j\neq l}^{k} p(i_j)\right\} .$$

By maximizing L with respect to $p(i_1, i_l)$'s and $p(i_j)$'s the maximum likelihood estimate of the joint probability is obtained by $\{n(i_1, i_l)/n^{k-1}\} \cdot \prod_{j=2, j\neq l}^{k} n(i_j)$. Since there are constraints

$$\sum_{i_1, i_l} p(i_1, i_l)=1 \qquad \text{and} \qquad \sum_{i_j} p(i_j)=1 ,$$

the model has $\{(C_{i_1}C_{i_l}-1)+\sum_{j=2, j\neq l}^{k} (C_{i_j}-1)\}$ parameters to be specified. Thus the statistic AIC for the model is given by

$$(2.3)\quad \mathrm{AIC}=(-2)\sum_{i_1,\cdots,i_k} n(i_1,\cdots,i_k)\log\left\{n(i_1,i_l)\prod_{j=2,j\neq l}^{k} n(i_j)/n^{k-1}\right\}$$
$$+2\left\{(C_{i_1}C_{i_l}-1)+\sum_{j=2,j\neq l}^{k}(C_{i_j}-1)\right\}$$
$$=(-2)\left[\sum_{i_1,i_l} n(i_1,i_l)\log\{n(i_1,i_l)/n\}+\sum_{j=2}^{k}\sum_{i_j} n(i_j)\log\{n(i_j)/n\}\right.$$
$$\left.-\sum_{i_l} n(i_l)\log\{n(i_l)/n\}\right]+2\left\{(C_{i_1}C_{i_l}-1)+\sum_{j=2}^{k}(C_{i_j}-1)-(C_{i_l}-1)\right\}.$$

For the purpose of comparison of models within the above set the common constant $(-2)\sum_{j=2}^{k}\sum_{i_j} n(i_j)\log\{n(i_j)/n\}+2\sum_{j=2}^{k}(C_{i_j}-1)$ is ignored and the statistic AIC is given by

$$(2.4)\quad \mathrm{AIC}=(-2)\left[\sum_{i_1,i_l} n(i_1,i_l)\log\{n(i_1,i_l)/n\}-\sum_{i_l} n(i_l)\log\{n(i_l)/n\}\right]$$
$$+2\{(C_{i_1}C_{i_l}-1)-(C_{i_l}-1)\}.$$

Tables 1.1, 1.2 and 1.3 of Section 1 clearly show that the response to the question 'which sex has more difficult life' depends most significantly on sex. It will be interesting to see if the MAICE procedure with the present model confirms this observation. The number of variables in our example is 4 and the necessary AIC's are obtained by putting $k=4$ in (2.3) or (2.4), where i_1 denotes a category of the opinion, i_2

Table 2

No.	Model No.	Model	AIC	No. of Parameters	χ^2-value	$1-F(\chi^2)$	Degrees of Freedom
1	(0, 1)	$p(i_1,i_2,i_3,i_4)$	25132.68	63	—	—	—
2	(1, 1)	$p(i_1,i_2,i_3)p(i_2,i_3,i_4)/p(i_2,i_3)$	25100.08	39	15.329	0.911	24
3	(1, 2)	$p(i_1,i_2,i_4)p(i_2,i_3,i_4)/p(i_2,i_4)$	25102.18	39	17.461	0.828	24
4	(1, 3)	$p(i_1,i_3,i_4)p(i_2,i_3,i_4)/p(i_3,i_4)$	25203.70	47	101.937	0.000	16
5	(2, 1)	$p(i_1,i_2)p(i_2,i_3,i_4)/p(i_2)$	25097.44*	33	24.686	0.740	30
6	(2, 2)	$p(i_1,i_3)p(i_2,i_3,i_4)/p(i_3)$	25187.96	35	110.096	0.000	28
7	(2, 3)	$p(i_1,i_4)p(i_2,i_3,i_4)/p(i_4)$	25187.20	35	109.520	0.000	28
8	(3, 1)	$p(i_1)p(i_2,i_3,i_4)$	25187.05	32	115.272	0.000	31
Model 1°		$p(i_1,i_2)p(i_3)p(i_4)$	25102.34	9	78.504	0.016	54
Model 2°		$p(i_1,i_3)p(i_2)p(i_4)$	25192.86	11	166.561	0.000	52
Model 3°		$p(i_1,i_4)p(i_2)p(i_3)$	25192.10	11	166.463	0.000	52

i_1: The question 'which sex has more difficult life?'
i_2: Sex
i_3: Age
i_4: Rural vs. urban breakdown
*: MAICE among all the models

of sex, i_3 of age and i_4 of urban vs. rural breakdown. To search for the predictor on which the opinion has the strongest dependence we have only to calculate AIC's for the two-way contingency tables shown in Tables 1.1, 1.2 and 1.3 and pick the one with the minimum AIC. From (2.3) we get 25102.34, 25192.86 and 25192.10, which are shown at the bottom of Table 2, as the AIC's of the models. Apparently the MAICE procedure suggests that we should adopt sex as the most effective predictor, which is identical to our empirical judgement. The detail of the dependence depends on how we categorize each predictor. This aspect will be discussed elsewhere [8].

3. More general models and their AIC's

A useful model for the search of an optimal combination of predictors can be obtained by using the multiplicative models of contingency tables which have previously been discussed by many authors such as Darroch [4], Bishop [3], Goodman [5], and Wermuth [9], [10]. A multiplicative model is a model such that the joint distribution of several variables is factored into the product of marginal distributions of subgroups of variables. One example of multiplicative model with $k=5$ is given by

$$(3.1)\quad p(i_1,\cdots,i_5)=p(i_1,i_4,i_5)p(i_2,i_4,i_5)p(i_3,i_4,i_5)/\{p(i_4,i_5)p(i_4,i_5)\}\ .$$

This model can be written as

$$(3.2)\qquad p(i_1,\cdots,i_5)=p(i_1|i_4,i_5)p(i_2|i_4,i_5)p(i_3|i_4,i_5)p(i_4,i_5)\ ,$$

where $p(i_1|i_4,i_5)$ denotes the conditional probability of i_1 given (i_4,i_5). This shows that in the model each of the variable pairs (i_1,i_2), (i_1,i_3) and (i_2,i_3) has zero partial association, that is, the variables in a pair is conditionally mutually independent, given the remaining three variables. Each multiplicative model is characterized by the variable groups in the parentheses of the numerator and denominator of the representation of its probability as in (3.1) and can be derived by successively assuming zero partial associations among various variable pairs. Following Wermuth [10], a multiplicative model is constructed as follows: Given a multiplicative model, choose a variable pair (i_j,i_l), which is to have zero partial association, from a variable group in the numerator. Here (i_j,i_l) is a variable pair that is not contained in any one of the variable groups in the denominator. Denote by (i_j,i_l,i_K) the variable group in the numerator that includes (i_j,i_l), where i_K denotes the variables other than i_j and i_l. To get the desired model, we have only to replace $p(i_j,i_l,i_K)$ in the numerator by $p(i_j,i_K)p(i_l,i_K)$ and multiply the denominator by $p(i_K)$ and cancel the common factors. For instance, if we assume the zero partial association of pair (i_1, i_4) in the above model, the application of the rule to (3.1) leads to the following model

$$(3.3)\quad \begin{aligned} p(i_1,\cdots,i_5) &= p(i_1,i_5)p(i_4,i_5)p(i_2,i_4,i_5)p(i_3,i_4,i_5) \\ &\quad /\{p(i_5)p(i_4,i_5)p(i_4,i_5)\} \\ &= p(i_1,i_5)p(i_2,i_4,i_5)p(i_3,i_4,i_5)/\{p(i_5)p(i_4,i_5)\}\,. \end{aligned}$$

To search for an optimal combination of predictors of a variable i_1, we want to eliminate those variables which will show zero partial association with the variable i_1. For this purpose we define a particular sequence of models as follows:

$$(3.4)\quad \begin{aligned}
&\text{MODEL}\,(0,1):\ p(i_1,\cdots,i_k)=p(i_1,\cdots,i_k) \\
&\text{MODEL}\,(1,1):\ p(i_1,\cdots,i_k)=p(i_1,\cdots,i_{k-1})p(i_2,\cdots,i_k)/p(i_2,\cdots,i_{k-1}) \\
&\qquad (1,2):\ p(i_1,\cdots,i_k)=p(i_1,\cdots,i_{k-2},i_k)p(i_2,\cdots,i_k) \\
&\qquad\qquad /p(i_2,\cdots,i_{k-2},i_k) \\
&\qquad \cdots\cdots\cdots\cdots \\
&\qquad (1,{}_{k-1}C_1):\ p(i_1,\cdots,i_k)=p(i_1,i_3,\cdots,i_k)p(i_2,\cdots,i_k) \\
&\qquad\qquad /p(i_3,\cdots,i_k) \\
&\text{MODEL}\,(2,1):\ p(i_1,\cdots,i_k)=p(i_1,\cdots,i_{k-2})p(i_2,\cdots,i_k)/p(i_2,\cdots,i_{k-2}) \\
&\qquad (2,2):\ p(i_1,\cdots,i_k)=p(i_1,\cdots,i_{k-3},i_{k-1})p(i_2,\cdots,i_k) \\
&\qquad\qquad /p(i_2,\cdots,i_{k-3},i_{k-1}) \\
&\qquad \cdots\cdots\cdots\cdots \\
&\qquad (2,{}_{k-1}C_2):\ p(i_1,\cdots,i_k)=p(i_1,i_4,\cdots,i_k)p(i_2,\cdots,i_k) \\
&\qquad\qquad /p(i_4,\cdots,i_k) \\
&\qquad\qquad \vdots \\
&\text{MODEL}\,(k-2,1):\ p(i_1,\cdots,i_k)=p(i_1,i_2)p(i_2,\cdots,i_k)/p(i_2) \\
&\qquad (k-2,2):\ p(i_1,\cdots,i_k)=p(i_1,i_3)p(i_2,\cdots,i_k)/p(i_3) \\
&\qquad \cdots\cdots\cdots\cdots \\
&\qquad (k-2,{}_{k-1}C_{k-2}):\ p(i_1,\cdots,i_k)=p(i_1,i_k)p(i_2,\cdots,i_k)/p(i_k) \\
&\text{MODEL}\,(k-1,{}_{k-1}C_{k-1}):\ p(i_1,\cdots,i_k)=p(i_1)p(i_2,\cdots,i_k)\,.
\end{aligned}$$

These model are generated by successively assuming zero partial associations and applying the above rule. MODEL (0, 1) means unconstrained model. MODEL (1, 1) represents the zero partial association between the variable i_1 and i_k in the sense that they are independent given the remaining $k-2$ variables. Similarly, MODEL (2, 1) represents the zero partial association between the variable i_1 and the set of variables $\{i_{k-1}, i_k\}$. Other models can be interpreted analogously. The variables appearing in the denominators of these equations define the candidates of the optimal combination of predictors of the variable i_1. The number 'l' of MODEL (l, m) denotes the number of zero partial associations to be assumed and the number 'm' denotes that the model is the mth with respect to some proper ordering of the models belonging to the class of models with one and the same 'l'. Therefore, m does not exceed ${}_{k-1}C_l$. The total number of models belonging to the above sequence of models is given by

$$ {}_{k-1}C_0 + {}_{k-1}C_1 + \cdots + {}_{k-1}C_{k-1} = 2^{k-1} . $$

For $k=2$ we get the unrestricted and the independence model discussed in Introduction.

Consider the set of variables defined by $I=\{i_2, \cdots, i_k\}$. Denote by E a subset of I. Taking into account that MODEL (0, 1) can be written as $p(i_1, \cdots, i_k) = p(i_1, \cdots, i_k)p(i_2, \cdots, i_k)/p(i_2, \cdots, i_k)$, or, using the above notations, $p(i_1, I) = p(i_1, I)p(I)/p(I)$, a model in the above sequence (3.4) can be represented in the form

$$ (3.5) \qquad p(i_1, I) = p(i_1, E)p(I)/p(E) , $$

where we assume that $p(E)=1$ for $E=\phi$, an empty set. The AIC for the model (3.5) is given by

$$ (3.6) \qquad \begin{aligned} \mathrm{AIC} = (-2) \sum_{i_1, I} n(i_1, I) \log [n(i_1, E)n(I)/\{n \cdot n(E)\}] \\ +2\{(C_{i_1}C_E - 1) + (C_I - 1) - (C_E - 1)\} , \end{aligned} $$

where C_E and C_I denotes the number of categories of the corresponding sets of variables and we assume that $n(E)=n$ and $C_E=1$ for $E=\phi$. In calculating AIC's it is assumed that $0 \log 0 = 0$. For the purpose of comparison of models within the above sequence, the common constant $(-2) \sum_I n(I) \log \{n(I)/n\} + 2(C_I - 1)$ can be ignored and the AIC is given by

$$ (3.7) \quad \mathrm{AIC} = (-2) \sum n(i_1, E) \log \{n(i_1, E)/n(E)\} + 2\{(C_{i_1}C_E - 1) - (C_E - 1)\} . $$

This shows that we can compare these models without using the full-dimensional table. Further we note that from the point of view of AIC the comparison of models belonging to (2.2) reduces to that of

Table 3 Which sex

	S_1															
	A_1				A_2				A_3				A_4			
	R_1	R_2	R_3	R_4	R_1	R_2	R_3	R_4	R_1	R_2	R_3	R_4	R_1	R_2	R_3	R_4
W_1	50	57	77	55	38	58	61	51	38	39	59	58	42	40	95	86
W_2	28	27	40	26	17	28	40	35	16	20	39	37	21	23	45	49
Total	78	84	117	81	55	86	101	86	54	59	98	95	63	63	140	135

* See Tables 1.1, 1.2 and 1.3 about notations.

the models, MODEL $(k-2, m)$, $m=1,\cdots, {}_{k-1}C_{k-2}$, of (3.4) since the statistic (3.7) is identical to (2.4) when $E=\{i_l\}$, $l=2,\cdots, k$.

Table 3 is the four-way contingency table of the question 'which sex has more difficult life' and the three demographic factors. It will be interesting to see what combination of predictors the MAICE procedure adopt as the optimal one. The necessary eight models and their AIC's are obtained by putting $k=4$ in (3.4) and using (3.6). The results are given in Table 2. The MAICE is MODEL (2, 1) and shows that still a single factor sex defines the best combination to define the predictor. The result of Table 2 gives a finer description of the interdependence relation between the opinion and other demographic predictors than the result of the simplified analysis of the preceding section. Nevertheless, the result shows that we have only to pay our attention to sex in the case of the analysis of the interaction between the opinion and other demographic predictors.

The survey of Japanese national character has been conducted every five years since 1953. We used questionnaires which were common to all five surveys for the purpose of detection of changes in people's way of thinking. We applied the MAICE procedure proposed in this paper to the analysis of all questions of the 1973 survey. The results are quite assuring. In almost all the cases the MAICE lead to the same conclusion as that obtained by a careful analysis of the data formerly reported in [6].

The analysis of a multidimensional contingency table has been a difficult and very much time-consuming task. This was mainly due to the inappropriate modeling and the lack of an objective criterion for the evaluation of the badness of a fitted model. By applying the procedure of this paper we can easily find what combination of predictors is the most important as a factor and list up the predictors in order of the dependence of the variable on the predictors. The use of the statistic (3.7) also facilitate the search for the optimal combination of predictors for a high dimensional table. This last aspect will be discussed in more detail in a future paper.

has more difficult life?

S_2																Total
A_1				A_2				A_3				A_4				
R_1	R_2	R_3	R_4	R_1	R_2	R_3	R_4	R_1	R_2	R_3	R_4	R_1	R_2	R_3	R_4	
47	57	63	48	40	50	57	53	32	45	47	45	35	42	66	63	1694
32	56	63	52	29	57	67	46	37	44	61	49	37	49	90	101	1361
79	113	126	100	69	107	124	99	69	89	108	94	72	91	156	164	3055

4. Discussion of statistical characteristics of the procedure

Suppose that a four-dimensional probability distribution is defined by

$$p(i_1, i_2, i_3, i_4)=p(i_1, i_2)p(i_2, i_3, i_4)/p(i_2) .$$

We assume the values of these probabilities shown in Table 4. The question is whether we can detect the true structure by the MAICE procedure. To answer this question we generated 100 sets of data each composed of 3000 random samples from the above probability distribution. The frequencies of the models chosen as the MAICE's are shown

Table 4

i_1, i_2	Probability	i_1, i_2	Probability
1, 1	0.2959	2, 1	0.1607
1, 2	0.2586	2, 2	0.2848

i_2, i_3, i_4	Probability	i_2, i_3, i_4	Probability
1, 1, 1	0.0255	2, 1, 1	0.0259
1, 1, 2	0.0275	2, 1, 2	0.0370
1, 1, 3	0.0383	2, 1, 3	0.0412
1, 1, 4	0.0265	2, 1, 4	0.0327
1, 2, 1	0.0180	2, 2, 1	0.0226
1, 2, 2	0.0281	2, 2, 2	0.0350
1, 2, 3	0.0331	2, 2, 3	0.0406
1, 2, 4	0.0282	2, 2, 4	0.0324
1, 3, 1	0.0177	2, 3, 1	0.0226
1, 3, 2	0.0193	2, 3, 2	0.0291
1, 3, 3	0.0321	2, 3, 3	0.0353
1, 3, 4	0.0311	2, 3, 4	0.0308
1, 4, 1	0.0206	2, 4, 1	0.0236
1, 4, 2	0.0206	2, 4, 2	0.0298
1, 4, 3	0.0458	2, 4, 3	0.0511
1, 4, 4	0.0442	2, 4, 4	0.0537

Table 5

Estimated Distribution	Frequency	Number of Free Parameters	Frequency Accepted by χ^2-test	Degrees of Freedom
$p(i_1, i_2, i_3)p(i_2, i_3, i_4)/p(i_2, i_3)$	6	39	95	24
$p(i_1, i_2, i_4)p(i_2, i_3, i_4)/p(i_2, i_4)$	9	39	97	24
$p(i_1, i_2)p(i_2, i_3, i_4)/p(i_2)$	85	33	95	30
Other distributions	0	—	0	—
Total	100	—	—	—

in Table 5. The result tells that the MAICE procedure produced correct answer 85 times out of 100. Needless to say the performance of the procedure depends on the sample size and the structure of the true distribution. The present result is a typical example as is expected from the definition of AIC statistic.

Consider that the chi-square goodness of fit tests are applied to our example. We regard the situation as the fitting of each of the models described in Section 3 to the observations $n(i_1, i_2, i_3, i_4)$. For example, for the above model we get the chi-square test statistic

$$\chi^2 = \sum_{i_1,\cdots,i_4} \{n(i_1, i_2, i_3, i_4) - n(i_1, i_2)n(i_2, i_3, i_4)/n(i_2)\}^2$$
$$/\{n(i_1, i_2)n(i_2, i_3, i_4)/n(i_2)\} .$$

The figures in the right half of Table 5 give the frequency for each model accepted at the level of 5%. The results show that three models including the true one were accepted about 95 times out of 100. This means that the test procedure can not discriminate more complicated models from the true structure.

The figures in the right half of Table 2 give χ^2, $1-F(\chi)^2$ and the degrees of freedom for each of the seven models of the data given by the four-way contingency table shown in Table 3. Here F denotes a cumulative distribution function of a chi-square variable. If the test is applied only to those models within MODEL (2, m), the χ^2 for the MODEL (2, 1) is insignificant, with respect to the 5% level of significance. This result shows that the model is acceptable, or at least not rejected, and coincides with the conclusion by MAICE for this case. However, if the test is applied to models defined by (2, 2), every model is rejected at the level of 5%, as is shown in the three lines from the bottom of Table 2. The MAICE is Model No. 1° for this case too, but by the test procedure MODEL (0, 1) is the only choice.

The relation between the MAICE and classical test procedures can be understood by considering the fact that the log likelihood ratio test statistic takes the form $\chi^2 = \mathrm{AIC}\,(k) - \mathrm{AIC}\,(K) + 2(K-k)$, where AIC ($k$) denotes the AIC of a model with k free parameters. K is usually the highest possible value of k and χ^2 is tested as a chi-square with the degrees of freedom d.f. $= K-k$. Taking into account that the expectation of χ^2 is equal to its degrees of freedom, we can understand that the MAICE procedure applied to each pair of models in the above example means the comparison of the value of χ^2 with twice its expectation. The values of $1-F(2\,\mathrm{d.f.})$ for various values of d.f. are given in Table 6. The table clearly shows that by AIC the "level of significance" is adjusted in such a way that the corresponding probability of rejection of the simpler model decreases as the degrees of freedom

Table 6

d.f.	1−*F*(2d.f.)	d.f.	1−*F*(2d.f.)
1	0.1572989537	10	0.0292526881
2	0.1353352832	15	0.0119215009
3	0.1116101347	20	0.0049954123
4	0.0915781944	25	0.0021311519
5	0.0752352001	30	0.0009206824
6	0.0619688044	40	0.0001763029
7	0.0511816101	50	0.0000345493
8	0.0423801120	60	0.0000068763
9	0.0351737134	70	0.0000013839

* Calculated by expansion formula for the χ^2-distribution function

increase. The MAICE procedure, therefore, has a tendency to adopt simpler models compared with the chi-square test procedure as the degrees of freedom increase. This characteristic of the MAICE seems to be in better agreement with our intuitive choice when a complex model is fitted than the one by the chi-square test. Now if a modification of a test procedure considered so that the significance level is adjusted in accordance with the degrees of freedom, one has to provide a rule for the adjustment. Even if this adjustment is made possible, it is still impossible to compare a model with every possible choice of the alternative. For example, it is impossible to compare MODEL (1, 1) with MODEL (1, 2) in Table 2 by the classical chi-square test. The salient feature of AIC is that it is an estimate of a clearly defined universal measure of fit, the entropy defined in Section 1. This fact justifies the comparison of AIC's among every possible model which cannot necessarily be compared by the classical goodness of fit test.

5. Concluding remarks

Generally there are two different types of analysis of survey results. The one is the case where the purpose of the analysis is to evaluate the dependence between a specific variable to be predicted and a specific predictor, such as the answer to the question "Which political party the youth has been supporting?" The other is the case where the object is to seek an explanation of phenomenon, exemplified by the question "What has caused the changes in political party support?" We are sure that the procedure proposed in this paper will be of great help to solve the latter problem. By our procedure, as was shown in preceding sections, the comparison of various models is very simple and under certain circumstances the search procedure for the optimal combination of predictors can be done without the use of the full-dimensional contingency table.

The definition of AIC will draw researcher's attention to the relation between the number of free parameters within a model and the sample size of the survey data. This aspect of statistical analysis was not clearly recognized in the application of classical tests. We are tempted to think that classical tests, such as the chi-square test of goodness of fit and that of independence, are merely approximate realizations of our procedure. However, our procedure needs further refinement of the basic model so as to take care of the situation where many cells are lacking observations. This will be the subject of further study.

A Fortran program for the entire procedure is available from the authors.

Acknowledgement

We wish to thank the referees for their helpful comments. Thanks are due to Mr. K. Katsura of the Institute of Statistical Mathematics for programming and testing the algorithms. We are grateful to Mr. G. Kitagawa and Mr. M. Ishiguro for their helpful suggestions.

THE INSTITUTE OF STATISTICAL MATHEMATICS

REFERENCES

[1] Akaike, H. (1973). Information theory and an extension of the maximum likelihood principle, *2nd International Symposium on Information Theory*, B. N. Petrov and F. Csaki, Eds., Akademiai Kiado, Budapest, 267-281.

[2] Akaike, H. (1976). On entropy maximization principle, *Applications of Statistics*, P. R. Krishnaiah, ed., North-Holland, Amsterdam, 27-41.

[3] Bishop, Y. M. M. (1969). Full contingency tables, logits and split contingency tables, *Biometrics*, **25**, 383-400.

[4] Darroch, J. N. (1962). Interactions in multifactor contingency tables, *J. R. Statist. Soc.*, B**24**, 251-263.

[5] Goodman, L. A. (1970). The multivariate analysis of qualitative data: interactions among multiple classifications, *J. Amer. Statist. Ass.*, **65**, 226-256.

[6] Research Committee on the Study of Japanese National Character (1976). *Nipponjin no Kokuminsei, sono san* (A study of the Japanese National Character, Part III), Shiseido, Tokyo. (In Japanese)

[7] Sakamoto, Y. (1974). A study of the Japanese National Character—Part V, *Ann. Inst. Statist. Math., Supplement* 8, 1-58.

[8] Sakamoto, Y. (1977). A model for the optimal pooling of categories of the predictor in a contingency table, *Research Memorandum*, No. 119, The Institute of Statistical Mathematics, Tokyo.

[9] Wermuth, N. (1976). Analogies between multiplicative models in contingency tables and covariance analysis, *Biometrics*, **32**, 95-108.

[10] Wermuth, N. (1976). Model search among multiplicative models, *Biometrics*, **32**, 253-263.

J. R. Statist. Soc. B (1982),
44, *No.* 1, *pp.* 102–107

On Linear Intensity Models for Mixed Doubly Stochastic Poisson and Self-exciting Point Processes

By Yosihiko Ogata and Hirotugu Akaike

The Institute of Statistical Mathematics, Tokyo, Japan

[Received August 1980. Final revision May 1981]

Summary

A flexible family of parametric models for intensity processes is introduced to represent a causal relationship between a point process and another stochastic process. Algorithms for the maximum likelihood computation and the procedure of model selection are discussed.

Keywords: PARTIAL LOG LIKELIHOOD; INTENSITY FUNCTION; RESPONSE FUNCTIONS; LAGUERRE TYPE POLYNOMIAL; RECURSIVE RELATIONS; AIC

1. Introduction

In 1978 Vere-Jones (1978) investigated the causal relationship between the eruption of a volcano and deep earthquakes by applying a doubly stochastic Poisson process model. In the present paper we intend to extend the idea to the case with a more general input process and develop a procedure of modelling which will be useful even for the analysis of a causal relation between two series of data of different kind, such as the continuous record of a geophysical quantity and the record of earthquake occurrences. The importance of this type of modelling was recognized during our discussion of some microearthquake data with Professor Oike of the Disaster Prevention Research Institute of Kyoto University.

For the observation over $[0, T]$ of a point process N_t the partial log likelihood in the sense of Cox (1975) is given by

$$L_T = \int_0^T \log \Lambda_\theta(t \mid H_t)\, dN_t - \int_0^T \Lambda_\theta(t \mid H_t)\, dt, \quad \theta \in \Theta, \tag{1.1}$$

where $\Lambda_\theta(t \mid H_t)$ denotes the intensity function parameterized by θ, and the family of conditioning events H_t consists of the past histories of N_t itself and those of another observable process X_t. It is assumed that the observation N_t is generated by a point process whose intensity is specified by $\Lambda_{\theta_0}(t \mid H_t)$ for some $\theta_0 \in \Theta$. For this case it can be shown, analogously to the proof of the asymptotic properties of the maximum likelihood estimate of a point process discussed in Ogata (1978), that the standard large sample theory holds under usual regularity conditions, such as the stationarity and ergodicity of the joint process $\{(X_t, N_t), t \geqslant 0\}$.

The practical applicability of the present model is heavily dependent on the availability of some proper parameterization of the intensity function $\Lambda_\theta(t \mid H_t)$. In the present paper we propose a system of parametric families of $\Lambda_\theta(t \mid H_t)$ which allows efficient calculation of the (partial) likelihoods. The model selection is then realized by the minimum AIC procedure.

2. Linear Intensity Models

Throughout the present paper we consider the point process such that the relations

$$P\{N_{t+\Delta t} - N_t = 1 \mid H_t\} = \Lambda(t \mid H_t)\Delta t + o(\Delta t)$$

and

$$P\{N_{t+\Delta t} - N_t \geqslant 2 \mid H_t\} = o(\Delta t) \tag{2.1}$$

 0035-9246/82/44102$2.00

hold for small Δt. This implies that

$$E\{N_{t+\Delta t} - N_t \mid H_t\} = \Lambda(t \mid H_t)\Delta t + o(\Delta t). \tag{2.2}$$

Since H_t consists of the past histories of the processes X_s and N_s before time t, we are interested in the linear model defined by

$$\Lambda(t \mid H_t) = \mu + \int_0^t g(t-s)\,dN_s + \int_0^t h(t-s)\,dX_s. \tag{2.3}$$

This type of model was first discussed by Hawkes (1971) for the case where $\{X_t\}$ is a point process. In our present model the process $\{X_t\}$ may either be a point process or a cumulative process

$$X_t = \int_0^t x(s)\,ds \tag{2.4}$$

of some stochastic process $\{x(t)\}$.

When $h(t) \equiv 0$ holds, this means that there is no causal relation between the input $\{X_t\}$ and the output $\{N_t\}$. Also $g(t) \equiv 0$ means that the output process is a doubly stochastic Poisson, while $g(t) \equiv h(t) \equiv 0$ means that the output process is a homogeneous Poisson process of rate μ.

3. PARAMETERIZATION AND THE LIKELIHOOD COMPUTATION

For the parameterization of $g(t)$ we propose the use of the Laguerre type polynomial

$$g(t) = \sum_{k=0}^{K} a_k t^k e^{-ct}. \tag{3.1}$$

We also adopt a similar parameterization for the response function $h(t)$ given by

$$h(t) = \sum_{k=0}^{L} b_k t^k e^{-ct}, \tag{3.2}$$

where the exponential coefficient c is assumed to be equal to that of $g(t)$. The assumption of the same c in both $g(t)$ and $h(t)$ is for the sake of convenience of the likelihood computation.

Given the occurrence times of two types of events $\{t_i, \tau_m;\ i = 1, \ldots, I,\ m = 1, \ldots, M\}$ during the time interval $[0, T]$, we fit the model (2.3) to the data, regarding the series $\{t_i\}$ as the output and $\{\tau_i\}$ as the input. The partial log likelihood (1.1) of the model thus defined can easily be obtained in terms of

$$\int_0^T \log \Lambda_\theta(t \mid H_t)\,dN_t = \sum_{i=1}^{I} \log \Lambda_\theta(t_i \mid H_{t_i}) = \sum_{i=1}^{I} \log\left\{\mu + \sum_{k=0}^{K} a_k P_k(i) + \sum_{k=0}^{L} b_k Q_k(i)\right\},$$

where

$$\begin{aligned} P_k(i) &= \sum_{t_j < t_i} (t_i - t_j)^k \exp\{-c(t_i - t_j)\} \\ Q_k(i) &= \sum_{\tau_m < t_i} (t_i - \tau_m)^k \exp\{-c(t_i - \tau_m)\}, \end{aligned} \tag{3.3}$$

and

$$\begin{aligned} \int_0^T \Lambda_\theta(t \mid H_t)\,dt &= \mu T + \int_0^T dt \int_0^t g(t-s)\,dN_s + \int_0^T dt \int_0^t h(t-s)\,dX_s \\ &= \mu T + \sum_{k=0}^{K} a_k \sum_{i=1}^{I} R_k(T - t_i) + \sum_{k=0}^{L} b_k \sum_{m=0}^{M} R_k(T - \tau_m), \end{aligned}$$

where

$$R_k(t) = \int_0^t t^k e^{-ct}\,dt. \tag{3.4}$$

The following recursive relations are useful for the computation of the likelihood:

$$P_k(i+1) = (t_{i+1}-t_i)^k \exp\{-c(t_{i+1}-t_i)\} + \sum_{j=0}^{k} \binom{k}{j} (t_{i+1}-t_i)^{k-j} \exp\{-c(t_{i+1}-t_i)\}\, P_j(i),$$

$$Q_k(i+1) = D_k(t_i, t_{i+1}) + \sum_{j=0}^{k} \binom{k}{j} (t_{i+1}-t_i)^{k-j} \exp\{-c(t_{i+1}-t_i)\}\, Q_j(i)$$

and

$$R_{k+1}(t) = \{(k+1)\,R_k(t) - t^{k+1} e^{-ct}\}/c, \tag{3.5}$$

where

$$D_k(t_i, t_{i+1}) = \sum_{t_i \leqslant \tau_m < t_{i+1}} (t_{i+1}-\tau_m)^k \exp\{-c(t_{i+1}-\tau_m)\}.$$

By definition $t_0 = 0$, we get $P_0(0) = Q_0(0) = 0$ and $R_0(t) = (1-e^{-ct})/c$.

If the input is defined by a cumulative process $dX_t = x(t)\,dt$, $0 \leqslant t \leqslant T$, we approximate it by

$$dX_t^* = (T/M) \sum_{m=1}^{M} x(t)\,\delta(t-\sigma_m)\,dt,$$

where M is a properly chosen large integer at least of the order of the sample size I, $\sigma_m = (m/M)T - (1/2M)T$, $m = 1, 2, \ldots, M$, and $\delta(t)$ denotes Dirac's delta function. With this approximation, we get an approximate partial log likelihood

$$L_T^*(\theta) = \sum_{i=1}^{I} \log\{\mu + \sum_{k=0}^{K} a_k P_k(i) + \sum_{k=0}^{L} b_k U_k(i)\} - \mu T$$
$$- \sum_{k=0}^{K} a_k \sum_{i=1}^{I} R_k(T-t_i) - \sum_{k=0}^{L} b_k \sum_{m=0}^{M} x(\sigma_m)\,R_k(T-\sigma_m),$$

where $P_k(i)$ and $R_k(t)$ are the same as in the previous case and $U_k(i)$ is given recursively by

$$U_k(i+1) = F_k(t_i, t_{i+1}) + \sum_{j=0}^{k} \binom{k}{j} U_j(i)\,(t_{i+1}-t_i)^{k-j} \exp\{-c(t_{i+1}-t_i)\}$$

and

$$F_k(t_i, t_{i+1}) = \sum_{t_i \leqslant \sigma_m < t_{i+1}} x(\sigma_m)\,(t_{i+1}-\sigma_m)^k \exp\{-c(t_{i+1}-\sigma_m)\}.$$

The gradient of each partial log likelihood function can easily be obtained by differentiating the above functions. Once the gradient is obtained the maximization of the likelihood function is performed by using a standard non-linear optimization technique developed by Fletcher and Powell (1963).

4. Model Selection

We can estimate the coefficients $a_k (k = 1, \ldots, K)$, $b_k (k = 1, \ldots, L)$ and c of the polynomials (3.1) and (3.2) by the maximum likelihood method, provided that the orders K and L are given. By the minimum AIC procedure (Akaike, 1977), we select (K, L) which minimizes

$$\mathrm{AIC}(K, L) \equiv (-2)\max(\text{log likelihood}) + 2(\text{number of parameters})$$
$$= (-2)\max L_T(\mu, c, a_0, \ldots, a_k, b_0, \ldots, b_L) + 2(K+L+4). \tag{4.1}$$

The AIC is an estimate of the expected negentropy which is a natural measure of discrimination between the true and estimated probability law of the data. The use of the minimum AIC procedure is justified under the assumption of the standard large sample theory of the maximum likelihood estimate which is briefly touched in Introduction.

It is certainly possible to get the maximum likelihood estimates of all the parameters. However this is too much time-consuming when the sample size is large or the orders K and L are high. This is due to the fact that the functions $P_k(i)$, $Q_k(i)$, $U_k(i)$ and $R_k(t)$ contain the exponential parameter c which changes in every step of the non-linear optimization. Also very frequently there are more than one maxima of the likelihood function.

However, once the parameter c is fixed, the log likelihood function has at most one maximum. This is seen from the fact that the Hessian is everywhere non-positive definite with respect to all the other parameters, as the intensity function $\Lambda_\theta(t)$ is linearly parameterized (Ogata, 1978, p. 255). Thus for the computational convenience, it is quite advisable to repeat the maximum likelihood computation for a finite number of coarsely distributed fixed values of c. Accordingly we adopt the following algorithms.

For a fixed exponential coefficient c, compute the following statistics

$$P(k,i) \equiv P_k(i), \quad Q(k,i) \equiv Q_k(i),$$

$$U(k,i) \equiv U_k(i), \quad V(k) \equiv \sum_{i=1}^{I} R_k(T-t_i),$$

$$W(k) \equiv \sum_{m=0}^{M} R_k(T-\tau_m) \tag{4.2}$$

and

$$S(k) \equiv \sum_{m=0}^{M} x(\sigma_m)\, R_k(T-\sigma_m)$$

for $i = 0, 1, \ldots, I$ and $k = 1, 2, \ldots, \bar{K}$ where $\bar{K}$ is the upper bound of the orders. Compute the partial log likelihood for the point process input by

$$L_T(\theta) = \sum_{i=1}^{I} \log\left\{\mu + \sum_{k=0}^{K} a_k\, P(k,i) + \sum_{k=0}^{L} b_k\, Q(k,i)\right\} - \mu T$$
$$- \sum_{k=0}^{K} a_k\, V(k) - \sum_{k=0}^{L} b_k\, W(k), \tag{4.3}$$

or that for the accumulated process input by

$$L_T^*(\theta) = \sum_{i=1}^{I} \log\left\{\mu + \sum_{k=0}^{K} a_k\, P(k,i) + \sum_{k=0}^{L} b_k\, Q(k,i)\right\} - \mu T$$
$$- \sum_{k=0}^{K} a_k\, V(k) - \sum_{k=0}^{L} b_k\, S(k), \tag{4.4}$$

where θ stands for $(\mu, \{a_k\}, \{b_k\})$. Keep the statistics (4.2) and maximize the function (4.3) or (4.4) sequentially for each pair (K, L) such that $K \leqslant \bar{K}$ and $L \leqslant \bar{K}$. Repeat the whole process with several other possible choices of c and find the maximum of the likelihood with respect to these values of c.

Suppose we have some candidate values of c (say c_j, $j = 1, 2, \ldots, J$). Then we can compare

$$\mathrm{AIC}(c_j, K, L) = (-2) \max_{\theta} L_T(c_j, \theta) + 2(K + L + 3),$$

where $\theta = (\mu, a_0, \ldots, a_K, b_0, \ldots, b_L)$, $K, L = 0, 1, 2, \ldots, \bar{K}$ and $j = 1, 2, \ldots, J$. Thus we can perform the minimum AIC procedure quite systematically up to some considerably high degree models.

The selection of the values c_j can be realized in the following manner. Firstly we assess the range of t where the response functions $g(t)$ and $h(t)$ are significantly different from zero. This may be obtained by some prior information about the data or by some preliminary analysis of the data such as the auto and cross-correlogram analysis. Let a rough estimate of the range be R, and assume the response function of the form $ax^k e^{-ct}$ which has its peak at $x = k/c$. Then, we assume that the peak is attained in the middle of the range, i.e. we assume the relation $k/c = R/2$. If we assume $k = 2$, for example, we get $c_0 = 4/R$ as an initial guess of the exponential coefficient. Another practical way of getting an initial estimate of c is to apply the direct maximum likelihood estimation procedure to the second-order polynomial model using a short subset of the data. We successively try $c_j = 2^j c_0$ or $2^{-j} c_0, j = 1, 2, ..., J$ and compare the AIC values sequentially until a minimum is obtained. Further search for the optimum value of c can be continued within the interval defined by the two c_j's, adjacent to the one which has given the minimum of AIC. By our experience the estimated response functions with different c_j's but with similar AIC values are quite similar to each other in spite of their possible differences of the orders.

5. DISCUSSION AND REMARKS

We checked the feasibility of the procedure described in Sections 3 and 4 by some artificially generated data. Also the model was applied to a pair of earthquake series which occurred in certain different seismic regions in Japan over the period 1924–74. The results of our analysis show that the earthquakes in one region are not only self-exciting but also significantly receive one-way stimulation from earthquake occurrences in the other region. The details of this earthquake example can be seen in a companion paper (Ogata, Akaike and Katsura, 1981) to which the interested reader is referred. Here we would like only to give some feeling of the estimation procedure of the response functions by Fig. 1 obtained by applying the

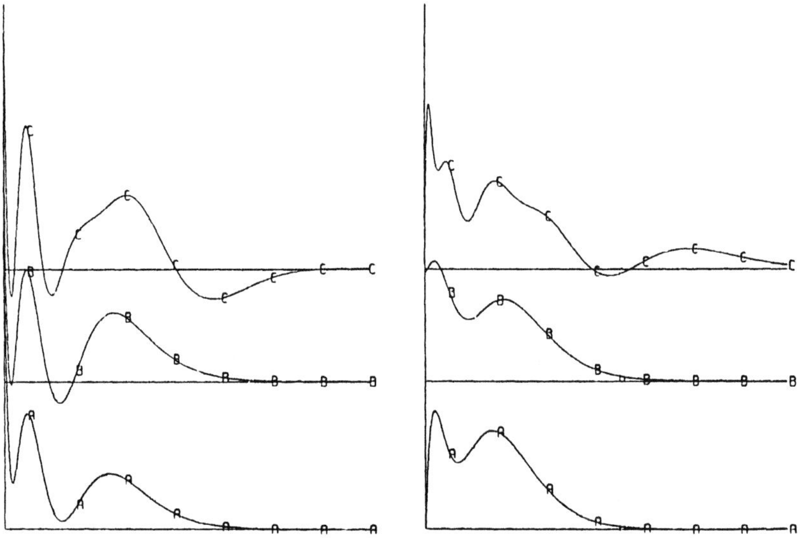

FIG. 1.
–A–A–A–: The true response functions; $C = 1{\cdot}1, K = L = 5$.
–B–B–B–: The estimated response functions with the minimum $\text{AIC}(2^{1/4}, 5, 5) = 3281{\cdot}1$.
–C–C–C–: The estimated response functions with $\text{AIC}(2^{1/4}, 10, 10) = 3289{\cdot}9$.

procedure of Section 4 to artificially generated bivariate series of events. The numbers of the events in the input and output series were 992 and 769 over the time interval [0, 2000], respectively. The result clearly shows the practical utility of the procedure.

Hawkes (1971) assumed the response functions $g(t)$ and $h(t)$ in (2.3) to be non-negative to ensure the non-negativity of the intensity process with probability one. The point process of this type has a certain kind of clustering property (see Hawkes and Oakes, 1974). However, there are examples of series of events with some inhibitory property like spike trains of nerve-system data. If we fit the parametric model of Section 3 to such data, we often get response functions having negative values in some parts of t (Nakamura, Ogata and Oomura, 1980). If the negativity of the response functions does not affect the non-negativity of the estimated intensity process throughout the observed time interval, the negativity could be understood as an indication of the existence of an inhibitory range. To handle such possibility properly we will have to replace the parametric intensity function in (1.3) by $\Lambda_\theta^+(t \mid H_t) = \max\{\Lambda_\theta(t \mid H_t), 0\}$ where $\Lambda_\theta(t \mid H_t)$ is given in (2.3).

If $\mu > 0$ holds in (2.3), it is easily seen that the existence of the negative valued parts in the estimated response functions does not always imply the existence of negative parts of the intensity process. Given a series of events $\{t_i, \tau_m, i = 1, \ldots, I, m = 1, \ldots, M\}$ over the time interval $[0, T]$, if the intensity $\Lambda_\theta(t \mid H_t)$ with θ obtained by the method of maximum likelihood is non-negative in $[0, T]$, then that estimate also maximizes the log likelihood function

$$L_T(\theta) = \int_0^T \log \Lambda_\theta^+(t \mid H_t)\, dN_t - \int_0^T \Lambda_\theta^+(t \mid H_t)\, dt.$$

ACKNOWLEDGEMENTS

The authors would like to thank Professor K. Oike of Kyoto University for valuable suggestions about earthquakes data. The authors are also grateful to Professor D. Vere-Jones and Dr T. Ozaki for very helpful suggestions during the preparation of this paper. The idea of using Laguerre type polynomials for modelling the response functions was obtained during the discussions with Ozaki. K. Katsura and G. Kitagawa generously helped us with the programming of the necessary computations. The authors are particularly grateful to the referees whose comments led to a significant revision of the paper.

REFERENCES

AKAIKE, H. (1977). On Entropy Maximization principle. In *Application of Statistics* (P. R. Krishnaiah, ed.), Amsterdam: North-Holland. 27–41.

COX, D. R. (1975). Partial likelihood. *Biometrika*, **62**, 269–276.

FLETCHER, R. and POWELL, M. J. D. (1963). A rapidly convergent descent method for minimization. *Computer J.*, **6**, 163–168.

HAWKES, A. G. (1971). Spectra of some mutually exciting point processes. *Biometrika*, **58**, 83–90.

HAWKES, A. G. and OAKES, D. (1974). A cluster process representation of a self-exciting process. *J. Appl. Prob.*, **11**, 493–503.

NAKAMURA, M., OGATA, Y. and OOMURA, Y. (1980). Evaluation of neuronal spike train data by new methodologies of wave form discrimination and point process analysis. In *Proceedings of the 7th International CODATA Conference* (Y. Mashiko, ed.). Oxford: Pergamon.

OGATA, Y. (1978). The asymptotic behaviour of maximum likelihood estimators for stationary point processes. *Ann. Inst. Statist. Math.*, **30**, A, 243–261.

OGATA, Y., AKAIKE, H. and KATSURA, K. (1981). The application of linear intensity models to the investigation of causal relations between a point process and another stochastic process, *Research Memo.*, No. 203, The Inst. Statist. Math., Tokyo (Submitted to *Ann. Inst. Statist. Math. B.*)

VERE-JONES, D. (1978). Personal communication.

Ann. Inst. Statist. Math.
30 (1978), Part A, 9-14

A BAYESIAN ANALYSIS OF THE MINIMUM AIC PROCEDURE

HIROTUGU AKAIKE

(Received Oct. 15, 1977; revised Apr. 24, 1978)

Summary

By using a simple example a minimax type optimality of the minimum AIC procedure for the selection of models is demonstrated.

1. Introduction

Akaike [1] introduced an information criterion which is by definition

$$\text{AIC} = (-2)\log(\text{maximum likelihood}) + 2(\text{number of parameters}) \tag{1.1}$$

as an estimate of minus twice the expected log likelihood of the model whose parameters are determined by the method of maximum likelihood. Here log denotes the natural logarithm. The simple procedure which selects a model with the minimum AIC among a set of models defines the minimum AIC estimate (MAICE) (Akaike [2]). The introduction of AIC helped the recognition of the importance of modeling in statistics and many practically useful statistical procedures have been developed as minimum AIC procedures; see, for example, Akaike [2], [3].

In spite of the accumulation of successful results in practical applications the logical foundation of MAICE has been continuously questioned by theoretically minded statisticians. The purpose of the present paper is to provide a Bayesian interpretation of the MAICE procedure and show that the procedure provides a minimax type solution to the problem of model selection under the assumption of equal prior probability of the models.

Our analysis starts with a brief review of the statistic of the form

$$(-2)\log(\text{maximum likelihood}) + (\log N)(\text{number of parameters}) + C\,, \tag{1.2}$$

where N is the sample size used for the computation of the maximum likelihood estimates. We will call this type of statistic by the generic name BIC. Two types of BIC have been introduced by Akaike [3] and Schwarz [4].

2. A review of BIC

Both Akaike and Schwarz based the introduction of BIC on some Bayesian arguments. Schwarz derived the statistic for models from a Koopman-Darmois family. Akaike introduced the statistic for the problem of selection of variables in linear regression. Here we will restrict our attention to the problem of selection of a multivariate Gaussian distribution. This is a special case of the model treated by Schwarz but may serve as a simplified model of the general situation where the use of the maximum likelihood estimates is contemplated.

Consider the situation where a set of observations $Y=\{y(n);\ n=1, 2, \cdots, N\}$ of L-dimensional vector random variables $y(n)=(y_1(n), y_2(n), \cdots, y_L(n))$ is given. It is assumed that $y(n)$'s are independently identically distributed as Gaussian $N(\theta, I)$, where $\theta=(\theta_1, \theta_2, \cdots, \theta_L)$ is the vector of the unknown means and I is an $L\times L$ identity matrix. We consider the set of models $N({}_k\theta, I)$ $(k=0, 1, \cdots, L)$ specified by assuming $\theta_{k+1}=\theta_{k+2}=\cdots=\theta_L=0$, i.e., ${}_k\theta=({}_k\theta_1, {}_k\theta_2, \cdots, {}_k\theta_k, 0, \cdots, 0)$ and ${}_k\theta_1, {}_k\theta_2, \cdots, {}_k\theta_k$ are unknown. Following Schwarz, we assume a prior distribution $\pi(k)$ over k, or the set of models $N({}_k\theta, I)$, $(k=0, 1, \cdots, L)$. Further we assume a prior distribution $N(0_k, \sigma^2 I_k)$ for $({}_k\theta_1, {}_k\theta_2, \cdots, {}_k\theta_k)$, where 0_k denotes a k-dimensional zero-vector and I_k a $k\times k$ identity matrix. Now the marginal posterior distribution of k is given by $C\pi(k)p(k|Y)$ $(k=0, 1, \cdots, L)$ with $p(k|Y)$ defined by

$$p(k|Y)=\exp\left\{-\frac{1}{2}\sum_{n=1}^{N}\sum_{i=k+1}^{L} y_i^2(n)\right\}\int \exp\left\{-\frac{1}{2}\sum_{n=1}^{N}\sum_{i=1}^{k}(y_i(n)-{}_k\theta_i)^2\right\}$$
$$\cdot\left(\frac{1}{2\pi\sigma^2}\right)^{k/2}\exp\left\{-\frac{1}{2\sigma^2}\sum_{i=1}^{k}{}_k\theta_i^2\right\}\prod_{i=1}^{k} d_k\theta_i\ .$$

By choosing the k that maximizes the posterior probability one can maximize the probability of correct decision on k. Consider the situation where σ^2 is sufficiently large so that we get an approximate equality

$$p(k|Y)=\exp\left\{-\frac{1}{2}\sum_{n=1}^{N}\sum_{i=k+1}^{L} y_i^2(n)\right\}\int \exp\left\{-\frac{1}{2}\sum_{i=1}^{k}\sum_{n=1}^{N}(y_i(n)-{}_k\theta_i)^2\right\}$$
$$\cdot\left(\frac{1}{2\pi\sigma^2}\right)^{k/2}\prod_{i=1}^{k} d_k\theta_i\ .$$

For this case we have

$$p(k|Y)=\exp\left\{-\frac{1}{2}S(k)\right\}\left(\frac{1}{N\sigma^2}\right)^{k/2},$$

where

$$S(k)=\sum_{i=1}^{L}\sum_{n=1}^{N}(y_i(n)-\bar{y}_i)^2+\sum_{i=k+1}^{L}N\bar{y}_i^2$$

and $\bar{y}_i$ denotes the sample mean $\sum y_i(n)/N$. The BIC statistic (1.2) of the kth model is obtained as minus twice the log posterior probability $(-2)\log\{C\pi(k)p(k|Y)\}$ and is given by

$$\text{BIC}(k)=S(k)+k\log N+R(k)\ , \tag{2.1}$$

where $R(k)=k\log\sigma^2-2\log\{C\pi(k)\}$. The corresponding AIC statistic (1.1) of the model is given by

$$\text{AIC}(k)=S(k)+2k+R\ , \tag{2.2}$$

where R is independent of k and may be ignored in the following discussion.

Taking into account the relation

$$S(k)=S(L)+\sum_{i=k+1}^{L}N\bar{y}_i^2\ , \tag{2.3}$$

it is easy to see that the MBICE, the k that minimizes BIC (k), provides a consistent estimate of the correct model. In contrast to this the MAICE, the k that minimizes AIC (k), does not have this consistency property. This is obvious, because when the k_0th model is correct the distribution of the differences of AIC (k)'s with $k=k_0, k_0+1,\cdots, L$ tends to a non-degenerate stationary distribution as N tends to infinity. Schwarz argues in Section 4 of his paper that this is a shortcoming of MAICE. Does this mean that MAICE is generally inferior to MBICE? Schwarz carefully qualifies his statement by saying that if the assumptions made in Section 2 of his paper, which are essentially equivalent to the assumptions of our present Bayesian model, are accepted MAICE cannot be optimal. In the next section it will be shown that MAICE is optimal under some assumptions which are quite different from and often more natural than those of MBICE.

3. Minimax property of MAICE

In this section we assume that the prior distribution $p({}_k\theta|k)$ of ${}_k\theta=({}_k\theta_1, {}_k\theta_2,\cdots, {}_k\theta_L)$ of the kth model is given by

$$p({}_k\theta|k)=\left(\frac{1}{2\pi\sigma^2}\right)^{k/2}\exp\left\{-\frac{1}{2\sigma^2}\sum_{i=1}^{k}{}_k\theta_i^2\right\}\left(\frac{1}{2\pi\delta^2}\right)^{(L-k)/2}$$

$$\cdot \exp\left\{-\frac{1}{2\delta^2}\sum_{i=k+1}^{L} {}_k\theta_i^2\right\} .$$

It is assumed that $N\sigma^2$ is greater than 1 and $N\delta^2$ is smaller than 1. The logarithm of the posterior distribution $p\{({}_k\theta, k)|Y\}$ of $({}_k\theta, k)$ is then given by

$$\log p\{({}_k\theta, k)|Y\} = -\frac{1}{2}\left\{\sum_{i=1}^{L} N(\bar{y}_i - {}_k\theta_i)^2 + \frac{1}{\sigma^2}\sum_{i=1}^{k} {}_k\theta_i^2 + \frac{1}{\delta^2}\sum_{i=k+1}^{L} {}_k\theta_i^2 \right.$$
$$\left. + k\log\left(\frac{\sigma^2}{\delta^2}\right)\right\} + \log \pi(k) + R ,$$

where $\pi(k)$ is the prior probability of the kth model and R denotes a quantity which is independent of k. Here we assume equal probability for $\pi(k)$ and the term $\log \pi(k)$ will be ignored. The mode of the posterior distribution is then given by $({}_k\hat{\theta}, k)$ with k that maximizes $p\{({}_k\hat{\theta}, k)|Y\}$, where ${}_k\hat{\theta}_i = \{N/(N+1/\sigma^2)\}\bar{y}_i$ for $i=1, 2, \cdots, k$, $\{N/(N+1/\delta^2)\}\bar{y}_i$ for $i=k+1, k+2, \cdots, L$. Now, minus twice $\log p\{p({}_k\hat{\theta}, k)|Y\}$ is given by

$$\text{(3.1)} \qquad \frac{1}{N\sigma^2+1}\sum_{i=1}^{k} N\bar{y}_i^2 + \frac{1}{N\delta^2+1}\sum_{i=k+1}^{L} N\bar{y}_i^2 + k\log\left(\frac{\sigma^2}{\delta^2}\right) ,$$

where a common additive constant is ignored. This formula tells that if we allow δ^2 diminish to zero the mode of the posterior distribution can only be attained at $k=0$, which is a nonsensical result and suggests the necessity of marginalization, or the integration of the posterior distribution with respect to $d_k\theta$.

By integrating $p\{({}_k\theta, k)|Y\}$ with respect to $d_k\theta$ the posterior probability $p(k|Y)$ of the kth model is obtained and minus twice its logarithm is given by

$$\text{(3.2)} \quad \mathrm{LOG}\,(k) = \frac{1}{N\sigma^2+1}\sum_{i=1}^{k} N\bar{y}_i^2 + \frac{1}{N\delta^2+1}\sum_{i=k+1}^{L} N\bar{y}_i^2 + k\log\left(\frac{N\sigma^2+1}{N\delta^2+1}\right) ,$$

where a common additive constant is ignored. For the purpose of comparison of models we use $\mathrm{CIC}\,(k) = \mathrm{LOG}\,(k) - \mathrm{LOG}\,(L)$ which is given by

$$\mathrm{CIC}\,(k) = \left(\frac{1}{N\delta^2+1} - \frac{1}{N\sigma^2+1}\right)\sum_{i=k+1}^{L} N\bar{y}_i^2 + k\log\left(\frac{N\sigma^2+1}{N\delta^2+1}\right) ,$$

where again a common additive constant is ignored. If N is large compared with σ^2 and $N\sigma^2$ is large but $N\delta^2$ is small compared with 1, $\mathrm{CIC}\,(k)$ will be approximated by $\mathrm{BIC}\,(k)$ of (2.1). These conditions can be satisfied by increasing the sample size N when δ^2 is sufficiently close or exactly equal to zero and represents the situation where the difference between the magnitudes of the elements in the set $({}_k\theta_1, {}_k\theta_2, \cdots, {}_k\theta_k)$ and those in $({}_k\theta_{k+1}, {}_k\theta_{k+2}, \cdots, {}_k\theta_L)$ is clearly visible through the observations $\bar{y}_i$ $(i=1, 2, \cdots, L)$. This is the situation which we cannot expect to hold very often in ordinary exploratory data analyses. Thus it must be concluded that MBICE will find rather limited applications.

The decision on the choice of k will be difficult when the difference between $({}_k\theta_1, {}_k\theta_2, \cdots, {}_k\theta_k)$ and $({}_k\theta_{k+1}, {}_k\theta_{k+2}, \cdots, {}_k\theta_L)$ cannot be clearly recognized through the observations $\bar{y}_i$ $(i=1, 2, \cdots, L)$. The most critical will then be the situation where $N\sigma^2$ (>1) and $N\delta^2$ (<1) are both very close to 1. For this critical situation we get

$$\text{(3.3)} \qquad \lim_{\substack{N\sigma^2 \downarrow 1 \\ N\delta^2 \uparrow 1}} \left(\frac{1}{N\delta^2+1} - \frac{1}{N\sigma^2+1} \right)^{-1} \mathrm{CIC}\,(k) = \sum_{i=k+1}^{L} N\bar{y}_i^2 + 2k\ .$$

Taking into account (2.2) and (2.3) one can see that the right-hand side of the above equation can be used as the definition of the statistic AIC (k) to be used in the definition of the minimum AIC procedure for the decision on k. Thus we get a proof of optimality of MAICE under this limiting condition. For the case with $\sigma^2=e^2N^{-1}$ and $\delta^2=e^{-2}N^{-1}$ we get a statistic with $2.63k$ in place of $2k$ of AIC (k). This result shows that for a fairly wide range of values of σ^2 and δ^2 the minimum AIC procedure will provide a reasonable approximation to the Bayes solution of the decision problem of k under the assumption of equal probability $\pi(k)$. Thus we have obtained a surprisingly simple proof of the minimax type optimality of MAICE and its robustness.

4. Discussion

In the discussion of Section 3, N was retained only to clarify the relation between AIC and BIC. For the discussion of MAICE N could have been put equal to 1. If we consider $\bar{y}_i$'s as the maximum likelihood estimates of the parameters of a distribution after a proper change of coordinate we can see that the result of the preceding section holds generally for MAICE and characterizes the procedure as optimal for the detection of k where the ratio of the signal, the bias squared, to the noise, the variance, crosses the critical value 1. This justifies the original intension of introduction of MAICE by Akaike [1], [2].

The formulas (3.1) and (3.2) demonstrate the close relation between the maximum and marginal, or integrated, likelihoods of each model. It is only when $N\sigma^2$ and $N\delta^2$ are both significantly greater than 1 that these two formulas become approximately equal, which is not a very interesting situation.

The optimality of MAICE discussed in the preceding section is only concerned with the probability of coincidence of the MAICE with the correct k. If an estimate of the correct value of k is required the mean of the approximate posterior distribution $p(k|Y)=C\exp\{(-1/2)\cdot\mathrm{AIC}(k)\}$ would be useful.

Obviously the minimum AIC procedure discussed in the preceding section is a direct extension of the classical method of maximum likelihood to the multi-model situation and is not free from the defect of ordinary point estimate. Although the AIC (k)'s are defined in terms of the maximum likelihood estimates $\bar{y}_i$ $(i=1, 2, \cdots, L)$ the discussion in the preceding section does not tell which estimate of ${}_k\theta$ should eventually be used. The use of the maximum likelihood estimates of ${}_k\theta$ under the assumption of $\sigma^2=\infty$ and $\delta^2=0$ has been customary but a further analysis in necessary when there is no such clear separation of θ_i's. If the choice of one single model is not the sole purpose of the analysis of the data the average of the models with respect to the approximate posterior probability $C\exp\{(-1/2)\,\mathrm{AIC}(k)\}$ will provide a better estimate of the true distribution of Y. In this type of application the $2k$ in the definition of AIC (k) may be adaptively modified by using the model of the prior distribution defined by $\pi(k)=C\lambda^k$ and adjusting the parameter λ by the data. The feasibility of this type of procedure has been confirmed by numerical experiments and the results will be discussed in a separate paper.

Acknowledgement

The present author is grateful to Professor Gideon Schwarz of Hebrew University for providing a chance to study his paper before its publication. The paper was one of the main stimulae which lead the present author to the present investigation. The work reported in this paper was partly supported by a grant from the Ministry of Education, Science and Culture.

THE INSTITUTE OF STATISTICAL MATHEMATICS

REFERENCES

[1] Akaike, H. (1973). Information theory and an extension of the maximum likelihood principle, *2nd International Symposium of Information Theory*, B. N. Petrov and F. Csaki, eds., Akademiai Kiado, Budapest, 267-281.

[2] Akaike, H. (1974). A new look at the statistical model identification, *IEEE Trans. Automat. Contr.*, AC-**19**, 716-723.

[3] Akaike, H. (1977). On entropy maximization principle, *Applications of Statistics*, P. R. Krishnaiah, ed., North-Holland, Amsterdam, 27-41.

[4] Schwarz, G. (1976). Estimating the dimension of a model, *Ann. Statist.*, **6**, 461-464.

Reprinted by permission of the Biometrika Trustees
Biometrika (1978), 65, 1, *pp.* 53–9
Printed in Great Britain

A new look at the Bayes procedure

BY HIROTUGU AKAIKE
Institute of Statistical Mathematics, Tokyo

SUMMARY

In developing an estimate of the distribution of a future observation it becomes natural and necessary to consider a distribution over the space of parameters. This justifies the use of Bayes procedures in statistical inference. An objective procedure of evaluation of the prior distribution in a Bayesian model is developed and the classical ignorance prior distribution is newly interpreted as the locally impartial prior distribution.

Some key words: Bayes; Entropy; Foundations of inference; Ignorance distribution; Likelihood; Predictive distribution; Prior distribution.

1. INTRODUCTION

Akaike (1974, 1977) advocates the formulation of statistical inference as the problem of determining the probability distribution $f(y)$ of a future observation y. From this point of view, the problem of parameter estimation is seen as that of determining a probability distribution specified by the parameter. As a natural choice of the gain function for taking a distribution $g(y)$ as an estimate of the true distribution $f(y)$ we adopt the entropy of $f(y)$ with respect to $g(y)$ defined by

$$B(f,g) = -\int \frac{f(y)}{g(y)} \log \left\{\frac{f(y)}{g(y)}\right\} g(y)\, dy,$$

where $f(y)$ and $g(y)$ are the density functions with respect to a properly chosen measure dy. The entropy $B(f,g)$ is asymptotically proportional to the logarithm of the probability of getting a sample distribution closely approximated by the distribution $f(y)$ when a very large number of observations are independently drawn from the distribution $g(y)$ (Sanov, 1961); thus this choice of the gain function seems reasonable. We are interested in finding $g(y)$ which will maximize the entropy $B(f,g)$. Since $B(f,g) = \int f(y) \log g(y)\, dy - \int f(y) \log f(y)\, dy$, we are interested in maximizing $E_y\{\log g(y)\} = \int f(y) \log g(y)\, dy$, the expected log likelihood of $g(y)$.

When a parametric family $\{g(y \mid \theta)\}$ is given, assume that a randomized decision procedure specifies a parameter θ by using a distribution $p(\theta \mid x)$ based on the observation x. We call $p(\theta \mid x)$ the inferential distribution. By Jensen's inequality we have

$$\log [E_{\theta|x}\{g(y \mid \theta)\}] \geqslant E_{\theta|x}\{\log g(y \mid \theta)\},$$

and thus

$$E_x E_{y|x}[\log \{E_{\theta|x}\, g(y \mid \theta)\}] \geqslant E_x E_{y|x} E_{\theta|x}\{\log g(y \mid \theta)\},$$

where $E_{\theta|x}$ and $E_{y|x}$ respectively denote the expectations with respect to the distributions $p(\theta \mid x)$ and $f(y \mid x)$, the conditional distribution of y on x, and E_x denotes the expectation with respect to the distribution of x. Thus, on average, the 'predictive distribution' defined by $E_{\theta|x}\{g(y \mid \theta)\}$ provides a better estimate of $f(y \mid x)$ than a particular $g(y \mid \theta)$ with θ randomly chosen by the inferential distribution $p(\theta \mid x)$. This explains why we should concentrate on the use of the inferential distribution rather than a point estimate. Hereafter we assume that x has density $h(x)$ and limit our attention to the case where $f(y \mid x) = f(y)$.

By applying Jensen's inequality again we get

$$E_y(\log[E_x E_{\theta|x}\{g(y\,|\,\theta)\}]) \geqslant E_y E_x(\log[E_{\theta|x}\{g(y\,|\,\theta)\}])$$

and this leads to the decomposition

$$E_y\{\log f(y)\} - E_y E_x(\log[E_{\theta|x}\{g(y\,|\,\theta)\}]) = B\{p(\theta\,|\,x)\} + V\{p(\theta\,|\,x)\},$$

where

$$B\{p(\theta\,|\,x)\} = E_y\{\log f(y)\} - E_y(\log[E_x E_{\theta|x}\{g(y\,|\,\theta)\}]),$$

$$V\{p(\theta\,|\,x)\} = E_y(\log[E_x E_{\theta|x}\{g(y\,|\,\theta)\}]) - E_y E_x(\log[E_{\theta|x}\{g(y\,|\,\theta)\}]).$$

The term $B\{p(\theta\,|\,x)\}$ represents the deviation of the best possible predictive distribution $E_x E_{\theta|x}\{g(y\,|\,\theta)\}$ from $f(y)$, and $V\{p(\theta\,|\,x)\}$ represents the further deterioration due to the variation of $E_{\theta|x}\{g(y\,|\,\theta)\}$ around $E_x E_{\theta|x}\{g(y\,|\,\theta)\}$. By obvious analogy with the classical concepts these two terms $B\{p(\theta\,|\,x)\}$ and $V\{p(\theta\,|\,x)\}$ may be called the bias and the variance of the predictive distribution $E_{\theta|x}\{g(y\,|\,\theta)\}$, or simply of the inferential distribution $p(\theta\,|\,x)$. With our present formulation the problem of statistical inference reduces to the choice of an inferential distribution $p(\theta\,|\,x)$ that will produce a good balance between the 'bias' and the 'variance'.

Consider the idealized situation where the distributions $f(y)$ and $h(x)$ are chosen randomly from the families $\{g(y\,|\,\theta)\}$ and $\{i(x\,|\,\theta)\}$, respectively, by a prior distribution $p(\theta)$. We want to find a distribution $p(\theta\,|\,x)$ which maximizes the average gain

$$\iiint g(y\,|\,\theta)\, i(x\,|\,\theta) \log\left\{\int g(y\,|\,\theta')\, p(\theta'\,|\,x)\, d\theta'\right\} dy\, dx\, p(\theta)\, d\theta.$$

This average is equal to

$$\iiint g(y\,|\,\theta)\, i(x\,|\,\theta)\, p(\theta)\, d\theta \log\left\{\int g(y\,|\,\theta')\, p(\theta'\,|\,x)\, p(x)\, d\theta'\right\} dy\, dx$$

$$- \iiint g(y\,|\,\theta)\, i(x\,|\,\theta)\, p(\theta)\, d\theta \log p(x)\, dy\, dx,$$

where $p(x) = \int i(x\,|\,\theta)\, p(\theta)\, d\theta$. As noted by Aitchison (1975), the desired maximum is attained with $p(\theta\,|\,x)\, p(x) = i(x\,|\,\theta)\, p(\theta)$. Thus the optimal choice of $p(\theta\,|\,x)$ is realized by following the Bayes procedure. It looks as if this result proves that an ultimate solution to the problem of statistical inference is produced by the Bayes procedure. Unfortunately this is not the case. The basic assumption here is that the true distributions $f(y)$ and $h(x)$ are obtained by random sampling from known families $\{g(y\,|\,\theta)\}$ and $\{i(x\,|\,\theta)\}$. Even when this assumption is accepted the problem of specification of $p(\theta)$ remains.

In spite of the apparent utility of Bayesian models in statistical inference no systematic analysis of the role played by the prior distributions has been developed to the point that it can give practising statisticians perfect confidence in using their own prior distributions. Even in the very useful review of the Bayesian method by Lindley (1972) the problem of probability assessments, essentially of the prior distribution, is apparently dismissed as a matter more for a psychologist than a mathematician. Since the subjectivist approach to Bayesian modelling apparently has lost ground (Hacking, 1967), we specify the purpose of the Bayes procedure as the construction of a posterior distribution which will be useful as an inferential distribution and try to develop procedures of evaluation and selection of the prior distribution.

2. Objective evaluation of prior distributions

The result given towards the end of the preceding section does not really justify the general use of the Bayes procedure for the situation assumed there, since the averaged gain function may not be an adequate criterion for the evaluation of a statistical decision. Specifically, when an observation x is made, we know that the corresponding parameter θ is already fixed at some value θ_0, say, and we are only interested in $g(y\,|\,\theta_0)$ and not in the possibilities of other $g(y\,|\,\theta)$'s. We want some security for each individual possible value of θ_0, rather than assurance on the average. Usually this objective is not fully attainable and we have to satisfy ourselves by choosing a procedure which produces satisfactory results for those θ_0's within some limited range. This observation suggests the use of the performance characteristic function of an inferential distribution $p(\theta\,|\,x)$ defined by

$$U(\theta_0) = \iint g(y\,|\,\theta_0)\, i(x\,|\,\theta_0) \log\left\{\int g(y\,|\,\theta)\, p(\theta\,|\,x)\, d\theta\right\} dy\, dx - \int g(y\,|\,\theta_0) \log\{g(y\,|\,\theta_0)\}\, dy.$$

If the posterior distribution obtained by the Bayes procedure is to be used as an inferential distribution $p(\theta\,|\,x)$ we compare various possible prior distributions by using the information supplied by their performance characteristic functions.

To see how this idea works suppose that $g(y\,|\,\theta) = {}_nC_y\, \theta^y(1-\theta)^{n-y}$ $(0 \leqslant \theta \leqslant 1)$ and that the true distribution is $f(y) = {}_nC_y\, \theta_0^y(1-\theta_0)^{n-y}$. We assume that $i(x\,|\,\theta) = g(x\,|\,\theta)$ and $h(x) = f(x)$. If we denote by $\theta(x)$ the maximum likelihood estimate of θ_0 based on the observation x, we have

$$E_x E_y(\log[g\{y\,|\,\theta(x)\}]) = E_x[n\theta_0 \log \theta(x) + n(1-\theta_0)\log\{1-\theta(x)\} + E_y(\log {}_nC_y)].$$

If $n = 2$, we have $\theta(0) = 0$, $\theta(1) = 0{\cdot}5$ and $\theta(2) = 1$. Thus unless $\theta_0 = 0$ or 1 the gain function takes the value minus infinity. This demonstrates the limitation of the method of maximum likelihood in this situation. By assuming the uniform prior distribution $p(\theta)\,d\theta = d\theta$ $(0 \leqslant \theta \leqslant 1)$ the Bayes procedure produces an inferential distribution $p(\theta\,|\,x)$ which is the beta distribution $B(x+1, n-x+1)^{-1}\theta^x(1-\theta)^{n-x}$. The corresponding predictive distribution is

$$E_{\theta|x}\{g(y\,|\,\theta)\} = \int g(y\,|\,\theta)\, p(\theta\,|\,x)\, d\theta = ({}_{x+y}C_x)\,({}_{2n-x-y}C_{n-x})\,({}_{2n+1}C_n)^{-1}.$$

Table 1 shows, for the case $n = 2$, the values of the performance characteristic function at various values of the true parameter θ_0. Contrary to our vague notion that the uniform prior distribution will represent noncommitment about the true parameter θ_0, the Bayes procedure based on the uniform prior distribution is definitely acting unfavourably towards those θ_0's which are close to 0 or 1. We now apply the same analysis to Jeffreys's ignorance prior distribution $\{B(\frac{1}{2},\frac{1}{2})\}^{-1}\theta^{-\frac{1}{2}}(1-\theta)^{-\frac{1}{2}}$ discussed by Good (1965, p. 18). Table 2 shows the performance characteristic. It can be seen that, compared with the uniform prior distribution,

Table 1. *Performance characteristic of the inferential distribution generated by the uniform prior distribution; binomial experiment, $n = 2$*

	$\theta_0 = 0{\cdot}5$	$\theta_0 = 0{\cdot}75$ or $0{\cdot}25$	$\theta_0 = 0{\cdot}95$ or $0{\cdot}05$	$\theta_0 = 0{\cdot}99$ or $0{\cdot}01$
$U(\theta_0)$	$-0{\cdot}143$	$-0{\cdot}154$	$-0{\cdot}313$	$-0{\cdot}440$

Table 2. *Performance characteristic of the inferential distribution generated by Jeffreys's ignorance prior distribution; binomial experiment, $n = 2$*

	$\theta_0 = 0{\cdot}5$	$\theta_0 = 0{\cdot}75$ or $0{\cdot}25$	$\theta_0 = 0{\cdot}95$ or $0{\cdot}05$	$\theta_0 = 0{\cdot}99$ or $0{\cdot}01$
$U(\theta_0)$	$-0{\cdot}364$	$-0{\cdot}210$	$-0{\cdot}181$	$-0{\cdot}259$

the ignorance prior distribution is acting definitely unfavourably towards those θ_0's which are close to 0·5. The results illustrated in Tables 1 and 2 do not exhibit any special features which uniquely characterize these two prior distributions as the representations of our 'ignorance' of θ_0.

Hereafter, as in the present example, we will consider only the special situation where $h(x) = f(x)$ and $i(x \mid \theta) = g(x \mid \theta)$.

3. Construction of impartial prior distributions

When a parameterized prior distribution $p(\theta \mid \sigma)$ is used in a Bayes procedure a choice of the optimum value of σ may be realized by the method of type II maximum likelihood of Good (1965), i.e. by maximizing the averaged likelihood $g(x \mid \sigma) = \int g(x \mid \theta)\, p(\theta \mid \sigma)\, d\theta$. It is natural to expect that this procedure will work if $\log g(x \mid \sigma)$ is a good estimate of $E_x\{\log g(x \mid \sigma)\}$. Usually the averaging will reduce the variability of $\log g(x \mid \sigma)$ as an estimate of $E_x\{\log g(x \mid \sigma)\}$, but the maximum expected log likelihood is reduced from $\max_\theta E_x\{\log g(x \mid \theta)\}$ to $\max_\sigma E_x\{\log g(x \mid \sigma)\}$. Thus $p(\theta \mid \sigma)$ must be chosen in such a way that, while the reduction of the statistical variability is significant, the reduction of the expected log likelihood will be kept to a minimum. Such a choice of $p(\theta \mid \sigma)$ is sometimes realized by using the prior information of the physical nature of θ; see, for example, Lindley & Smith (1972).

If the variability of this procedure is still expected to be significant we can further smooth $g(x \mid \sigma)$ by $p(\sigma \mid \tau)$, a probability distribution of σ with a parameter τ. By repeating the process we come to the point where no further meaningful parameterization of the distribution of the final parameter is conceivable. At this point two procedures are possible to stop this process of regression. The first is to fix the value of the final parameter at the value which maximizes the final averaged likelihood. The second is to assume explicitly some reasonably defined prior distribution of the parameter. The first procedure is essentially the type II maximum likelihood procedure and its use can be evaluated by the performance characteristic function of the overall procedure. We will discuss the second procedure in this section.

As shown in § 2, a uniform prior distribution does not necessarily assure the uniform performance of the corresponding Bayes procedure. The implicit requirement of an ignorance prior distribution is the uniformity or the impartiality of the performance of the corresponding Bayes procedure with respect to the values of the true parameter. This observation suggests the following procedure for the construction of a practically impartial prior distribution. Find a prior distribution which will maximize the minimum of the performance characteristic function within a prescribed set of parameters and use it as the impartial prior distribution. It seems that at least theoretically there is no serious problem in implementing the above procedure in a particular situation. Also with the present objective procedure of evaluation of the prior distribution it seems that objectivist statisticians can now use Bayesian models without the sense of dishonesty pointed out by Pearson (1964). Our attempt here has been to place the Bayes procedure under the control of what Cox & Hinkley (1974, p. 45) call the strong repeated sampling principle.

As is well known, the concept of ignorance is closely related to the choice of the coordinate system for the parameterization. The coordinate system where the uniform distribution exactly represents an impartial prior distribution is the one, if any, where the structure of the neighbourhood of each possible value θ of the true parameter is homogeneous everywhere. Here the deviation of a parameter θ' from θ is measured by the negative-entropy $N\{g(.\mid \theta'), g(.\mid \theta)\} = -B\{g(.\mid \theta'), g(.\mid \theta)\}$. To see the possibility of existence of such a homogeneous coordinate system we consider the case where $g(y \mid \theta) = (2\pi\sigma^2)^{-\frac{1}{2}} \exp\{-\frac{1}{2}(y-m)^2/\sigma^2\}$

with $\theta = (m, \sigma)$. We have

$$2N\{g(.\,|\,\theta'), g(.\,|\,\theta)\} = 2\int g(y\,|\,\theta')\log\left\{\frac{g(y\,|\,\theta')}{g(y\,|\,\theta)}\right\}dy$$

$$= \log\sigma^2 - \log\sigma'^2 - 1 + \frac{\sigma'^2}{\sigma^2} + \frac{(m'-m)^2}{\sigma^2}.$$

Because $\sigma'/\sigma = \exp(\log\sigma' - \log\sigma)$, the deviation of θ' from θ due to the deviation of σ' from σ is measured by a monotonic function of the difference $\log\sigma' - \log\sigma$. The deviation of θ' from θ due to the deviation of m' from m is measured by a monotonic function of the difference $(m'/\sigma) - (m/\sigma)$ and is evaluated additively. Thus it is clear that if we use

$$\phi = (\phi_1, \phi_2) = (m/\sigma, \log\sigma)$$

for the parametric representation of $g\{y\,|\,(m, \sigma)\}$ the space of ϕ exhibits a homogeneous structure, where the uniform measure $d\phi_1 d\phi_2$ will specify an impartial improper prior distribution.

It should be mentioned here that we are only interested in the use of the posterior distribution as an inferential distribution and the use of the improper prior distribution is admitted only if the formal application of the Bayes procedure produces a meaningful posterior distribution. Obviously the present $d\phi_1 d\phi_2$ corresponds to $(dm/\sigma)(d\sigma/\sigma)$ in the original parameterization and the gain function of the corresponding posterior distribution is given in the form

$$E_x E_y(n\log[\{S_x + S_y + \tfrac{1}{2}n(\bar{x}-\bar{y})^2\}(\sigma_0 S_x^{\frac{1}{2}})^{-1}]) + \text{const},$$

where S_x and $\bar{x}$ denote $\Sigma(x_i - \bar{x})^2$ and $\Sigma x_i/n$, respectively, and σ_0 denotes the true value of σ. The gain function is independent of the values of the true parameters. This shows that $dm\, d\sigma/\sigma^2$ satisfies the basic requirement of an impartial prior distribution. It is of some interest to note that $dm\, d\sigma/\sigma$ also provides an impartial result. This suggests the necessity of further analysis.

The fact that a perfectly impartial prior distribution does not always exist can easily be recognized by the analysis of the simple binomial experiment with $g(y\,|\,\theta) = {}_nC_y\,\theta^y(1-\theta)^{n-y}$. For this case we have the negative entropy

$$N\{g(.\,|\,\theta'), g(.\,|\,\theta)\} = n\{\theta'\log\theta' + (1-\theta')\log(1-\theta') - \theta'\log\theta - (1-\theta')\log(1-\theta)\}$$

and we cannot expect a neat separation of θ' and θ which assures a homogeneous reparameterization. When both θ' and θ are close to 1 or 0 the variation of the negative entropy is dominated by $n(1-\theta')\{\log(1-\theta') - \log(1-\theta)\}$ or by $n\theta'(\log\theta' - \log\theta)$. This suggests the reparameterizations by $\log(1-\theta)$ and $\log\theta$ with the measures $(1-\theta)^{-1}d\theta$ and $\theta^{-1}d\theta$, respectively. These prior distributions are discussed by Thatcher (1964) for prediction in binomial samples. Our present derivation shows that the prior distributions $(1-\theta)^{-1}d\theta$ and $\theta^{-1}d\theta$ may act as impartial prior distributions only when θ is close to 1 and 0, respectively. This is rather remarkable, since it is quite contrary to the common interpretation that the prior distributions $(1-\theta)^{-1}d\theta$ and $\theta^{-1}d\theta$ represent the prior knowledge that the true value of θ is close to 1 and 0, respectively. This latter interpretation obviously contradicts the use of these prior distributions as ignorance prior distributions. It seems that the nonexistence of the globally impartial prior distribution is the cause of the dispute of the prediction problem of binomial samples discussed by Thatcher (1964).

The difficulty of getting a globally impartial prior distribution suggests the necessity of introducing locally impartial prior distributions. As a function of θ' the negative entropy

$N\{g(.|\theta'), g(.|\theta)\}$ attains the minimum value 0 at $\theta' = \theta$. Thus for $\theta = (\theta_1, \ldots, \theta_p)$ we have asymptotically for $N = N\{g(.|\theta'), g(.|\theta)\}$,

$$2N\{g(.|\theta'), g(.|\theta)\} = \sum_{i=1}^{p} \sum_{j=1}^{p} \Delta\theta_i \Delta\theta_j \left(\frac{\partial^2 N}{\partial\theta_i' \partial\theta_j'}\right)_{\theta'=\theta},$$

where $\Delta\theta_i = \theta_i' - \theta_i$ and the derivative is taken with respect to θ' and evaluated at $\theta' = \theta$. A locally homogeneous coordinate system ϕ could then be defined in such a way that $\partial^2 N/\partial\phi_i \partial\phi_j = 1$, for $i = j$, and 0, otherwise. The corresponding locally and sometimes globally impartial prior distribution would then be given by

$$d\phi_1 \ldots d\phi_p = \left| \left(\frac{\partial^2 N}{\partial\theta_i' \partial\theta_j'}\right)_{\theta'=\theta} \right|^{\frac{1}{2}} d\theta_1 \ldots d\theta_p,$$

where $|A|$ denotes the determinant of a matrix A. Because the negative entropy is identical to Kullback's information quantity, we know that $(\partial^2 N/\partial\theta_i' \partial\theta_j')_{\theta'=\theta}$ is equal to Fisher's information matrix (Kullback, 1959, pp. 26–8). Thus the locally impartial prior distribution obtained above is identical to the ignorance prior distribution introduced by Jeffreys (1961, § 3·10) by invariance theory. Our present approach shows that generally the impartial property of an ignorance prior distribution can be expected only locally and thus its overall performance must be checked through the analysis of the performance characteristic function of the corresponding posterior distribution. The ignorance prior distribution for the binomial experiment with $n = 2$ is given by $B(\frac{1}{2}, \frac{1}{2})^{-1} \theta^{-\frac{1}{2}}(1-\theta)^{-\frac{1}{2}}$ and its partial characteristic is already noted in § 2.

4. Discussion

Although ignorance prior distributions may only be locally impartial yet by adhering to the present interpretation of the ignorance prior distribution as an impartial prior distribution, some of the arguments previously developed against ignorance prior distributions can be dismissed rather simply. Mitchell (1967) discusses the difficulty of introducing a prior distribution of the parameters of an exponential regression $E(y)_x = \alpha + \beta\rho^x$, with $x = x_0 + is$; $i = 0, \ldots, k-1$; $s > 0$ and known. Here y_x is assumed to be independently normally distributed with mean $E(y)_x$ and variance σ^2. By assuming the prior distribution $d\rho\, d\alpha\, d\beta\, d\sigma/\sigma$ she obtains a peculiar posterior distribution $\{\rho^{x_0}(1-\rho^s) f(\rho, y)\}^{-1}$, where $f(\rho, y)$ is bounded and has no zeros. It is mentioned that Jeffreys's invariance theory leads to answers which are unacceptable on commonsense ground. Now it is easy to see that the square root of the determinant of the corresponding Fisher's information matrix contains $\rho^{x_0}(1-\rho^s)$ as its factor. Thus if Jeffreys's invariance theory was followed consistently and the use was made of the above defined ignorance prior distribution, that peculiar posterior distribution with infinite concentration of probability at $\rho = 0$ and 1 would not have appeared. This clearly demonstrates the superiority of the objectively defined ignorance prior distribution to the intuitively chosen prior distribution $d\rho\, d\alpha\, d\beta\, d\sigma/\sigma$.

Another problem with the ignorance prior distribution is that it is often improper. The un-Bayesian implication of the use of an improper prior distribution is discussed by Dawid, Stone & Zidek (1973) in relation to the so-called marginalization paradox, i.e. the inconsistency between the marginal posterior distribution and the posterior distribution based on the marginal likelihood. From the present point of view of the use of the Bayes procedure the role of the posterior distribution is to provide an inferential distribution to be used for the definition of a predictive distribution. Obviously the marginal posterior distribution is useless for this purpose and we can see that the process of marginalization itself is illegitimate for the present use of the Bayesian model.

It may also be argued that ignorance prior distributions often do not produce very useful results. In our present approach to Bayesian modelling the ignorance or impartial prior distribution is to be used only at the last stage of the modelling. By definition it cannot, and in fact should not, modify significantly the information provided by the corresponding likelihood. With this understanding of the use of the locally or globally impartial prior distribution the classical concept of ignorance prior distribution will be able to survive in the future use of Bayesian models.

The author is grateful to Professor Y. C. Ho of Harvard University for drawing his attention to the use of the minimax strategy in a stochastic situation. Thanks are due to Professor D. R. Cox and the referees for their suggestions which led to a drastic revision of the paper. The present work was partly supported by the Vinton Hayes Senior Fellowship at the Division of Engineering and Applied Physics, Harvard University, and by a grant from the Ministry of Education, Science and Culture, at the Institute of Statistical Mathematics, Tokyo.

References

AITCHISON, J. (1975). Goodness of prediction fit. *Biometrika* **62**, 547–54.

AKAIKE, H. (1974). A new look at the statistical model identification. *I.E.E.E. Trans. Auto. Control* AC-**19**, 716–23.

AKAIKE, H. (1977). On entropy maximization principle. *Applications of Statistics*, Ed. P. R. Krishnaiah, pp. 27–41. Amsterdam: North-Holland.

COX, D. R. & HINKLEY, D. V. (1974). *Theoretical Statistics*. London: Chapman & Hall.

DAWID, A. P., STONE, M. & ZIDEK, J. V. (1973). Marginalization paradoxes in Bayesian and structural inference (with discussion). *J. R. Statist Soc.* B **35**, 189–233.

GOOD, I. J. (1965). *The Estimation of Probabilities*. Massachusetts Institute of Technology Press.

HACKING, I. (1967). Slightly more realistic personal probability. *Phil. of Science* **34**, 311–25.

JEFFREYS, H. (1961). *Theory of Probability*, 3rd edition. Oxford University Press.

KULLBACK, S. (1959). *Information Theory and Statistics*. New York: Wiley.

LINDLEY, D. V. (1972). *Bayesian Statistics: A Review*. Philadelphia: Society for Industrial and Applied Mathematics.

LINDLEY, D. V. & SMITH, A. F. M. (1972). Bayes estimates for the linear model (with discussion). *J. R. Statist. Soc.* B **34**, 1–41.

MITCHELL, A. F. S. (1967). Contribution to discussion of the paper by I. J. Good. *J. R. Statist. Soc.* B **39**, 423–4.

PEARSON, E. S. (1964). Contribution to discussion of the papers by J. Aitchison and A. R. Thatcher. *J. R. Statist. Soc.* B **26**, 196.

SANOV, I. N. (1961). On the probability of large deviations of random variables. *Selected Transl. Math. Statist. Prob.* **1**, 213–44.

THATCHER, A. R. (1964). Relationships between Bayesian and confidence limits for predictions (with discussion). *J. R. Statist. Soc.* B **26**, 176–210.

[*Received November* 1976. *Revised September* 1977]

Reprinted from *The Statistician*, Vol. 27, 1978, 217-235
by permission from Blackwell Publishers

On the Likelihood of a Time Series Model

HIROTUGU AKAIKE
Institute of Statistical Mathematics, 4-6-7 Minami-Azabu, Minato-ku, Tokyo 106, Japan

The conventional approach to parametric model fitting of time series is realized through the comparison of various competing models by some ad hoc *criterion. Since each of the models is usually specified by the parameters determined by the information from the data, the extension of the classical concept of likelihood to this situation is not obvious. By asking the log likelihood of a model to be an unbiased estimate of the expected log likelihood of the model, a reasonable definition of the likelihood is obtained and this allows us to develop a systematic approach to parametric time series modelling. Practical utility of this approach is demonstrated by numerical examples.*

Introduction

A typical difficulty in time series analysis is the choice of an appropriate model. Although this difficulty is common to other statistical analyses, it is especially significant in the case of time series analysis. The historical success of the book by Box and Jenkins (1970) is undoubtedly due to its contribution to this art of time series modelling through the elucidation of the stages of identification, estimation and diagnostic checking.

By the Box and Jenkins approach the process of model building is iterative. Thus, at some stage, we have at hand several models which are the candidates for our final choice. In this situation each model is specified by the parameters determined from the data under consideration. If we have at hand a set of models, each of which is specified by a fixed vector of parameters, then, assuming equal prior weight for each model, the choice of the best model will simply be realized by picking the one with the highest value of the likelihood. Thus it is obvious that the difficulty in making the decision on the best choice of a model is due to the unavailability of the "likelihood" of a model which is specified by the information from the data.

Recently Akaike (1978d) introduced a definition of the likelihood of a model which is determined by data. This definition is based on the observation that the log likelihood is the basic quantity for measuring the goodness of fit of a model. The likelihood of a model is so defined that its logarithm will be an unbiased estimate of the expected log likelihood of the model with respect to a future observation, where the expectation is taken with respect to the distribution of the present and future observations.

The purpose of the present paper is to show the practical use of this concept of likelihood of a model in the case of time series modelling. Since the definition of the likelihood is based on a heuristic argument, the only justification for its use will come from its performance in applications. The use of the likelihood is therefore demonstrated by applying it to several real and artificial time series. The results will prove that the introduction of the likelihood enables us to develop systematic procedures of time series model fitting, which will relieve time series analysts from much of the burden of making subjective judgements at the final stage of model selection.

Likelihood of a Model

Assume that the data set x is given. We specify the purpose of statistical analysis of x as the prediction of future observations y whose distribution is identical to that giving the elements of x. The prediction is realized by specifying a distribution $p(y|x)$, the predictive distribution of y as a function of the available data x. Assume that the true distribution of y is given by $p(y)$. The goodness of $p(y|x)$ as an estimate of $p(y)$ is measured by the entropy of $p(y)$, with respect to $p(y|x)$, which is defined by

$$B\{p(\cdot); p(\cdot|x)\} = -\int \left\{\frac{p(y)}{p(y|x)}\right\} \log \left\{\frac{p(y)}{p(y|x)}\right\} p(y|x)\,dy$$

For a justification of the use of entropy as the measure of goodness of fit, see the Appendix. On rewriting

$$B\{p(\cdot); p(\cdot|x)\} = \int p(y) \log p(y|x)\,dy - \int p(y) \log p(y)\,dy$$

we have the first term on the right-hand side determining the goodness of $p(y|x)$ as an estimate of $p(y)$. This term can be represented as $E_y \log p(y|x)$, where E_y denotes the expectation with respect to the distribution of y. The goodness of the estimation procedure specified by $p(y|x)$ is then measured by $E_x E_y \log p(y|x)$ which is the expected log likelihood of the model $p(\cdot|x)$ with respect to a future observation y.

Naturally it is impossible to evaluate $E_x E_y$ and we want to use $\log p(x|x)$ as an estimate of $E_x E_y \log p(y|x)$. When $p(\cdot|x) = p(\cdot|\boldsymbol{\theta})$, a distribution specified by a fixed parameter vector $\boldsymbol{\theta}$, we have $\log p(x|x) = \log p(x|\boldsymbol{\theta})$. This is exactly the classical definition of the log likelihood of the model specified by $p(\cdot|\boldsymbol{\theta})$, and we have $E_x \log p(x|\boldsymbol{\theta}) = E_x E_y \log p(y|x)$. For a general $p(\cdot|x)$, $E_x \log p(x|x)$ is not usually equal to

$E_xE_y \log p(y|x)$, and we propose that the log likelihood of the model $p(\cdot|x)$ be defined by

$$l\{p(\cdot|x)\} = \log p(x|x) + C$$

where C is a constant such that

$$E_x l\{p(\cdot|x)\} = E_xE_y \log p(y|x)$$

Obviously the definition is meaningful only when we restrict our attention to a family of possible $p(y)$'s such that C is a constant for the members of that family. Although the present definition of the likelihood of a model can be applied to any $p(y|x)$, including the predictive distribution obtained by a Bayes procedure, we are particularly interested in the case where the predictive distribution $p(y|x)$ is given in the form $p(y|x) = p(y|\theta(x))$ and $\theta(x)$ is the maximum likelihood estimate of θ defined by

$$p(x|\theta(x)) = \operatorname*{Max}_{\theta} p(x|\theta)$$

When the true distribution $p(y)$ is given by $p(y|\theta_0)$ we have, under certain regularity conditions, asymptotic equalities

$$2 \log p(x|\theta(x)) - 2 \log p(x|\theta_0) \sim \chi_k^2 \tag{1}$$

and

$$E_y\{2 \log p(y|\theta_0) - 2 \log p(y|\theta(x))\} \sim \chi_k^2 \tag{2}$$

where $\sim$ denotes the asymptotic equality of the distributions of the variables and χ_k^2 denotes a chi-squared with the degrees of freedom equal to k, the dimension of θ. Asymptotically, the two variables defined by the left-hand sides of the above formulae tend to be numerically equal. Assuming the existence of the expectations, we have asymptotic equalities

$$2\{E_x \log p(x|\theta(x)) - E_x \log p(x|\theta_0)\} = k$$

and

$$2\{E_y \log p(y|\theta_0) - E_xE_y \log p(y|\theta(x))\} = k$$

By adding the two equalities, and taking into account the relation $E_y \log p(y|\theta_0) = E_x \log p(x|\theta_0)$, we get

$$-2E_xE_y \log p(y|\theta(x)) = -2E_x \log p(x|\theta(x)) + 2k$$

This shows that asymptotically

$$l\{p(\cdot|\theta(x))\} = \log p(x|\theta(x)) - k$$

satisfies the condition of the log likelihood of the model $p(\cdot|\theta(x))$. The right-hand side is identical to $-(1/2)$ AIC, where AIC stands for an information criterion (Akaike, 1974) defined by

$$\text{AIC} = -2 \log (\text{maximum likelihood}) + 2 (\text{number of parameters})$$

This result suggests that exp $\{-(1/2)\ \text{AIC}\}$ will asymptotically be a reasonable definition for the likelihood of a model specified by the parameters determined by the method of maximum likelihood. For the original derivation of AIC the reader is referred to Akaike (1973).

Stationary Gaussian Models

In the case of a time series model the computation of the exact likelihood is usually time consuming. This is due to the fact that, even for a stationary model, the finiteness of data demands non-homogeneous treatment of each data point. Here we consider the fitting of stationary Gaussian models and adopt an approximation to the exact likelihood. This approximation is discussed in detail in Akaike (1976).

Assume that a set of scalar observations $\{y(n);\ n=1, 2, \ldots, N\}$ is given. For the sake of simplicity we assume that the $y(n)$'s are measured as deviations from the sample mean of the original record. Define the sample autocovariances $C(k)$, $k=0, \pm 1, \pm 2, \ldots,$ by

$$C(k)=\begin{cases} \dfrac{1}{N}\sum_{n=1}^{N-k} y(n+k)y(n) & (0 \le k \le N-1) \\ C(-k) & (-N+1 \le k < 0) \\ 0 & (N \le |k|) \end{cases}$$

Obviously $\{C(k)\}$ defines a positive definite sequence and so there exists a stationary Gaussian process with zero mean and autocovariance sequence equal to $\{C(k)\}$. When a stationary ARMA model

$$y(n)+b_1y(n-1)+\ldots+b_My(n-M) \\ =x(n)+a_1x(n-1)+\ldots+a_Lx(n-L)$$

is considered, where $\{x(n)\}$ is a white noise process, this relation defines a constant linear filter which transforms the sequence $\{y(n)\}$ into $\{x(n)\}$. The power, or equivalently the variance, of the output $x(n)$ when we apply the filter to the stationary Gaussian process with zero mean and autocovariance sequence $\{C(k)\}$ will be denoted by $S(a_1, a_2, \ldots, a_L, b_1, b_2, \ldots, b_M)$. The approximate maximum likelihood estimates of the coefficients are obtained by minimizing $\log S(a_1, \ldots, a_L, b_1, \ldots, b_M)$, with respect to $a_1, \ldots, a_L, b_1, \ldots, b_M$, and the estimate of the variance of the white noise is given by the corresponding value of $S(a_1, \ldots, a_L, b_1, \ldots, b_M)$. The AIC for the comparison of ARMA models is then defined by

$$\text{AIC}=N \log [\text{minimum } S(a_1, \ldots, a_L, b_1, \ldots, b_M)]+2(L+M)$$

A smaller value of AIC means a better fit for the model. A computer program based on this type of approximate likelihood is included in the time series analysis and control program package TIMSAC-74 developed by Akaike *et al* (1975, 1976).

To show the behaviour of AIC, the present procedure was applied to a simulated realization of length $N = 1\,000$ from an ARMA process defined by

$$y(n) - 0{\cdot}80\, y(n-1) = x(n) - 0{\cdot}85\, x(n-1)$$

The spectra corresponding to the fitted models are illustrated in Figure 1 together with the theoretical spectrum. We can see significant drops in the values of AIC between ARMA(0, M) and ARMA(1, M) and also between ARMA(L, 0) and ARMA(L, 1). Within the 25 models, AIC attains the minimum $-56{\cdot}28$ at ARMA(1, 1) which is the true structure incidentally. We can see the wiggly behaviour of the estimated spectra for the more over-parametrized models. The inferiority of these over-parametrizations is reflected by the increases of their AICs or, equivalently, by the decreases in the likelihoods for these models. The effect of under-parametrization is reflected by the very significant increases in the AICs for those models which lack either an AR or MA part. Although the attainment of the minimum for the AIC at the correct structure may be considered as only really due to the very large sample size chosen, the overall behaviour of the AIC in Figure 1 is the sort which might be considered typical for smaller choices of N.

To check the performance of AIC with real data, it was applied to the ARMA model fitting of Wölfer's sunspot number series for the years 1749–1924, as given in Anderson (1971, p. 660). The data length N is 176. The spectra of the ARMA(M, L) models with minimum AICs for $L = 0, 2, 4, 6, 8$ are illustrated in Figure 2, along with the periodogram or the Fourier transform of the autocovariance sequence $\{C(k)\}$. The minimum of the AIC is attained by ARMA(7, 2) which, on comparison with the periodogram, seems to be a reasonable choice. The application of AIC to the examples given in the book by Box and Jenkins is discussed by Ozaki (1977).

The definition of the approximate likelihood can be extended to the case of a multivariate Gaussian process (Akaike, 1976). Figure 3 illustrates the spectra for the AR and Markovian models fitted to the pair of Series J, from Box and Jenkins (1970), which give the minimum values for the AIC. A Markovian model of a stationary (vector) Gaussian process $y(n)$ is given by

$$z(n) = Fz(n-1) + Gx(n)$$
$$y(n) = Hz(n)$$

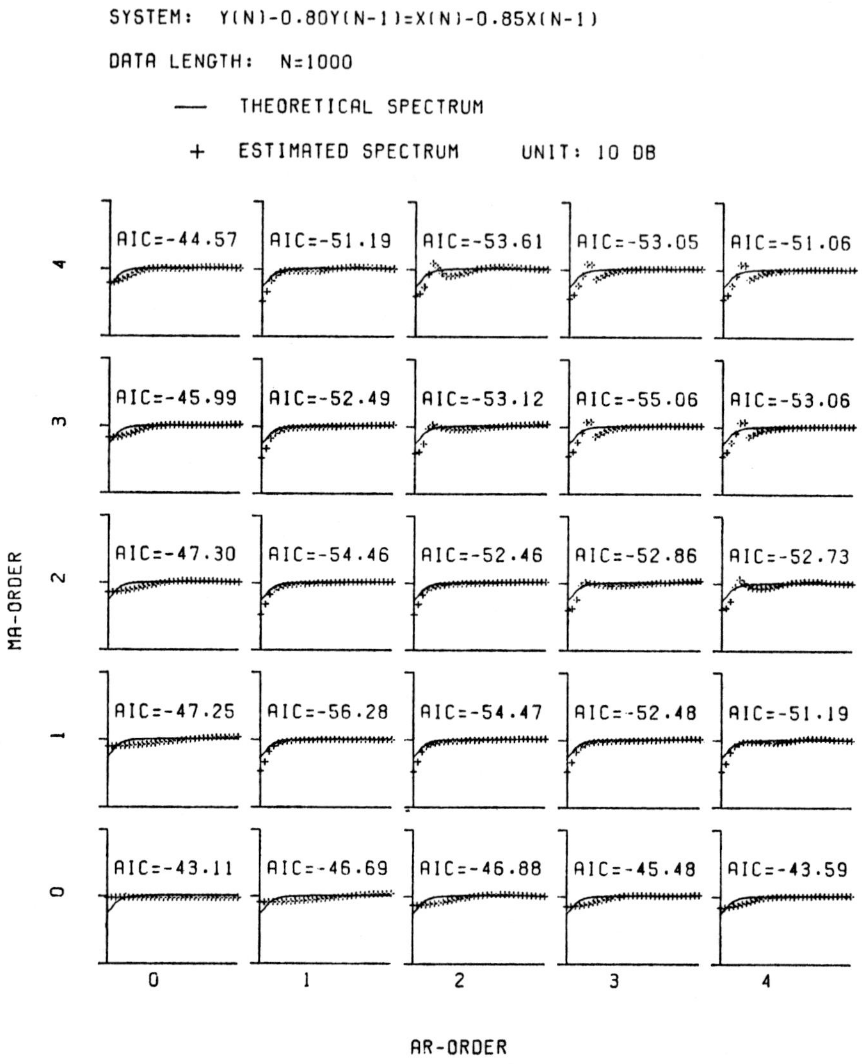

Figure 1

Behaviour of the AIC in ARMA model fitting

where $x(n)$ is independent of $z(n-1)$, $z(n-2)$, ... and is given by $x(n)=y(n)-HFz(n-1)$. If $z(n)$, the state vector, is properly defined, this representation is free from the identifiability problem which is rather serious for multivariate ARMA model fitting; see Akaike (1976) for details. There is a unique correspondence between Markovian and ARMA models. In the present example, with $y(n)=\{y_1(n), y_2(n)\}'$, the Markovian model is defined with a five-dimensional state vector $z(n)$, and the corresponding ARMA model is a two-dimensional ARMA(3, 2). The values

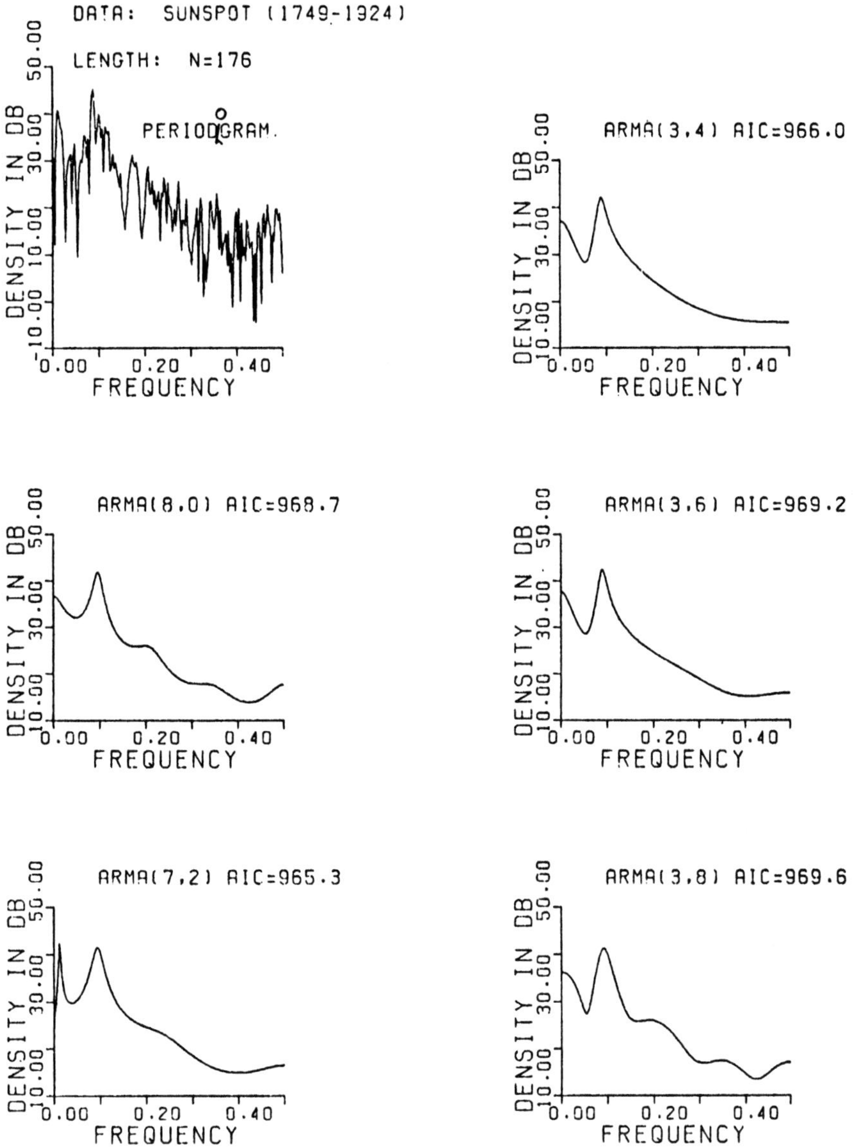

Figure 2

Estimates of the power spectrum of Wölfer's sunspot series.

of the AICs are $-1653{\cdot}4$ and $-1657{\cdot}8$ for the AR and Markovian models, respectively, and these suggest a consistent determination of the models. The numbers of parameters in these models are both equal to 16, ignoring those required to define the coveriance matrix of $x(n)$. The estimates of the frequency response function, between the output variable y_1 and the input variable y_2, were obtained as the ratios of the cross spectra between y_1 and y_2 to the corresponding power spectra of y_2. In this example, as was expected from the small difference between the AICs, the difference in the spectra for the two models is not so significant. A computer program for the fitting of Markovian models is also included in TIMSAC-74.

Transformation of Variable

In the analysis of time series it is common to try some kind of transformation on the variable. The decision on the choice of the transformation can be realized very simply by using the likelihoods of the models. Figure 4 illustrates the sunspot series for the years 1749–1924. For the original series, $y(n)$, two transformations $\{y(n)+1\}^{1/2}$ and $\log\{y(n)+1\}$ were applied. The choice of the additive constant 1 was quite arbitrary and the results were intended only for the purpose of illustration.

AIC attains minimum values at ARMA(7, 2), for the original $y(n)$, ARMA(7, 3) for $\{y(n)+1\}^{1/2}$ and ARMA(5, 4) for $\log\{y(n)+1\}$. The effect of transforming the variable is represented simply by the multiplication of the likelihood by the corresponding Jacobian and thus by the addition of minus twice the logarithm of the Jacobian to the AIC. For the case of $\{y(n)+1\}^{1/2}$, this last quantity is equal to $\sum \log\{y(n)+1\} + 2N \log 2$ and, for the case of $\log\{y(n)+1\}$, it is $2\sum \log\{y(n)+1\}$, where the summation extends over $n=1, 2, \ldots, N$.

The AICs obtained after the corrections for the Jacobians are shown in Figure 5. The minimum of the AIC is attained by the transformation $\{y(n)+1\}^{1/2}$, which shows that the then resulting series is the one closest to being Gaussian, from among the three. By looking at Figure 4, we notice significant asymmetries of the series $y(n)$ and $\log\{y(n)+1\}$, and we see that the conclusion from the AIC is in good agreement with this observation.

Obviously the Gaussian model can never be exact for these variables, since in all cases they are bounded below. It is just because of the convenience of its approximation that we use this model. It is quite probable that a model based on some physical consideration of the generating mechanism would produce a better fit to the data. At least the present results suggest the wisdom of considering non-linear predictors, when attempting to predict future values of the raw sunspot series. For a deeper discussion of the transformation of variable in time series analysis, see Anderson (1977).

Non-stationarity

On looking at Figure 4, one might suspect that there was some non-stationarity in the series during the period 1789–1829. To investigate

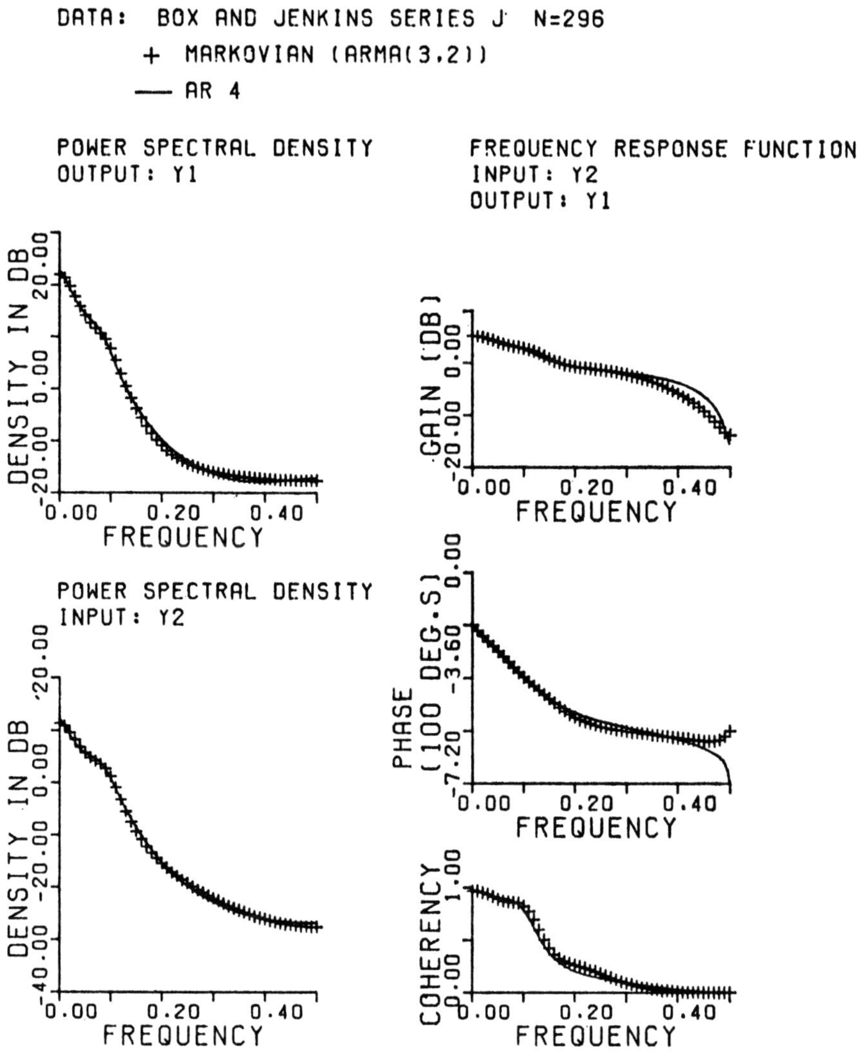

Figure 3

Estimates of the spectra for the pair of Series J from Box and Jenkins (1970)

this, we divided the data $\{(y(n)+1)^{1/2};\ n=1, 2, \ldots, 176\}$ into four consecutive segments, each of length 44, and fitted AR models to various possible combinations of consecutive segments. The first 14 data points, from the period 1749–92, were used to define the initial state of the model, and the fitting of the AR models was done by the least squares method using the Householder transformation. This procedure is discussed in Kitagawa and Akaike (1978) and is an improved version of the procedure first developed by Ozaki and Tong (1975). The computer program for

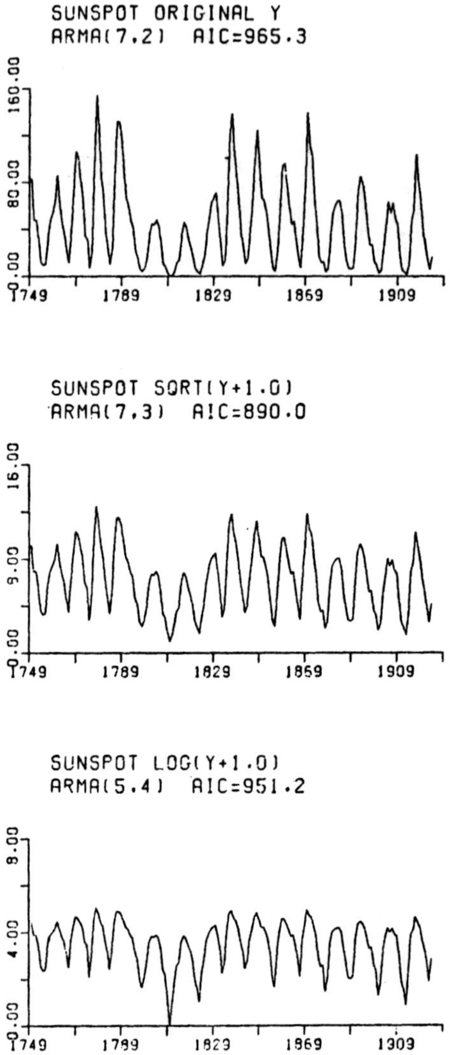

Figure 4

Effects of transformation on Wölfer's sunspot series

the original version is given in the program package TIMSAC-74, and programs based on the revised version will be published in a future issue of Computer Science Monographs from the Institute of Statistical Mathematics.

For the present use, to allow the comparison of models with possibly different structures in different segments, AIC is defined by

$$\mathrm{AIC} = N \log S + 2(M+2)$$

where N denotes the number of data points within a segment, S is the average of the squared residuals and M is the order of the AR model. The number 2, in $(M+2)$, takes into account the effect of estimating the mean of the process and the variance of the residual white noise.

In Figure 5 the resulting estimates for the spectra and the AICs are illustrated. During the period 1749–1836, the model defined by the minimum AIC is AR(9) for the subperiod 1749–92 and AR(2) for the subperiod 1793–1836, which together show a lower value of AIC= 15·69+(−1·95), than the minimum AIC estimate of 29·17 for the whole period AR(2) choice. The minimum AIC procedure suggested the use of a new model for the period 1837–80. For the period 1837–1924, the minimum AIC estimate for the whole period, an AR(10), gives AIC=4·06, which is lower than the AIC=8·65+(−1·90) obtained from the minimum AIC estimates, AR(7) and AR(6), for the subperiods 1837–80 and 1881–1924, respectively. The models and their spectra, chosen by these analyses, are denoted by asterisks in Figure 5. Obviously our procedure has detected some change of the spectrum during the period 1793–1836, which again is in good agreement with the impression obtained from observing the behaviour of the original time series.

The analysis of non-stationarity of time series has usually been realized by somewhat *ad hoc* procedures. The present result shows that it is only the introduction of some appropriate models of non-stationarity that is required in order that a practical and systematic procedure for the analysis of non-stationarity may be developed.

A Bayesian Modelling

Once we get a feeling for the behaviour of the AIC as minus twice the log likelihood of the model, determined from the maximum likelihood estimates of the parameters, we can proceed to using $\exp\{-(1/2)\text{AIC}\}$, $=L$ say, as the likelihood of the model.

To show the necessity of this type of consideration, the results from fitting AR models to our simulated data are illustrated in Figure 6. The likelihood L is normalized so that it takes the value 1·0 for the model with the minimum AIC. Due to the difficulty of tracing the dip of the power spectrum at zero frequency by an AR model, the order tends to assume higher values to attain higher likelihoods.

From the analysis of its statistical behaviour, we know that the AIC tends to show low values at higher order models when in fact the model is poor. This fact can be explained by equations (1) and (2); see, for more details, Shimizu (1978). To reduce the risk due to this behaviour of the AIC, we introduce a parametric prior distribution $\{\pi(k)\}$ over the set

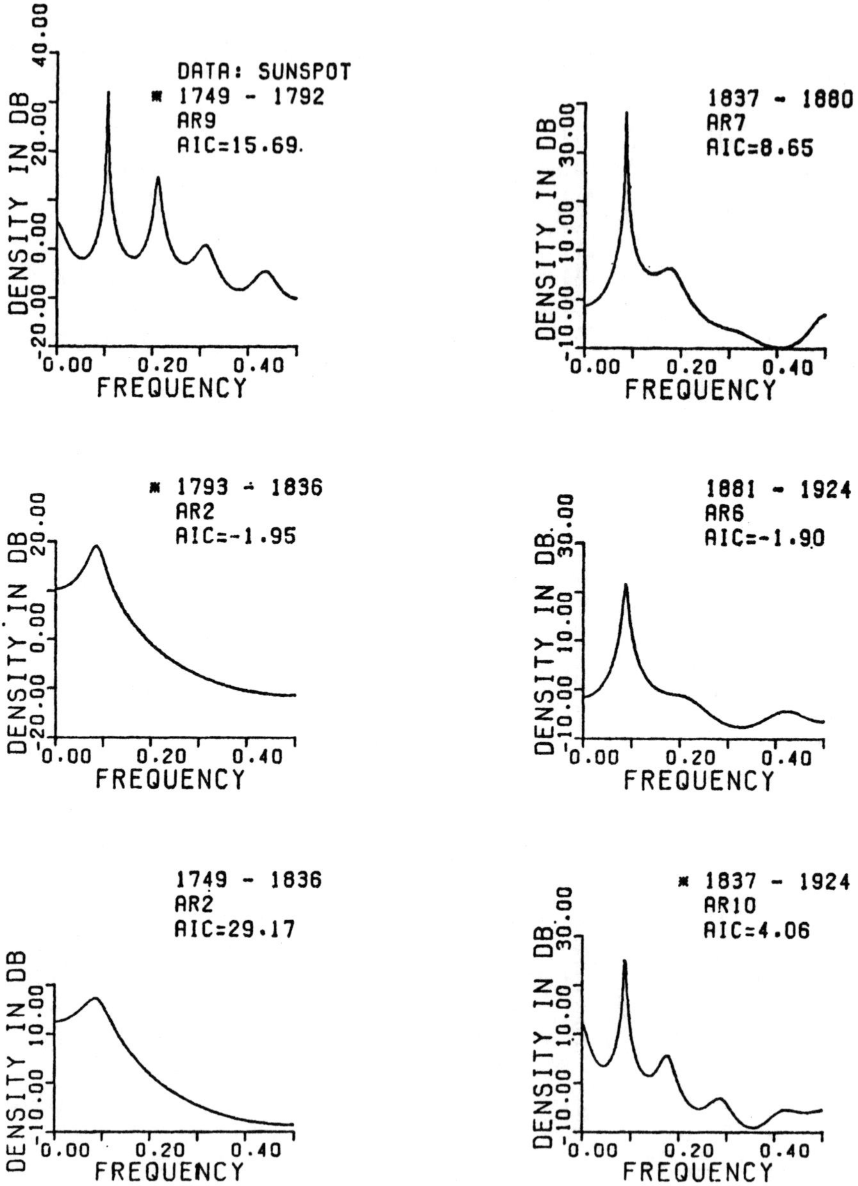

Figure 5

Non-stationary autoregressive spectrum analysis of Wölfer's sunspot series

$\{AR(k), k=0, 1, 2, \ldots\}$ which is defined by

$$\pi(k)=(1-\rho)\rho^k \qquad (0<\rho<1)$$

To realize an appropriate choice of the parameter ρ, we maximize the entropy of the distribution specified by the likelihoods with respect to the distribution $\{\pi(k)\}$; that is we choose the value of ρ which maximizes

$$\sum_{k=0}^{\infty} \exp\{-(1/2)\mathrm{AIC}(k)\} \log \pi(k)$$

where AIC(k) denotes the AIC of AR(k), and we put $\exp\{-(1/2)\mathrm{AIC}(k)\} = 0$ for those k's which are larger than some preassigned upper limit to the AR order. The optimum choice of ρ is then given by $\rho = 1/(1+\bar{k})$, where

$$\bar{k} = \frac{\sum_{k=0}^{\infty} k \exp\{-(1/2)\mathrm{AIC}(k)\}}{\sum_{k=0}^{\infty} \exp\{-(1/2)\mathrm{AIC}(k)\}}$$

The present method of maximum entropy, for the fitting of a parametric prior distribution, will avoid a gross misfit of this prior distribution to the data under observation. Also, since $\bar{k}$ is the average of the distribution specified by the likelihoods, we can expect rather small statistical variability of ρ.

Once we get the value of ρ we can define the "posterior probability" $p(k)$ of AR(k) by

$$p(k) = \frac{\exp\{-(1/2)\mathrm{AIC}(k)\}\pi(k)}{\sum_{k=0}^{\infty} \exp\{-(1/2)\mathrm{AIC}(k)\}\pi(k)}$$

Our Bayesian, strictly speaking quasi-Bayesian, estimate of the process is then obtained by taking the average of the Gaussian AR models with respect to the distribution $\{p(k)\}$. The corresponding estimate of the spectrum is given by the average of the spectra for the AR models. A computationally more efficient procedure is obtained through the averaging of the partial autocorrelation coefficients and the estimates of the variance of the white noise to define an averaged estimate; for details, see Akaike (1977b).

In Figure 6, next to the spectrum of AR(23), is shown the spectrum of the Bayesian estimate obtained by the averaging of the partial autocorrelation coefficients. The averaging process is acting effectively in reducing the fluctuations caused by the fitting of an individual AR model. The results of similar applications to real and artificial data suggest that the present definition for the likelihood of a model shows a good agreement with what we expect of the notion of likelihood, i.e. it can be useful in constructing a Bayesian model over a set of models which are determined by data. The minimum AIC procedure for the selection of a model is obviously the maximum likelihood procedure for the models determined

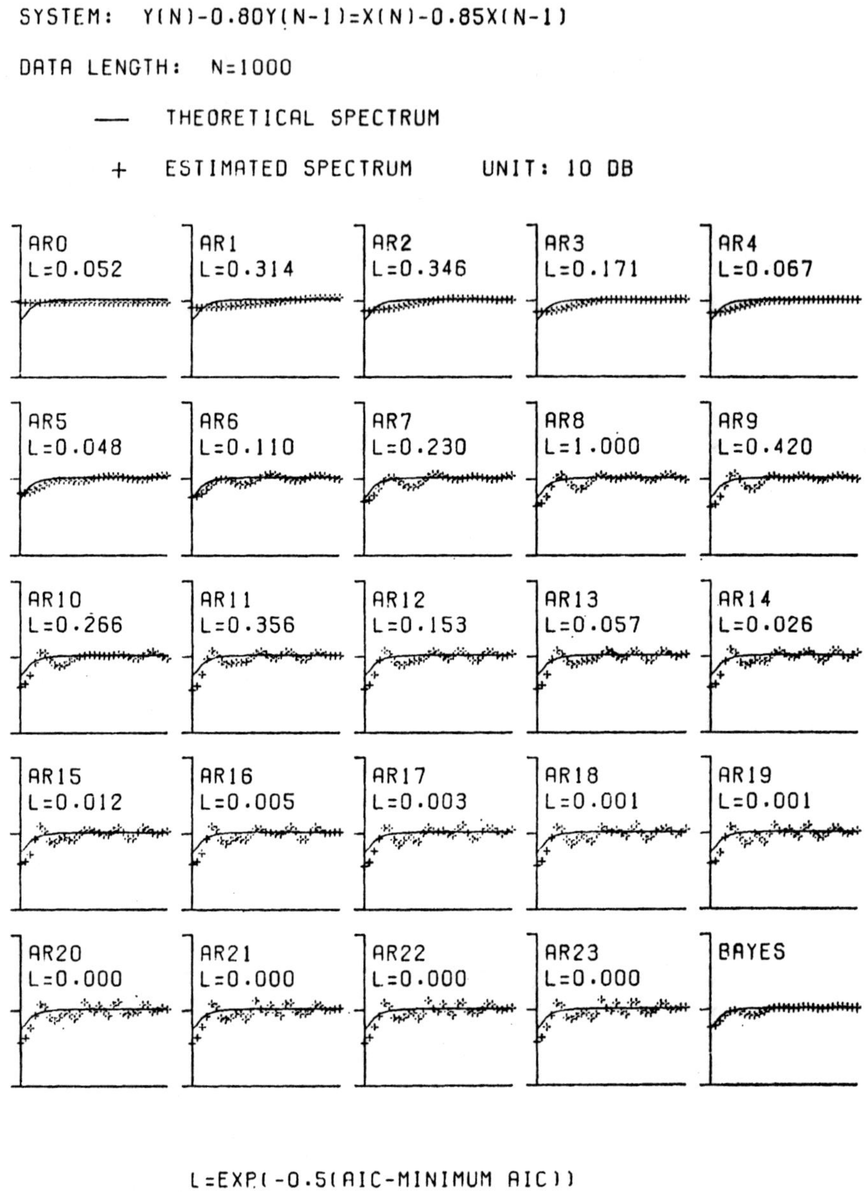

Figure 6

Use of a Bayesian model in autoregressive spectrum estimation

by the maximum likelihood estimates of the parameters. Thus the minimum AIC procedure is not free from the shortcomings of the method of maximum likelihood. Our experience of application of the present Bayesian type procedure suggests that the procedure is effective in decreasing the maximum loss compared with the minimum AIC procedure. This effect can be seen in the numerical examples given in Akaike (1977b, 1978b).

Although the present result is interesting as a proof of the use of $\exp\{-(1/2)\text{AIC}\}$ as a definition of the likelihood of a model determined by the maximum likelihood estimates of the parameters, the comparison of Figures 1 and 6 clearly shows that the choice of the set of basic models is much more fundamental in getting a good result than the sophistication of the estimation procedure. Nevertheless the confirmation of the use of $\exp\{-(1/2)\ \text{AIC}\}$ as the likelihood will significantly help the development of appropriate applications of the AIC.

Conclusion

The definition of the likelihood for a time series model is based on the recognition that the expected log likelihood, $E_y \log p(y\,|\,\boldsymbol{\theta})$, or the entropy of the true distribution $p(y)$ with respect to the assumed distribution $p(y\,|\,\boldsymbol{\theta})$, is the basic quantity with which to measure the goodness of $p(y\,|\,\boldsymbol{\theta})$ as an approximation to $p(y)$. The likelihood $p(\boldsymbol{x}\,|\,\boldsymbol{\theta})$, which Fisher (1956, p. 68) considered as a "measure of rational belief" in the vector parameter $\boldsymbol{\theta}$, has the clearly defined objective characterization that its logarithm is a natural estimate of $E_y \log p(y\,|\,\boldsymbol{\theta})$. We consider that the use of the likelihood in statistical inference stems from this fact. A typical use of the likelihood can be seen in the Bayes procedure. It suggests that any reasonable definition of the likelihood must prove its utility in Bayesian type applications.

The AIC statistic was originally introduced as an asymptotically unbiased estimate of minus twice the expected log likelihood of a model determined by the maximum likelihood estimates of the parameters; and the practical utility of $\exp\{-(1/2)\text{AIC}\}$, as the likelihood of the model, has been confirmed by numerical examples. Thus the definition of AIC as minus twice the log likelihood of the model, determined by the maximum likelihood estimates of the parameters, seems to be a reasonable one.

With the assumption of equal prior probability for the models, the distribution of $\exp\{-(1/2)\text{AIC}\}$ defines the posterior distribution of the models. The practical implication of this result is enormous when the analysis of a multivariate time series is undertaken. It was only with the aid of the minimum AIC or the maximum posterior probability procedure, or its equivalent, that we could develop a practically useful procedure of statistical model identification for implementation to multivariable feedback control systems (Otomo *et al* 1972; Akaike, 1978b, c). The high dimensionality of the vector of parameters defies any attempt at *ad hoc* procedures for model identification. Incidentally, Fisher's definition of the theory of estimation was limited to a single model situation

(Fisher, 1936, p. 250), and the selection of the model was realized by a test of significance. This formulation of the process of statistical model building led to the introduction of various *ad hoc* pretest estimation type procedures, and excluded the possibility of developing a unifying procedure of model building, as exemplified by the minimum AIC procedure.

The conceptual clarification obtained by the use of Bayesian models in statistical inference is quite remarkable, yet the main difficulty in their practical use is the choice of appropriate prior distributions. Since we place the greatest faith in the log likelihood as an estimate of the absolute criterion of the model, the expected log likelihood, we tend to depend on the information provided by the likelihood to resolve the uncertainty in choosing a prior distribution. The choice of the family of possible prior distributions must reflect all the relevant prior information for the case under consideration. Thus the choice must reflect the results of the analysis of the statistical characteristics of the related likelihoods. This point has not been sufficiently stressed by the conventional approach to Bayesian modelling, where the prior information is vaguely considered to be given independently of the choice of the data distribution $p(\cdot \mid \boldsymbol{\theta})$. The entropy maximization principle developed by Akaike (1977a, 1978a) provides a point of view which suggests the necessity of this type of consideration.

Although the discussion in this paper is limited to the case where the models are determined by the maximum likelihood estimates of the parameters, the definition of the likelihood of a model can be extended to other models, like those defined by some Bayesian type procedure. Thus there is almost unlimited scope for introducing new procedures of time series model fitting. However, it must be recognized that the most important constituent of a Bayesian type procedure is the set of basic models. When the choice of the basic models is poor, no sophisticated procedure can produce good results. The likelihood of a model discussed in this paper has an objective meaning as the measure of the model's adequacy, which is independent of the construction of the prior probabilities of the models, and should prove most useful in the preliminary search for satisfactory models. Typically, the second term in the definition of AIC shows the necessity of parsimony in the parameterization of the models, and the first term in the definition suggests the importance of models which will adequately represent the structure of the stochastic process under consideration. It is only when the first term is sufficiently controlled that the second term becomes relevant.

The introduction of the concept of the likelihood of a model prepared a systematic approach to the statistical model fitting of time series. It is the introduction of new models that is most urgently needed to produce rapid progress in practical applications of time series analysis. For this end, close cooperation between people with real problems and statisticians is a *sine qua non.*

Appendix: Statistical Characterization of Entropy

Consider a distribution $(q_1, q_2, \ldots, q_k)$ with $q_i > 0$ $(i=1, \ldots, k)$ and $q_1 + \ldots + q_k = 1$. Assume that N independent drawings are made from the distribution and that the resulting frequency distribution is given by $(N_1, N_2, \ldots, N_k)$, where $N_1 + \ldots + N_k = N$. Then the probability of getting this same $(N_1, \ldots, N_k)$ by sampling from $(q_1, \ldots, q_k)$ is given by

$$W = \frac{N!}{N_1! \ldots N_k!} q_1^{N_1} \ldots q_k^{N_k} \qquad \text{(A.1)}$$

By taking logarithms and using the asymptotic equality

$$\log N! = N \log N - N,$$

we get the asymptotic equality

$$\log W = -N \sum_{i=1}^{k} \frac{N_i}{N} \log \left(\frac{N_i}{Nq_i} \right)$$

If we next put $p_i = N_i/N$, we get

$$\begin{aligned} \log W &= -N \sum_{i=1}^{k} p_i \log \left(\frac{p_i}{q_i} \right) \\ &= NB(p; q) \end{aligned} \qquad \text{(A.2)}$$

where $B(p; q)$ is the entropy of the distribution $\{p_i\}$ with respect to the distribution $\{q_i\}$.

The historical probabilistic interpretation of thermodynamic entropy by Boltzmann (1877) is symbolically represented by the formula

$$S = k \log W \qquad \text{(A.3)}$$

where S denotes the thermodynamic entropy and k is a constant. From (A.2) and (A.3) the analogy between S and $B(p; q)$ is obvious. The relation (A.2) suggests the interpretation of the entropy $B(p; q)$ as the logarithm of the probability of getting the distribution $\{p_i\}$ by sampling from the assumed distribution $\{q_i\}$. The distinction between the hypothetical $\{q_i\}$ and the factual $\{p_i\}$ is very important in developing a proper use of the concept of entropy $B(p; q)$.

The neg-entropy $-B(p; q)$ is equal to the Kullback–Leibler information number $I(p; q)$ (Kullback and Leibler, 1951). A characterization of $I(p; q) = -B(p; q)$ as a measure of discrepancy between $\{p_i\}$ and $\{q_i\}$

can be seen in Savage (1971, p. 794). Good (1971, p. 132), who attributes the first use of the quantity to Turing, prefers to call $I(p; q)$ the expected weight of evidence. The extension of the interpretation of $B(p; q)$, as the logarithm of the probability of getting $\{p_i\}$ from $\{q_i\}$, to more general distributions is made possible by the work of Sanov (1961).

Acknowledgement

This work was partly supported by a grant from the Ministry of Education, Science and Culture.

REFERENCES

AKAIKE, H. (1973). Information theory and an extension of the maximum likelihood principle. In *2nd International Symposium on Information Theory* (eds B. N. Petrov and F. Csaki), pp. 267–81. Akademaii-Kiado, Budapest.

AKAIKE, H. (1974). A new look at the statistical model identification. *IEEE Transactions on Automatic Control*, **AC-19**, 716–23.

AKAIKE, H. (1976). Canonical correlation analysis of time series and the use of an information criterion. In *System Identification: Advances and Case Studies* (eds R. K. Mehra and D. G. Lainotis), pp. 27–96. Academic Press, New York.

AKAIKE, H. (1977a). On entropy maximization principle. In *Applications of Statistics* (ed. P. R. Krishnaiah), pp. 27–41. North-Holland, Amsterdam.

AKAIKE, H. (1977b). A Bayesian extension of the minimum AIC procedure of autoregressive model fitting. *Research Memorandum No. 126.* Institute of Statistical Mathematics, Tokyo. (To appear in *Biometrika*, **66.**)

AKAIKE, H. (1978a). A new look at the Bayes procedure. *Biometrika*, **65**, 53–9.

AKAIKE, H. (1978b). On the identification of state space models and their use in control. A paper presented at the *IMS Special Topics Meeting on Time Series Analysis*, Ames, Iowa.

AKAIKE, H. (1978c). On newer approaches to parameter estimation and structure determination. A paper presented at the *7th IFAC World Congress*, Helsinki.

AKAIKE, H. (1978d). Likelihood of a model. *Research Memorandum No. 127.* Institute of Statistical Mathematics, Tokyo.

AKAIKE, H., ARAHATA, E. and OZAKI, T. (1975, 1976). TIMSAC-74 – a time series analysis and control program package - (1) & (2). *Computer Science Monographs, Nos 5 & 6.* Institute of Statistical Mathematics, Tokyo.

ANDERSON, O. D. (1977). A Box-Jenkins analysis of the coloured fox data from Nain, Labrador. *The Statistician*, **26**, 51–75.

ANDERSON, T. W. (1971). *The Statistical Analysis of Time Series.* John Wiley, New York.

BOLTZMANN, L. (1877). Uber die Beziehung zwischen dem zweiten Hauptsatze der mechanischen Warmetheorie und der Wahrscheinlichkeitsrechnung respective den Satzen uber das Warmegleichgewicht. *Wiener Berichte*, **76**, 373–435.

BOX, G. E. P. and JENKINS, G. M. (1970). *Time Series Analysis, Forecasting and Control.* Holden Day, San Francisco.

FISHER, R. A. (1936). Uncertain inference. *Proceedings of the American Academy of Arts and Science*, **71**, 245–58.

FISHER, R. A. (1956). *Statistical Methods and Scientific Inference*. Oliver and Boyd, London.

GOOD, I. J. (1971). The probabilistic explication of information, evidence, surprise, causality, explanation, and utility. In *Foundation of Statistical Inference* (eds V. P. Godambe and D. A. Sprott), pp. 108–27. Holt, Rinehart and Winston, Toronto, Montreal.

KITAGAWA, G. and AKAIKE, H. (1978). A procedure for the modelling of non-stationary time series. *Annals of the Institute of Statistical Mathematics*, **30B**, 351–63.

KULLBACK, S. and LEIBLER, R. A. (1951). On information and sufficiency. *Annals of Mathematical Statistics*, **22**, 79–86.

OTOMO, T., NAKAGAWA, T. and AKAIKE, H. (1972). Statistical approach to computer control of cement rotary kilns. *Automatica*, **8**, 35–48.

OZAKI, T. (1977). On the order determination of ARIMA models. *Applied Statistics*, **26**, 290–301.

OZAKI, T. and TONG, H. (1975). On fitting of non-stationary autoregressive models in time series analysis. *Proc. 8th Hawaii International Conference on System Sciences*, pp. 225–6. Western Periodicals, North Hollywood, California.

SANOV, I. N. (1961). On the probability of large deviations of random variable. *IMS and AMS Selected Translations in Mathematical Statistics and Probability*, **1**, 213–44.

SAVAGE, L. J. (1971). Elicitation of personal probabilities and expectations. *Journal of American Statistical Association*, **66**, 783–801.

SHIMIZU, R. (1978). Entropy maximization principle and selection of the order of an autoregressive Gaussian process. *Anmals of the Institute of Statistical Mathematics*, **30A**, 263–70.

Reprinted from *Bayesian Statistics*, J.M. Bernardo, M.H. De Groot, D.V. Lindley and A.F.M. Smith, eds., University Press, Valencia, Spain, 1980, 1-13 by permission from J. M. Bernardo

Likelihood and the Bayes procedure

HIROTUGU AKAIKE

The Institute of Statistical Mathematics, Tokyo

SUMMARY

In this paper the likelihood function is considered to be the primary source of the objectivity of a Bayesian method. The necessity of using the expected behavior of the likelihood function for the choice of the prior distribution is emphasized. Numerical examples, including seasonal adjustment of time series, are given to illustrate the practical utility of the common-sense approach to Bayesian statistics proposed in this paper.

Keywords: LIKELIHOOD; BAYES PROCEDURE; AIC; SEASONAL ADJUSTMENT.

1. INTRODUCTION

The view that the Bayesian approach to statistical inference is useful, practically as well as conceptually, is now widely accepted. Nevertheless we must also accept the fact that there still remain some conceptual confusions about the Bayes procedure. Although many strong impetuses for the use of the procedure came from the subjective theory of probability, it seems that the confusions are also caused by the subjective interpretation of the procedure.

By looking through the works on the Bayes procedure by subjectivists, it quickly becomes clear that there is not much discussion of the concept of likelihood. The subjective theory of probability is used only to justify the use of the prior distribution of the parameters of a data distribution. It is almost trivial to see that no practically useful Bayes procedure is defined without the use of the likelihood function, while the likelihood function can be defined without the prior distribution. Thus the data distribution represents the basic part of our prior information and the Bayes procedure gives only one specific way of utilizing the information supplied by data through the likelihood function.

From this point of view there is nothing special about the choice of prior distributions to differentiate it from the design of ordinary statistical procedures such as the choice of the sampling procedure in a sample survey and the choice of the spectrum window in the spectrum analysis of a time series.

In this paper we first discuss some conceptual confusions with the Bayes procedure which we believe to be due to the subjective interpretation of the procedure. We argue that it is necessary to recognize the limitation of the subjective theory and put more emphasis on the concept of likelihood. We take the position of regarding the Bayes procedure as one possible way of utilizing the information provided by the likelihood function. Once such an attitude towards the Bayes procedure is accepted we can freely develop Bayesian models simply by representing a particular preference of the parameters by a prior distribution. The goodness of the prior distribution can then be checked by evaluating expected performances of the corresponding Bayes procedure in various conceivable situations.

We demonstrate the use of this type of approach by developing a general Bayesian model for the analysis of linear relations between variables. The model contains as special cases the basic models of those estimation procedures such as the Stein estimator, ridge regression, Shiller's distributes lag estimator and O'Hagan's localized regression. Numerical examples are given to illustrate the practical utility of some quasi-Bayesian procedures developed for these models and for a more conventional model of polynomial regression. The result of application to the seasonal adjustment of time series seems particularly interesting as the model contains twice as many parameters as the number of the observations.

2. CONCEPTUAL DIFFICULTIES OF THE SUBJECTIVE APPROACH

Significant impetus for the advancement of Bayesian statistics has come from the side of the subjective theory of probability. This is natural as every statistical procedure may be viewed as a formulation of the psychological process of information processing and evaluation by a skilful researcher. In spite of the significant contribution of the subjective theory of probability to clarifying the nature of the psychological aspect of this process, several conceptual difficulties remain with the theory. Here we discuss some difficulties, which we believe to be misconceptions, related to the Bayes procedure and clear the way for the development of practically useful Bayesian methods.

2.1. *Rationality and Savage's axiom*

It is sometimes said that a rational person must behave as if he has a clearly defined system of subjective probabilities of uncertain events. This is often ascribed to Savage (1954) who developed a theory of personal probability by axiomatizing the preference behavior of a person under uncertainty. Unfortunately the very first postulate P1 of Savage, which assumes the linear ordering of the preference, excludes the real difficulty of preference. This can be explained by the following simple example.

Consider a young boy who wants to choose a girl as his wife. His preference is based on the three characteristics, H, I and L. Here H stands for health, I for intelligence and L for looks. Each characteristic is ranked by the numbers 1, 2, and 3, with higher number denoting higher rank. The difference of ranks by 1 is marginal and the difference by 2 means a significant difference. Denote by $R_i = (H_i, I_i, L_i)$ the vector of the ranks of the characteristics of the ith girl. Being uncertain about the relative importance of these characteristics in his future life, he ignores the marginal differences and pays attention only to the significant differences. Thus his preference is defined by the following scheme:

$$R_i \leq R_j, \text{ i.e., the } j^{th} \text{ girl is preferred to the } i^{th} \text{ girl,}$$

$$\text{iff } C_i \leq C_j \text{ for the characteristic } C$$

$$\text{for which } |C_j - C_i| \text{ is maximum.}$$

Now he has three girl friends (i = 1, 2, 3) whose R_i's are defined by $R_1 = (1, 2, 3)$, $R_2 = (3, 1, 2)$ and $R_3 = (2, 3, 1)$, respectively. Obviously it holds that

$$R_1 \leq R_2, R_2 \leq R_3 \text{ and } R_3 \leq R_1,$$

which shows that his natural preference scheme does not satisfy the postulate P1 of Savage.

It is the difficulty of this type of preference that make us feel the need of a horoscope or some other help in making the decision in a real life situation. Since Savage's system excludes the possibility of this type of difficulty, the corresponding theory of personal probability cannot tell how we should treat the difficulty. The exact characterization of Savage's theory is then a theory of one particular aspect of preference and there is no compelling reason to demand that a rational person's preference should be represented by a single system of subjective probability. Wolfowitz (1962) presents a pertinent discussion of this point. Thus to justify the use of a system of personal probability one must prove its adequacy by some means. Certainly the proof cannot be found within the particular system of personal probability itself.

2.2. *The role of parameters in a Bayesian modeling*

The subjective theory of probability of De Finetti demands that the probability distribution or the expectations of the uncertain events of interest should completely be specified (de Finetti, 1974b, p. 87). If we accept this demand and decide to use the Bayes procedure, all we have to do is to compute $p(y|x)$, the probability of an event y conditional on a given set of data x. The theory only asserts that the necessary probability distribution should be there and does not consider the special role played by the parameters in constructing a statistical model or the probability distribution. De Finetti (1974a, p. 125) even rejects the concept of a parameter as metaphysical, unless it is a decidable event.

That the concept of parameter cannot be eliminated is shown by the simple example of the binomial experiment where the probability of occurence of a head in a coin tossing is considered. The concept of independent trials with a fixed probability of head is unacceptable by the subjective theory of probability of de Finetti and the solution is sought in the concept of exchangeability (de Finetti, 1975, pp. 211-218). The difficulty is caused by the fact that the probability of a head, which must be decided, plays the role of a parameter that is not actually decidable (Akaike, 1979b).

We may use the theory of probability to develop some understanding of what we psychologically expect of the parameters of a statistical model. Consider a random variable x and the observations x_1, x_2, ...of some related events. We expect that a parameter θ exhausts the information about x to be gained through the observations x_1, x_2... The probabilistic expression of this expectation is given by

$$p(x|\theta, x_1, x_2, \ldots) = p(x|\theta), \tag{2.1}$$

where $p(x|z_1, z_2, \ldots)$ denotes the distribution of x conditional on z_1, z_2, ... To allow this type of discussion we must consider θ as a random variable as is advocated by Kudo (1973). The formula (2.1) then gives a very natural characterization of the parameters as a condensed representation of the information contained in the observations, i.e., once θ is known no further observations can improve our predictions on x. Thus we want to know the value of θ. Actually de Finetti's discussion of the exchangeable distribution of the binomial experiment has given a proof of the existence of such a variable.

Although the above characterization of a parameter is interesting, in the statistical model building for inference the order of reasoning is reserved. The prior information first suggests what type of parameterization of the data distribution $p(x|\theta)$ should be used. The prior distribution $\pi(\theta)$, if at all specified, represents only a part of the prior information. To take the parameters as something prespecified and assume that the prior distribution can or should be determined independently of the data distribution constitutes a serious misconception about the inferencial use of the Bayes procedure.

2.3. *Likelihood principle and the Bayes procedure*

It has often been claimed that the likelihood principle, which demands that the statistical inference should be identical if the likelihood function is identical, is a direct consequence of the Bayesian approach; see, for example, Savage (1962, p. 17). In the example of coin tossing, if we denote the probability of head by θ and assume the independence and homogeneity of the tossings, we have

$$p(x|\theta) = C\,\theta^{x}(1-\theta)^{n-x}$$

as the likelihood of θ when x heads appeared in n tossings. It is argued that there is no difference in the inference through the Bayes procedure if the above likelihood is obtained as the result of n tosses, with n predetermined, or as the result of tossing continued until x heads appeared, with x predetermined.

This seemingly innocuous argument is against the principle of rationality of the subjective theory of probability which suggests that the choice of a statistical decision be based on its expected utility. The expected behavior of the likelihood function $p(x|\theta)$ is certainly different for the two schemes of the coin tossing and it is irrational to adopt one and the same prior distribution $\pi(\theta)$, irrespectively of the expected difference of the statistical behavior of the likelihood functions.

To clarify the nature of the confusion by a concrete example, consider the use of the posterior distribution

$$C\,\theta^{x}(1-\theta)^{n-x}\,\pi(\theta)$$

as an estimate of the probability distribution of the result y of the next toss, where $y = 1$ for head and 0 otherwise. The predictive distributions are defined as the averages of the data distribution $p(y|\theta)$ with respect to the posterior distributions of θ. These will be denoted by $p(y|x)$ and $p(y|n)$ to indicate that x and n are the realizations of the random variables, respectively. They are defined by

$$p(y|*) = C\int_0^1 \theta^{x+y}(1-\theta)^{n+1-x-y}\pi(\theta)d\theta,$$

where $*$ stands for either x or n. When the "true" value of θ is θ_o the goodness of $p(y|*)$ as an estimate of the true distribution $p(y|\theta_o) = \theta_o^y(1-\theta_o)^{1-y}$ can be measured by the entropy of $p(y|\theta_o)$ with respect to $p(y|x)$ or $p(y|n)$ which is defined by

$$\mathrm{B}\,\{p(\cdot|\theta_o),\, p(\cdot|*)\}$$
$$= -\Sigma_{y=0}^{1}\left\{\frac{p(y|\theta_o)}{p(y|*)}\right\}\log\left\{\frac{p(y|\theta_o)}{p(y|*)}\right\}p(y|*)$$

The larger the entropy the better is the approximation of $p(\cdot|*)$ to $p(\cdot|\theta_o)$. Before we observe x or n we evaluate $\mathrm{E}_*\,\mathrm{B}\{p(\cdot|\theta_o),\, p(\cdot|*)\}$ for some possible values of θ_o, where E_* denotes the expectation with respect to the distribution of $*$ defined with $\theta = \theta_o$. We have

$$\mathrm{E}_x\,\mathrm{B}\{p(\cdot|\theta_o),\, p(\cdot|x)\}$$
$$= \Sigma_{y=0}^{1}\, p(y|\theta_o)\, \Sigma_{x=0}^{n}\, \log\left\{\frac{p(y|x)}{p(y|\theta_o)}\right\}{}_nC_x\, \theta_o^x\, (1-\theta_o)^{n-x}$$

and

$$\mathrm{E}_n\,\mathrm{B}\{p(\cdot|\theta_o),\, p(\cdot|n)\}$$
$$= \Sigma_{y=0}^{1}\, p(y|\theta_o)\, \Sigma_{n=x}^{\infty}\, \log\left\{\frac{p(y|n)}{p(y|\theta_o)}\right\}{}_{n-1}C_{x-1}\, \theta_o^x\, (1-\theta_o)^{n-x}.$$

Obviously we have no reason to expect that these two quantities will take one and the same value and, at least for that matter, there is no reason for us to assume one and the same prior distribution $\pi(\theta)$ for both cases.

3. LIKELIHOOD AS THE SOURCE OF OBJECTIVITY

The discussion in the preceding section illustrates both the subjective and objective elements in the Bayesian approach to statistical inference. It is subjective because a statistical inference procedure is designed to satisfy a subjectively chosen objective. The choice of the data distribution is particularly subjective and the prior distribution reflects the object of the inference which is often expressed in the form of a psychological expectation.

What is then objective with the procedure ? The objectivity stems from the dependence on the data which is a production of the outside world. This objectivity is fed into the Bayes procedure through the likelihood function. Since $\mathrm{B}\{p_o(\cdot),\, p(\cdot|\theta)\} = \mathrm{E}_x \log p(x|\theta) - \mathrm{E}_x \log p_o(x)$, we can see that, ignoring the additive constant $\mathrm{E}_x \log p_o(x)$, the log likelihood $\log p(x|\theta)$ is a natural estimate of the entropy of $p_o(\cdot)$ with respect to $p(\cdot|\theta)$. Here E_x denotes the expectation with respect to the distribution $p_o(\cdot)$ of x. Thus the likelihood $p(x|\theta)$ represents an objective measure of the goodness, as measured by x, of

$p(\cdot|\theta)$ as an approximation to $p_o(\cdot)$. This fact forms the basis of the practical utility of the Bayes procedure even for the family $\{p(\cdot|\theta)\}$ which is chosen subjectively and does not contain the true distribution of x.

The likelihood function $p(x|\theta)$ is the basic device for the extraction or condensation of the information supplied by the data x. The role of the prior distribution $\pi(\theta)$ is to aid further condensation of the information supplied by the likelihood function $p(x|\theta)$ through the introduction of some particular preference of the parameters. By evaluating the expected entropy of the true distribution with respect to the predictive distribution specified by a posterior distribution we can extend the concepts of bias and variance to the posterior distribution (Akaike, 1978a). If we try to keep a balance between the bias and variance, we cannot ignore the influence of the statistical behavior of the likelihood function on the choice of our prior distribution. Some of the conflicts between the conventional and Bayesian statistics are caused by ignoring the possible dependence of the choice of the prior distribution, or even the choice of the basic data distribution, on the number of available observations which influences the behaviour of the likelihood function; see, for example, Lindley (1957), Schwarz (1978) and Akaike (1978b).

4. A GENERAL BAYESIAN MODELING FOR LINEAR PROBLEMS

In this section we demonstrate the practical utility of the point of view discussed in the preceding section through the discussion of a general Bayesian model for the analysis of linear problems. The basic idea here may be characterized as the common-sense approach to Bayesian statistics.

Consider the analysis of the linear relation between the vector of observations $y = [y(1),\ldots,y(N)]'$ and the vectors of the independent variables $x_i = [x_i(1), x_i(2),\ldots x_i(N)]'$ $(i = 1,2,..,K)$, where ' denotes transposition. The method of least squares leads to the minimization of

$$L(a) = \Sigma_{j=1}^{N} [y(j) - \Sigma_{i=1}^{K} a_i x_i(j)]^2. \tag{4.1}$$

We know, when K is large compared with N or when the matrix $X = [x_1, x_2,\ldots,x_K]$ is ill-conditioned the least squares estimates behave badly. To control this we introduce some preference on the values of the parameters and try to minimize

$$L(a) + \mu \| a - a_0 \|_R^2 \tag{4.2}$$

where a_0 denotes a particular vector of parameters $[a_{01}, a_{02},\ldots,a_{0K}]'$, $\| \ \|_R^2$ the norm defined by a positive definite matrix R, and μ a positive constant. The use of this type of constrained least squares for the solution of an ill-posed problem is wellknown; see, for example, Tihonov (1965).

The difficulty with the application of this method of constrained least squares is in the choice of the value of μ. To solve this we transform the problem into the maximization of

$$\ell(a) = \exp\left\{ -(1/2\sigma^2)\,[L(a) + \mu\|a - a_0\|_R^2]\right\},$$

where temporarily σ^2 is assumed to be known. Since we have

$$\ell(a) = \exp[\,-(1/2\sigma^2)\,L(a)]\,\exp[\,-(\mu/2\sigma^2)\,\|a-a_0\|_R^2\,],$$

we can see that the solution of the constrained least squares problem is now given as the mean of the posterior distribution defined by the data distribution

$$f(y|\sigma^2,a) = (1/2\pi)^{N/2}(1/\sigma)^N \exp[\,-(1/2\sigma^2)\,L(a)], \qquad (4.3)$$

and the prior distribution

$$\pi(a|d) = (1/2\pi)^{K/2}(1/\sigma)^K \exp[\,-(d^2/2\sigma^2)\,\|a-a_0\|_R^2], \qquad (4.4)$$

where $d^2 = \mu$. By properly choosing X, a_0 and R, we can get many practically useful models. Particularly, we will restrict our attention to the case where $\| a- a_0\|_R^2$ is defined by

$$\|a - a\|_R^2 = \|c_0 - Da\|^2, \qquad (4.5)$$

where D is a properly chosen matrix, $c_0 = Da_0$ and $\|v\|^2$ denotes the sum of squares of the components of v. In this case the posterior mean of the vector parameter a is obtained by minimizing $\|z(a|d)\|^2$ of the vector $z(a|d)$ defined by

$$z(a|d) = \begin{bmatrix} y(1) \\ y(2) \\ \cdot \\ \cdot \\ \cdot \\ y(N) \\ dc_0(1) \\ dc_0(2) \\ \cdot \\ \cdot \\ dc_0(L) \end{bmatrix} - \begin{bmatrix} X \\ \\ aD \end{bmatrix} \begin{bmatrix} a(1) \\ a(2) \\ \cdot \\ \cdot \\ \cdot \\ \\ a(K) \end{bmatrix} \tag{4.6}$$

Examples.

a. *Stein type shrunken estimator*

This is defined by putting $L = N$, $D = X$ and $c_o = 0$, the zero vector. The case with $K = N$ and $X = I_{NxN}$ corresponds to the original problem of estimation of the mean vector of a multivariate Gaussian distribution treated by Stein. By putting c_o equal to the vector of the parameters obtained from some similar former observations, we can realize a reasonable use of the prior information.

b. *Ridge regression*

This is defined by putting $L = K$, $D = I_{KxK}$ and $c_o = 0$.

c. *Shiller's distributed lag estimator*

Shiller (1973) developed a procedure for the estimation of a smoothly changing impulse response sequence. In this case $[y(1), y(2), \ldots, y(N)]$ is-obtained as the time series of the output of a constant linear system under the input $u(j)$. X is defined by $x_i(j) = u(j-i+1)$ and $c_o = 0$.

D is put equal to

$$D_1 = \begin{bmatrix} \alpha & & & & & \\ -1 & 1 & & & & \\ & -1 & 1 & & 0 & \\ & & \cdot & \cdot & & \\ & 0 & & \cdot & \cdot & \\ & & & & \cdot & \cdot \\ & & & & -1 & 1 \end{bmatrix}$$

or

$$D_2 = \begin{bmatrix} \alpha & & & & & & & \\ -\beta & \beta & & & & & & \\ 1 & -2 & 1 & & & & & \\ & 1 & -2 & 1 & & 0 & & \\ & & \cdot & \cdot & \cdot & & & \\ & & & \cdot & \cdot & \cdot & & \\ & 0 & & & \cdot & \cdot & \cdot & \\ & & & & & 1 & -2 & 1 \end{bmatrix}$$

where α and β are properly chosen constants. D_1 controls the first order differences of $a(j)$ and D_2 the second order differences.

d. *Localized regression of O'Hagan.*

O'Hagan (1978) introduced an interesting Bayesian model for the estimation of the locally gradually changing regression of a time series $y(i)$ on-$x(i)$. Our model corresponding to O'Hagan's is given by putting $K = N$, $c_o = 0$ and

$$X = \begin{bmatrix} x(1) & & & & \\ & x(2) & & 0 & \\ & & \cdot & & \\ & & & \cdot & \\ & 0 & & \cdot & x(N) \end{bmatrix}.$$

D is put equal to D_1 or D_2 of the above example or

$$D_3 = \begin{bmatrix} \alpha & & & & & & & & \\ -\beta & \beta & & & & & & & \\ \gamma & -2\gamma & \gamma & & & & 0 & & \\ -1 & 3 & -3 & 1 & & & & & \\ & -1 & 3 & -3 & 1 & & & & \\ & & \cdot & \cdot & \cdot & \cdot & & & \\ & & & \cdot & \cdot & \cdot & \cdot & & \\ & 0 & & & \cdot & \cdot & \cdot & \cdot & \\ & & & & & -1 & 3 & -3 & 1 \end{bmatrix}$$

One particularly interesting model is obtained by putting $x(i) = 1$ $(i = 1,2,\ldots, N)$. The number of parameters in this model is equal to the number of observations $y(i)$.

e. *Locally smooth trend fitting*

For a time series $y(i)$, by putting $c_o = 0$ and $D = D_k X$ where D_k is as given in the preceding examples, we get a model for the fitting of a smooth trend curve. One special choice of X is given by $X = I_{NxN}$. We will call the model defined with $X = I_{NxN}$ and $D = D_k$ the model of locally smooth trend of kth order.

f. *Bayesian seasonal adjustment*

We consider the decomposition of the monthly observations $y(i)$ for M years, where $i = 12m + j$ $(j = 1,2, \ldots, 12, m = 0,1, \ldots, M\text{-}1)$, into the form

$$y(i) = T_i + S_i + I_i,$$

where T_i denotes the trend, S_i the seasonal and I_i the irregular component. For this problem we put $K = 2N$ $(N = 12M)$ and define $a = (T_1, T_2, \ldots, T_N, S_1, S_2, \ldots, S_N)$ and put $c_o = 0$.

The matrix X is defined by

$$X = N \updownarrow \left[\begin{array}{cccccc:cccccc} \multicolumn{6}{c:}{\longleftarrow N \longrightarrow} & \multicolumn{6}{c}{\longleftarrow N \longrightarrow} \\ 1 & & & & & & 1 & & & & & \\ & 1 & & & & & & 1 & & & & \\ & & 1 & 0 & & & & & 1 & 0 & & \\ & & & \ddots & & & & & & \ddots & & \\ & 0 & & & \ddots & & & 0 & & & \ddots & \\ & & & & & 1 & & & & & & 1 \end{array}\right]$$

and D by

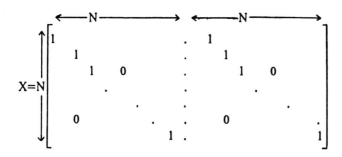

$$
D_{kp} = \begin{array}{c} \updownarrow N \\ \\ \\ \updownarrow N \\ \\ \\ \\ \updownarrow N \\ \\ \end{array}
\overset{\longleftarrow N \longrightarrow \qquad \longleftarrow\!\!-\!\!-\!\!-\!\!-\!\!-\!\!- N -\!\!-\!\!-\!\!-\!\!-\!\!-\!\!-\!\!\longrightarrow}{\underset{\longleftarrow N \longrightarrow \quad \leftarrow 12 \rightarrow \leftarrow 12 \rightarrow \;\cdots\; \longleftarrow 12 \longrightarrow}{\left[\begin{array}{c:ccccc}
D_k & & & 0 & & \\
\hdashline
 & eI & 0 & 0 & & 0 \\
 & fI & -fI & 0 & & 0 \\
0 & 0 & fI & -fI & & 0 \\
 & \vdots & \vdots & \ddots & \ddots & \vdots \\
 & 0 & 0 & & fI & -fI \\
\hdashline
 & g1' & 0 & & & 0 \\
 & 0 & g1' & & & 0 \\
0 & \vdots & \vdots & & \ddots & \vdots \\
 & 0 & 0 & & & g1'
\end{array}\right]}}
$$

where D_k is one of those defined in the preceding examples, $I = I_{12\mathrm{x}12}$, $1' = (1, 1, \ldots, 1)$, and e, f, g are properly chosen constants.

A notable characteristic of this model is that it has twice as many parameters as the number of observations. This constitutes a typical many parameters problem which cannot be handled by the ordinary unconstrained least squares or the method of maximum likelihood.

The fundamental problem in applying these models to real data is the choice of the constant d. Assuming that other constants are specified, the decision on d is equivalent to the decision on the prior distribution of a. From (4.2) the choice of d, or μ, determines the relative weight of the additional term $\| a - a_o \|_R^2$ against $L(a)$, the sum of squares of the residuals. When a_o is not exactly equal to the true value of a, we expect that the bias of the estimate increases as d is increased but the variance decreases. It is natural to try to keep a balance between these two factors. To realize this it is necessary not to specify d uniquely but use the information supplied by the likelihood function or $L(a)$.

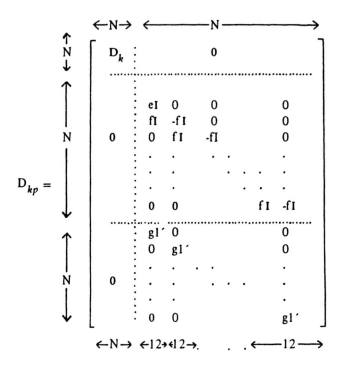

In the Bayesian terminology this is to consider d as a hyperparameter which has its own prior distribution. Now it is obvious that by considering d as a hyperparameter we are trying to use the information supplied by the likelihood function for the determination of d. This observation suggests that a proper choice of the prior distribution to be used in an inferential situation can only be realized through the analysis of the statistical characteristics of the related likelihood function. The infinite digression of considering the priors of priors can only be stopped by the analysis of the expected output at each stage, which is determined by the behavior of the likelihood function.

Incidentally, the present observation shows why the conventional subjectivist doctrine of assuming the determination of the prior distribution of the parameters independently of the related likelihood function was not strictly followed by the research workers dealing with real inference problems. This point is discussed as the Bayes / Non-Bayes compromise by Good (1965). We take here the very flexible attitude towards the Bayes procedure to consider it only as one possibility of utilizing the information supplied by the likelihood function. Thus we consider that any practically useful statistical procedure which utilizes the information supplied by the likelihood function should not be rejected only because it is non-Bayesian. It is not the dogmatic exclusion of other procedures but the explicit proposal of useful models that proves the advantage of the Bayesian approach over the conventional statistics.

5. NUMERICAL EXAMPLES

To show that our discussion in the preceding sections is not vacuous, here we show some numerical examples. These were obtained by Bayesian modelings but with the help of some procedures which are not strictly Bayesian. The first three examples are concerned with the models discussed in the preceding section. The last one is an example of polynomial fitting and is included to show the feasibility of a Bayesian modeling with the aid of an information criterion (AIC) to deal with the difficulty of choosing a prior distribution for a multimodel situation where the models are with different number of parameters.

For the first three examples the essential statistic used for the determination of the parameter d in (4.4) is the likelihood of the model specified by the prior distribution. We consider the marginal likelihood of (d,σ^2) defined by

$$L(d,\sigma^2) = \int f(y\,|\,\sigma^2,a)\,\pi\,(a\,|\,d)\,da,$$

where $f(y\,|\,\sigma^2,a)$ and $\pi(a\,|\,d)$ are given by (4.3) and (4.4), respectively. If we assume (4.5) and put $c_0 = 0$ we get

$$L(d,\sigma^2) = (1/2\pi)^{N/2}(1/\sigma)^N \exp\,[\,-(\,1/2\sigma^2)\,\|\,z(a_\cdot\,|\,d)\,\|^2]$$

$$\cdot\,|\,d^2D'D\,|^{1/2}\,|\,d^2D'D + X'X\,|^{-1/2},$$

where $\| z(a_{\bullet}|d) \|^2$ denotes the minimun of $\| z(a|d) \|^2$ with $z(a|d)$ defined by (4.6). Instead of developing a prior distribution of (d,σ^2) we consider the use of the procedure which chooses a model with the maximum marginal likelihood. This is called the method of type II maximum likelihood by Good (1965). For a given d, the maximum with respect to σ^2 is attained at

$$\sigma_d^2 = (1/N)\| z(a_{\bullet}|d)\|^2.$$

For the case of practical applications, we consider a finite set of possible values $(d_1, d_2, \ldots, d_l)$ of d and choose the one that maximizes $L(d,\sigma_d^2)$. Since we are familiar with the use of minus twice the log likelihood, we propose to minimize

$$\begin{aligned}\text{ABIC} &= (-2) \log L(d, \sigma_d^2) \\ &= N \log \left[1/N \| z(a_{\bullet}|d) \|^2 \right] + \log \| d^2D'D + X'X \| \\ &\quad - \log \| d^2D'D \| + \text{const},\end{aligned}$$

where ABIC stands for "a Bayesian information criterion". When different D's are not considered, the term $\log \| d^2D'D \|$ may be replaced by $2K \log d$, where K is the dimension of the vector a.

In the last example we demonstrate the practical utility of $\exp(-\frac{1}{2}\text{AIC})$ as the definition of the likelihood of a model specified by the maximum likelihood estimates of the parameters. Here AIC is by definition (Akaike, 1974)

$$\text{AIC} = (-2) \log (\text{maximum likelihood}) + 2\,(\text{number of free parameters}).$$

This definition allows a very practical procedure of developing a Bayesian type approach to the situation where several models with different numbers of parameters are considered.

The general definition of ABIC of a model with hyperparameters determined by the method of type II maximum likelihood would have been ABIC = (-2) log (maximum marginal likelihood) + 2 (number of adjusted hyperparameters). In the examples treated in this paper the numbers of the adjusted hyperparameters are identical within the models being compared and their influence on the maximum marginal likelihoods is ignored.

Examples

a. *Distributed lag estimation*

We did a simulation with the second example of Shiller (1973, p. 783). The result is illustrated in Table 1. This result was obtained by using the model

c of the precedin section with $N = 40$, $K = 20$ and $D = D_2$ with $\alpha = \beta = 0$. Considering that this is a limiting situation with non-zero α and β, ABIC was defined by

$$\mathrm{ABIC} = N \log [(1/N) \mid z(a_* | d) |^2]$$

$$+ \log \mid d^2 D'D + X'X \mid - 2K \log d,$$

and the ABIC was minimized over $d = 5.0, 2.5, 1.25, 0.625, 0.3125$. the values of the ABIC at these d's were -43.4, -51.5, -52.7, -45.0, -30.9, respectively. The minimum, -52.7, was attained at $d = 1.25$ and corresponding estimates of the parameters are given in Table 1 along with the theoretical values and the least squares estimates. By taking a properly weighted average of the results with different d's we may get a procedure which has smaller sampling variability, but it seems that the present simple procedure is almost sufficient for many practical applications.

TABLA 1
Example of distributed lag estimation

i	1	2	3	4	5
Theoretical	.000	.000	.001	.004	.018
Bayes	-.009	-.003	.004	.009	.017
Least squares	-.010	.021	-.045	.037	.078

i	6	7	8	9	10
Theoretical	.054	.130	.242	.352	.399
Bayes	-.051	.134	.242	.345	.395
Least squares	-.074	.255	.113	.462	.334

i	11	12	13	14	15
Theoretical	.352	.242	.130	.054	.018
Bayes	.362	.257	.134	.052	.012
Least squares	.359	.329	.046	.072	.042

i	16	17	18	19	20
Theoretical	.004	.001	.000	.000	.000
Bayes	-.001	-.015	.006	.035	-.018
Least squares	-.018	-.050	.065	-.008	-.008

b. *Locally smooth trend fitting*

In this example the original data $y(i)$ ($i = 1, 2, \ldots, 30$) were generated by the relation

$$y(i) = 4 \exp [- (1/2) ((i-5)/4)^2] + z(i),$$

where z(i)'s are independently and identically distributed as $N(0,1)$. Twelve models of locally smooth trend of kth order defined by the model *e* of the preceding section with $\dot{d} = 2^{8-j}$ (j = 1, 2, ... , 12) were tried with k = 1, 2, 3. The constants α, β, and γ of the D_k's were all put equal to 0.001. The ABIC was defined by

$$\text{ABIC} = N \log [(1/N) | z(a_* | \text{d}) |^2] + \log | \text{d}^2 D'D + X'X |$$

$$-\log | d^2 D'D |.$$

The minimum of ABIC was attained at k = 1 and d = 2.0. The original data, the theoretical trend and some of the estimated trends are illustrated in Fig. 1. In this figure SSDEV stands for the sum of squares of deviations of the estimates from the theoretical. It can be seen that the present procedure can produce meaningful results even with these rather noisy observations. In the figures *ID* stands for k.

c. *Seasonal adjustment*

In this case the model f was applied to various artificial and real time series of length six years, i.e., N = 72. The constants of D_k in D_{kp} were the same as in the preceding example and other constants were e = 0.001, f = 1.0 and g = 10.0. The set of twelve values of d used in the preceding example was also used here and k = 1, 2, 3 were tried. Results corresponding to the minima of the ABIC's are illustrated in Fig.'s 2—4.

Fig. 2 shows the result of application of the present procedure to an artificial series given in Abe, Ito, Maruyama et al (1971, pp. 250-251). The result shows a very good reproduction of the true trend curve which was disturbed by a fixed multiplicative seasonality and the addition of the irregular components to produce the observations denoted by original.

It is remarkable that by this procedure no special treatment is necessary at the end of the series. This point is a significant advantage over the conventional procedures which require various ad hoc adjustments at the beginning and end of the series (Shiskin and Eisenpress, 1957). Fig. 3 shows the result of application to the last six years of the series of the logarithms of the number of airline passengers, given as Series *G* in Box and Jenkins (1970). The result reveals a very reasonable gradual change of the seasonality. The procedure has also been applied to the time series of labor force given in Table

1 of Shiskin and Eisenpress (1957, p.442) and the result is given in Fig. 4. The adjusted series is simply defined by $y(i)$ - S_i and is compared with the series adjusted by the Method II by Shiskin and Eisenpress.

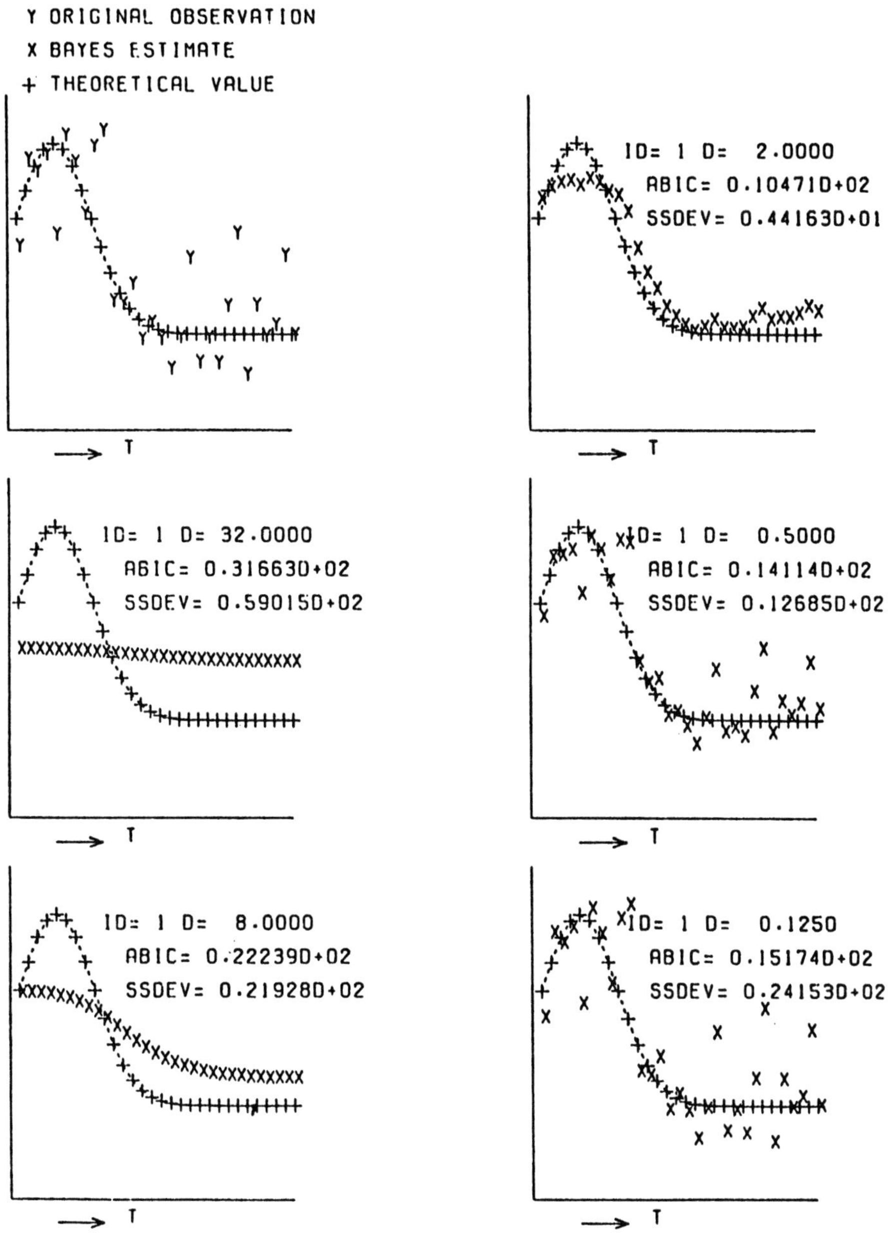

FIGURE 1

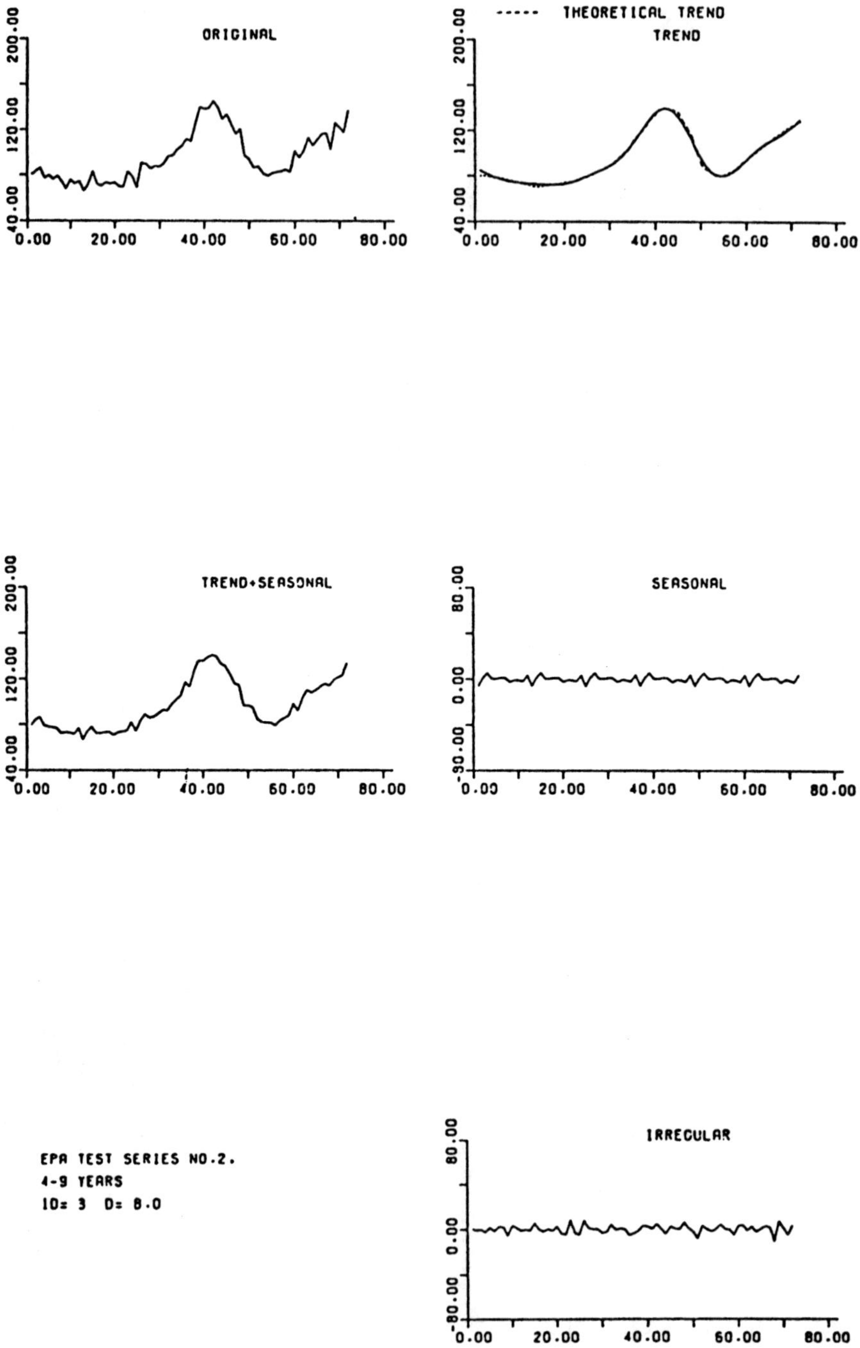
ORIGINAL
----- THEORETICAL TREND
TREND
TREND+SEASONAL
SEASONAL
IRREGULAR
EPA TEST SERIES NO.2.
4-9 YEARS
ID= 3 D= 8.0
200.00
120.00
40.00
80.00
0.00
-80.00
0.00
20.00
40.00
60.00
80.00

FIGURE 2

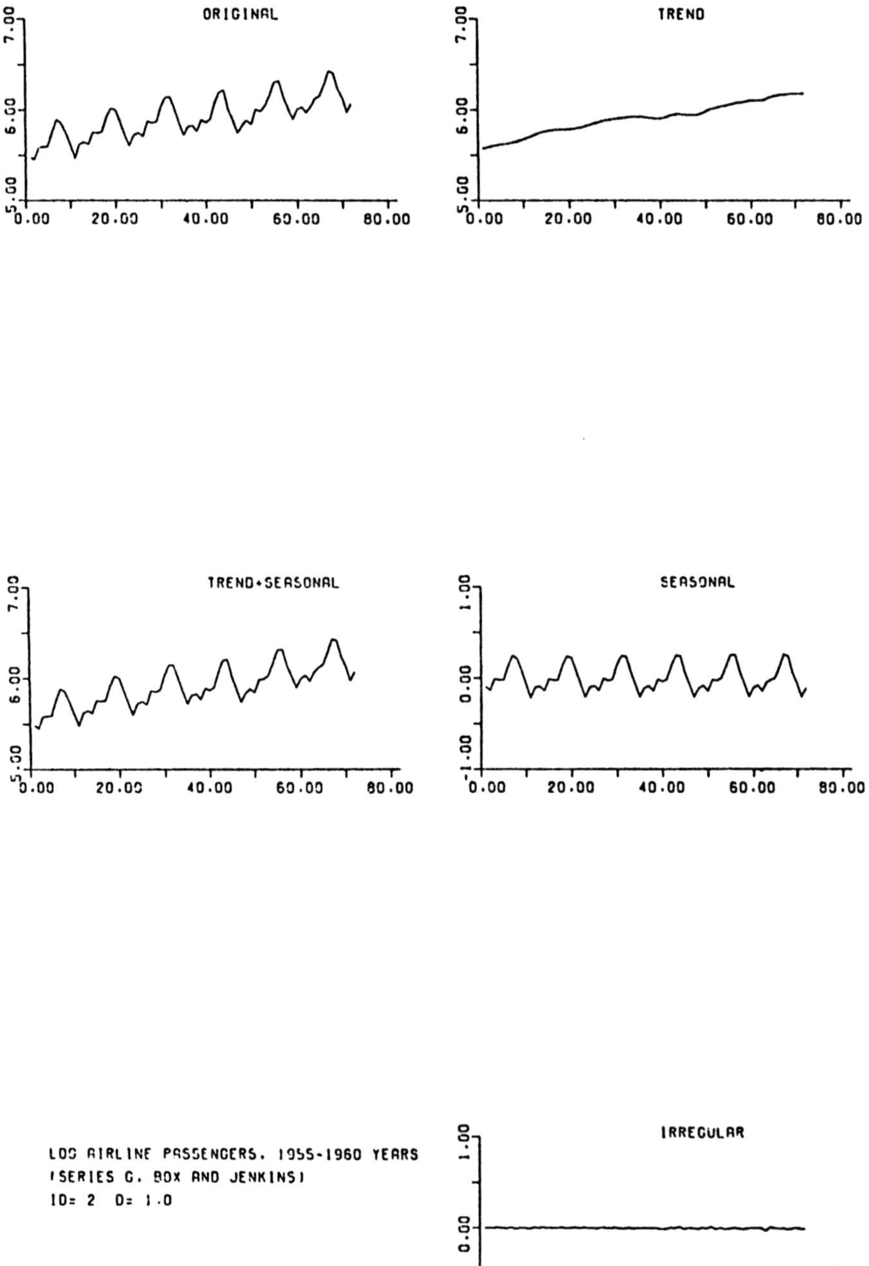
ORIGINAL
TREND
TREND+SEASONAL
SEASONAL
IRREGULAR
LOG AIRLINE PASSENGERS, 1955-1960 YEARS
(SERIES G, BOX AND JENKINS)
ID= 2 D= 1.0
7.00
6.00
5.00
1.00
0.00
-1.00
0.00
20.00
40.00
60.00
80.00

FIGURE 3

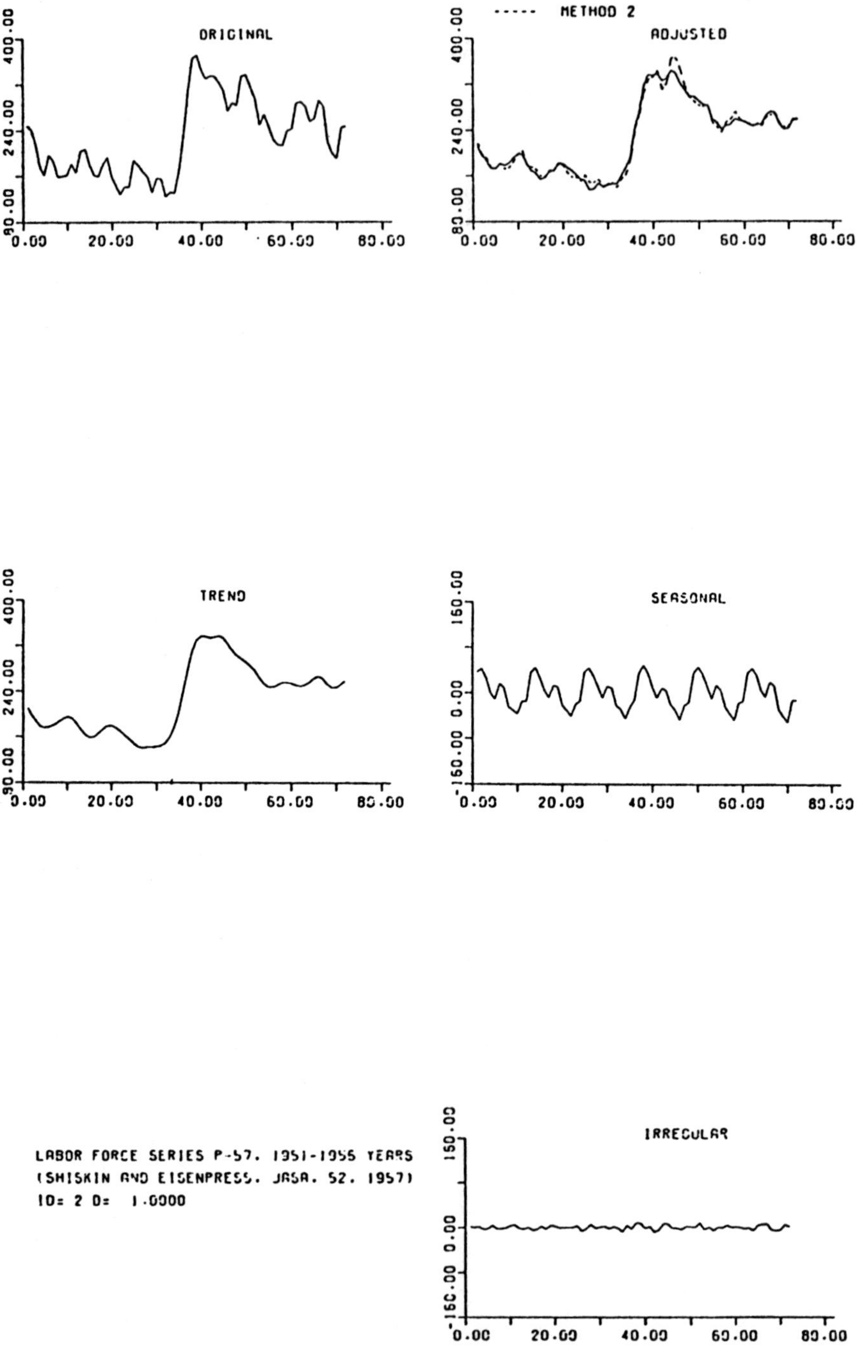
ORIGINAL
METHOD 2
ADJUSTED
TREND
SEASONAL
IRREGULAR
LABOR FORCE SERIES P-57. 1951-1956 YEARS
(SHISKIN AND EISENPRESS. JASA. 52. 1957)
ID= 2 D= 1.0000
0.00
20.00
40.00
60.00
80.00
400.00
240.00
80.00
150.00
0.00
-150.00

FIGURE 4

d. *Polynomial fitting*

By this example we wish to demonstrate that a reasonable definition of the likelihood of a model defined by the maximum likelihood estimates of the parameters can be given by exp (-(1/2) AIC) (Akaike, 1979a, c). The observations $y(i)$ are identical to those of the example *b* of this section and the polynomials of successively increasing order were fitted up to the 10th order by the method of maximum likelihood. Under the assumption of the Gaussian distribution, the AIC of the Mth order model is defined by

$$\text{AIC}\ (M) = N \log [\ (1/N)\ \text{S}(M)\ ] + 2M,$$

where $S(M)$ denotes the sum of squares of the residuals. Some of the estimated regression curves and the values of the AIC are illustrated in Fig. 5.

We smoothed these regression curves with the weight proportional to exp $[\ -\ (1/2)\ \text{AIC}\ (M)\ ]\ \pi(M)$ with $\pi(M) \propto (M + 1)^{-1}$. The result is denoted by "Bayes" in the figure. The same type of procedure has been applied to the fitting of autoregressive models by Akaike (1979a) where the choice of $\pi(M)$ is discussed.

The present result shows that the procedure is practically useful, although its performance depends on the choice of the system of the basic functions or the polynomials. Usually this choice produces significant effects at the beginning and end of the regression curve. This shows the advantage of the models used in the preceding examples *b* and *c* over the present model. Nevertheless the present result demonstrates the feasibility of a Bayesian modeling of a multi-model problem with models defined with different number of parameters.

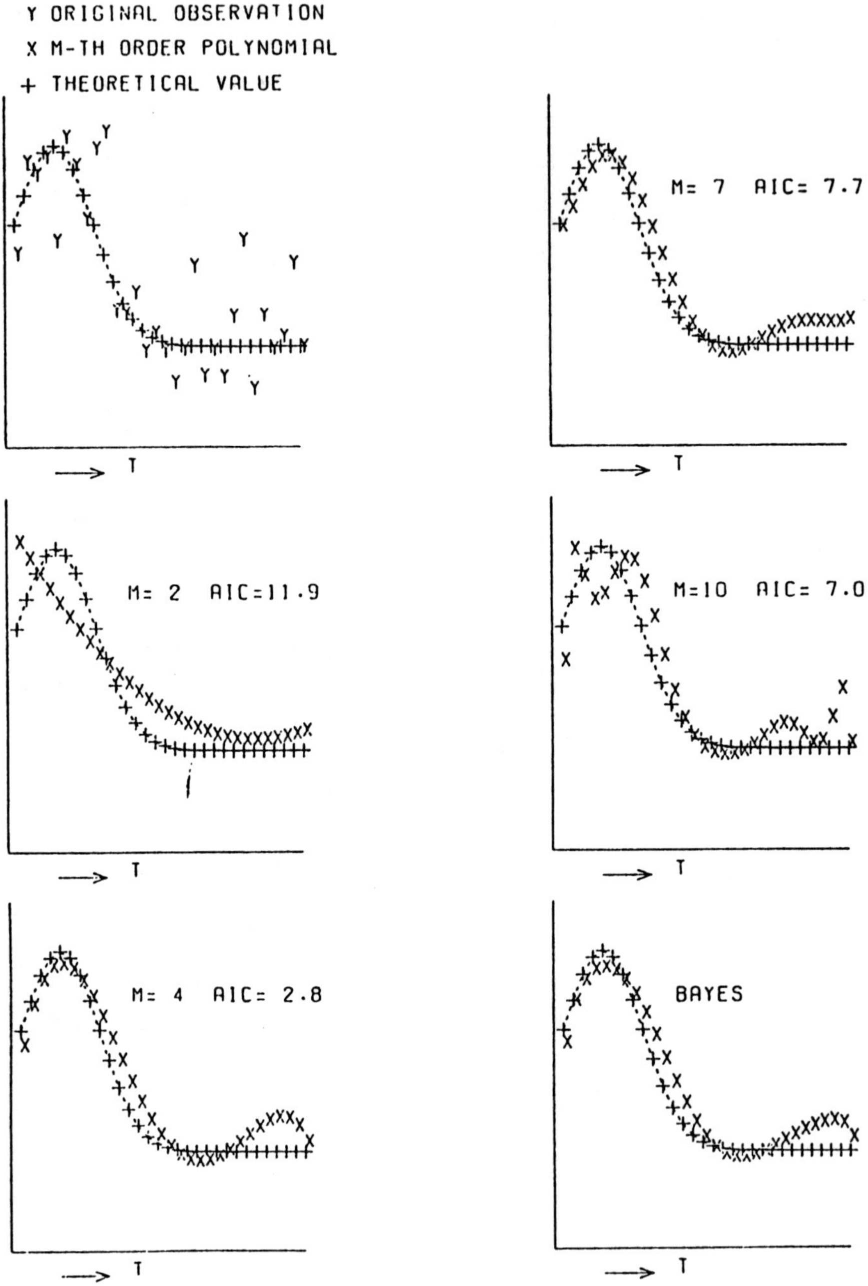
Y ORIGINAL OBSERVATION
X M-TH ORDER POLYNOMIAL
+ THEORETICAL VALUE
M= 7 AIC= 7.7
M= 2 AIC=11.9
M=10 AIC= 7.0
M= 4 AIC= 2.8
BAYES
T

FIGURE 5

6. DISCUSSION

The numerical results presented in the preceding section suggest the possibility of developing further applications of the general linear model to problems such as the gradually changing autoregression and the general trend analysis of time series. This possibility is pursued in Akaike (1979d). By choosing the set of d's properly the type II maximum likelihood method may be replaced by a procedure which takes an average of the models with respect to the weight proportional to the likelihood of each model. The performance of these procedures are controlled by the statistical characteristics of the related likelihood functions. One particular possibility is the extension of the concept of ignorance prior distribution to the prior distribution of a hyperparameter. This is discussed in Akaike (1980).

The application to seasonal adjustement is particularly interesting as it provides an example of the model which cannot be treated by the ordinary method of maximum likelihood. This example clearly demonstrates the practical utility of the Bayesian approach. It also shows that our present procedure may be characterized as a tempered method of maximum likelihood. The practical utility of the general linear model stems from the understandability and manipulability of the related prior distributions. This allows us to make proper judgement on how to temper the likelihood function through the choice of the values of the constants within the priors.

The subjective theory of probability is developed on the basis of our psychological reaction to uncertainty. Acordingly the final justification of the theory must be sought in the psychological satisfaction it can produce throught its application to real problems. It is only the accumulation of successful results of application that can really make the Bayesian statistics attractive.

The Bayes procedure provides a natural and systematic way of utilizing the information supplied by a likelihood function. The likelihood has a clearly defined objective meaning as the measure of the goodness of a model. It is this objectivity that provides the basis for the use of the subjective theory of probability as a guide in developing statistical procedures. Only this objectivity allows us to develop our confidence on the practical utility of the Bayes procedure, even when we know that the related model is our subjective construction.

ACKNOWLEDGEMENTS

The author is grateful to Dr K. Tanabe for the stimulating discussion of the linear model. The author is also indebted to Ms. E. Arahata for preparing the numerical and graphical outputs reported in this paper.

REFERENCES

ABE, K; ITO, M., MARUYAMA, A., YOSHIKAWA, J., ISUKADA, K. and IKEGAMI, M. (1971). *Methods of Seasonal Adjustements. Research Series No. 22.*, Tokyo: Economic Planning Agency Economic Research Institute (In Japanese).

AKAIKE, H. (1974). A new look at the statistical model identification. *IEEE Trans. Automat. Control*, **AC-19**, 716-723.

— (1978a). A new look at the Bayes procedure. *Biometrika*, **65**, 53-59.

— (1978b). A Bayesian analysis of the minimum AIC procedure. *Ann. Inst. Statist. Math.*, **30**, *A*, 9-14.

— (1979a). A Bayesian extension of the minimum AIC procedure of autoregressive model fitting. *Biometrika*, **66**, 53-59.

— (1979b). A subjective view of the Bayes procedure. *Research Memo. No. 117.* Tokyo: The Institute of Statistical Mathematics. Revised, February 1979.

— (1979c). On the use of the predictive likelihood of a Gaussian model. *Research Memo. No 159.* Tokyo: The Institute of Statistical Mathematics.

— (1979d). On the construction of composite time series models. *Research Memo. No 161.* Tokyo: The Institute of Statistical Mathematics.

— (1980). Ignorance prior distribution of a hyperparameter and Stein's estimator. *Ann. Inst. Statist. Math.*, **32**, *A*, 171-178.

BOX, G.E.P. and JENKINS, G.M. (1970). *Time Series Analysis, Forecasting and Control.* San Francisco: Holden Day.

DE FINETTI, B (1974a). Bayesianism: Its unifying role for both the foundation and application of statistics. *Int. Stat. Rev.*, **42**, 117-130.

— (1974b/1975) *The Theory of Probability, Volumes 1 and 2.* New York: Wiley.

GOOD, I.J. (1965) *The Estimation of Probabilities.* Cambridge, Massachusetts: M.I.T. Press.

KUDO, H. (1973). The duality of parameter and sample. *Proceedings of the Institute of Statistical Mathematics Symposium*, **6**, 9-15 (In Japanese).

LINDLEY, D.V. (1957). A statistical paradox. *Biometrika*, **44**, 187-192.

O'HAGAN, A. (1978). Curve fitting and optimal design for prediction. *J.R. Statist. Soc. B*, **40**, 1-42.

SAVAGE, L.J. (1962). Subjective probability and statistical practice. In *The Foundations of Statistical Inference.* (G.A. Barnard and D.R. Cox eds.) 9-35.London: Methuen.

— (1954) *The Foundations of Statistics.* New York: Wiley.

SCHWARZ, G. (1978) Estimating the dimension of a model. *Ann. Statist.*, **6**, 461-464.

SHILLER, R. (1973) A distributed lag estimator derived from smoothness priors. *Econometrica*, **41**, 775-778.

SHISKIN, J. and EISENPRESS, H. (1957). Seasonal adjustments by electronic computer methods. *J. Amer. Statist. Ass.*, **52**, 415-499.

TIHONOV, A.N. (1965). Incorrect problems of linear algebra and a stable method for their solution. *Soviet Math. Dokl.*, **6**, 988-991.

WOLFOWITZ, J. (1962). Bayesian inference and axioms of consistent decision. *Econometrica*, **30**, 470-479.

Reprinted from *Journal of Time Series Analysis*, Vol. 1, No. 1, 1980, 1-13 by permission from Blackwell Publishers

SEASONAL ADJUSTMENT BY A BAYESIAN MODELING

By Hirotugu Akaike

The Institute of Statistical Mathematics, Tokyo

Abstract. The basic ideas underlying the construction of a newly introduced seasonal adjustment procedure by a Bayesian modeling are discussed in detail. Particular emphasis is placed on the use of the concept of the likelihood of a Bayesian model for model selection. The performance of the procedure is illustrated by a numerical example.

Keywords. Seasonal adjustment; ill-posed problem; Bayesian model; likelihood of a model; X-11; BAYSEA.

1. INTRODUCTION

At present there are three main types of procedures of the seasonal adjustment of time series. The first is based on some moving average procedures for the extraction of trend and seasonal components. The well-known Census Method II X-11 variant (Shiskin, Young and Musgrave, 1967) belongs to this category. The second is based on the multiple regression technique, where some smooth elementary functions of time are used to define the independent variables; see, for example, Stephenson and Farr (1972). The third is by the direct time series modeling, particularly of ARIMA type, of the trend and seasonal components. This type of approach is discussed, for example, by Box, Hillmer and Tiao (1978) and Engle (1978).

The moving average type procedure, particularly the X-11 procedure, is now developed to such an extent that it is rather hard to contemplate any essential modifications. The X-11 procedure is composed of a large collection of sub-procedures and they are selectively applied by empirically established complex decision rules. Thus, although the procedure has been very successful, each time we encounter a new difficulty we have to develop some ad hoc modifications. A typical example of this is given by the recent discussion by Shiskin and Plewes (1978) of the U.S. unemployment rate.

The regression approach has not been considered to be particularly hopeful (Kallek, 1978; Shiskin and Plewes, 1978). This is due to the fact that the final result is quite sensitive, particularly at the beginning and end of the series, to the choice of the independent variables.

If one considers the prediction of a time series as the final object of the seasonal adjustment the time series modeling approach will appear as most natural and promising. However, this approach is more appropriate for the detailed analyses of some particular time series and, as is noted by Engle (1978), further developments of both computationally efficient procedures of estimation and criteria for model selection are required to establish the advantage of this approach over the X-11 type procedures.

0143-9782/80/01 001-13 $02.50/0

One practical way of improving an existing procedure belonging to one of the above three categories will be to partially incorporate the characteristics of procedures of other categories. The X-11 ARIMA variant developed by Dagum (1978) is such an example. However, we must pay our attention to the dichotomy stressed by Box, Hillmer and Tiao (1978), the dichotomy of empirical and model-based approach. This separates the X-11 type procedure from the procedures in other two categories. The significant advantage of the model-based approach is that it is based on a clearly defined statistical model of the time series and the modification of the procedure can be realized effectively by the modification of the basic model.

From the standpoint of practical application the crucial problem is the choice of one particular procedure for the adjustment of a given time series. The choice of the multiplicative or additive model in the application of the X-11 procedure is an example. If the procedure is model-based, we can, at least in principle, systematically apply the concept of the likelihood of a time series model discussed by Akaike (1978) for the selection of the best model.

The procedure we are going to discuss in this paper is based on a Bayesian model of the time series. Formally it may be viewed as a time series modeling procedure based on a very limited family of models. These models assume that the time differences of appropriate orders of the trend and seasonal components form mutually independent Gaussian white noises. Although at first sight this looks rather restrictive as a family of time series models the final estimates of the trend and seasonal components take the form of regression estimates based on an extremely flexible set of independent variables. Thus the procedure has the characteristics of the regression approach, particularly with its simplicity of estimation, and yet has the characteristics of the time series modeling approach which is free from the trouble of the ordinary regression approach at both ends of the time series.

The procedure is particularly flexible in handling the additional information such as the trading-day or leap year effect, the shifting of the base line of observations and abnormal or missing observations. An objective procedure of evaluation and selection of models can be realized by the consistent use of the concept of the likelihood of a Bayesian model. The feasibility of the seasonal adjustment procedure of this type was first demonstrated by Akaike (1979a) and its modifications were discussed by Akaike (1979b), Akaike and Ishiguro (1980b) and Ishiguro and Akaike (1980). The computer program, called BAYSEA (Bayesian seasonal adjustment) which realizes the basic part of the procedure is published in Akaike and Ishiguro (1980a). It is the purpose of this paper to provide a unifying view of the basic ideas underlying the development of this new seasonal adjustment procedure.

2. CONSTRAINED LEAST SQUARES AND ITS BAYESIAN INTERPRETATION

Consider the problem of decomposing a time series $y(i)$ $(i = 1, \ldots, N)$ into the form

$$y(i) = T_i + S_i + I_i,$$

where T_i denotes the trend, S_i the seasonal and I_i the irregular component. The fundamental period of the seasonal component is denoted by p. For simplicity, we assume that the length of the data is given by $N = Mp$ and the observations are taken at the time points $i = mp + j$ $(j = 1, \dots, p,\ m = 0, \dots, M-1)$.

In the regression approach T_i and S_i are represented respectively by

$$T_i = \sum_{k=1}^{K} a_k f_k(i) \quad \text{and} \quad S_i = \sum_{k=1}^{K} b_k g_k(i),$$

where $f_k(i)$ and $g_k(i)$ are properly chosen functions of i and the coefficients a_k and b_k are determined by minimizing the sum of squares

$$\sum_{i=1}^{N} \{y(i) - T_i - S_i\}^2.$$

The results are usually affected significantly at the beginning and end of the series by the choice of the functions $f_k(i)$ and $g_k(i)$. The most nonrestrictive choice of the functions is realized by putting

$$f_k(i) = g_k(i) = \delta(i-k) \text{ with } K = N,$$

where $\delta(i-k) = 1$, for $i = k$, and 0, otherwise. With this choice we have $T_i = a_i$ and $S_i = b_i$ and the direct application of the method of least squares cannot produce any meaningful result. To solve the difficulty, instead of trying to restrict the family of the functions $f_k(i)$ and $g_k(i)$, we put some constraints on the behavior of T_i and S_i based on the prior preference of the behavior of these components.

The requirement of the smoothness of the trend may be represented by constraining the sum of squares of the differences $T_i - T_{i-1}$, or of any other higher order differences such as $(T_i - T_{i-1}) - (T_{i-1} - T_{i-2})$, to a small value. Similarly the gradual change of the seasonal component may be realized by assuming the sum of squares of the differences $S_i - S_{i-p}$ to be small. The requirement that the sum of S_i's within a period be close to zero can similarly be handled. A solution which satisfies these requirements may be obtained, for example, by minimizing

$$\sum_{i=1}^{N} [\{y(i) - T_i - S_i\}^2 + d^2\{(T_i - 2T_{i-1} + T_{i-2})^2 + r^2(S_i - S_{i-12})^2 + z^2(S_i + S_{i-1} + \cdots + S_{i-11})^2\}],$$

where d, r and z are properly chosen constants.

In the above formulation of the problem our prior preference on the trend and seasonal components is given an explicit representation by the second term inside the square brackets. The usefulness of this type of constrained least squares method for the solution of the so-called ill-posed problem is well-known; see, for example, Tihonov (1965). However, the choice of the parameters d, r and z has been a serious problem. Here we solve this problem by a Bayesian modeling.

To clarify the basic structure of the model we consider in this section the linearly constrained least squares problem in its most general form. Denote by y the N-(column) vector of observations $y(i)$ and by a the M-vector of parameters, such as T_i's and S_i's. The linearly constrained least squares problem is to find the vector a that minimizes

$$L(a)=\|y-Xa\|^2+\|a-a_0\|_R^2,$$

where X is an $N\times N$ matrix and a_0 denotes a known M-vector and where $\|\ \|$ denotes the ordinary Euclidean norm and $\|\ \|_R$ the norm specified by a positive definite matrix R, i.e., $\|a\|_R^2=a'Ra$. Here $'$ denotes the transpose.

Obviously the minimization of $L(a)$ is equivalent to the maximization of

$$l(a)=\exp\left\{-\frac{1}{2\sigma^2}L(a)\right\},$$

where σ^2 is a positive constant. We have

$$l(a)=\exp\left(-\frac{1}{2\sigma^2}\|y-Xa\|^2\right)\exp\left\{-\frac{1}{2\sigma^2}(a-a_0)'R(a-a_0)\right\},$$

which shows that the constrained least squares problem is equivalent to the maximization of the posterior density of the vector a under the assumption of the data distribution

$$f(y|\sigma^2,a)=\left(\frac{1}{2\pi}\right)^{N/2}\left(\frac{1}{\sigma}\right)^N\exp\left\{-\frac{1}{2\sigma^2}\|y-Xa\|^2\right\}$$

and the prior distribution

$$p(a|R,\sigma^2,a_0)=\left(\frac{1}{2\pi}\right)^{M/2}\left(\frac{1}{\sigma}\right)^N|R|^{1/2}\exp\left\{-\frac{1}{2\sigma^2}\|a-a_0\|_R^2\right\},$$

where $|R|$ denotes the determinant of R.

We are particularly interested in the situation where we have

$$\|a-a_0\|_R^2=\|D(a-a_0)\|^2,$$

where D is an $L\times M$ matrix with rank M. In this case we have

$$L(a)=\left\|\begin{pmatrix}y\\c_0\end{pmatrix}-\begin{pmatrix}x\\D\end{pmatrix}a\right\|^2,$$

where $c_0=Da_0$. $L(a)$ attains its minimum at

$$a_*=(X'X+D'D)^{-1}(X'y+D'c_0),$$

and we have

$$L(a)=L(a_*)+(a-a_*)'(X'X+D'D)(a-a_*).$$

This shows that, as the posterior density of a is proportional to

$$f(y|\sigma^2,a)p(a|R,\sigma^2,a_0)=\left(\frac{1}{2\pi}\right)^{(N+M)/2}\left(\frac{1}{\sigma}\right)^{N+M}|R|^{1/2}\exp\left\{-\frac{1}{2\sigma^2}L(a)\right\},$$

where $R=D'D$, the solution a_* of the constrained least squares problem is identical to the posterior mean of a.

3. THE BASIC BAYESIAN MODEL FOR SEASONAL ADJUSTMENT

In this section we will specialize the general Bayesian model of the preceding section for the application to seasonal adjustment. To maintain a better correspondence of the model with the computational procedure of the program BAYSEA we define the vector y of observation by

$$y = (y(N), y(N-1), \dots, y(1))'.$$

A natural choice of the vector a of the parameters will then be

$$a = (S_N, S_{N-1}, \dots, S_1, T_N, T_{N-1}, \dots, T_1)'.$$

The matrix X is then defined by

$$X = [I_N, I_N],$$

where I_N denotes an $N \times N$ identity matrix. The matrix D is defined by

$$D = d\begin{pmatrix} rD_{11} & 0 \\ zD_{12} & 0 \\ 0 & D_{32} \end{pmatrix},$$

where 0 denotes an $N \times N$ zero-matrix. D_{11} is given by the Kronecker product $E_{11} \otimes I_p$, i.e., the (i, j)th block of D_{11} is given by $E_{11}(i, j) \cdot I_p$ where $E_{11}(i, j)$ denotes the (i, j)th element of E_{11}. Here I_p is a $p \times p$ identity matrix and E_{11} is an $M \times M$ matrix defined, for example, by

$$E_{11} = \begin{pmatrix} 1 & -1 & & & \\ & 1 & -1 & 0 & \\ & & \ddots & \ddots & \\ & 0 & & 1 & -1 \\ & & & & 1 \end{pmatrix},$$

where 0 denotes zero entries. Obviously D_{11} controls the smoothness of the seasonal component. D_{21} is an $N \times N$ matrix introduced to keep the running averages of S_i close to zero and is defined by

$$D_{21} = \begin{pmatrix} \overbrace{1 & 1 & \cdots & 1}^{\leftarrow p \rightarrow} & & & & & \\ & 1 & 1 & \cdots & 1 & & & & \\ & & 1 & & 1 & \cdots & 1 & & 0 \\ & & & & & \cdots & & & \\ & & & & & & 1 & 1 & \cdots & 1 \\ & & 0 & & & & & 1 & \cdots & 1 \\ & & & & & & & & & 1 \end{pmatrix} \begin{matrix} \\ \\ \\ \\ \uparrow \\ p \\ \downarrow \end{matrix}.$$

The matrix D_{32} controls the smoothness of the trend component and is defined, for example, by

$$D_{32} = \begin{pmatrix} 1 & -2 & 1 & & & & \\ & 1 & -2 & 1 & & 0 & \\ & & \ddots & \ddots & \ddots & & \\ & & & 1 & -2 & 1 & \\ & 0 & & & 1 & -2 & \\ & & & & & 1 & \end{pmatrix}.$$

This D_{32} controls the smoothness of the trend component through the control of the second order differences of T_i.

The final estimate a_* depends on the choice of the parameters d, r, z of D and the choice of D_{11} and D_{32}. Also the present definition of D suggests that there is a problem of the initial values of T_i and S_i at the beginning of the series. These problems will be discussed in the following sections.

A word of caution is in order. This is on the choice of d in the definition of D. From the definition of D_{32} it is obvious that if d is less than 1 the fluctuations of the second order differences of the trend component is less penalized than the residual errors, or the fluctuations of the irregular components. Thus it may happen that we get zero irregulars as the final result. This observation suggests that we should limit d to a range of values greater than or equal to 1.

4. LIKELIHOOD OF A BAYESIAN MODEL

The potential of the Bayesian model developed in the preceding section can be confirmed only when a proper procedure is developed for the determination of the necessary prior distribution. The concept of the likelihood of a Bayesian model is particularly useful for this purpose.

When a set of Bayesian models specified by the pairs of the data distribution $f_k(\cdot|\theta_k)$ and the prior distribution $p(\theta_k)$ $(k = 1, \ldots, K)$ is given the Bayesian modeling will be completed by specifying a prior distribution $\pi(k)$ for k. When y is observed the prior distribution $p(\theta_k)\pi(k)$ is transformed into the posterior distribution $p(\theta_k|y)\pi(k|y)$. Here $p(\theta_k|y)$ is the posterior distribution of θ_k conditional on k and $\pi(k|y)$ is the posterior probability of the kth model defined by the relation

$$\pi(k|y) \propto f_k(y)\pi(k),$$

where $f_k(y) = \int f_k(y|\theta_k)p(\theta_k)\, d\theta_k$ which we call the likelihood of the Bayesian model specified by $f_k(\cdot|\theta_k)$ and $p(\theta_k)$.

The definition of $\pi(k|y)$ clearly shows that $f_k(y)$ plays exactly the role of the likelihood function in the ordinary Bayes procedure. In particular, when the prior probabilities $\pi(k)$ are equal, the model with the maximum likelihood defines the model with the maximum posterior probability. This suggests the utility of selecting a model with maximum likelihood.

The likelihood of the Bayesian model developed in the preceding section is given in terms of the notations of Section 2 as the integral of $f(y|\sigma^2, a)\cdot p(a|R, \sigma^2, a_0)$. From the formula given at the end of Section 2 it is easy to see that the likelihood is given by

$$l(R, \sigma^2, a_0) = \left(\frac{1}{2\pi}\right)^{N/2}\left(\frac{1}{\sigma}\right)^{N}|R|^{1/2}|X'X+R|^{-1/2}\exp\left\{-\frac{1}{2\sigma^2}L(a_*)\right\},$$

where $R = D'D$ and $L(a_*)$ is the minimum of the sum of squares $\|y - Xa\|^2 + \|D(a-a_0)\|^2$. Numerically the values of $|R|$, $|X'X+R|$ and $L(a_*)$ can very simply be obtained by the application of the Householder transformation for the solution of the least squares problem (Golub, 1969).

For a given pair of R and a_0 we choose the model with $\sigma^2 = \sigma_0^2$ that maximizes $l(R, \sigma^2, a_0)$. Then from a set of finite number of R's we choose the one that maximizes $l(R, \sigma_0^2, a_0)$. To keep the similarity with the AIC criterion (Akaike, 1978, p. 219) we use

$$\text{ABIC} = (-2)\ln\{l(R, \sigma^2, a_0)\}$$

for the comparison of models. Here ABIC stands for a Bayesian information criterion. As a simple illustration of the use of this criterion we may consider the comparison of the additive and log additive model. We have only to multiply the likelihood of the model of the log transformed variables $z(i) = \ln y(i)$ by the Jacobian, in this case $\{\Pi y(i)\}^{-1}$, to get the likelihood in terms of the original variables. Thus the necessary modification of ABIC of the log-additive model for the comparison with the ABIC of the additive model is simply realized by adding $2\,\Sigma \ln y(i)$ to the ABIC of the additive model for z.

In the program BAYSEA the search for the model with the minimum ABIC is limited over a finite set of the values of the parameter d in the definition of D of section 3, with other parameters being fixed. Comparison between the models with different structures of D is then realized by comparing ABIC's each minimized with respect to d.

5. THE INITIAL VALUE PROBLEM

Before going into the discussion of the initial value problem first we note the extreme flexibility of the present Bayesian model. In particular, there is no constraint on the definition of the vector a except that Xa defines the mean of the observation y by a proper choice of X. We may even consider that a is composed of the trend and seasonal components extending from the infinite past to the infinite future. The Bayesian model allows the inference on these components based on the information supplied by the data y. This may suggest that the initial value problem can be avoided by the modeling of infinitely long past histories. However, at least computationally, we must limit the modeling to a finite span of time and we can never be free from the initial value problem.

The versatility of the model allows a simple handling of the missing value problem. We have only to discard the missing components of y and their corresponding rows of X from our model (Akaike and Ishiguro, 1980b). The initial value problem can be treated as a special case of this missing value problem. With the model of Section 3 the initial value problem can be handled simply by augmenting the original observation vector y by the imaginary observations $y(0)$, $y(-1), \ldots, y(-p+1)$ and then declaring them as missing. The corresponding vector a then takes the form

$$a = (s', s'_-, t', t'_-)',$$

where $s = (S_N, S_{N-1}, \ldots, S_1)'$ and $s_- = (S_0, S_{-1}, \ldots, S_{-p+1})$ and t and t_- are similarly defined.

For simplicity we assume $a_0 = 0$, a zero vector. This is equivalent to replacing $a - a_0$ by a. The prior distribution of a is then determined by $\|Da\|^2$. Now D can be represented in the form

$$D = \begin{pmatrix} \overset{\leftarrow N \leftarrow}{G_{11}} & \overset{\leftarrow p \rightarrow}{G_{12}} & \overset{\leftarrow N \rightarrow}{0} & \overset{\leftarrow p \rightarrow}{0} \\ 0 & 0 & G_{23} & G_{24} \end{pmatrix},$$

so that we have $\|Da\|^2 = \|G_{11}s + G_{12}s_-\|^2 + \|G_{23}t + G_{24}t_-\|^2$. Denote the projection of $G_{12}s_-$ onto the space spanned by the column vectors of G_{11} by $-G_{11}s_0$. Then we have

$$\|G_{11}s + G_{12}s_-\|^2 = \|G_{11}(s - s_0)\|^2 + \|G_{12}s_- + G_{11}s_0\|^2,$$

where $s_0 = -(G'_{11}G_{11})^{-1}G_{11}G_{12}s_-$. Thus we can see that the prior distribution of s conditional on s_- is determined by the first term on the right hand side of the above equation and that it has mean equal to s_0. The second term on the right hand side determines the distribution of s_-. As s_0 is obtained by a linear transformation of s_- the mean of this distribution is zero. The prior distribution of t can similarly be analyzed and the conditional mean t_0 of t is given by a linear transformation of t_-. Thus, when s_- and t_- are given the present a will be replaced by $a = (s', t')'$ and a_0 by $a_0 = (s'_0, t'_0)'$.

In the above discussion we could have used any s_- and t_-, if only they are composed of sufficiently long past histories of the seasonal and trend components. Thus in comparing a finite number of models defined with different D's we take s_- and t_- sufficiently long to cover the whole set of models. Our experience suggests that the final estimates of the trend and seasonal components are fairly sensitive to the choice of G_{12} and G_{24}, those parts of D that mainly control the distribution of s_- and t_-. As a practical solution to this problem we assume a common (improper) uniform prior distribution of s_- and t_- for the models and apply the method of maximum likelihood to determine the values of s_- and t_- for each model.

In the program BAYSEA these maximum likelihood estimates are further approximated by the estimates obtained by fitting the Bayesian model backward in time, starting with the zero initials and with reduced values of the entries in the non-zero triangular parts of the lower right ends of the matrices D_{11}, D_{21} and D_{32} defined in section 3. This reduction of values make the model less sensitive to the assumption of the zero initials. Also to reduce the effect of uncertainty of the absolute level of the trend component the observations $y(i)$ are replaced by $y(i)-y(N)$ for this backward model fitting.

6. HANDLING OF PRIOR INFORMATION

The present Bayesian model allows easy handling of various prior information by simple modification of the matrices X and D and the vector a_0. The handling of the missing value problem briefly described in the preceding section is one such example. Any prior information which can be expressed in the form of the regression of y can be handled by augmenting X by the column vectors of some appropriate independent variables and also introducing reasonable constraints on the values of the additional regression coefficients by augmenting D properly. The detection of a jump in the trend discussed by Akaike (1979b) and the evaluation of trading-day and leap year effects discussed by Ishiguro and Akaike (1980) are examples of this type of handling.

In the case of economic time series it is often considered that when three or four years of additional observations are available the result of seasonal adjustment of a particular year becomes fairly stable; see, for example, Kuiper (1978). Under these circumstances, and also from practical considerations, it becomes necessary to declare at one point that the adjustment for a particular year is final. By our present Bayesian model this situation can be handled very easily. Actually this is the situation where we have s_- and t_-, the trend and seasonal components for the preceding years, fixed. The necessary (conditional) mean vector a_0 for the succeeding years is then computed by the procedure described in the preceding section. By our experience with BAYSEA this procedure works very well for ordinary economic monthly data with additional two or three years span of observations.

7. NUMERICAL EXAMPLES

To show how our Bayesian seasonal adjustment procedure works one numerical example is presented here. The original data is the series C of the monthly soft drink data of Gersovitz and Mackinnon (1978, table 4). The results denoted by the cases 1, 2 and 3 in fig. 1a and b were obtained by the combinations

$$(\text{ORDER}=1, \text{SORDER}=1), (\text{ORDER}=2, \text{SORDER}=1)$$
$$\text{and } (\text{ORDER}=2, \text{SORDER}=2),$$

respectively, where ORDER and SORDER respectively denote the orders of differencing of the trend and seasonal component used for the definition of D. The results were obtained by applying the procedure to successive four years spans of data with each span shifted by one year, except for the initial span where the length of the span was seven years and the result for the first four years was used as the initial condition for the following span. The parameters r and z were fixed at 1.0 and $1/\sqrt{p}$, now with $p=12$, respectively. At the initial backward fitting of the

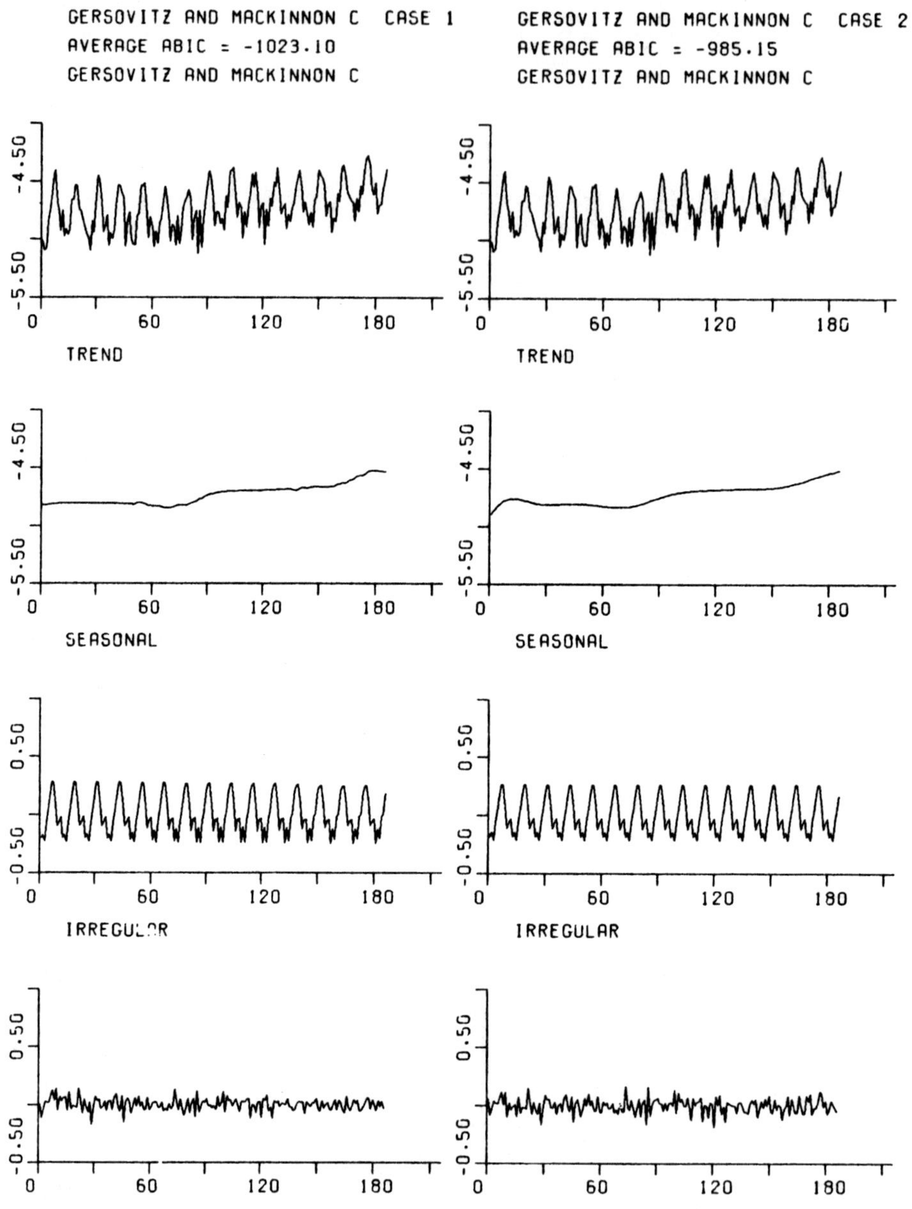

FIGURE 1a. Decomposition of soft drink data by BAYSEA, case 1 and 2.

model the incomplete rows, or the rows with smaller number of non-zero entries than others, of D_{11}, D_{21} and D_{32} were multiplied by 0.1, 0.1 and 0.01, respectively, to reduce the effect of the zero initials. The search for d was limited to integral multiples of $2^{1/4}$.

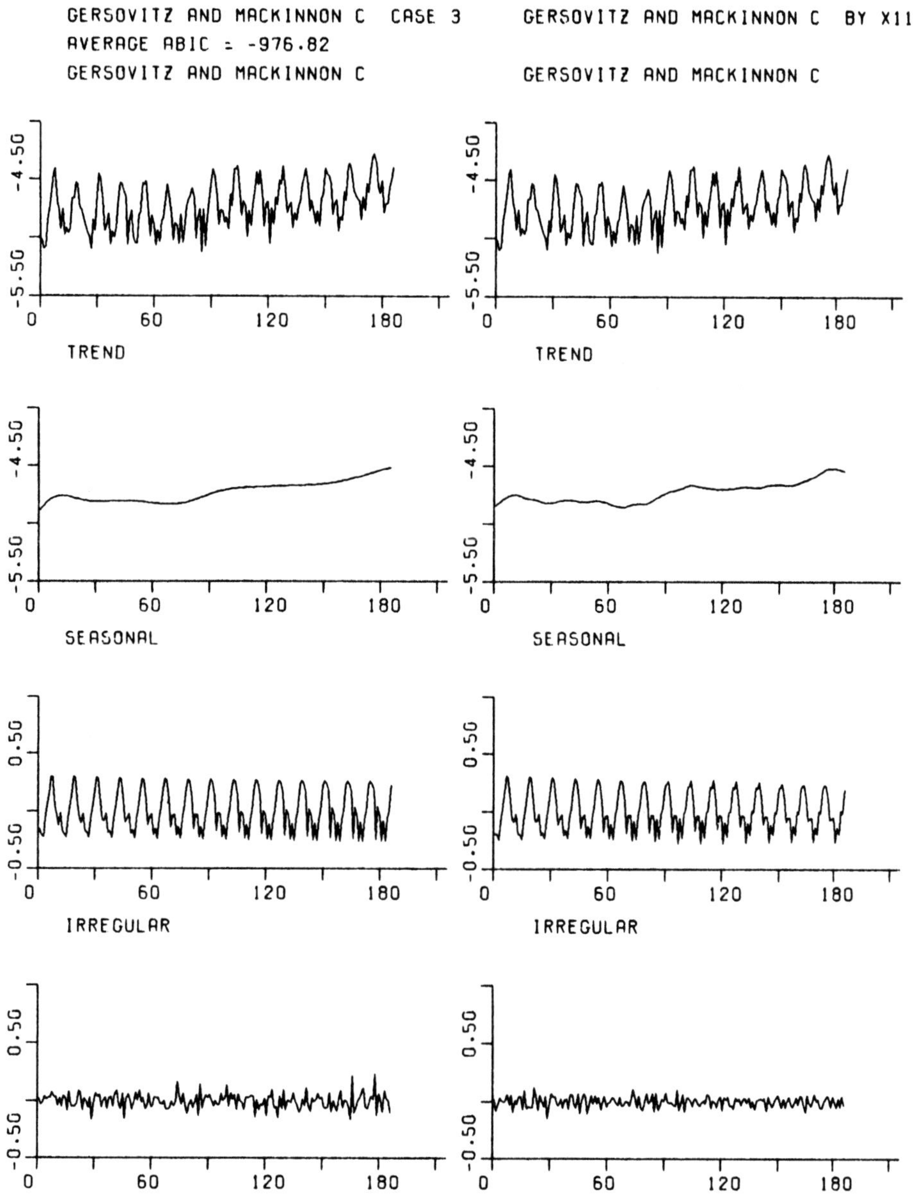

FIGURE 1b. Decomposition of soft drink data by BAYSEA, case 3, and by X-11.

The average ABIC within the figure stands for the sum of the ABIC's, minimized with respect to d for each consecutive span of data, times a scaling constant. It is seen that case 1 produces the best result by this criterion. The result obtained by the X-11 additive procedure is included in Fig. 1b for the purpose of comparison. In this experiment extreme value correction was not adopted in both BAYSEA and X-11. The X-11 procedure produces a trend component which is more flexible than those obtained by BAYSEA. Our numerical investigation with other examples suggests that the X-11 procedure produces trend and seasonal components which are too responsive to non-systematic fluctuations of original data. However we will not discuss this point here.

8. CONCLUDING REMARK

The simplicity and versatility of the basic model and the objectivity of the model selection criterion are the characteristics unique to our procedure. Our experience suggests that our procedure is a viable alternative to the X-11 procedure. Although we started with a rather modest expectation that the procedure may provide a reasonable approximation to the X-11 procedure, our experience suggests that to have the above stated characteristics is a definite advantage in a problematic situation. The discussion of the result of our comparative study of the X-11 and the BAYSE procedure is beyond the scope of the present paper and will be published elsewhere.

It has often been considered that the orthodox and Bayesian approach to statistics are incompatible. The development of our procedure clearly demonstrates the potential of Bayesian modeling in practical applications. Obviously our approach here is not purely subjective Bayesian. It may be characterized as an objective use of a Bayesian model. The construction of our model clearly demonstrates the subjective nature of the definition of the trend, seasonal and irregular components. They are characterized by what we expect them to be. This is a completely subjective procedure, but it is described objectively. Whether the resulting model provides a reasonable description of the data is then judged by an objectively defined criterion ABIC.

No procedure is available for the judgement of the absolute goodness of a model in a practical application. However, if we adopt the point of view of maximizing the entropy, or the expected log likelihood, of the model, as discussed in Akaike (1977), we may search for the best model by comparing the likelihoods of various models. However, to compare the present type of models with a model of different construction we will have to adjust ABIC for the bias caused by the maximum likelihood estimation of the necessary parameters, as in the case of AIC. This problem has not been pursued yet.

ACKNOWLEDGEMENTS

Many of the technical points discussed in this paper were clarified by the discussion with Makio Ishiguro during the process of implementation of BAYSEA. For this I am most grateful. Thanks are due to Sadao Naniwa of the Bank of Japan for providing useful information on the seasonal adjustment of economic time series. This work was facilitated by a grant from the Asahi Shimbun Publishing Company.

REFERENCES

AKAIKE, H. (1977) On Entropy Maximization Principle in *Applications of Statistics*, P. R. Krishnaiah, ed., Amsterdam: North-Holland, pp. 27–41.

AKAIKE, H. (1978) On the Likelihood of a Time Series Model *The Statistician* 27, 217–235.

AKAIKE, H. (1979a) Likelihood and the Bayes Procedure in *Proceedings of the International Meeting on Bayesian Statistics held at Valencia, Spain, 1979. A special issue of Trabajos de Estadistica* (to appear).

AKAIKE, H. (1979b) On the Construction of Composite Time Series Models *Proceedings of 42nd Session of the International Statistical Institute* (to appear).

AKAIKE, H. and M. ISHIGURO, (1980a) BAYSEA, A Bayesian Seasonal Adjustment Program *Computer Science Monographs* No. 13, Tokyo: The Institute of Statistical Mathematics.

AKAIKE, H. and M. ISHIGURO, (1980b) Trend Estimation with Missing Observations *Annals of the Institute of Statistical Mathematics* (to appear).

BOX, G. E. P., S. C. HILLMER, and G. C. TIAO, (1978) Analysis and Modeling of Seasonal Time Series in *Seasonal Analysis of Economic Time Series*, A. Zellner ed., U.S. Bureau of the Census, Economic Research Report ER-1, pp. 309–334.

DAGUM, E. B. (1978) Modelling, Forecasting and Seasonally Adjusting Economic Time Series with the X-11 ARIMA Method *The Statistician* 27, 203–216.

ENGLE, R. F. (1978) Estimating Structural Models of Seasonality in *Seasonal Analysis of Economic Time Series*, A. Zellner ed., U.S. Bureau of the Census, Economic Research Report ER-1, pp. 281–297.

GERSOVITZ, M. and J. G. MACKINNON, (1978) Seasonality in Regression: An Application of Smoothness Priors *Journal of the American Statistical Association* 73, 264–273.

GOLUB, G. H. (1969) Matrix Decompositions and Statistical Computations in *Statistical Computation*, R. C. Milton and J. A. Nelder eds., New York: Academic Press, pp. 365–397.

ISHIGURO, M. and H. AKAIKE, (1980) A Bayesian Approach to the Trading-day Adjustment of Monthly Data in *Time Series Analysis: Proceedings of the International Conference held at Houston, August 1980*, O. D. Anderson and M. R. Perryman eds., Amsterdam: North-Holland (to appear).

KALLEK, S. (1978) An Overview of the Objectives and Framework of Seasonal Adjustment in *Seasonal Analysis of Economic Time Series*, A. Zellner, ed., U.S. Bureau of the Census, Economic Research Report ER-1, pp. 3–25.

KUIPER, J. (1978) A Survey and Comparative Analysis of Various Methods of Seasonal Adjustment in *Seasonal Analysis of Economic Time Series*, A. Zellner ed., U.S. Bureau of the Census, Economic Research Report ER-1, pp. 59–76.

SHISKIN, J., A. H. YOUNG, and J. C. MUSGRAVE, (1976) The X-11 Variant of the Census Method II Seasonal Adjustment Program *Technical Paper No. 15*, Bureau of the Census, U.S. Department of Commerce.

SHISKIN, J. and T. J. PLEWES, (1978) Seasonal Adjustment of the U.S. Unemployment Rate *The Statistician* 27, 181–202.

STEPEHENSON J. A. and H. T. FARR, (1972) Seasonal Adjustment of Economic Data by General Linear Statistical Model *Journal of the American Statistical Association* 67, 37–45.

TIHONOV, A. N. (1965) Incorrect Problem of Linear Algebra and a Stable Method for Their Solution *Soviet Mathematics Doklady* 6, 988–991.

Ann. Inst. Statist. Math.
34 (1982), Part B, 389-398

A QUASI BAYESIAN APPROACH TO OUTLIER DETECTION

GENSHIRO KITAGAWA AND HIROTUGU AKAIKE

(Received Aug. 1, 1981; revised Dec. 14, 1981)

Summary

A quasi Bayesian procedure is developed for the detection of outliers. A particular Gaussian distribution with ordered means is assumed as the basic model of the data distribution. By introducing a definition of the likelihood of a model whose parameters are determined by the method of maximum likelihood, the posterior probability of the model is obtained for a particular choice of the prior probability distribution. Numerical examples are given to illustrate the practical utility of the procedure.

1. Introduction

The problem of outlier detection attracted much attention because of its practical and conceptual importance. Numerous papers treated this problem from the traditional testing point of view. However, in the detection of unknown number of multiple outliers, a severe difficulty is caused by the so-called masking effect (Tietjen and Moore [15]) and none of the solutions hitherto proposed is considered to be entirely satisfactory (Hawkins [12]). The work by Box and Tiao [7] demonstrated the importance of formulating the problem explicitly in terms of a Bayesian modeling. Recently Freeman [8] reviewed three Bayesian models of outliers in data from the linear model discussed by Box and Tiao [7], Abraham and Box [1] and Guttman, Dutter and Freeman [11]. In the paper Freeman noticed the difficulty of handling improper prior distributions of parameters when the numbers of the parameters are different among the models. Box and Tiao [7] and Abraham and Box [1] avoided this difficulty by considering only the models with a fixed number of parameters. In Guttman, Dutter and Freeman [11], an ad hoc procedure is developed to get information on the number of outliers.

Although Freeman [8] tried to avoid the difficulty by using proper priors throughout, the difficulty of choosing an appropriate set of priors is clearly demonstrated by his numerical examples. In the discussion of his paper it is mentioned that it might be that attempts like the AIC criterion to produce a standard way of answering a wide variety of questions regardless of their different contexts are doomed to failure. This comment sounds rather out of place as it is the importance of proper modeling and objective criterion for model selection that has been stressed by the introduction of AIC.

The purpose of the present paper is to show that by developing an appropriate family of models and using the concept of the likelihood of a model obtained by properly extending the basic idea of the AIC criterion (Akaike [4]) we can in fact develop a procedure which practically avoids the difficulty discussed by Freeman. The predictive log likelihood of a model determined by specifying the set of assumed outliers and applying the method of maximum likelihood, is defined as an approximately unbiased estimate of the expected log likelihood of the model. A quasi Bayesian procedure is then realized by using the predictive likelihoods and appropriately chosen prior probabilities of the models. Numerical results are given to show that the procedure produces reasonable results for some of the examples frequently discussed in the literature.

2. The model and its predictive likelihood

Let $x^t=(x_1,\cdots,x_n)$ be a vector of n observations. J is an ordered set of k integers $\{i_1,\cdots,i_k\}$ chosen from the integers $\{1,2,\cdots,n\}$. When J is specified, we consider that k observations $x_{i_1},\cdots,x_{i_k}$ are outliers, whereas the rests, x_j $(j\notin J)$, are normal observations drawn from a Gaussian distribution with unknown mean μ_0 and variance σ^2. The outliers are assumed to be obtained from Gaussian distributions with ordered means, $\mu_1\leqq\mu_2\leqq\cdots\leqq\mu_k$, and common variance σ^2. Thus the component model given J is specified by the data distribution

$$f(x\,|\,J,\theta)=\prod_{j\notin J}\frac{1}{\sigma}\phi\left[\frac{x_j-\mu_0}{\sigma}\right]\prod_{j=1}^{k}\frac{1}{\sigma}\phi\left[\frac{x_{i_j}-\mu_j}{\sigma}\right]$$

where $\theta=(\mu_0,\mu_1,\cdots,\mu_k,\sigma^2)$ and $\phi(x)$ denotes the standard Gaussian density function.

Given the observation, $x_1,\cdots,x_n$, the log likelihood of the component model is given by

$$\log f(x\,|\,\theta,J)=-\frac{n}{2}\log 2\pi\sigma^2-\frac{1}{2\sigma^2}\left\{\sum_{j\notin J}(x_j-\mu_0)^2+\sum_{j=1}^{k}(x_{i_j}-\mu_j)^2\right\}.$$

Thus the maximum likelihood estimate of the mean of normal observations, μ_0, is obtained by $\hat{\mu}_0=(1/(n-k))\sum_{j\notin J}x_j$. While those of the mean values of outliers, $\hat{\mu}_i$ $(i=1,\cdots,k)$, are obtained by maximizing the log likelihood function under order restrictions. This is equivalent to finding the solution to the problem of quadratic programming:

$$\text{minimize} \qquad F(\mu_1,\cdots,\mu_k)=\sum_{j=1}^{k}(\mu_j-x_{i_j})^2$$

$$\text{subject to} \qquad \mu_1 \leqq \mu_2 \leqq \cdots \leqq \mu_k .$$

They are easily obtained numerically by the pool-adjacent-violators algorithm (Barlow et al. [5]) which will be briefly described in Section 4. For the model specified by $J=\{i_1,\cdots,i_k\}$ which satisfies the natural ordering condition, $x_{i_1} \leqq x_{i_2} \leqq \cdots \leqq x_{i_k}$, they are simply given by $\hat{\mu}_j = x_{i_j}$. The maximum likelihood estimate of the variance σ^2 is then obtained by

$$\hat{\sigma}^2 = \frac{1}{n}\left\{\sum_{j\notin J}(x_j-\hat{\mu}_0)^2+\sum_{j=1}^{k}(x_{i_j}-\hat{\mu}_j)^2\right\} .$$

According to the entropy maximization principle [3], we evaluate the goodness of the model specified by these maximum likelihood estimates by its expected log likelihood

$$\text{(1)} \quad \mathrm{E}_y \log f(y|J,\hat{\theta}) = -\frac{n}{2}\log 2\pi\hat{\sigma}^2 - \frac{1}{2\hat{\sigma}^2}\left\{n\sigma^2+(n-k)(\mu_0-\hat{\mu}_0)^2+\sum_{j\in J}(\mu_j-\hat{\mu}_j)^2\right\} ,$$

where E_y denotes the expectation under the assumed distribution of y, $f(y|J,\theta)$. In a practical situation, the true parameter $\theta=(\mu_0,\mu_1,\cdots,\mu_k,\sigma^2)$ is unknown and thus the present form of the expected log likelihood is useless. Following the idea underlying the definition of AIC ([2], [3]) we try to correct the bias of the maximized log likelihood,

$$\text{(2)} \qquad \log f(x|J,\hat{\theta}) = -\frac{n}{2}\log 2\pi\hat{\sigma}^2 - \frac{n}{2} ,$$

as an estimate of the expected log likelihood. From (1) and (2), the average increase of the maximum log likelihood is obtained by

$$\begin{aligned} C_{k,n} &= \mathrm{E}_x\{\log f(x|J,\hat{\theta}) - \mathrm{E}_y \log f(y|J,\hat{\theta})\} \\ &= \mathrm{E}_x\left[\frac{1}{2\hat{\sigma}^2}\left\{n\sigma^2+(n-k)(\mu_0-\hat{\mu}_0)^2+\sum_{j=1}^{k}(\mu_j-\hat{\mu}_j)^2-\frac{n}{2}\right\}\right] \end{aligned}$$

where E_x denotes the expectation under the assumed distribution $f(x|J,\theta)$ of the data. However, since the maximum likelihood estimates $\hat{\mu}_1,\cdots,\hat{\mu}_k$ depend in a complicated way on the magnitudes of $x_{i_1},\cdots,x_{i_k}$, it is difficult to evaluate $C_{k,n}$ analytically. Therefore for the present purpose of evaluation of the bias, we will use the simple estimator $\hat{\mu}_j = x_{i_j}$ $(j=1,\cdots,k)$ irrespectively of the ordering of the magnitudes of $x_{i_1},\cdots,x_{i_k}$. This is equivalent to assuming that our model satisfies the natural ordering $x_{i_1} \leqq x_{i_2} \leqq \cdots \leqq x_{i_k}$ and will produce a reasonable approximation when $\sigma^{-1}(\mu_i - \mu_{i-1})$ $(i=2,\cdots,k)$ are sufficiently large. The effect of this assumption will be checked in Section 6.

Under the assumption that the original data x was drawn from the distribution $f(x|J,\theta)$, we have $\mathrm{E}_x[\hat{\sigma}^{-2}] = n(n-k-3)^{-1}\sigma^{-2}$, $\mathrm{E}_x[(\hat{\mu}_0 - \mu_0)^2] = (n-k)^{-1}\sigma^2$ and $\mathrm{E}_x[(\hat{\mu}_i - \mu_i)^2] = \sigma^2$, and it follows that

$$C_{k,n} = \frac{n(k+2)}{n-k-3} .$$

An unbiased estimate of the expected log likelihood of the estimated model is now obtained as

$$\log f(x|J,\hat{\theta}) - C_{k,n}$$

and the predictive likelihood of the estimated model under the assumption of J is defined by

$$p(x|J) = \exp\{\log f(x|J,\hat{\theta}) - C_{k,n}\} .$$

3. The prior and posterior probabilities

We assume that we have no information initially to say that a specific k, the number of outliers, is more likely than others. Hence we put $p(0) = p(1) = \cdots = p(n) = 1/(n+1)$, where $p(k)$ denotes the prior probability that there are k outliers. Given that there are k outliers, there are ${}_nC_k$ ways of specifying k observations as outliers out of the n. Thus the prior probability that a specific set of k observations are the outliers is given by

$$\frac{1}{n+1}\,{}_nC_k^{-1} = \frac{(n-k)!k!}{(n+1)!} .$$

This prior probability admits another derivation. We assume that an observation is an outlier with probability α. The probability that a particular set of k observations are the outliers is then given by $\alpha^k(1-\alpha)^{n-k}$. By integrating $\alpha^k(1-\alpha)^{n-k}$ over 0 through 1, we obtain the above prior probability.

For each set of k assumed outliers there are $k!$ ways of assigning them to the k distributions specified by the means μ_j $(j=1,\cdots,k)$. By assuming every configurations to be equally probable, we obtain the prior probability of the model specified by $J = \{i_1,\cdots,i_k\}$ as

$$\pi(j)=\frac{(n-k)!}{(n+1)!} .$$

The posterior probability of the model specified by J is then given by $\pi(J|x)=p(x)^{-1}p(x|J)\pi(J)$, where $p(x)=\sum_J p(x|J)\pi(J)$ and $p(x|J)$ denotes the predictive likelihood of the model defined in the preceding section. The posterior probability of $x_{i_1},\cdots,x_{i_k}$ being the outliers is given by $\sum \pi(J|x)$, where the summation extends over the set of $k!$ J's obtained by permuting $\{i_1,\cdots,i_k\}$.

4. Algorithm

In this section we will describe an algorithm for the computation of posterior probabilities.

(1) *Specification of outliers*

For $m=0,1,\cdots,2^n-1$, put $ind(i)$ $(i=1,\cdots,n)$ equal to the ith bit of the binary expansion of m, i.e., $\sum_{i=1}^n 2^{i-1}ind(i)=m$. The number of outliers is given by $k=\sum_{i=1}^n ind(i)$ and the set of outliers $\{x_{i_j};\ j=1,\cdots,k\}$ is specified by putting $i_j=i$ for the jth non-zero $ind(i)$.

(2) *Computation of the log posterior probability of the naturally ordered model*

For the given combination $\{i_1,\cdots,i_k\}$ of assumed outliers, compute the logarithm of the posterior probability of the naturally ordered model by

$$\log p(m|x)=-\frac{n}{2}\log\sigma^2(m)-\frac{n(k+2)}{n-k-3}-\log n!+\log(n-k)! ,$$

where a common additive constant is ignored and

$$\sigma^2(m)=\frac{1}{n}\sum_{ind(j)=0}(x_j-\mu(m))^2 , \qquad \mu(m)=\frac{1}{n-k}\sum_{ind(i)=0}x_j .$$

(3) *Computation of the posterior probability*

The posterior probability of another model, specified by $J(j)=\{jnd(1),\cdots,jnd(k)\}$ which is obtained by rearranging $\{i_1,\cdots,i_k\}$, is obtained by using the weight, $w(j,m)$, relative to that of the naturally ordered model

$$w(j,m)=\left\{\frac{n\sigma^2(m)+\sum_{i=1}^k(x_{jnd(i)}-\hat{\mu}_i)^2}{n\sigma^2(m)}\right\}^{-n/2} ,$$

where the maximum likelihood estimates $\hat{\mu}_i$ $(i=1,\cdots,k)$ are obtained by the following pool-adjacent-violators algorithm (Barlow et al. [5]).

The algorithm starts with the initial estimates $\mu_i=x_{jnd(i)}$ $(i=1,\cdots,k)$. If the initial estimates satisfy the order condition, they are the final estimates $\hat{\mu}_i$ $(i=1,\cdots,k)$. If not, select all the sequence of violators of the ordering; that is, select all the pairs of p and q such that $\mu_{p-1}\leqq\mu_p>\mu_{p+1}>\cdots>\mu_q\leqq\mu_{q+1}$. For every pair of p and q, replace the estimates μ_i $(i=p,\cdots,q)$ by the pooled one $\frac{1}{q-p+1}\sum_{r=p}^{q}\mu_r$. If the resulting μ_i's do not yet satisfy the order condition, repeat the above step until to produce the final estimates $\hat{\mu}_i$ $(i=1,\cdots,k)$.

The posterior probability of x_{i_j} $(j=1,\cdots,k)$ being the outliers is obtained by

$$\pi(m\,|\,x)\propto p(m\,|\,x)\sum_j w(j,m)\,,$$

where the summation extends over $k!$ possible models specified by $J(j)$'s.

5. Examples

To check the performance of the present procedure, we applied it to two sets of familiar data. The computation was performed by a computer program, OUTLAP, developed by Kitagawa [14].

Example 1 (Darwin's data).

Table 1. Darwin's data

−67	−48	6	8	14	16	23	24
28	29	41	49	56	60	75	

This data set has been discussed by many researchers, such as Box and Tiao [7], Abraham and Box [1] and Freeman [8]. Here the data $x_1, x_2, \cdots, x_n$ are ar-

Table 2. Prior and posterior probabilities of some combinations of possible outliers (Darwin's data)

Number of outliers	Possible outliers	Prior probability	Posterior probability	w
2	−67, −48	0.0016	0.515	1.787
0	none	0.1667	0.140	1.
1	−67	0.0111	0.107	1.
3	−67, −48, 75	3.7×10^{-4}	0.085	1.697
1	−48	0.0111	0.017	1.
3	−67, −48, 6	3.7×10^{-4}	0.014	1.988
3	−67, −48, 60	3.7×10^{-4}	0.012	1.761
3	−67, −48, 8	3.7×10^{-4}	0.011	1.973
3	−67, −48, 56	3.7×10^{-4}	0.009	1.772
4	−67, −48, 60, 75	1.2×10^{-4}	0.007	2.795

ranged in ascending order of magnitude. The lowest two observations look rather discrepant from the rest. The posterior probabilities of various possible combinations of outliers are obtained by multiplying the predictive likelihoods, $\exp\{\log f(x|J)-C_{k,n}\}$, by the prior probabilities $\pi(J)$. The ten largest posterior probabilities which together constitute 91.7% of the total probability are listed in Table 2. The highest posterior probability 0.515 occurs at $J=\{1, 2\}$, indicating that -67 and -48 are probably the outliers. The posterior probabilities of $J=\phi$ (no outliers), {1} and {1, 2, 15} are also considerable. The probabilities of other combinations of assumed outliers are almost negligible. The marginal posterior probabilities $\pi(i|x)$ that the observation x_i is an outlier are obtained as $\pi(1|x)=0.812$, $\pi(2|x)=0.705$, $\pi(15|x)=0.120$, $\pi(14|x)=0.030$ and $\pi(i|x)<0.022$ for $i=3, 4, \cdots, 13$. The meaning of the values in the last column denoted by w will be explained in Section 6.

Example 2 (Herndon's data).

The second example is Herndon's data discussed by Grubbs [9], Tietjen and Moore [15] and Kitagawa [13]: Here the extreme values

Table 3. Herndon's data

−1.40	−0.44	−0.30	−0.24	−0.22	−0.13	−0.05	0.06
0.10	0.18	0.20	0.39	0.48	0.63	1.01	

on both sides are suspicious. The ten largest posterior probabilities which together constitute 85.1% of the total probability is listed in Table 3. The highest posterior probability 0.414 occurs at $J=\{1\}$, indicating the -1.40 is the outlier. The posterior probabilities that -1.40 and 1.01 are the outliers and that there is no outliers are not negligible but those of other cases are very small. The marginal posterior prob-

Table 4. Prior and posterior probabilities of some combinations of possible outliers (Herndon's data)

Number of outliers	Possible outliers	Prior probability	Posterior probability	w
1	−1.40	0.0111	0.414	1.0
2	−1.40, 1.01	0.0016	0.223	1.0
0	none	0.1667	0.152	1.0
3	−1.40, 0.63, 1.01	3.7×10^{-4}	0.035	1.553
2	−1.40, −0.44	0.0016	0.019	1.175
1	1.01	0.0111	0.018	1.0
2	−1.40, 0.63	0.0016	0.013	1.003
3	−1.40, −0.44, 1.01	3.7×10^{-4}	0.010	1.060
2	−1.40, −0.30	0.0016	0.010	1.126
3	−1.40, 0.48, 1.01	3.7×10^{-4}	0.009	1.388

abilities $\pi(i|x)$ that the observation x_i is an outlier are respectively given by $\pi(1|x)=0.802$, $\pi(15|x)=0.326$, $\pi(14|x)=0.064$, $\pi(2|x)=0.040$, $\pi(13|x)=0.027$ and $\pi(i|x)<0.021$ for $i=3, 4, \cdots, 12$.

6. Discussion

In Section 2, we evaluated the average increase of the maximum log likelihood due to the increase of the number of parameters under the simplifying assumption that the condition $x_{i_1} \leqq x_{i_2} \leqq \cdots \leqq x_{i_k}$ always holds. Now we will present a result of our empirical study on the effect of this assumption.

The following two situations were considered:

(1) $k=2$, $J=\{1, 2\}$

$$f(x|J, \mu_0, \mu_j, \sigma)=\sigma^{-n}\phi\left(\frac{x_1-\mu_1}{\sigma}\right)\phi\left(\frac{x_2-\mu_2}{\sigma}\right)\prod_{i=3}^{n}\phi\left(\frac{x_i-\mu_0}{\sigma}\right)$$

with $a=\mu_2-\mu_1 \geqq 0$, $\mu_0=0$ and $\sigma=1$.

(2) $k=3$, $J=\{1, 2, 3\}$

$$f(x|J, \mu_0, \mu_j, \sigma)=\sigma^{-n}\phi\left(\frac{x_1-\mu_1}{\sigma}\right)\phi\left(\frac{x_2-\mu_2}{\sigma}\right)\phi\left(\frac{x_3-\mu_3}{\sigma}\right)\prod_{i=4}^{n}\phi\left(\frac{x_i-\mu_0}{\sigma}\right)$$

with $a=\mu_2-\mu_1=\mu_3-\mu_2 \geqq 0$, $\mu_0=0$ and $\sigma=1$.

In Fig. 1 each dot shows the sample means of $C^*_{k,n}=M^{-1}\sum_x \{\log f(x|J,$

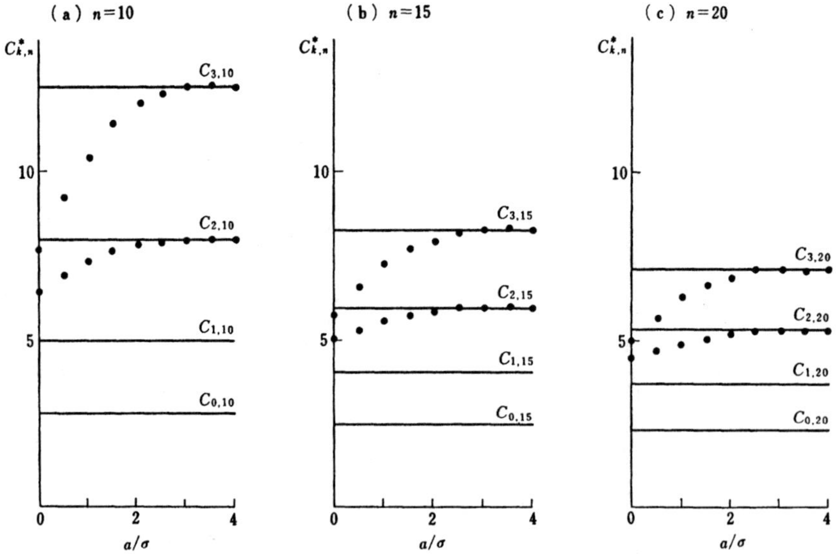

Fig. 1. Sample means of the average increase of the maximum log likelihood due to the increase of the number of parameters

$\hat{\mu}_0, \hat{\mu}_j, \hat{\sigma}) - E_y \log f(y|J, \hat{\mu}_0, \hat{\mu}_j, \hat{\sigma})\}$ obtained from $M=40000$ simulations for each combination of $k=2$ and 3, $n=10$, 15 and 20 and $a=0.0, 0.5, 1.0, \cdots, 4.0$. The solid lines show $C_{k,n}$ for $k=0$, 1, 2 and 3.

From the figure we can see that if $a/\sigma = \mu_i - \mu_{i-1} \geqq 2$, the difference of $C^*_{k,n}$ and $C_{k,n}$ is negligible compared with that of $C^*_{k,n}$ and $C^*_{k+1,n}$. On the other hand, if a/σ is much smaller than 2 the difference of $C^*_{k,n}$ and $C_{k,n}$ becomes considerable. It shows that if some of the outliers are located closely each other then the posterior probability that they are simultaneously outliers is underestimated. This indicates that our procedure may not work well for the detection of a cluster of several outliers, particularly when they are located close to the main distribution. For the detection of these outliers, we should use a modified model which allows the situation where some of the outliers are from a common distribution.

There are some practical ways of simplifying the algorithm to obtain reasonable approximations to the posterior probabilities. In an actual computation, we first rearrange the data in order of increasing magnitude as $x_{(1)} \leqq \cdots \leqq x_{(n)}$ and specify the maximum K of the number of possible outliers, so that, if $K < i \leqq n-K$, the posterior probability that $x_{(i)}$ is an outlier is very small. Then we set $\pi(J|x)=0$ if $\{x_i;\ i \in J\}$ contains any of $x_{(i)}$ with $K < i \leqq n-K$. With this approximation, the number of combinations of possible outliers is reduced to 2^{2K} from 2^n. This will avoid the combinatorial explosion of the number of models.

Generally $w(m)$ $(=\sum w(j, m))$ takes a value between 1 and $k!$. But we can see from Tables 2 and 4 that the values of $w(m)$ of models with significant posterior probabilities are close to 1. Thus we may put $w=1$ to get a reasonable approximation. With this approximation the algorithm can be simplified greatly. The value of $w(m)$ takes a significant value only when there are outliers with nearly equal values.

It is easy to extend the present quasi Bayesian approach to the detection of outliers of large variance type. The model in this case is specified by the conditional data distribution

$$p(x|J, \mu, \sigma, \tau) = \prod_{i \notin J} \frac{1}{\sigma} \phi\left(\frac{x_i - \mu}{\sigma}\right) \prod_{j \in J} \frac{1}{\tau} \phi\left(\frac{x_i - \mu}{\tau}\right),$$

and the prior probability

$$\pi(J) = \frac{(n-k)!k!}{(n+1)!}.$$

By our experience, however, the procedure based on this model is generally insensitive to outliers compared with the procedure discussed in this paper.

The advantage of the present quasi Bayesian approach to the outlier problem is that it allows a natural definition of the outlier correction. When an observation x_i is considered to be an outlier a natural correction will be to replace it with the sample mean of the normal observations. Since the posterior probability of the model specified by J is given by $\pi(J|x)$, the final corrected values are obtained by $z_i=\sum_J \pi(J|x)x_i(J)$ $(i=1,\cdots,n)$, where $x_i(J)$ is the corrected value of x_i under the assumption of the model specified by J. This will find a wide application in the area of automatic extreme value correction. Computer program based on this idea is already given in Kitagawa [14].

Acknowledgement

The work reported in this paper was partly supported by a grant from the Asahi Shimbun Publishing Company.

THE INSTITUTE OF STATISTICAL MATHEMATICS

REFERENCES

[1] Abraham, B. and Box, G. E. P. (1978). Linear models and spurious observations, *Appl. Statist.*, **27**, 131-138.
[2] Akaike, H. (1973). Information theory and an extension of the maximum likelihood principle, *2nd Int. Symp. on Information Theory* (eds. B. N. Petrov and F. Csaki), Akademiai Kiado, Bdapest, 267-281.
[3] Akaike, H. (1977). On entropy maximization principle, *Applications of Statistics* (ed. P. R. Krishnaiah), North-Holland, 27-41.
[4] Akaike, H. (1980). On the use of the predictive likelihood of a Gaussian model, *Ann. Inst. Statist. Math.*, **32**, A, 311-324.
[5] Barlow, R. E., Bartholomew, D. J., Bremner, J. M. and Brunk, H. D. (1972). *Statistical Inference under Order Restrictions*, John Wiley and Sons.
[6] Barnett, V. (1978). The study of outliers: Purpose and model, *Appl. Statist.*, **27**, 242-250.
[7] Box, G. E. P. and Tiao, G. G. (1968). A Bayesian approach to some outlier problems, *Biometrika*, **55**, 119-129.
[8] Freeman, F. R. (1979). On the number of outliers in data from a linear model, Int. Meeting on Bayesian Statistic, May 28-June 2, Valencia, Spain, to be published in a special issue of *Trab. Estadist.*
[9] Grubbs, F. E. (1969). Procedures for detecting outlying observations in samples, *Technometrics*, **11**, 1-21.
[10] Guttman, I. (1973). Care and handling of univariate or multivariate outliers in detecting spuriosity—A Bayesian approach, *Technometrics*, **15**, 723-738.
[11] Guttman, I., Dutter, R. and Freeman, P. R. (1978). Care and handling of univariate outliers in the general linear model to detect spuriosity—A Bayesian approach, *Technometrics*, **20**, 187-193.
[12] Hawkins, D. M. (1980). *Identification of Outliers*, Chapman and Hall.
[13] Kitagawa, G. (1979). On the use of AIC for the detection of outliers, *Technometrics*, **21**, 193-199.
[14] Kitagawa, G. (1980). OUTLAP, an outlier analysis program, *Computer Science Monographs*, No. 15, Institute of Statistical Mathematics, Tokyo.
[15] Tiejen, G. L. and Moore, R. H. (1972). Some Grubbs-type statistics for the detection of several outliers, *Technometrics*, **14**, 583-597.

Reprinted with kind permission from Elsevier Science-NL, Sara Burgerhartstraat 25, 1055 KV Amsterdam, The Netherlands

Statistics & Probability Letters 1 (1982) 75–78
North-Holland Publishing Company

November 1982

On the Fallacy of the Likelihood Principle *

Hirotugu Akaike

The Institute of Statistical Mathematics, Tokyo, Japan

Received June 1982

Abstract. By using the direct and inverse binomial experiments it is shown that there is a situation where Birnbaum's basic axiom of mathematical equivalence and the likelihood principle is a tautology. This observation disqualifies Birnbaum's proof of the likelihood principle based on the axioms of mathematical equivalence and conditionality. The implication of this disproof of Birnbaum's argument for Bayesian statistics is briefly discussed.

Keywords. Likelihood principle, conditionality, sufficiency, Bayesian statistics, prior distribution.

1. Introduction

There has been a continuous discussion of the likelihood principle which demands that the statistical inferences based on the data with identical likelihood functions should be identical (see, for example, Birnbaum (1962, 1972) and references therein).

The likelihood principle has been considered to be providing a strong support for Bayesian statistics (de Finetti, 1972; Lindley, 1972; Savage, 1962). In particular, the direct and inverse binomial experiments with unknown probability of occurrence is often quoted as a typical example where the conventional and Bayesian approaches differ significantly.

The likelihood principle gained much support by its 'proof' given by Birnbaum (1962) who derived the principle from the principles of conditionality and sufficiency. The 'proof' was later improved by replacing the sufficiency principle by a logically weaker axiom, or principle, of mathematical equivalence (Birnbaum, 1972). The force of Birnbaum's 'proof' lies in the fact that the two axioms of conditionality and mathematical equivalence are undeniably 'obvious' or 'natural'. The definitions of these axioms are given in the next section.

Many authors have discussed the likelihood principle. A good summary of the interplay between various principles of inference related with the likelihood principle is given by Dawid (1977). Particularly notable in these works is the one by Durbin (1970) who developed a strong argument against Birnbaum's simultaneous application of the conditionality and sufficiency principles. Apparently Birnbaum's introduction of the axiom of mathematical equivalence was intended to replace the axiom of sufficiency by one that is more 'obvious' and thus will be more independent of the conditionality principle.

The purpose of the present paper is to show, by using the example of direct and inverse binomial experiments, that the axiom of mathematical equivalence is by no means obvious as claimed by Birnbaum. In this example, under the assumption

* This work was supported by the United States Army under Contract No. DAAG 29-80-C-0041 at Mathematics Research Center, University of Wisconsin, Madison.

0167-7152/82/0000-0000/$02.75 © 1982 North-Holland

of the axiom of conditionality, the relation between the axiom of mathematical equivalence and that of likelihood reduces to a tautology, which shows that the difficulty pointed out by Durbin still persists.

The relation between the likelihood principle and Bayesian statistics is discussed briefly at the end of this paper.

2. Birnbaum's axioms and the likelihood principle

Following Birnbaum (1972), an experiment E is defined by $E=(\Omega, S, f)$, where $S=\{x\}$ is the discrete sample space, $\Omega=\{\theta\}$ is the parameter space and $f=f(x, \theta)=\mathbf{P}(X=x \mid \theta)$. For each x, (E, x) is used to represent (an instance of) statistical evidence. For E and E^* with common parameter space the judgement of the equivalence of the statistical evidences (E, x) and (E^*, x^*) is represented by $E_v(E, x)=E_v(E^*, x^*)$.

A statistic $h(x)$ is called ancillary if it satisfies the relation

$$f(x, \theta)=g(h) f(x \mid h, \theta)$$

where $g(h)=\mathbf{P}\{(x;\ h(x)=h)\}$ is independent of θ. For each possible value h of an ancillary statistic $h(x)$, E_h is defined by $E_h=(\Omega, S_h, f_h)$ where $S_h=\{x;\ h(x)=h\}$ and $f_h=f(x \mid h, \theta)$. Then E may be considered as a mixture experiment composed of components E_h having respective probabilities $g(h)$. Birnbaum's axioms required for the recent discussion are the following.

Conditionality (C). $E_v(E, x)=E_v(E_h, x)$ where h is an ancillary statistic and $h=h(x)$.

Mathematical equivalence (M). If $f(x, \theta)=f(x', \theta)$ for all $\theta \in \Omega$, then $E_v(E, x)=E_v(E, x')$.

Likelihood (L). If for some $c>0$, $f(x, \theta)=c f^*(x^*, \theta)$ for all $\theta \in \Omega$, then $E_v(E, x)=E_v(E^*, x^*)$.

The likelihood principle dictates that the evidence produced by a valid statistical inference procedure must satisfy (L). A support for the principle comes from Birnbaum's proof that (C) and (M) jointly imply (L). This is often considered a 'proof' of the likelihood principle.

Birnbaum also showed that (L) implies (C) and (M). Thus, under the assumption of (C), (M) and (L) are mathematically equivalent.

3. The fallacy

The force of support for the likelihood principle rendered by Birnbaum's 'proof' is based on the observation that (C) and (M) are 'natural' or 'obvious'. When (C) is accepted as one of the premises, then, due to the logical equivalence of (M) and (L), the force of the support depends entirely on the 'obviousness' of (M).

Birnbaum adopted (M) simply by assuming that it is the simplest formalization of the equivalent statistical evidence. This argument is obviously extra-mathematical and is concerned with the meaning of (M). We will show that this extra-mathematical argument is unfounded.

Consider the direct binomial experiment $E=\{\Omega, S, f\}$ defined by

$$f(x, \theta)={}_nC_x \theta^x(1-\theta)^{n-x},$$

$S=\{0, 1, \ldots, n\}$ and $\Omega=\{\theta;\ 0 \leqslant \theta \leqslant 1\}$. Also consider the inverse binomial experiment $E'=\{\Omega, S', f'\}$ defined by

$$f'(y, \theta)={}_{y-1}C_{m-1} \theta^m(1-\theta)^{y-m},$$

$S'=\{m, m+1, \ldots\}$ and with the same Ω as that of E. We assume that n and m are positive integers such that $n>m$.

For $x=m$ and $y=n$ we have $f(m, \theta)=c f'(n, \theta)$ $(c>0)$. Following Birnbaum we generate the mixture experiment $E^*=\{\Omega, S^*, f^*\}$ where $\Omega=\{\theta\}$, $S^*=\{(u, v);\ u, v=0, 1, 2, \ldots\}$ and f^* is defined by

$$f^*((u, v), \theta)=\begin{cases} k f(x, \theta) & \\ \qquad \text{for } (u, v)=(n, x), & \\ (1-k) f'(y, \theta) & \\ \qquad \text{for } (u, v)=(m, y), & \\ 0 \quad \text{otherwise} & \end{cases}$$

where $k=1/(1+c)$. For this experiment we have

$$\begin{aligned} f^*((n, m), \theta) &= k f(m, \theta)=(1-k) f'(n, \theta) \\ &= f^*((m, n), \theta). \end{aligned}$$

In this case (M) says that $E_v(E^*, (n, m)) = E_v(E^*, (m, n))$.

Birnbaum's justification of (M) rests on the argument that by relabelling (n, m) by (m, n) and vice versa, no observable change is introduced into the probabilistic structure of the experiment (see the discussion of the example at the beginning of p. 859 of Birnbaum (1972)). However, in the present example, by observing (n, m) we know that the data were generated by the direct binomial experiment. Also, by observing (m, n) we know that it came from the inverse binomial experiment. *The practical meaning of the relabelling is to make a false report* (m, n) *when actually* (n, m) *is observed. Whether such reporting is acceptable or not is not an obvious matter.* It is acceptable only when we are willing to ignore the difference between (m, n) and (n, m).

When (C) is invoked the acceptance of the false reporting procedure is equivalent to accepting the equality $E_v(E, m) = E_v(E', n)$, i.e., the likelihood axiom. Thus, under the assumption of (C), the relabelling assumed by (M) is neither more nor less compelling than the likelihood axiom (L) itself. In this situation the relation between (M) and (L) is a tautology. A tautology is logically a truth. However, it does not provide any new information that adds to our knowledge. This observation constitutes a semantic disproof of Birnbaum's 'proof' of the likelihood principle.

4. Discussion

Birnbaum's 'proof' has produced significant effect on Bayesian statisticians. Savage's reaction to Birnbaum's 'proof' (Savage, 1962) typically represents the interpretation by a Bayesian statistician of the relation between the likelihood principle and Bayesian statistics. Lindley (1972) changed his interpretation of the principle from that stated in the unquestionable definition given in Lindley (1965), where the equality of prior distributions was an explicit prerequisite of the principle to that of Savage who simply assumed the uniqueness of the prior distribution. Apparently, Lindley (1972) accepted Birnbaum's 'proof' as a confirmation of the principle.

It is an unlucky coincidence that the 'proof' strengthened the then-growing misconception that the principle automatically holds in Bayesian statistics. Many text books adopt this view. Such a view is obviously in contradiction with the original position of the subjective theory of probability which tells that every prior information should be taken into account when developing a probability distribution of an unknown. It is rather surprising to see that even the typical subjectivist de Finetti violates this basic position by simply admitting the likelihood principle (de Finetti, 1972). Such a breach of the basic teaching has unfailingly lead Bayesian statistics to a stalemate.

Strongly negative reactions have been shown against the present author's discussion of the necessity of including the sampling scheme into prior information when developing prior distributions for direct or inverse binomial experiments (Akaike, 1980). This looks natural when we recognize that Birnbaum's argument attracted even philosophers' interest (Kyburg, 1974; Seidenfeld, 1979). However, scientifically minded users of Bayesian statistics do not show any doubt in taking into account the property of sampling scheme in developing a prior distribution. A typical example is Jeffreys (1946) who defines an ignorance prior distribution by using the information supplied by the specification of the likelihood function. Box and Tiao (1973) convincingly discuss the rationality of using different ignorance priors for direct and inverse binomial experiments.

It seems fair to say that the likelihood principle is unconditionally supported only by those Bayesian statisticians who tend to see some uniquely defined objective meaning in the parameter or its probability distribution. An argument for the use of such an 'objective' prior distribution as our 'subjective' prior distribution, when it does exist, is developed in Akaike (1974). However, when we do not have such a strong prior information, we cannot deny the possibility of using the information about the expected behavior of the posterior distribution in choosing a prior distribution. Such a procedure of selection naturally induces the dependence of the prior distribution on the sampling scheme.

The semantic disproof of Birnbaum's derivation of the likelihood principle presented in this paper will emancipate statisticians from the spell of the yet unfounded likelihood principle. Statisticians are still free in developing their prior distributions for each particular problem. They may use the information on the assumed sampling scheme in choosing a prior distribution. This freedom will give Bayesian statisticians more chance to appreciate the common sense embodied in conventional statistics and thus eventually enhance the vitality of Bayesian statistics.

Acknowledgement

The author is grateful to G.E.P. Box, A.P. Dawid, T. Leonard and C.F. Wu for helpful discussions and comments.

References

Akaike, H. (1974), An objective use of Bayesian models, *Ann. Inst. Statist. Math.* **29**, 711–720.

Akaike, H. (1980), Likelihood and the Bayes procedure, with discussion, in: J.M. Bernardo, M.H. DeGroot, D.V. Lindley and A.F.M. Smith, eds., *Bayesian Statistics* (University Press, Valencia) pp. 143–166, 185–203.

Birnbaum, A. (1962), On the foundation of statistical inference, with discussion, *J. Amer. Statist. Assoc.* **57**, 269–326.

Birnbaum, A. (1972), More on concepts of statistical evidence, *J. Amer. Statist. Assoc.* **67**, 858–861.

Box, G.E.P. and G.C. Tiao (1973), *Bayesian Inference in Statistical Analysis* (Addison-Wesley, Reading, MA).

Dawid, A.P. (1977), Conformity of inference patterns, in: J.R. Barra, B. van Cutsen, F. Brodeau and G. Romier, eds., *Recent Developments in Statistics* (North-Holland, Amsterdam) pp. 245–256.

de Finetti, B. (1972), *Probability, Induction and Statistics* (Wiley, New York).

Durbin, J. (1970), On Birnbaum's theorem on the relation between sufficiency, conditionality and likelihood, *J. Amer. Statist. Assoc.* **65**, 395–398.

Jeffreys, H. (1946), An invariant form for the prior probability in estimation problems, *Proc. Roy. Soc. London Ser. A* **186**, 453–561.

Kyburg, H.E. (1974), *The Logical Foundation of Statistical Inference* (Reidel, Dordrecht).

Lindley, D.V. (1965), *Introduction to Probability and Statistics* Part 2 (Cambridge University Press, London).

Lindley, D.V. (1972), *Bayesian Statistics, A Review* (Society for Industrial and Applied Mathematics, Philadelphia).

Savage, L.J. (1962), Discussion on Birnbaum's paper, *J. Amer. Statis. Assoc.* **57**, 307–308.

Seidenfeld, T. (1979), *Philsophical Problems of Statistical Inference* (Reidel, Dordrecth).

Reprinted from *Proceedings of the Nineth International Symposium on Earth Tides*, J.T. Kuo, ed., 1983, 283-292 by permission from E. Schweizerbart'sche Verlagsbuchhandlung

A Bayesian Approach to the Analysis of Earth Tides

M. ISHIGURO, H. AKAIKE, Tokyo, M. OOE and S. NAKAI, Mizusawa

with 4 figures and 2 tables

Ishiguro M., Akaike, H., Ooe, M. & Nakai, S., 1983: A Bayesian Approach to the Analysis of Earth Tides. Proceedings of the Ninth International Symposium on Earth Tides, pp. 283-292.

Abstract: A new Bayesian method for tidal analysis is proposed. In contrast with the conventional filtering approach, this method is based on a time domain model which includes the response of the Earth to theoretical tidal input and other associated meteorological variables. It also includes the term which represents the drift of the record.

The basic assumption of this new procedure is the smoothness of the drift, and this requirement is represented in the form of probability of the Bayesian model. The parameters of the model are given as the mean of the posterior distribution defined by the data distribution and the prior distribution of the parameters.

The method allows an objective decision on the choice of the lag of the response functions and of the grouping of the tidal waves as a problem of statistical model selection. The problem of missing observation data and unexpected steps of the drift can be easily handled.

Practical applicability of this model to the analysis of earth tide records is demonstrated by using synthetic and actual earth tides data.

Keywords: Bayesian model, grouping, response model, drift estimation, meteorological disturbances, Esashi earth tides station.

1. Introduction

The objective of the earth tide analysis is to decompose the data y_i ($i=1, 2, \ldots, N$) into the form

$$y_i = \sum R_n \cos(\omega_n i - p_n) + r_i + d_i + e_i \ , \qquad (1)$$

where y_i is the observed data at time 'i'; R_n, ω_n, and p_n are the amplitude, angular velocity and initial phase of the n-th tidal constituent, respectively; r_i represents non-tidal effects such as those of atmospheric pressure and/or temperature; d_i and e_i denote the drift and the observation error, respectively. It is important

Publisher: E. Schweizerbart'sche Verlagsbuchhandlung, D-7000 Stuttgart

to estimate not only R_n and p_n but also r_i and d_i (BAKER, 1978a,b).

Currently used tidal analysis methods assume that $r_i + d_i$ can be approximated, at least for a short time span, by low order polynomials (MELCHIOR, 1973). Based on this assumption, numerical filters were designed to eliminate the term $r_i + d_i$ (VENEDIKOV, 1966). Based on the same assumption, a preprocessing procedure was developed to eliminate the drift and to adjust occasional steps and/or missing observation data (NAKAI, 1977, 1979).

Other widely-used devices for tidal analyses are the grouping of tidal waves and response modeling (VENEDIKOV, 1966; MUNK & CARTWRIGHT, 1966; LAMBERT, 1974). The tidal part of Eq. 1 is expressed as, using parameters a_{mk} and q_m to be estimated,

$$\sum R_n \cos(\omega_n i - p_n) = \sum_{m=1}^{M} \sum_{k=-L_m}^{L_m} a_{mk} \sum_{j=1}^{J_m} R^*_{mj} \cos\left(\omega^*_{mj}(i-\tau_m k) - p^*_{mj} + q_m\right), \tag{2}$$

where R^*_{mj}, ω^*_{mj} and p^*_{mj} are the theoretical amplitude, angular velocity and initial phase of the j-th wave of the m-th group, respectively. τ_m are suitably chosen lag intervals. We obtain the response model of MUNK & CARTWRIGHT (1966) by putting $q_m=0$, and the Venedikov grouping model by putting lags $L_m=0$. Apparently, the most important and difficult problem of Eq. 2 is the choice of the grouping and of the L_m. Although LAMBERT (1974) employs a statistical test for this decision, there remains the choice of the levels of significance.

The purpose of this paper is to propose a new Bayesian method to fit and evaluate models of the form of Eq. 1. This is a particular realization of the general Bayesian modeling for linear problems discussed by AKAIKE (1980a). The features of the method are 1) r_i and d_i can be estimated separately; 2) the drift d_i is only assumed in its smoothness; 3) occasional steps in the drift curve are allowable; 4) missing observations are allowable; 5) the decisions on the grouping of tidal waves and the lags of the response function are based on objective criterion.

2 Bayesian model of earth tides data

2.1 Model

We assume models of the form

$$
\begin{aligned}
y_i &= F(i,\{\alpha_m\},\{\beta_m\},\{d_i\},\{\gamma_k\};M,\{J_m\},K) + e_i \\
&= \sum_{m=1}^{M} \{\alpha_m \sum_{j=1}^{J_m} R^*_{mj} \cos(\ \omega^*_{mj} i - p^*_{mj}\ ) \\
&\qquad + \beta_m \sum_{j=1}^{J_m} R^*_{mj} \sin(\ \omega^*_{mj} i - p^*_{mj}\ )\} \\
&\quad + \sum_{k=0}^{K} \gamma'_k \chi_{i-k} + d_i + e_i \qquad (i=1,2,\cdots,N)
\end{aligned} \tag{3}
$$

for the earth tides data, where $\{\chi_i\}$ denote vectors of associated records such as atmospheric pressure and/or temperature, $\{\gamma_k\}$ represents the vectors of regression coefficients and ' denotes the transpose , and e_i is assumed to be identically and independently as Gaussian with mean 0 and variance σ^2. The parameters of Eq. 3 to be estimated are $\{\alpha_m\}$, $\{\beta_m\}$, $\{\gamma_k\}$ $\{d_i\}$ and σ^2. This model also contains other parameters M, $\{J_m\}$ and K, which are hereafter referred to as hyperparameters.

We further assume that parameters $\{\alpha_m\}$, $\{\beta_m\}$ and $\{d_i\}$ <u>approximately</u> satisfy relations

$$
\begin{aligned}
& d_i - 2d_{i-1} + d_{i-2} = 0 && (i=1,2,\cdot\cdot,N), \\
& \alpha_m - \alpha_{m-1} = 0 && (m=2,\cdot\cdot,M) \\
\text{and}\quad & \beta_m - \beta_{m-1} = 0 && (m=2,\cdot\cdot,M),
\end{aligned} \tag{4}
$$

where d-1 and d_o are assumed to be given.

We estimate the parameters in Eq. 3 which approximately satisfy Eq. 4 by minimizing

$$
\sum_{i=1}^{N} |\ y_i - F(i,\{\alpha_m\},\{\beta_m\},\{d_i\},\{\gamma_k\};M,\{J_m\},K)\ |^2
$$
$$
+\ v^2 \sum_{i=1}^{N} |\ d_i - 2d_{i-1} + d_{i-2}\ |^2
$$

$$+ \sum_{m=2}^{M} w_m^2 \{ | \alpha_m - \alpha_{m-1} |^2 + | \beta_m - \beta_{m-1} |^2 \} \qquad (5)$$

with M, $\{J_m\}$, K, v^2 and $\{w_m^2\}$ properly chosen. And v and $\{w_m\}$ are also considered as hyperparameters.

2.2 Criterion

We adopt a Bayesian model for choosing the values of the hyperparameters. Here, we briefly review the procedure described by AKAIKE (1980a, or 1980b).

When observations $\{y_i\}$ and variables $\{z_{ik}\}$ are given, the least squares method leads to minimization of

$$L(a) = \sum_{i=1}^{N} | y_i - \sum_{k=1}^{K} a_k z_{ik} |^2 .$$

The least squares estimate of the parameter vector $a=(a_1, a_2, \ldots a_k)'$ becomes meaningless or unreliable when K is large as compared with N. One way to avoid this difficulty is to introduce some preference on the values of the vector a and try to minimize

$$L(a) + \mu^2 || a - a_0 ||_R^2 , \qquad (6)$$

where a_0 denotes a particular vector of parameters, $|| \; ||_R$ the norm defined by a positive definite matrix R. The most important problem here is the choice of μ .

The minimization of Eq. 6 is the maximization of

$$\ell(a) = \exp[-\frac{1}{2\sigma^2}\{ L(a) + \mu^2 || a - a_0 ||_R^2 \}] , \qquad (7)$$

where σ^2 is a positive constant. The value of a which maximizes Eq. 7 can be regarded as the mean of the posterior distribution defined by the data distribution

$$f(y|\sigma^2,a) = (\frac{1}{2\pi})^{\frac{N}{2}} (\frac{1}{\sigma})^N \exp\{-\frac{1}{2\sigma^2}L(a)\} \tag{8}$$

and the prior distribution

$$\pi(a|\mu) = (\frac{1}{2\pi})^{\frac{K}{2}} (\frac{1}{\sigma})^K ||\mu^2 R||^{\frac{1}{2}} \exp\{-\frac{\mu^2}{2\sigma^2}||a-a_0||_R^2\}, \tag{9}$$

where $||\ ||$ denotes the determinant. Thus, in this context the choice of μ in interpreted as the choice of the parameter of the prior distribution. Now it is natural to choose μ such that the marginal likelihood defined by

$$L(\mu,\sigma^2) = \int f(y|\sigma^2,a)\pi(a|\mu)da \tag{10}$$

is maximized. AKAIKE (1980a) proposed the use of ABIC (A Bayesian Information Criterion) is defined by

$$ABIC = -2\log L(\mu,\sigma^2) \tag{11}$$

for the choice of μ . The best choice of μ is that which minimizes ABIC.

Note that Eq. 5 is a generalization of Eq. 6. Instead of the single μ in Eq. 6, we have the hyperparameters M, $\{J_m\}$, K, v and $\{w_m\}$ to be adjusted, and we can also define ABIC for the present case and choose these values, objectively. This means that we can objectively decide on the grouping of tidal waves and the choice of the lag of the response function of non-tidal effect and even on the inclusion or exclusion of some specific non-tidal effects. If there is a step between $i=i_\ell$ and $i_\ell+1$, we have only to add a term $\delta z_{i_\ell i}$ to the right-hand side of Eq. 3, where $z_{i_\ell i}=0$ if $i \leq i_\ell$, 1 otherwise. The amount of step δ can be estimated. In handling missing observations, we have only to omit the terms which contain the missing observations from Eq. 5.

3. Numerical results

3.1 Analysis of synthetic data

Synthetic data of NS component of the extensometer for the period from June 1, 1979 to July 31, 1979 were generated using Eq. 3, where $\{R^*_{mj}\}$, $\{w^*_{mj}\}$, and $\{p^*_{mj}\}$ were set to be theoretical values. The grouping and the tidal factors f_m and phase lag ℓ_m (in degree) of each group 'm' are given as 'real value' in Table 1. These are the roughly estimated real values at Esashi earth tide station (SATO et al., 1983). The relations between (f_m, ℓ_m) and (α_m, β_m) are $\alpha_m=f_m\cos\ell_m$ and $\beta_m=-f_m\sin\ell_m$. Drift $\{d_i\}$ was generated by an integrated first order auto-regressive process. Several steps were added at random time points. The variance σ^2 of the noise e_i was set equal to $(0.3)^2$. Data were generated for two months at half-hourly readings.

Table 1. Analyses of synthetic data.

<table>
<tr><th rowspan="2">group</th><th rowspan="2">waves*</th><th colspan="2">Real Value**</th><th colspan="2">Model 6***</th><th colspan="2">Model 7</th><th colspan="2">Model 12</th></tr>
<tr><th>f_m</th><th>ℓ_m</th><th>f_m</th><th>ℓ_m</th><th>f_m</th><th>ℓ_m</th><th>f_m</th><th>ℓ_m</th></tr>
<tr><td>Q1</td><td>1-62</td><td rowspan="2">0.23</td><td rowspan="2">-170.</td><td rowspan="4">0.204</td><td rowspan="4">-179.</td><td rowspan="2">0.231</td><td rowspan="2">-170.</td><td>0.231</td><td>-171.</td></tr>
<tr><td>O1</td><td>63-88</td><td>0.231</td><td>-170.</td></tr>
<tr><td>M1</td><td>89-110</td><td rowspan="2">0.2</td><td rowspan="2">180.</td><td rowspan="2">0.200</td><td rowspan="2">180.</td><td>0.212</td><td>-178.</td></tr>
<tr><td>P1S1K1</td><td>111-143</td><td>0.199</td><td>180.</td></tr>
<tr><td>J1</td><td>144-165</td><td rowspan="2">0.5</td><td rowspan="2">-150.</td><td rowspan="2">0.550</td><td rowspan="2">-149.</td><td rowspan="2">0.509</td><td rowspan="2">-149.</td><td>0.507</td><td>-149.</td></tr>
<tr><td>OO1</td><td>166-197</td><td>0.531</td><td>-148.</td></tr>
<tr><td>2N2</td><td>198-236</td><td rowspan="3">0.47</td><td rowspan="3">170.</td><td rowspan="3">0.471</td><td rowspan="3">170.</td><td rowspan="3">0.470</td><td rowspan="3">170.</td><td>0.484</td><td>170.</td></tr>
<tr><td>N2</td><td>237-260</td><td>0.471</td><td>170.</td></tr>
<tr><td>M2</td><td>261-286</td><td>0.471</td><td>170.</td></tr>
<tr><td>L2</td><td>287-300</td><td>0.45</td><td>165.</td><td>0.445</td><td>164.</td><td>0.441</td><td>164.</td><td>0.470</td><td>164.</td></tr>
<tr><td>S2K2</td><td>301-347</td><td>0.52</td><td>180.</td><td>0.519</td><td>-180.</td><td>0.520</td><td>180.</td><td>0.442</td><td>180.</td></tr>
<tr><td>M3</td><td>348-363</td><td>0.27</td><td>-170.</td><td>0.258</td><td>-163.</td><td>0.262</td><td>-165.</td><td>0.263</td><td>-165.</td></tr>
<tr><td>σ</td><td></td><td colspan="2">0.3</td><td colspan="2">0.293</td><td colspan="2">0.295</td><td colspan="2">0.295</td></tr>
<tr><td>ABIC</td><td></td><td colspan="2"></td><td colspan="2">-5916.2</td><td colspan="2">-6508.2</td><td colspan="2">-6455.8</td></tr>
</table>

* Group and waves follow the definition given in the FORTRAN program initially coded at ICET in 1974.
** Amplitude factors for groups Q1 and O1 are set equal to 0.23.
*** Model 6 groups Q1, O1, M1 and P1S1K1 together.

Three models with different groupings were fitted to the synthetic data. Results are summarized in Table 1. Note that the value of ABIC is minimum for Eq. 7, which is with the true structure of 7 groups.

3.2 Analyses of strains observed at Esashi

SATO et al. (1983) have discussed in detail the analyses of strains observed in Esashi.

We check to see if the effect of atmospheric pressure is present in the NS component of the extensometer. The grouping of all models fitted are the same with that of Model 12 of Table 1. Models with lags K= -1,0,2,4 were fitted, where K= -1 means that this model does not take into account the effect of atmospheric pressure. The results show that the effect of atmospheric pressure should be included in the model and that the best choice of the lag among the values tested is 4. Table 2 gives the results of analyses of real strain data.

Table 2. Analyses of real strain data.

K	-1	0	2	4	6
ABIC	-4212.5	-4643.4	-4981.4	-4990.8	-4979.8

3.3 Response model

We have fitted the model

$$y_i = \sum_{m=1}^{3} \sum_{k=-3}^{3} a_{mk} \sum_{j=1}^{J_m} R^*_{mj} \cos\{\omega^*_{mj}(i-k) - p^*_{mj}\}$$
$$+ d_i + \sum_{\ell} \delta_\ell z_{i\ell i} + e_i$$

to the EW component of the extensometer observed from June 5, 1979 to July 31, 1979 at Esashi (see Figure 1). The three groups assumed in this model were those of diurnal, semidiurnal and ter-diurnal waves. The results are shown in Figures 2, 3 and 4.

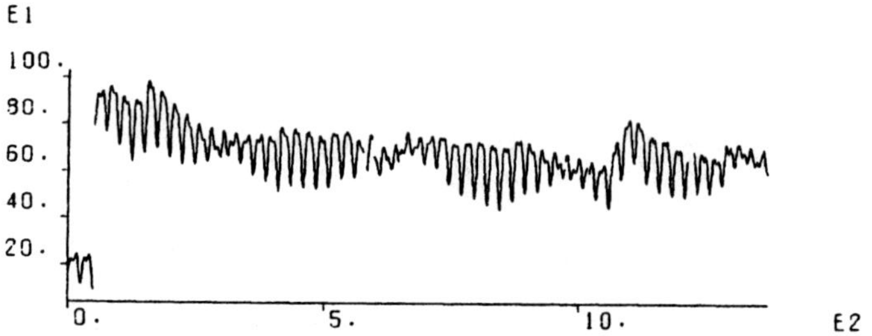

Figure 1. EW component of the extensometer.

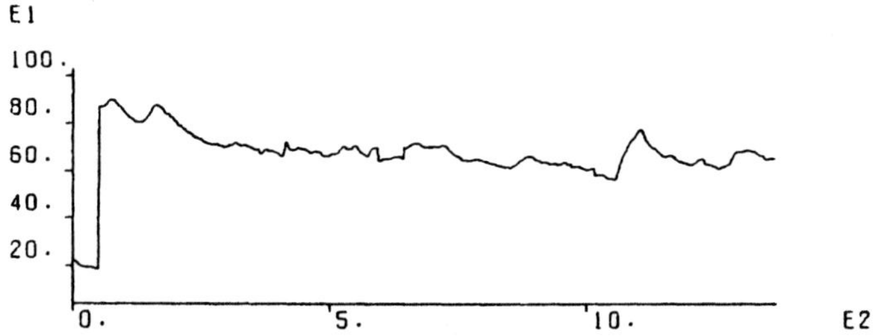

Figure 2. Estimated drift and step.

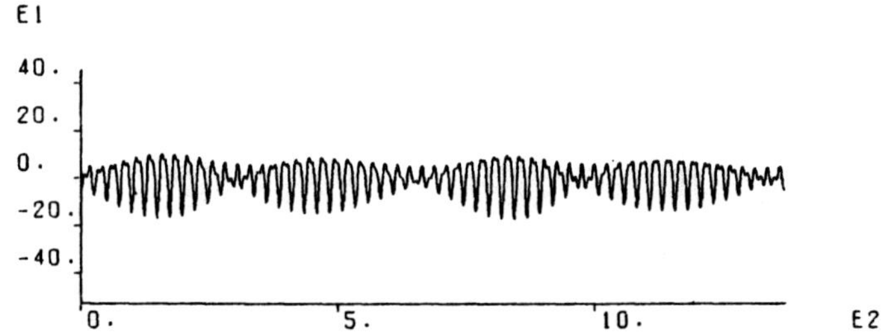

Figure 3. Estimated tidal component.

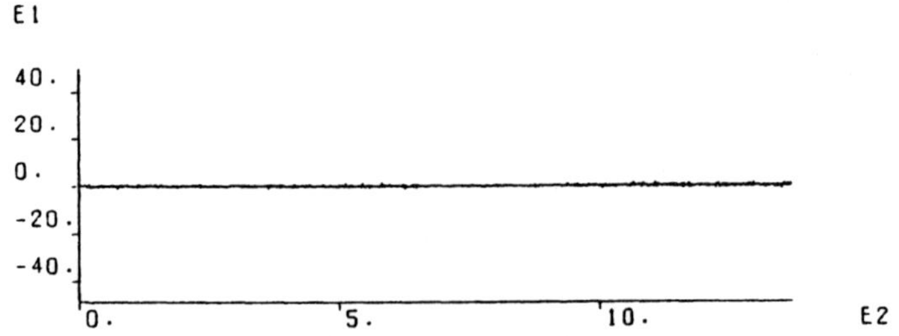

Figure 4. Estimated observation error.

Figure 3 shows the sum of two terms d_i and $\sum_{\ell}\delta_{\ell}z_{i\ell i}$.

4. Discussion

By a suitable choice of $\{w_m\}$ in Eq. 5, the model can follow the critical behavior of the tidal admittances around the frequency of the fluid-core resonance.

The tidal admittances are not necessarily constant. Besides, tidal recordings by an astatized instrument commonly show sensitivity changes which cannot be controlled by calibration (DUCARME, 1979). It is not difficult to generalize Eq. 3 to simulate the case. We should only replace α_m and β_m in Eq. 3 by α_{mi} and β_{mi}, and

$$\sum_m w_m^2\{|\alpha_m - \alpha_{m-1}|^2 + |\beta_m - \beta_{m-1}|^2\} \text{ in (5) by}$$

$$\sum_{i,m} w_m^2 \{|\alpha_{mi} - \alpha_{m-1,i}|^2 + |\beta_{mi} - \beta_{m-1,i}|^2\}$$

$$+ \sum_{i,m} u_m^2 \{|\alpha_{mi} - 2\alpha_{m,i-1} + \alpha_{m,i-2}|^2 + |\beta_{mi} - 2\beta_{m,i-1} + \beta_{m,i-2}|^2\},$$

where $\{u_m\}$ are additional hyperparameters.

ACKNOWLEDGEMENT

This research was partly supported by a grant from the Asahi Shunbun Publishing Company.

REFERENCES

Akaike, H., 1980a. Likelihood and the Bayes procedure, in Bayesian Statistics, University Press, Valencia (Spain).

Akaike, H., 1980b. Seasonal adjustment by a Bayesian modeling, J. of Time Series Analyses, 1:1-13.

Baker, T.F., 1978a. BIM, No. 78:4571-4578.

Baker, T.F., 1978b. BIM, No. 78:4596-4610.

Ducarme, B., 1979. Sensitivity smoothing before the analyses of the tidal data, BIM, No. 76: 4962-4981.

Lambert, A., 1974. Earth tide analysis and prediction by the response method, J. Geophys. Res., 79, No. 32: 4952-4960.

Melchior, P., 1973. Harmonic analysis of earth tides, in Methods in Computational physics, vol. 13, Academic Press, New York.

Munk, W.H. & Cartwright, D.E., 1966. Tidal spectroscopy and prediction. Phil. Trans. Roy. Soc. London, Ser. A, Vol. 259:553-581.

Nakai, S., 1979. Revised method of the preprocessing of tidal data, BIM, No. 81:4955-4961.

Sato, T., Ooe, M. & Sato, N., 1983. Tidal Tilt and Strain Measurements

and analyses at the Esashi earth tides station, in this volume.
Venedikov, A.P., 1966. Une method pour l'analyse des marées terrestres à partir d'enregistrements de longueur arbitraire, Obs. Roy. Bel., Comm. No. 250, Sr. Geophys., No. 71: 437-459.

Ishiguro, M., Akaike, H.
The Institute of Statistical Mathematics
4-6-7 Minami-Azabu,Minato-ku
Tokyo, Japan

Ooe, M., Nakai, S.
International Latitude Observatory of Mizusawa
2-12 Hoshigaoka-cho
Mizusawa-shi, Iwate-ken
023 Japan

DISCUSSION

Q: How much computer time does this method request as compared to the standard Venedikov method? (Ducarme)

A: I cannot give you an exact estimation of computer time now, but it may be about 2 - 3 times compared with Venedikov method. (Ooe)

Q: The method allows the correction of jumps. Do you have to introduce the position of the jump or can you detect it automatically? (Ducarme)

A: The position of the jumps is assigned in these results. We, however, can search automatically the position if needed. In this case the computer time gets longer. (Ooe)

Q: It is a good idea to include a term different from the drift. But if the associated process has tidal frequencies it will be not possible to separate from the tide. (Venedikov)

A: The associated process doesn't have the same construction with the tidal one. This is especially true if the wide frequency band is treated. The present method provides an objective choice of the regression coefficients optimal to the data. In this sense, such as S_2 wave of the atmospheric pressure could be separated from the earth tide data. (Ooe)

PSYCHOMETRIKA—VOL. 52, NO. 3, 317–332
SEPTEMBER 1987
SPECIAL SECTION

FACTOR ANALYSIS AND AIC

HIROTUGU AKAIKE

THE INSTITUTE OF STATISTICAL MATHEMATICS

The information criterion AIC was introduced to extend the method of maximum likelihood to the multimodel situation. It was obtained by relating the successful experience of the order determination of an autoregressive model to the determination of the number of factors in the maximum likelihood factor analysis. The use of the AIC criterion in the factor analysis is particularly interesting when it is viewed as the choice of a Bayesian model. This observation shows that the area of application of AIC can be much wider than the conventional i.i.d. type models on which the original derivation of the criterion was based. The observation of the Bayesian structure of the factor analysis model leads us to the handling of the problem of improper solution by introducing a natural prior distribution of factor loadings.

Key words: factor analysis, maximum likelihood, information criterion AIC, improper solution, Bayesian modeling.

1. Introduction

The factor analysis model has been producing thought provoking statistical problems. The model is typically represented by

$$\mathbf{y}(n) = A\mathbf{x}(n) + \mathbf{u}(n), \qquad n = 1, 2, \cdots, N$$

where $\mathbf{y}(n)$ denotes a p-dimensional vector of observations, $\mathbf{x}(n)$ a k-dimensional vector of factor scores, $\mathbf{u}(n)$ a p-dimensional vector of specific variations. It is assumed that the variables with different n's are mutually independent and that $\mathbf{x}(n)$ and $\mathbf{u}(n)$ are mutually independently distributed as Gaussian random variables with variance covariance matrices $I_{k \times k}$ and Ψ, respectively, where Ψ is a diagonal matrix. The covariance matrix Σ of $\mathbf{y}(n)$ is then given by

$$\Sigma = AA' + \Psi.$$

This model is characterized by the use of a large number of unknown parameters, much larger than the number of unknown parameters of a model used in the conventional multivariate analysis. The empirical principle of parsimony in statistical model building dictates that the increase of the number of parameters should be stopped as soon as it is observed that a further increase does not produce significant improvement of fit of the model to the data. Thus the control of the number of parameters has usually been realized by applying a test of significance.

The author would like to express his thanks to Jim Ramsay, Yoshio Takane, Donald Ramirez and Hamparsum Bozdogan for helpful comments on the original version of the paper. Thanks are also due to Emiko Arahata for her help in computing.

Requests for reprints should be sent to Hirotugu Akaike, The Institute of Statistical Mathematics, 4-6-7 Minami-Azabu, Minato-Ku, Tokyo 106, Japan.

0033-3123/87/0900-SS01$00.75/0

In the case of the maximum likelihood factor analysis this is done by adopting the likelihood ratio test. However, in this test procedure, the unstructured saturated model is always used as the reference and the significance is judged by referring to a chi-square distribution with a large number of degrees of freedom equal to the difference between the number of parameters of the saturated model and that of the model being tested. As will be seen in section 3, an example discussed by Jöreskog (1978) shows that direct application of such a test to the selection of a factor analysis model is not quite appropriate. There the expert's view clearly contradicts the conventional use of the likelihood ratio test.

In 1969 the present author introduced final prediction error (FPE) criterion for the choice of the order of an autoregressive model of a time series (Akaike, 1969, 1970). The criterion was defined by an estimate of the expected mean square one-step ahead prediction error by the model with parameters estimated by the method of least squares. The successful experience of application of the FPE criterion to real data suggested the possibility of developing a similar criterion for the choice of the number of factors in the factor analysis. The choice of the order of an autoregression controlled the number of unknown parameters in the model, that controlled the expected mean square one-step ahead prediction error. By analogy it was easily observed that the control of the number of factors was required for the control of the expected prediction error by the fitted model. However, it was not easy to identify what the prediction error meant in the case of the factor analysis.

In the case of the autoregressive model an estimate of the expected predictive performance was adopted as the criterion; in the case of the maximum likelihood factor analysis it was the fitted distribution that was evaluated by the likelihood. The realization of this fact quickly led to the understanding that our prediction was represented by the fitted model in the case of the factor analysis, which then led to the understanding that the expectation of the log likelihood with respect to the "true" distribution was related to the Kullback-Leibler information that defined the amount of deviation of the "true" distribution from the assumed model.

The analogy with the FPE criterion then led to the introduction of the criterion

$$\text{AIC} = (-2) \log \text{maximum likelihood} + 2\,(\text{number of parameters}),$$

as the measure of the badness of fit of a model defined with parameters estimated by the method of maximum likelihood, where log denotes a natural logarithm (Akaike, 1973, 1974). We will present a simple explanation of AIC in the next section and illustrate its use by applying it to an example in section 3.

Although AIC produces a satisfactory solution to the problem of the choice of the number of factors, the application of AIC is hampered by the frequent appearance of improper solutions. This shows that successive increase of the number of factors quickly lead to models that are not quite appropriate for the direct application of the method of maximum likelihood.

In section 4 it will be discussed that the factor analysis model may be viewed as a Bayesian model and the choice of a factor analysis model by minimizing the AIC criterion is essentially concerned with the choice of a Bayesian model. This recognition encourages the use of further Bayesian modeling for the elimination of improper solutions. In section 5 a natural prior distribution for the factor loadings is introduced through the analysis of the likelihood function. Numerical examples will be given in Section 6 to show that the introduction of the prior distribution suppresses the appearance of improper solutions and that the indefinite increase of a communality caused by the conventional maximum likelihood procedure may be viewed as of little practical significance.

The paper concludes with brief remarks on the contribution of factor analysis to the development of general statistical methodology.

2. Brief Review of AIC

The fundamental ideas underlying the introduction of AIC are:

1. The predictive use of the fitted model.
2. The adoption of the expected log likelihood as the basic criterion.

Here the concept of parameter estimation is replaced by the estimation of a distribution and the accuracy is measured by a universal criterion, the expected log likelihood of the fitted model.

The relation between the expected log likelihood and the Kullback-Leibler information number is given by

$$I(f; g) = E \log f(x) - E \log g(x),$$

where $I(f; g)$ denotes the Kullback-Leibler information of the distribution f relative to the distribution g, and E denotes the expectation with respect to the "true" distribution $f(x)$ of x. The second term on the right-hand side represents the expected log likelihood of an assumed model $g(x)$ with respect to the "true" distribution $f(x)$. Since $I(f; g)$ provides a measure of the deviation of f from g and since $\log g(x)$ provides an unbiased estimate of $E \log g(x)$ the above equation provides a justification for the use of log likelihoods for the purpose of comparison of statistical models.

Consider the situation where the model $g(x)$ contains unknown parameter θ, that is, $g(x) = g(x \mid \theta)$. When the data x are observed and the maximum likelihood estimate $\theta(x)$ of θ is obtained, the predictive point of view suggests the evaluation of $\theta(x)$ by the goodness of $g(\cdot \mid \theta(x))$ as an estimate of the true distribution $f(\cdot)$. By adopting the information $I(f; g)$ as the basic criterion we are led to the use of $E_y \log g(y \mid \theta(x))$ as the measure of the goodness of $\theta(x)$, where E_y denotes the expectation with respect to the true distribution $f(y)$ of y. To relate this criterion to the familiar log likelihood ratio test statistic we adopt $2E_y \log g(y \mid \theta)$ as our measure of the goodness of $g(y \mid \theta)$ as an estimate of $f(y)$.

Here we consider the conventional setting where the true distribution $f(y)$ is given by $g(y \mid \theta_0)$, that is, θ_0 is the true value of the unknown parameter, the data x are a realization of the vector of i.i.d. random variables $x_1, x_2, \cdots, x_N$, and the log likelihood ratio test statistic asymptotically satisfies the relation

$$2 \log g(x \mid \theta(x)) - 2 \log g(x \mid \theta_0) = \chi^2_m,$$

where χ^2_m denotes a chi-squared with degrees of freedom m which is equal to the dimension of the parameter vector θ. Under this setting it is expected that the curvature of the log likelihood surface provides a good approximation to that of the expected log likelihood surface. This observation leads to another asymptotic equality

$$2E_y \log g(y \mid \theta(x)) - 2E_y \log g(y \mid \theta_0) = -\chi^2_m,$$

where it is assumed that y is another independent observation from the same distribution as that of x and the chi-squared variable is identical to that defined by the log likelihood ratio test statistic.

The above equations show that the amount of increase of $2 \log g(x \mid \theta(x))$ from $2 \log g(x \mid \theta_0)$ obtained by adjusting the parameter value by the method of maximum likelihood is asymptotically equal to the amount of decrease of $2E_y \log g(y \mid \theta(x))$ from $2E_y$ log

$g(y \mid \theta_0)$. Thus, to measure the deviation of $\theta(x)$ from θ_0 in terms of the basic criterion of twice the expected log likelihood, χ_m^2 must be subtracted twice from 2 log $g(x \mid \theta(x))$ to make the difference of twice the log likelihoods an unbiased estimate of that of twice the expected log likelihoods.

Since χ_m^2 is unobservable, as we do not know θ_0, we consider the use of its expected value m. The negative of the quantity thus obtained defines

$$\text{AIC} = (-2) \log g(x \mid \theta(x)) + 2m.$$

When several different g's are compared the one that gives the minimum of AIC represents the best fit. Such an estimate is denoted as MAICE (minimum AIC estimate). For more detailed discussion of the predictive point of view of statistics and the use of the information criterion readers are referred to Akaike (1985).

3. How AIC Works With The Factor Analysis Model

Given a set of observations $y = (y(n); n = 1, 2, \cdots, N)$ the maximum likelihood factor analysis starts with the definition of the log likelihood function given by

$$\log L(k) = -\tfrac{1}{2}N[\log |\Sigma_k| + \text{tr } \Sigma_k^{-1}S],$$

where S denotes the sample covariance matrix of y and k the number of factors and Σ_k is given by

$$\Sigma_k = A_k A_k' + \Psi,$$

where A_k denotes the matrix of factor loadings and Ψ the uniqueness variance matrix. The diagonal elements of $A_k A_k'$ define the communalities. The AIC statistic for the k-factor model is then defined by

$$\text{AIC}(k) = (-2) \log L(k) + [2p(k + 1) - k(k - 1)].$$

To show the use of AIC in the maximum likelihood factor analysis and to illustrate the difference between the AIC and conventional test approach in particular here we will discuss an example treated by Jöreskog (1978, p.457). This examle is concerned with the analysis of Harman's example of twenty-four psychological variables. The unrestricted four factor model was first fitted which produced

$$\chi_{186}^2 = 246.36.$$

This model was considered to be "representing a reasonably good fit" but a further restriction of parameters produced a simple structure model with

$$\chi_{231}^2 = 301.42.$$

This model was accepted as the best fitting simple structure.

Now we have

$$\text{Prob } \{\chi_{186}^2 \geqq 246.36 \mid H_0\} \approx 0.0009,$$

and

$$\text{Prob}\{\chi_{231}^2 \geqq 301.42 \mid H_0'\} \approx 0.0005,$$

where H_0 and H_0' denote the hypotheses of the four factor and the simple structure, respectively, and the chi-squared variables stand for the random variables with respective degrees of freedom. By the standard of conventional tests these figures show that the results are extremely significant and both H_0 and H_0' should be rejected. In spite of this, the expert judgment of Jöreskog was to accept the four factor model as a reasonable fit and prefer the simple structure model to the unrestricted. This conclusion suggests that the large values of the degrees of freedom appearing in the chi-squared statistics preclude the application of conventional levels of significance, such as 0.05 or 0.01, in making the final judgment of models in this situation.

The chi-squared statistic is defined by

$$\chi^2 = (-2) \max \log L(H) - (-2) \max \log L(H_\infty),$$

where max log $L(H)$ denotes the maximum log likelihood under the hypothesis H and H_∞ denotes the saturated or completely unconstrained model. Since AIC for an hypothesis H is defined by

$$\mathrm{AIC}(H) = (-2) \max \log L(H) + 2 \dim \theta,$$

where dim θ denotes the dimension of the vector of unknown parameters θ, we have

$$\mathrm{AIC}(H) - \mathrm{AIC}(H_\infty) = \chi^2_{\mathrm{d.f.}} - 2(\mathrm{d.f.}),$$

where d.f. denotes the difference between the number of unknown parameters of H_∞ and that of H. By neglecting the common additive constant AIC (H_∞) we may define AIC(H) simply by

$$\mathrm{AIC}(H) = \chi^2_{\mathrm{d.f.}} - 2(\mathrm{d.f.}).$$

For the models discussed by Jöreskog we get

$$\mathrm{AIC}(H_0) = 246.36 - 2 \times 186$$
$$= -125.64,$$

and

$$\mathrm{AIC}(H'_0) = 301.42 - 2 \times 231$$
$$= -160.58.$$

Since AIC(H_∞) = 0, these AIC's show that both H_0 and H'_0 are by far better than H_∞ and that the simple structure model H'_0 is showing a better fit than the unrestricted four factor model H_0.

This result by AIC is in complete agreement with Jöreskog's conclusion. The conventional theory of statistics does not tell how to evaluate the significance of a test in each particular application and there is no hope of arriving at a similar conclusion. Obviously the objective procedure of model selection by an information criterion can be fully implemented to define an automatic factor analysis procedure. Such a possibility is discussed by Bozdogan and Ramirez (1987).

4. Factor Analysis Model Viewed as a Bayesian Model

As was demonstrated by the application to Jöreskog's example the AIC approach produced a satisfactory solution to the model selection problem in factor analysis. In spite of this success the use of AIC in the maximum likelihood factor analysis has been severely limited by the frequent occurrence of improper solutions, that is, by the appearance of zero estimates of specific variances. Apparently this is caused by the overparametrization of the model.

The introduction of AIC is motivated by the desire to control the effect of overparametrization and the minimum AIC procedure for model selection is considered to be a realization of the well-known empirical principle of parsimony in statistical modeling. However the application of the minimum AIC procedure assumes the existence of proper maximum likelihood estimates of the models considered. The frequent occurrence of improper solutions in the maximum likelihood factor analysis means that the models are often too much overparametrized for the application of the method of maximum likelihood. This suggests the necessity of further control of the likelihood function. This can be realized by the use of some proper Bayesian modeling.

Before going into the discussion of this Bayesian modeling we will first notice the essentially Bayesian characteristic of the factor analysis model and point out that the minimum AIC procedure is concerned with the problem of the selection of a Bayesian model. In the basic factor analysis model $\mathbf{y} = A\mathbf{x} + \mathbf{u}$ the vector of observations $\mathbf{y}$ is assumed to be distributed following a Gaussian distribution with mean $A\mathbf{x}$ and unique variance Ψ. The vector of factor scores x is unobservable but is assumed to be distributed following a Gaussian distribution with zero mean and variance $I_{k \times k}$. Since $\mathbf{x}$ is never observed this distribution is simply a psychological construction for the explanation of the behavior of y. Under the assumption that A is fixed the distribution of $\mathbf{x}$ specifies the prior distribution of the mean of the observation $\mathbf{y}$. Thus we can see that the choice of k, the number of factors, is essentially concerned with the choice of a Bayesian model. Incidentally, the recognition of the Bayesian characteristic of the factor analysis model also suggests the use of the posterior distribution of $\mathbf{x}$ for the estimation of the factor scores as is discussed by Bartholomew (1981).

The basic problem in the use of a Bayesian model is how to justify the use of a subjectively constructed model. Our belief is that it is possible only by considering various possibilities as alternative models and comparing them with an objectively defined criterion. In particular we propose the use of the log likelihood, or the AIC when some parameters are estimated by the method of maximum likelihood, as the criterion of fit.

Let us consider the likelihood of a factor analysis model as a Bayesian model. For a Bayesian model specified by the data distribution $p(\cdot \mid \theta)$ and prior distribution $p(\theta)$ its likelihood with respect to the observed data y is given by

$$\int p(y \mid \theta) p(\theta)\, d\theta.$$

From the representation $\mathbf{y}(n) = A\mathbf{x}(n) + \mathbf{u}(n)$, $n = 1, 2, \cdots, N$, and the assumption of the mutual independence among the variables the likelihood of the Bayesian model defined with $\theta = (x(1), x(2), \cdots, x(N))$ is given by

$$L = \prod_{n=1}^{N} \left(\frac{1}{2\pi}\right)^{p/2} |\Sigma|^{-1/2} \exp\left\{-\frac{1}{2} \operatorname{tr} \Sigma^{-1}\mathbf{y}(n)\mathbf{y}(n)'\right\}$$

$$= \left(\frac{1}{2\pi}\right)^{Np/2} |\Sigma|^{-N/2} \exp\left\{-\frac{N}{2} \operatorname{tr} \Sigma^{-1} S\right\},$$

where $\Sigma = AA' + \Psi$, $|\Sigma|$ denotes the determinant, and

$$S = \frac{1}{N} \sum_{n=1}^{N} \mathbf{y}(n)\mathbf{y}(n)'.$$

For simplicity the mean of $\mathbf{y}(n)$ is assumed to be zero. Thus we get

$$\log L = -\frac{N}{2} [\log |\Sigma| + \operatorname{tr} S\Sigma^{-1}] + \text{const.}$$

This is exactly the likelihood function used in the conventional maximum likelihood factor analysis. Thus the maximum likelihood estimates of A and Ψ in the classical sense are the maximum likelihood estimates of the unknown parameters of a Bayesian model.

The above result shows that the AIC criterion defined for the factor analysis model is actually the ABIC criterion for the evaluation of a Bayesian model with parameters estimated by the method of maximum likelihood, where ABIC is defined by (Akaike, 1980)

$$\begin{aligned}\text{ABIC} &= (-2) \text{ maximum log likelihood of a Bayesian model} \\ &\quad + 2 \text{ (number of estimated parameters).}\end{aligned}$$

In the case of the factor analysis model we have

$$\text{ABIC} = \text{AIC}.$$

This identity clearly shows that there is no essential distinction between the classical and Bayesian models when they are viewed from the point of view of the information criterion.

5. Control of Improper Solutions by a Bayesian Modeling

The appearance of improper solutions suggests the necessity of the reduction of the number of parameters to be estimated by the method of maximum likelihood. The recognition of the Bayesian structure of the factor analysis model suggests that further modeling of the prior distribution of the unknown parameters in A and Ψ is possible. The use of the Bayesian approach for the control of improper solutions is already discussed in an earlier paper by Martin and McDonald (1975). These authors point out the importance of choosing a prior distribution that does not have the appearance of arbitrariness and discuss the use of a reasonably defined prior distribution of specific variances.

The informational approach to statistics puts very much faith in the information supplied by the log likelihood. Hence in the present paper we try to develop a prior distribution without using outside information except for the knowledge of the likelihood function of the data distribution. In the present situation this is particularly appropriate as the prior distribution is considered only for the purpose of tempering the likelihood function to clarify the nature of improper solutions.

By this approach we need a detailed analysis of the likelihood function. For the convenience of the analysis let us consider

$$q = \left(-\frac{2}{N}\right) \log L - \log |S|,$$

where the log likelihood $\log L$ is defined in the preceding section. By ignoring the additive constant we have

$$q = -\log |\Sigma^{-1}S| + \operatorname{tr} \Sigma^{-1}S.$$

By putting $\Psi = D^2$, where D is a diagonal matrix with positive diagonal elements, we get

$$\begin{aligned}\Sigma &= AA' + D^2 \\ &= D(I + CC')D,\end{aligned}$$

where A is $p \times k$, I is a $p \times p$ identity matrix and $C = D^{-1}A$, the matrix of standardized factor loadings. We have

$$\mathrm{tr}\,\Sigma^{-1}S = \mathrm{tr}\,(I + CC')^{-1}D^{-1}SD^{-1},$$

and

$$|\Sigma^{-1}S| = |(I + CC')^{-1}|\,|D^{-1}SD^{-1}|.$$

The modified negative log likelihood q can conveniently be expressed by using the eigenvectors z_i and eigenvalues ζ_i of $D^{-1}SD^{-1}$, the standardized sample covariance matrix. Define the matrix Z by

$$Z = [z_1, z_2, \cdots, z_p].$$

It is assumed that Z is normalized so that $Z'Z' = I$ holds. Represent C by Z in the form

$$C = ZF.$$

Adopt the representation

$$FF' = \sum_{i=1}^{p} \mu_i m_i m_i',$$

where $\mu_i > 0$, for $i = 1, 2, \cdots, k$, $= 0$, otherwise, and $m_i' m_i = \delta_{ij}$, where $\delta_{ij} = 1$, for $i = j$, 0, otherwise. Then we get

$$CC' = ZFF'Z' = \sum_{i=1}^{p} \mu_i l_i l_i',$$

where $l_i = Zm_i$ with $l_i' l_j = \delta_{ij}$, and

$$I + CC' = \sum_{i=1}^{p} \lambda_i l_i l_i',$$

where $\lambda_i = 1 + \mu_i$. From this representation we get

$$(I + CC')^{-1} = \sum_{i=1}^{p} \lambda_i^{-1} l_i l_i',$$

and

$$\begin{aligned}\mathrm{tr}\,(I + CC')^{-1}D^{-1}SD^{-1} &= \mathrm{tr} \sum_i \lambda_i^{-1} l_i l_i' \sum_j \zeta_j z_j z_j' \\ &= \sum_i \sum_j \lambda_i^{-1} \zeta_j m_i^2(j),\end{aligned}$$

where $m_i(j)$ denotes the j-th element of m_i. The last relation is obtained from the equation

$$z_j' l_i = m_i(j).$$

We also have

$$|(I + CC')^{-1}|\,|D^{-1}SD^{-1}| = \prod_{i=1}^{p} \lambda_i^{-1} \prod_{j=1}^{p} \zeta_j.$$

Thus we get the following representation of the modified negative likelihood function as a function of $\lambda = (\lambda_1, \lambda_2, \cdots, \lambda_p)$ and $m = (m_1, m_2, \cdots, m_p)$:

$$q(\lambda, m) = -\sum_{i=1}^{p} \log \lambda_i^{-1}\zeta_i + \sum_{i=1}^{p} \sum_{j=1}^{p} \lambda_i^{-1} \zeta_j m_i^2(j).$$

Assume that ζ_i and λ_i are arranged in the descending order, that is, $\zeta_1 \geqq \zeta_2 \geqq \cdots \geqq \zeta_p$ and $\lambda_1 \geqq \lambda_2 \geqq \cdots \geqq \lambda_p$, where $\lambda_{k+1} = \cdots = \lambda_p = 1$. Then the successive minimization of $q(\lambda, m)$ with respect to $m_p, m_{p-1}, \cdots, m_1$ leads to

$$q(\lambda) = \sum_{i=1}^{k} [\lambda_i^{-1}\zeta_i - \log(\lambda_i^{-1}\zeta_i)] + \sum_{i=k+1}^{p} (\zeta_i - \log \zeta_i).$$

As a function of λ, $(\lambda^{-1}\zeta) - \log(\lambda^{-1}\zeta)$ attains its minimum at $\lambda = \zeta$, for $\zeta > 1$, and at $\lambda = 1$, otherwise. Thus we get

$$\operatorname*{Min}_{\lambda} q(\lambda) = k^* + \sum_{i=k^*+1}^{p} (z_i - \log \zeta_i),$$

where $\zeta_i > 1$, for $i \leqq k^*$, $\leqq 1$, otherwise. This last quantity is equal to the quantity given by the Equation (18) of Jöreskog (1967, p.448) and is $(-2/N)$ times the maximum log likelihood of the factor analysis model when D is given.

In maximizing the likelihood we would normally hope that a too small value of some of the diagonal elements of D will reduce the maximum likelihood of the corresponding model. However, that this is not the case is shown by the above result which explains that the value of the maximum likelihood is sensitive only to the behavior of smaller eigenvalues of $D^{-1}SD^{-1}$. A very small diagonal element of D will only produce a very large eigenvalue. Thus the process of maximizing the likelihood with respect to the elements of D does not eliminate the possibility of some of these elements going down to zero.

The form of $q(\lambda)$ shows that if we introduce an additive term $\rho\Sigma\mu_i$ with $\rho > 0$ then the minimization of

$$q(\lambda) = \sum_{i=1}^{p} [\lambda_i^{-1}\zeta_i - \log(\lambda_i^{-1}\zeta_i)] + \rho \sum_{i=1}^{p} \mu_i,$$

with respect to λ does not allow any of λ_i $(= 1 + \mu_i)$ going to infinity. Taking into account the relations $C = ZF$ and $FF' = \Sigma\mu_i m_i m_i'$ we get

$$\sum_{i=1}^{p} \mu_i = \text{tr } FF' = \text{tr } CC'.$$

Since $C = D^{-1}A$ the minimization of $q(\lambda)$ produces an estimate that is given as the posterior mode under the assumption of the prior distribution given by

$$K \exp\left\{-\frac{N}{2} \rho \text{ tr } D^{-1}AA'D^{-1}\right\},$$

where K denotes the normalizing constant and N the sample size. This prior distribution is defined by a spherical normal distribution of the standardized factor loadings and will be referred to as the standard spherical prior distribution of the factor loadings.

For the complete specification of the Bayesian model it is necessary to define the prior distribution of D. However, an arbitrarily defined prior distribution of the elements of D can easily eliminate improper solutions if only it penalizes smaller values sufficiently. Since our interest here is mainly in the clarification of the nature of improper solutions obtained by the conventional maximum likelihood procedure we will not proceed to the modeling of the prior distribution of D and simply adopt the uniform prior.

TABLE 1

Communality estimates*

Harman : eight physical

$\rho = 0$ (MLE)

$k \backslash i$	1	2	3	4	5	6	7	8
1	842	865	810	813	240	171	123	199
2	830	893	834	801	911	636	584	463
3	872	1000	806	844	909	641	589	509

$\rho = 0.1$

$k \backslash i$	1	2	3	4	5	6	7	8
1	837	858	804	810	241	172	124	200
2	828	881	828	800	855	647	591	476
3	858	910	830	832	859	650	590	523
4	865	910	832	843	851	689	649	521
5	same as above							

$\rho = 1.0$

$k \backslash i$	1	2	3	4	5	6	7	8
1	763	768	725	739	252	181	134	204
2	766	781	742	743	590	486	440	409
3								
4	same as above							
5								

* In this and following tables maximum possible communality is normalized to 1000.

6. Numerical Examples

The Bayesian model defined with the standard spherical prior distribution of the factor loadings was applied to six published examples of improper solutions. These examples are Harman's eight physical variables data (Harman, 1960, p.82), with $p = 8$ and improper at $k = 3$, Davis data (Rao, 1955, p.110), with $p = 9$ and improper at $k = 2$, Maxwell's normal children data (Maxwell, 1961, p.55), with $p = 10$ and improper at $k = 4$, Emmett data (Lawley & Maxwell, 1971, p.43), with $p = 9$ and improper at $k = 5$, Maxwell's neurotic children data (p.53), with $p = 10$ and improper at $k = 5$, and Harman's twenty-four psychological variables data (Harman, 1960, p.137), with $p = 24$ and improper at $k = 6$.

The informational point of view suggests that hyperparameter ρ of the prior distri-

TABLE 2

Communality estimates

Davis data

$\rho = 0$ (MLE)									
$k \backslash i$	1	2	3	4	5	6	7	8	9
1	658	661	228	168	454	800	705	434	703
2	652	1000	243	168	464	816	704	435	701
3	1000	661	220	204	451	1000	701	488	696
$\rho = 0.1$									
$k \backslash i$	1	2	3	4	5	6	7	8	9
1	653	656	226	167	451	790	700	431	697
2	694	689	227	171	470	800	698	434	696
3	701	695	251	197	470	801	698	444	696
4	same as above								
$\rho = 1.0$									
$k \backslash i$	1	2	3	4	5	6	7	8	9
1	596	598	210	156	415	702	633	400	631
2									
3	same as above								
4									

bution may be "estimated" by maximizing the likelihood of the Bayesian model with respect to ρ. However, for this purpose integration in a high-dimensional space is required. In this paper we will limit our attention to the analysis of solutions with some fixed values of ρ.

The estimation of specific variances under the present Bayesian model was realized by the following procedure. Given the initial estimate D_1^2 of D^2 the sample covariance matrix S is replaced by $S_1 = D_1^{-1}SD_1^{-1}$ and the next estimate D_2^2 of D^2 is obtained by the relation $D_2^2 = \text{diag}(S - D_1B_1B_1'D_1)$, where B_1 is a $p \times k$ matrix such that B_1B_1' provides a least squares fit

$$2\sum_{i=1}^{p-1}\sum_{j=i+1}^{p}[S_1(i,j) - \sum_{l=1}^{k}B_1(i,l)B_1(j,l)]^2 + \rho\sum_{i=1}^{p}\sum_{j=1}^{k}B_1^2(i,j) = \text{Min.},$$

where $B(i, j)$ denotes (i, j)th element of B. The estimates of communalities are defined by diag $(D_1B_1B_1'D_1)$. The process is repeated until convergence is established.

When $\rho = 0$ the above procedure produced maximum likelihood solutions that were confirmed by a procedure based on the result of Jennrich and Robinson (1969). When $\rho > 0$ the solution may only be considered as an arbitrary approximation to the posterior

TABLE 3

Three factor maximum likelihood solution of Emmett data

i	A(·1)	A(·2)	A(·3)	Ψ
1	.664	.321	.074	.450
2	.689	.247	-.193	.427
3	.493	.302	-.222	.617
4	.837	-.292	-.035	.212
5	.705	-.315	-.153	.381
6	.819	-.377	.105	.177
7	.611	.396	-.078	.400
8	.458	.296	.491*	.462
9	.766	.427	-.012	.231

* The value suggests singular increase of the 8th communality.

mode. Nevertheless it will be sufficient for the purpose of confirming of the effect of the tempering of the likelihood function. For convenience we will call the solution the Bayesian estimate.

In the case of the above six examples the choice of $\rho = 1.0$ produced solutions with signficant overall reduction of communalities, or increase of specific variances. With the choice of $\rho = 0.1$ solutions were usually close to the conventional maximum likelihood estimates but with the improper estimates of communalities suppressed. Improper estimates disappeared completely, unless ρ was made extremely small. For a fixed ρ estimates of communalities usually stabilized as k, the number of factors, was increased.

It was generally observed that when the maximum likelihood method produced an improper solution first at $k = k_0$ the corresponding Bayesian estimate with $\rho = 0.1$ was proper but with only one communality estimate inflated compared with the estimate at $k = k_0 - 1$. Such a singular increase of the communality means the reinterpretation of a part of the specific variation as an independent factor. This fact and the result of our analysis of the likelihood function suggest that the singular increase of the communality is usually caused by the overparametrization that makes the estimate sensitive to the sampling variability of the data rather than by the structural change of the best fitting model at $k = k_0$. This is in agreement with the earlier observation of Tsumura and Sato (1981) on the nature of improper solutions.

Tables 1 and 2 provide estimates of communalities of Harman's eight physical variables data and of Davis' data, respectively, for various choices of the order, k, and ρ. In the

TABLE 4

Communality estimates by various procedures

Emmett data

$\rho = 0$ (MLE)

$k \backslash i$	1	2	3	4	5	6	7	8	9
1	510	537	300	548	390	481	525	224	665
2	538	536	332	809	592	778	597	256	782
3	550	573	384	788	619	823	600	538	769
4	554	666	379	772	663	856	648	480	759
5	556	868	1000	780	664	836	666	464	743

$\rho = 0.1$

$k \backslash i$	1	2	3	4	5	6	7	8	9
1	502	529	296	545	388	478	516	221	652
2	535	531	330	790	588	762	590	252	753
3	549	561	378	783	611	786	590	399	750
4									
5				same as above					

$\rho = 1.0$

$k \backslash i$	1	2	3	4	5	6	7	8	9
1	425	448	254	478	344	422	434	189	540
2	433	450	261	522	391	472	445	196	551
3									
4				same as above					
5									

case of the Harman data the result in Table 1 shows that the improper value 1000 at $i = 2$ with $k = 3$, obtained with $\rho = 0$, disappeared for the positive values of ρ. In particular, with $\rho = 0.1$, the solutions with $k = 2$, 3 and 4 are all mutually very close and they are close to the solutions with $\rho = 0$ and $k = 2$ and 3, except for the improper component at $k = 3$. This suggests that the two-factor model is an appropriate choice, which is in agreement with Harman's original observation. The soltuion with $\rho = 1.0$ conforms with this observation.

For the Davis data with $k_0 = 2$ the non-uniqueness of the convergence of iterative procedures for the maximum likelihood was first reported by Tsumura, Fukutomi, and Asoo (1968). With $k = 2$, Jöreskog (1967, p.474) reported improper estimate of specific variance for the 1st component and Tsumura et al. (p.57) found one for the 8th component. As is shown in Table 2 our procedure found one at the 2nd component. The result

TABLE 5

Suggested choices of dimensionalities*

Harman: eight physical		
$p = 8$	$k_o = 3$	$k_s = 2$
	MAICE $= \infty$**	
Davis		
$p = 9$	$k_o = 2$	$k_s = 1$
	MAICE $= \infty$**	
Maxwell: normal		
$p = 10$	$k_o = 4$	$k_s = 3$
	MAICE $= \infty$**	
Emmett		
$p = 9$	$k_o = 5$	$k_s = 2$
	MAICE $= 3$	
Maxwell: neurotic		
$p = 10$	$k_o = 5$	$k_s = 2$
	MAICE $= 3$	
Harman : 24 variables		
$p = 24$	$k_o = 6$	$k_s = 5$
	MAICE $= 5$	

* p : dimension of observation
k_o : lowest order with improper solution
k_s : suggested order by the Bayesian analysis

** ∞ denotes saturated model.

given in Table 4 of Martin and McDonald (1975, p.515) also suggests the existence of improper solution with zero unique variance for the 2nd component. These results suggest the existence of local maxima of the likelihood function. Table 2 also gives improper estimates for the 1st and 6th components with $k = 3$, which is in agreement with the result reported by Jöreskog.

The estimates obtained with $\rho = 0.1$ may be viewed as practically identical and are close to the solution with $\rho = 0$, the maximum likelihood estimate, for $k = 1$. This result strongly suggests that the improper solutions are spurious in the sense that they can be suppressed by mild tempering of the likelihood function. The one-factor model seems a reasonable choice in this case. The solution with $\rho = 1.0$ conforms with the present observation.

The phenomenon of the singular increase of a communality estimate is observed even with $k < k_0$. Such an example is given by the three-factor maximum likelihood solution of the Emmett data. The maximum likelihood solution by Lawley and Maxwell (1971, p.43) is reproduced in Table 3 which suggests the singular increase of the communality of the 8th component at $k = 3$. In Table 4 the estimate with $\rho = 0.1$ shows substantial increase of communality at only the 8th component at $k = 3$, compared with the estimate at $k = 2$. The increase is completely suppressed with $\rho = 1.0$. This result suggests that the high value of the communality estimate of the 8th component at $k = 3$ obtained with $\rho = 0$ is spurious. Similar phenomenon was observed with Maxwell's data of neurotic children for the 2nd component at $k = 3$.

Tsumura and Sato (1981, p.163) report that, by their experience, improper solutions were always with "quasi-specific factors" that respectively showed singular contributions to some specific variances. The above example shows that our present Bayesian approach can detect the appearance of such a factor even before one gets a definitely improper solution. Thus we can expect that the present approach will realize a reasonable control of improper solutions.

Table 5 summarizes the suggested choices of the number of factors for the six examples where the choices by the minimum AIC procedure, MAICE, are also included. The suggested choices are based on subjective judgments of the numerical results. It is quite desirable to develop a numerical procedure for the evaluation of the likelihood of each Bayesian model to arrive at an objective judgment.

It is interesting to note here that by a proper choice of ρ the Bayesian approach can produce estimate of A even with $k = p$. This explains the drastic change of the emphasis between the modelings by the conventional and Bayesian approach. By the Bayesian approach there is no particular meaning in trying to reduce the number of factors. To avoid unnecessary distortion of the model it is even advisable to adopt a large value of k and control the estimation procedure by a proper choice of ρ.

7. Concluding Remarks

It is remarkable that the idea of factor analysis has been producing so much stimulus to the development of statistical modeling. In terms of the structure of the model it is essentially Bayesian. Nevertheless, the practical use of the model was realized by the application of the method of maximum likelihood and this eventually led to the introduction of AIC.

The concept of the information measure underlying the introduction of AIC leads our attention from parameters to the distribution. This then provides a conceptual framework for the handling of the Bayesian modeling as a natural extension of the conventional statistical modeling. The occurence of improper solutions in the maximum likelihood factor analysis is a typical example that explains the limitation of the conventional modeling. The introduction of the standard spherical prior distribution of factor loadings provided an example of overcoming the limitation by a proper Bayesian modeling.

This series of experiences clearly explains the close dependence between the factor analysis and AIC, or the informational point of view of statistics, and illustrates their contribution to the development of general statistical methodology. It is hoped that this close contact between psychometrics and statistics will be maintained in the future and contribute to the advancement of both fields.

References

Akaike, H. (1969). Fitting autoregressive models for prediction. *Annals of the Institute of Statistical Mathematics, 21,* 243–247.

Akaike, H. (1970). Statistical predictor identification. *Annals of the Institute of Statistical Mathematics, 22,* 203–217.

Akaike, H. (1973). Information theory and an extension of the maximum likelihood principle. In B. N. Petrov & F. Csaki (Eds.), *2nd International Symposium on Information Theory* (pp. 267–281). Budapest: Akademiai Kiado.

Akaike, H. (1974). A new look at the statistical model identification. *IEEE Transactions on Automatic Control, AC-19,* 716–723.

Akaike, H. (1980). Likelihood and the Bayes procedure. In J. M. Bernardo, M. H. De Groot, D. V. Lindley, & A. F. M. Smith (Eds.), *Bayesian Statistics* (pp. 143–166). Valencia: University Press.

Akaike, H. (1985). Prediction and entropy. In A. C. Atkinson & S. E. Fienberg (Eds.), *A Celebration of Statistics* (pp. 1–24). New York: Springer-Verlag.

Bartholomew, D. J. (1981). Posterior analysis of the factor model. *British Journal of Mathematical and Statistical Psychology, 34,* 93–99.

Bozdogan, H., & Ramirez, D. E. (1987). An expert model selection approach to determine the "best" pattern structure in factor analysis models. Unpublished manuscript.

Harman, H. H. (1960). *Modern Factor Analysis.* Chicago: University Press.

Jennrich, R. I., & Robinson, S. M. (1969). A Newton-Raphson algorithm for maximum likelihood factor analysis. *Psychometrika, 34,* 111–123.

Jöreskog, K. G. (1967). Some contributions to maximum likelihood factor analysis. *Psychometrika, 32,* 443–482.

Jöreskog, K. G. (1978). Structural analysis of covariance and correlation matrices. *Psychometrika, 43,* 443–477.

Lawley, D. N., & Maxwell, A. E. (1971). *Factor Analysis as a Statistical Method, 2nd Edition.* London: Butterworths.

Martin, J. K., & McDonald, R. P. (1975). Bayesian estimation in unrestricted factor analysis: a treatment for Heywood cases. *Psychometrika, 40,* 505–517.

Maxwell, A. E. (1961). Recent trends in factor analysis. *Journal of the Royal Statistical Society, Series A, 124,* 49–59.

Rao, C. R. (1955). Estimation and tests of significance in factor analysis. *Psychometrika, 20,* 93–111.

Tsumura, Y., Fukutomi, K., & Asoo, Y. (1968). On the unique convergence of iterative procedures in factor analysis. *TRU Mathematics, 4,* 52–59. (Science University of Tokyo).

Tsumura, Y., & Sato, M. (1981). On the convergence of iterative procedures in factor analysis. *TRU Mathematics, 17,* 159–168. (Science University of Tokyo).

Reprinted from *A CELEBRATION OF STATISTICS, The ISI Centenary Volume*, A.C. Atkinson and S.E. Fienberg, eds., @Sringer-Varlag, New York, 1985, 1-24, by permission from Springer-Verlag, New York

CHAPTER 1

Prediction and Entropy

Hirotugu Akaike

Abstract

The emergence of the magic number 2 in recent statistical literature is explained by adopting the predictive point of view of statistics with entropy as the basic criterion of the goodness of a fitted model. The historical development of the concept of entropy is reviewed, and its relation to statistics is explained by examples. The importance of the entropy maximization principle as a basis of the unification of conventional and Bayesian statistics is discussed.

1. Introduction and Summary

We start with an observation that the emergence of a particular constant, the magic number 2, in several statistical papers is inherently related to the predictive use of statistics. The generality of the constant can only be appreciated when we adopt the statistical concept of entropy, originally developed by a physicist L. Boltzmann, as the criterion to measure the deviation of a distribution from another.

A historical review of Boltzmann's work on entropy is given to provide a basis for the interpretation of statistical entropy. The negentropy, or the negative of the entropy, is often equated to the amount of information. This review clarifies the limitation of Shannon's definition of the entropy of a probability distribution. The relation between Boltzmann entropy and the asymptotic theory of statistics is discussed briefly.

The concept of entropy provides a proof of the objectivity of the log likelihood as a measure of the goodness of a statistical model. It is shown that this observation, combined with the predictive point of view, provides a simple explanation of the generality of the magic number 2. This is done through the explanation of the AIC statistic introduced by the present author. The use of AIC is illustrated by its application to multidimensional contingency table analysis.

The discussion of AIC naturally leads to the entropy maximization principle which specifies the object of statistics as the maximization of the expected

Key words and phrases: AIC, Bayes procedure, Entropy, entropy maximization principle, information, likelihood, model selection, predictive distribution.

entropy of a true distribution with respect to the fitted predictive distribution. The generality of this principle is demonstrated through its application to Bayesian statistics. The necessity of Bayesian modeling is discussed and its similarity to the construction of the statistical model of thermodynamics by Boltzmann is pointed out. The principle provides a basis for the unification of Bayesian and conventional approaches to statistics. Referring to Boltzmann's fundamental contribution to statistics, the paper concludes by emphasizing the importance of the research on real problems for the development of statistics.

2. Emergence of the Magic Number 2

Around the year 1970, the constant 2 appeared in a curious fashion over and over again in a series of papers. This represents the emergence of what Stone (1977a) symbolically calls the magic number 2.

The number appears in Mallow's C_p statistic for the selection of independent variables in multiple regression, which is by definition

$$C_p = \frac{1}{s^2}\text{RSS}_p - n + 2p,$$

where RSS_p denotes the residual sum of squares after regression on p independent variables, n the sample size, and s^2 an estimate of the common variance σ^2 of the error terms (Mallows, 1973). The final prediction error (FPE) introduced by Akaike (1969, 1970) for the determination of the order of an autoregression is an estimate of the mean squared error of the one-step prediction when the fitted model is used for prediction. It satisfies asymptotically the relation

$$n \log \text{FPE} = n \log S_p + 2p,$$

where n denotes the length of the time series, S_p the maximum-likelihood estimate of the innovation variance obtained by fitting the pth-order autoregression. Both Leonard and Ord (1976) and Stone (1977a) noticed the number 2 as the asymptotic critical level of F-tests when the number of observations is increased.

An explanation of the multiple appearances of this number 2 can easily be given for the case of the multiple regression analysis. The effect of regression is usually evaluated by the value of RSS_p. A smaller RSS_p may be obtained by increasing the number of independent variables p. However, we know that after adding a certain number of independent variables further addition of variables often merely increases the expected variability of the estimate. When the increase of the expected variability is measured in terms of the mean squared prediction error, it will be seen that the increase is exactly equal to the expected amount of decrease of the sample residual variance RSS_p/n. Thus to convert RSS_p into an unbiased estimate of the mean squared error of prediction we must apply *twice* the correction that is required to convert RSS_p into an unbiased estimate of $n\sigma^2$.

The appearance of the critical value 2 for the F-test discussed by Leonard and Ord (1976) is more instructive. The F-test is considered as a preliminary test of significance in the estimation of the one-way ANOVA model where K independent observations y_{jk} ($k = 1, 2, \ldots, K$) are taken from each group j ($j = 1, 2, \ldots, J$). Under the assumption that y_{jk} are distributed as normal with mean θ_j and variance σ_W^2, the F-statistic for testing the hypothesis $\theta_1 = \theta_2 = \cdots = \theta_J$ is given by

$$F = \frac{(J-1)^{-1}S_B^2}{(J(K-1))^{-1}S_W^2},$$

where $S_B^2 = K\sum_j (y_{j\cdot} - y_{\cdot\cdot})^2$ and $S_W^2 = \sum_j\sum_k (y_{jk} - y_{j\cdot})^2$, and where $y_{j\cdot}$ and $y_{\cdot\cdot}$ denote the mean of the jth group and the grand mean, respectively. The final estimate of θ_j is defined by

$$\tilde{\theta}_j = \begin{cases} y_{j\cdot} & \text{if the hypothesis is rejected,} \\ y_{\cdot\cdot} & \text{otherwise.} \end{cases}$$

Consider the loss function $L(\tilde{\theta}, \theta) = \sum(\tilde{\theta}_j - \theta_j)^2$. For the simpler estimates defined by $\tilde{\theta}_j = y_{\cdot\cdot}$ and $\tilde{\theta}_j = y_{j\cdot}$ it can easily be shown that the difference of the risks of these estimates has one and the same sign as that of $E(J(K-1))^{-1}S_W^2(F-2)$, where E denotes expectation. Thus when the sample size K is sufficiently large, the choice of the critical value 2 for the F-test to select $\tilde{\theta}_j$ is appropriate.

The one characteristic that is clearly common to these papers is that the authors considered some predictive use of the models. An early example of the use of the concept of future observation to clarify the structure of an inference procedure is given by Fisher (1935, p. 393). The concept is explicitly adopted as the philosophical motivation in a work by Guttman (1967). In the present paper, the predictive point of view adopts as the purpose of statistics the realization of appropriate predictions.

In the example of the ANOVA model above, if the number of groups, J, is increased indefinitely, the test statistic F converges to 1 under the null hypothesis. Thus the critical value of the F-test for any fixed level of significance must also converge to 1 instead of 2. As is observed by Leonard and Ord, this dramatically demonstrates the difference between the conventional approach to model selection by testing with a fixed level of significance and the predictive approach. Thus the emergence of the magic number 2 must be considered as a sign of the impending change of the paradigm of statistics. However, to fully appreciate the generality of the number, we have first to expand our view of the statistical estimation.

3. From Point to Distribution

The risk functions considered in the preceding section were the mean squared errors of the predictions. Such a choice of the criterion is conventional but quite arbitrary. The weakness of the ad hoc definition becomes apparent when we try to extend the concept to multivariate problems.

A typical example of multivariate analysis is factor analysis. At first sight it is not at all clear how the analysis is related to prediction. In 1971, in an attempt to extend the concept of FPE to solve the problem of determination of the number of factors, the present author recognized that in factor analysis a prediction was realized through the specification of a distribution (Akaike, 1981). This quickly led to the observation that almost all the important statistical procedures hitherto developed were concerned, either explicitly or implicitly, with the realization of predictions through the specification of distributions.

Stigler (1975) remarks that the interest of statisticians shifted from point estimation to distribution estimation towards the end of the 19th century. However, it seems that Fisher's very effective use of the concept of parameter drew the attention of statisticians back to the estimation of a point in a parameter space. We are now in a position to return to distributions, and here the basic problem is the introduction of a natural topology in the space of distributions. The probabilistic interpretation of thermodynamic entropy developed by Boltzmann provides, historically, one of the most successful examples of a solution to this problem.

4. Entropy and Information

The statistical interpretation of the thermodynamic entropy, a measure of the unavailable energy within a thermodynamic system, was developed in a series of papers by L. Boltzmann in the 1870's. His first contribution was the observation of the monotone decreasing behavior in time of a quantity defined by

$$E = \int_0^\infty f(x, t) \log\left[\frac{f(x, t)}{\sqrt{x}}\right] dx,$$

where $f(x, t)$ denotes the frequency distribution of the number of molecules with energy between x and $x + dx$ at time t (Boltzmann, 1872). Boltzmann showed that for a closed system, under proper assumptions on the collision process of the molecules, the quantity E can only decrease. When the distribution f is defined in terms of the velocities and positions of the molecules, the above quantity takes the form

$$E = \iint f \log f \, dx \, d\xi,$$

where x and ξ denote the vectors of the position and velocity, respectively. Boltzmann showed that for some gases this quantity, multiplied by a negative constant, was identical to the thermodynamic entropy.

The negative of the above quantity was adopted by C. E. Shannon as the definition of the entropy of a probability distribution:

$$H = -\int p(x) \log p(x)\, dx,$$

where $p(x)$ denotes the probability density with respect to the measure dx (Shannon and Weaver, 1949).

Almost uncountably many papers and books have been written about the use of the Shannon entropy, where the quantity H is simply referred to as a measure of information, or uncertainty, or randomness. One departure from this definition of entropy, known as the Kullback–Leibler information (Kullback and Leibler, 1951), is defined by

$$I(q; p) = \int q(x) \log\left(\frac{q(x)}{p(x)}\right) dx$$

and relates the distribution $q(x)$ to another distribution $p(x)$.

Much interest has been shown in the use of these quantities as measures of statistical information. However, it seems that their potential as statistical concepts has not been fully evaluated. It seems to the present author that this is due to the neglect of Boltzmann's original work on the probabilistic interpretation of thermodynamic entropy. Karl Pearson (1929, p. 205) cites the words of D. F. Gregory: "... we sacrifice many of the advantages and more of the pleasure of studying any science by omitting all reference to the history of its progress." It seems that this has been precisely the case with the development of the statistical concept of entropy or information.

5. Distribution and Entropy

The work of Boltzmann (1872) produced a demonstration of the second law of thermodynamics, the irreversible increase of entropy in an isolated closed system. In answering the criticism that the proof of irreversibility is based on the assumption of a reversible mechanical process, Boltzmann (1877a) pointed out the necessity of probabilistic interpretation of the result.

At that time Meyer, a physicist, produced a derivation of the Maxwell distribution of the kinetic energy among gas molecules at equilibrium as the "most probable" distribution. Pointing out the error in Meyer's proof, Boltzmann (1877b) established the now well-known identity

$$\text{entropy} = \log(\text{probability of a statistical distribution}).$$

His reasoning was based on the asymptotic equality

$$\log \frac{n!}{n_0!n_1!\cdots n_p!} = -n \sum_{i=0}^{p} \frac{n_i}{n} \log \frac{n_i}{n}, \tag{5.1}$$

where n_i denotes the frequency of the molecules at the ith energy level and $n = n_0 + n_1 + \cdots + n_p$. If we put $p_i = n_i/n$, then the right-hand side is equal to $nH(p)$, where

$$H(p) = -\sum_{i=0}^{p} p_i \log p_i,$$

i.e., the Shannon entropy of the distribution $p = (p_0, p_1, \ldots, p_p)$.

Following the idea that the frequency distribution f of molecules at thermal equilibrium is the distribution which is the most probable under the assumption of a given total energy, Boltzmann maximized

$$H(f) = -\int_0^\infty f \log f\, dx$$

under the constraints

$$\int_0^\infty f\, dx = N \quad \text{and} \quad \int_0^\infty x f(x)\, dx = L,$$

where x denotes the energy level, N the total number of molecules, and L the total energy. The maximization produces as the energy distribution $f(x) = Ce^{-hx}$ with a proper positive constant h. Boltzmann discussed in great detail that this result could be physically meaningful only for a proper definition of the energy level x, a point commonly ignored by later users of the Shannon entropy. Incidentally, we notice here an early derivation of the exponential family of distributions by the constrained maximization of $H(f)$, a technique of probability-distribution generation later named the maximum-entropy method (Jaynes, 1957).

The change in Boltzmann's view of the energy distribution between 1872 and 1877 is quite significant. In the 1872 paper the distribution $f(x, t)$ represented a unique entity. In the 1877b paper the distribution was considered as a random sample and its probability of occurrence was the main subject.

Boltzmann (1878) further extended the discussion of this point. Noting that the probability of geting a sample frequency distribution $(w_0, w_1, \ldots, w_p)$ from a probability distribution $(f_0, f_1, \ldots, f_p)$ is given by

$$\Omega = f_0^{w_0} f_1^{w_1} \cdots f_p^{w_p} \cdot \frac{n!}{w_0! w_1! \cdots w_p!},$$

Boltzmann derived an asymptotic equality

$$l\Omega = w_0 lf_0 + w_1 lf_1 + \cdots + w_p lf_p - w_0 lw_0 - w_1 lw_1 \cdots - w_p lw_p + \text{const}, \tag{5.2}$$

where $n = w_1 + w_2 + \cdots + w_p$ and l denoted the natural logarithm. He pointed out that the former formula (5.1) is a special case of (5.2) where it is assumed that $f_0 = f_1 = \cdots = f_p$. If the additive constant is ignored, the present formula (5.2) can be rearranged in the form

$$l\Omega = -n \sum_{i=0}^{p} g_i l\left(\frac{g_i}{f_i}\right),$$

where $g_i = w_i/n$. Thus to retain the interpretation that the entropy is the log probability of a distribution we have to adopt, instead of $H(p)$, the quantity

$$B(g; f) = -\sum_i g_i \log\left(\frac{g_i}{f_i}\right)$$

as the definition of the entropy of the secondary distribution g with respect to the primary distribution f. When the distributions g and f are defined in terms of densities $f(x)$ and $g(x)$, the entropy is defined by

$$B(g; f) = -\int g(x) \log\left(\frac{g(x)}{f(x)}\right) dx.$$

When it is necessary to distinguish this quantity from the thermodynamic entropy or the Shannon entropy, we will call it the Boltzmann entropy. It is now obvious that $B(g; f)$ provides a natural measure of the deviation of g from f.

The equality of the above quantity to the thermodynamic entropy holds only when the former is maximized under the assumption of a given mean energy for an appropriately chosen "primary distribution" f and then multiplied by a proper constant. Thus it can be seen that the Shannon entropy $H(g) = -\sum g_i \log g_i$ yields the physical meaning of the entropy contemplated by Boltzmann only under very limited circumstances. Obviously $l\Omega$ or $B(g; f)$ is the more fundamental concept. This point is reflected in the fact that in Shannon and Weaver (1949) essential use is made not of $H(f)$ but of its derived quantities taking the form of $B(g; f)$.

The Kullback–Leibler (KL) information number is defined by $I(g; f) = -B(g; f)$. Kullback (1959) describes this quantity as the mean information per observation from $g(x)$ for discrimination in favor of $g(x)$ against $f(x)$ and simultaneously considers $I(f; g)$ to define the Jeffreys divergence:

$$J(g; f) = I(g; f) + I(f; g).$$

Contrary to the formal definition of $I(g; f)$ by Kullback, the present derivation of $B(g; f)$ based on Boltzmann's $l\Omega$ clearly explains the difference of the roles played by g and f. The primary distribution f is hypothetical, while the secondary g is factual. It is the fictitious sampling from $f(x)$ that provides the probabilistic meaning of $B(g; f)$. This may be seen more clearly by the representation

$$B(g; f) = -\int \frac{g(x)}{f(x)} \log\left(\frac{g(x)}{f(x)}\right) f(x)\, dx.$$

Boltzmann (1878) also arrived at a generalization of the exponential family of distributions by maximizing the entropy under certain constraints. These results demonstrate the fundamental contribution of Boltzmann to the science of statistics. A good summary of mathematical properties of the Boltzmann entropy or the Kullback–Leibler information is given by Csiszar (1975).

6. Entropy and the Asymptotic Theory of Statistics

The Boltzmann entropy appears, sometimes implicitly, in many basic contributions to statistics, particularly in the area of asymptotic theory. For a pair of distributions $p(\cdot \mid \theta_1)$ and $p(\cdot \mid \theta_2)$ from a parametric family $\{p(\cdot \mid \theta); \theta \in \Theta\}$ the deviation of the former from the latter can be measured by $B(\theta_1; \theta_2) = B(p(\cdot \mid \theta_1); p(\cdot \mid \theta_2))$. This induces a natural topology in the space of parameters.

When θ_1 and θ_2 are k-dimensional parameters given by $\theta_1 = (\theta_{11}, \theta_{12}, \ldots, \theta_{1k})$ and $\theta_2 = (\theta_{21}, \theta_{22}, \ldots, \theta_{2k})$, under appropriate regularity conditions we have

$$B(\theta_1; \theta_2) = -\tfrac{1}{2}(\theta_2 - \theta_1)' E\left[\frac{\partial^2}{\partial\theta' \, \partial\theta} \log p(x \mid \theta_1)\right](\theta_2 - \theta_1) + o(\|\theta_2 - \theta_1\|^2),$$

where $(\partial^2/\partial\theta' \, \partial\theta) \log p(x \mid \theta_1)$ denotes the Hessian evaluated at $\theta = \theta_1$, E denotes the expectation with respect to $p(\cdot \mid \theta_1)$, and $o(\|\theta_2 - \theta_1\|^2)$ is a term of order lower than $\|\theta_2 - \theta_1\|^2 = \sum(\theta_{1i} - \theta_{2i})^2$. The quantity $-E[(\partial^2/\partial\theta' \, \partial\theta) \log p(x \mid \theta_1)]$ is the Fisher information matrix. The fact that the Fisher information matrix is just minus twice the Hessian of the entropy clearly shows that it is related to the local property of the topology induced by the entropy.

The likelihood-ratio test statistic for testing a specific model, or hypothesis, defined by $\theta = \theta_0$ is given by

$$\lambda_n = \frac{\prod p(x_i \mid \theta_0)}{\sup\{\prod p(x_i \mid \theta); \theta \in \Theta\}},$$

where $(x_1, x_2, \ldots, x_n)$ denotes the sample. If the true distribution is defined by $p(\cdot \mid \theta)$, we expect that

$$T_n = -\frac{1}{n} \log \lambda_n$$

will converge stochastically to $-B(\theta; \theta_0)$ as n is increased to infinity. The result of Bahadur (1976) shows that under certain regularity conditions

$$\lim_{n \to \infty} \frac{1}{n} \log P(T_n > t_n \mid \theta_0) = B(\theta; \theta_0),$$

where t_n denotes the sample value of the test statistic T_n for a particular realization $(x_1, x_2, \ldots, x_n)$. This means that if one calculates the probability of the statistic T_n being larger than t_n, assuming that the data have come from the hypothetical distribution $p(\cdot \mid \theta_0)$, it will asymptotically be equal to $\exp(nB(\theta; \theta_0))$, where θ denotes the true distribution.

In a practical application the hypothesis will never be exact, and the above result says that by calculating the P-value of the log likelihood-ratio test we are actually measuring the entropy $nB(\theta; \theta_0)$. It is often argued that the test is logically meaningless, since the falsity of θ_0 is almost always certain. The present observation partially clarifies the confusion in this argument.

The concept of second-order efficiency was introduced by Rao (1961). In that paper he discussed the performance of an estimator obtained by minimizing the Kullback–Leibler information number $\sum \pi_r \log(\pi_r/p_r)$, where π_r denotes the probability of the rth cell in a multinomial distribution, defined as a function of a parameter θ, and p_r the observed relative frequency. This estimator can also be characterized as the one that maximizes $B(\pi; p)$, while the maximum-likelihood estimate maximizes $B(p; \pi)$.

If we carefully follow the derivation of $B(g;f)$, we can see that the primary distribution f is always hypothetical, while the secondary distribution g is factual. It is interesting to note that Rao has shown that the minimum KL number estimator, defined by the entropy with a factual primary distribution and an hypothetical secondary, is less efficient than the maximum-likelihood estimator defined by the more natural definition of the entropy. A similar relation has been observed between the estimators defined by minimizing the chi-square and the modified chi-square that are approximations to $-2B(p; \pi)$ and $-2B(\pi; p)$, respectively. These results suggest that the present interpretation of entropy can produce useful insights not available from the use of Fisher information which does not discriminate between the primary and secondary distributions.

The relation between the entropy and the asymptotic distribution of the corresponding sample distribution function is discussed by Sanov (1957) and Stone (1974). Other standard references on the relation between the entropy and large-sample theory are Chernoff (1956) and Rao (1962).

7. Likelihood, Entropy and the Predictive Point of View

Obviously, one of the most significant contributions to statistics by R. A. Fisher is the development of the method of maximum likelihood. However, there is a definite limitation to the applicability of the idea of maximizing the likelihood.

The limitation can most clearly be seen by the following model selection problem. Consider a set of nested parametric families $\{p(\cdot \mid \theta_k)\}$ $(k = 1, 2, \ldots, K)$, defined by $\theta_k = (\theta_{k1}, \theta_{k2}, \ldots, \theta_{kk}, \theta_{0k+1}, \ldots, \theta_{0K})$. In the kth family, only the first k components of the parameter vector θ_k are allowed to vary; the rest are fixed at some preasssigned values $\theta_{0\,k+1}, \ldots, \theta_{0K}$. When data x are given, if we simply maximize the likelihood over the whole family, we always end up with the choice of $p(\cdot\,|\,\theta_K^*)$, where θ_K^* denotes the maximum-likelihood estimate that maximizes $p(x\,|\,\theta_K)$. This means that the method of maximum likelihood always leads to the selection of the unconstrained model. This is obviously against our expectation. If a statistician suggests the choice of the highest possible order whenever fitting a polynomial regression, he will certainly lose the trust of his clients.

Fisher was clearly aware of the limitation of his theory of estimation. After pointing out the necessity of the knowledge of the functional form of the distribution as the prerequisite of his theory of estimation, Fisher (1936, p. 250) admits the possibility of a wider type of inductive argument that would discuss methods of determining the functional form by data. However he also states, "At present it is only important to make clear that no such theory has been established." This clearly suggests the necessity of extending the theory of statistical estimation to the situation where several possible parametric models are involved. Such an extension is possible with a proper combination of the predictive point of view and the concept of entropy.

The predictive point of view generalizes the concept of estimation from that of a parameter to that of the distribution of a future observation. We refer to such an estimate as a predictive distribution. The basic criterion in this generalized theory of estimation is then the measure of the "goodness" of the predictive distribution. One natural choice of such a measure is the expected deviation of the true distribution from the predictive distribution as measured by the expected entropy EB(true; predictive). Here, the expectation E is taken with respect to the true distribution of the data used to define the predictive distribution.

Except for data obtained by an artificial sampling scheme, we do not know exactly what is meant by the true distribution. Indeed, the concept of the true distribution obtains a practical meaning only through the specification of an estimation procedure or a model. The true distribution may thus be viewed as a conceptual construct that provides a basis for the design of an estimation procedure for a particular type of data. Since the concept of the true distribution is quite personal, the validity of an estimation procedure based on such a concept must be judged by the collective experience of its use by human society. In such a circumstance it becomes crucial to find the objectivity of a statistical inference procedure to make it a vehicle for the communication of our experiences.

When a parametric model $\{p(\cdot\,|\,\theta);\ \theta \in \Theta\}$ of the distribution of a future observation y is given, the goodness of a particular model $p(\cdot\,|\,\theta)$ as the predictive distribution of y is evaluated by the entropy

$$B(f; p(\cdot\,|\,\theta)) = -E_y \log\left(\frac{f(y)}{p(y\,|\,\theta)}\right),$$

where E_y denotes the expectation with respect to the true distribution denoted by $f(y)$. Here the true distribution is unspecified; only its existence is assumed. Since it holds that

$$B(f; p(\cdot \mid \theta)) = E_y \log p(y \mid \theta) - E_y \log f(y),$$

we may restrict our attention to $E_y \log p(y \mid \theta)$ for the comparison of possible choices of θ.

We further specify the predictive point of view by assuming that *the future observation y is another independent sample taken from the same distribution as that of the present data x*. The accuracy of our inference is evaluated only in its relation to the prediction of an observation similar to the present one.

One of the important consequences of the present specification of the predictive point of view is that it leads to the observation that the log likelihood, $\log p(x \mid \theta)$, is a natural estimate of $E_y \log p(y \mid \theta)$. Obviously, by the present predictive point of view, *the log likelihood* $\log p(x \mid \theta)$ *provides an unbiased estimate of* $E_y \log p(y \mid \theta)$, *irrespective of the form of the true distribution* $f(y)$. The log likelihood provides an unbiased estimate of the basic criterion $E_y \log p(y \mid \theta)$ to everyone who accepts the concept of the true distribution, irrespective of the form of $f(y)$. In this sense, objectivity is imparted to statistical inference through the use of log likelihoods. We can see that the range of the validity of the concept of likelihood is not restricted to one particular parametric family of distributions. This observation constitutes the basis for the solution of the model selection problem considered at the beginning of this section.

8. Model Selection and an Information Criterion (AIC)

We will first show that our basic criterion, the expected entropy, provides a natural extension of the mean-squared-error criterion. The quality of a predictive distribution $f(y \mid x)$ is evaluated by the expected negentropy defined by

$$-E_x B(f; f(\cdot \mid x)) = E_y \log f(y) - E_x E_y \log f(y \mid x),$$

where $f(y)$ denotes the true distribution of y, which is assumed to be independent of x, and E_x and E_y denote the expectations with respect to the true distributions of x and y, respectively. By Jensen's inequality we have $E_x \log f(y \mid x) \leqslant \log E_x f(y \mid x)$, and we get the additive decomposition

$$\begin{aligned} -E_x B(f; f(\cdot \mid x)) &= \{E_y \log f(y) - E_y \log E_x f(y \mid x)\} \\ &\quad + \{E_y \log E_x f(y \mid x) - E_y E_x \log f(y \mid x)\}. \end{aligned}$$

The term inside the first braces on the right-hand side represents the amount of increase of the expected negentropy due to the deviation of $f(y)$ from $E_x f(y \mid x)$. This term corresponds to the squared bias in the case of ordinary estimation of a parameter. The term inside the second braces represents the increase of the expected negentropy due to the sampling fluctuation of $f(y \mid x)$

around $E_x f(y|x)$. This quantity corresponds to the variance. The present result shows why the two different concepts, squared bias and variance, can be added together in a meaningful way.

Having observed that the expected negentropy provides a natural extension of the mean-squared-error criterion, we recognize that the main problem is the estimation of the entropy or the expected log likelihood $E_y \log f(y|x)$ of the predictive distribution. In the case of the ANOVA model discussed by Leonard and Ord, the F-test was used for the selection of the model underlying the definition of the final estimate. For the present general model we consider the use of the log likelihood-ratio test. The test statistic for the testing of $\{p(\cdot|\theta_k)\}$ against $\{p(\cdot|\theta_K)\}$ of the preceding section is defined y

$$(-2)\{\log p(x|\theta_k^*) - \log p(x|\theta_K^*)\},$$

where θ_k^* denotes the maximum-likelihood estimate determined by the date x, and the statistic is taken to follow a chi-square distribution with K-k degrees of freedom.

We consider that the test is developed to make a reasonable choice between $p(y|\theta_k^*)$ and $p(y|\theta_K^*)$. From our present point of view this means that the test must be in good correspondence to the choice by $(-2)E_y\{\log p(y|\theta_k^*) - \log p(y|\theta_K^*)\}$. The result of Wald (1943) on the asymptotic behavior of the log-likelihood-ratio test shows that, when x is a vector of observations of independently identically distributed random variables with the likelihood functions satisfying certain regularity conditions, we have asymptotically

$$E_x^*[-2\{\log p(x|\theta_k^*) - \log p(x|\theta_K^*)\}] = \|\theta_k^0 - \theta_K^0\|_I^2 + (K-k),$$

where E_x^* denotes the mean of the limiting distribution, $\| \ \|_I$ is the Euclidean norm defined by the Fisher information matrix, and θ_k^0 denotes the value of θ_k that maximizes $E_x \log p(x|\theta_k)$, where E_x denotes the expectation with respect to the true distribution under the assumption that it is given by $p(x|\theta_K^0)$.

Similarly, from the analysis of the asymptotic behavior of the maximum-likelihood estimates we have asymptotically

$$E_x^*[-2E_y\{\log p(y|\theta_k^*) - \log p(y|\theta_K^*)\}] = \|\theta_k^0 - \theta_K^0\|_I^2 - (K-k),$$

where the restricted predictive point of view is adopted, and x and y are assumed to be independently identically distributed.

From these two results it can be seen that as a measurement of $(-2)E_y\{\log p(y|\theta_k^*) - \log p(y|\theta_K^*)\}$ the log-likelihood-ratio test statistic $(-2)\{\log p(x|\theta_k^*) - \log p(x|\theta_K^*)\}$ shows an upward bias by the amount $2(K-k)$. If we correct for this bias, then we get $\{-2\log p(x|\theta_k^*) + 2k\} - \{-2\log p(x|\theta_K^*) + 2K\}$ as a measurement of the difference of the entropies of the models specified by $p(\cdot|\theta_k^*)$ and $p(\cdot|\theta_K^*)$. This observation leads to the conclusion that the statistic $-2\log p(x|\theta_k^*) + 2k$ should be used as a measure of the badness of the model specified by $p(\cdot|\theta_k^*)$ (Akaike, 1973). The acronym AIC adopted by Akaike (1974) for this statistic is an abbreviation of "an information criterion" and is symbolically defined by

$$\text{AIC} = -2\log(\text{maximum likelihood}) + 2(\text{number of parameters}),$$

where log denotes natural logarithm.

If the log-likelihood-ratio test is considered as a measurement of the entropy difference, then the above observation suggests that from our present point of view *we should choose the model with smaller value of* AIC. If we follow this idea, we get an estimation procedure which simultaneously realizes the model selection and parameter estimation. An estimate thus obtained is called a minimum AIC estimate (MAICE). Now it is a simple matter to see that the critical level 2 of the F-test by Leonard and Ord corresponds to the factor 2 of the second term in the definition of AIC.

One important observation about AIC is that it is defined without specific reference to the true model $p(\cdot \mid \theta_K^0)$. Thus, for any finite number of parametric models, we may always consider an extended model that will play the role of $p(\cdot \mid \theta_K^0)$. This suggests that AIC can be useful, at least in principle, for the comparison of models which are nonnested, i.e., the situation where the conventional log likelihood-ratio test is not applicable.

We will demonstrate the practical utility of AIC by its application to the multidimensional contingency-table analysis discussed by Goodman (1971). Observing the frequency f_{ijkl} in the cell (i, j, k, l) of a 4-way contingency table ($i = 1, 2, \ldots, I; j = 1, 2, \ldots, J; k = 1, 2, \ldots, K; l = 1, 2, \ldots, L$) with $\sum_{ijkl} f_{ijkl} = n$, the basic model is specified by the parametrization

$$\log F_{ijkl} = \theta + \lambda_i^A + \cdots + \lambda_l^D + \lambda_{ij}^{AB} + \cdots + \lambda_{kl}^{CD} + \lambda_{ijk}^{ABC} + \cdots + \lambda_{jkl}^{BCD} + \lambda_{ijkl}^{ABCD},$$

where F_{ijkl} denotes the expected frequency and the λ's satisfy the condition that any sum with respect to one of the suffixes is equal to zero. The characters A, B, C, D symbolically denote the group of parameters that are related to the factors denoted by these characters. Hypotheses are defined by putting some of the parameters equal to zero.

Goodman discussed the application to the analysis of detergent-user data which included information on the following four factors: the softness of the water used (S), the previous use of a brand (U), the temperature of the water used (T), and the preference for one brand over the other (P). In Table 1 the initial portion of Goodman's Table 3 is shown with the corresponding AIC's. In Goodman's modeling, when a higher-order effect is considered, all the corresponding lower-order effects are included in the model.

Goodman asserts that H_1 and H_2 do not fit the data but H_3 and H_4 do, where H_i denotes hypothesis number i. By the present definition of AIC the negative signs of AIC for H_3 and H_4 mean that the corresponding models are preferred to the saturated nonrestricted model. This corresponds to Goodman's assertion. The AIC already suggests that H_4 is an overfit, and Goodman actually proceeds to the detailed analysis of H_3 and arrives at H_5.

The significances of S and T are then respectively checked by comparing H_6 and H_7 with H_5. The hypothesis H_8 is then judged to be an improvement over

Table 1. Goodman's Analysis of Consumer Data

Hypothesis	Estimated group of parameters	Degrees of freedom	$-2x$ log likelihood ratio)	AIC[a]
1	None	23	118.63	72.63
2, (a)[b]	S, P, T, U	18	42.93	6.93
3	All the pairs	9	9.85	−8.15
4	All the triplets	2	0.74	−3.26
5, (b)[b]	PU, S, T	17	22.35	−11.65
6	PU, S	18	95.56	59.56
7	PU, T	19	22.85	−15.15
8	PU, PT	18	18.49	−17.51
9	PT, U	19	39.07	1.07
10, (d)[b]	PU, PT, ST	14	11.89	−16.11
11, (c)[b]	PU, PT, S	16	17.99	−14.01
—[c], (e)[b]	PTU, ST	12	8.4	−15.6
28, (f)[b]	PTU, STU	8	5.66	−10.34

[a] AIC = −2(log likelihood ratio) −2(number of degrees of freedom) = AIC(i) − AIC(∞), where AIC(i) denotes the original AIC of H_i, and AIC(∞) denotes that of the saturated model with all the parameters unrestricted.

[b] Models considered in Table 5-4 of Fienberg (1980, p. 77).

[c] Missing in Goodman (1971). Numbers obtained from Fienberg (1980).

H_7. The effect of PU is then confirmed by comparing H_8 with H_9. Further elaboration of H_8 leads to H_{10}. However, its improvement over H_8 is not considered to be significant, although the effect ST is judged to be significant by the comparison of H_{10} with H_{11}. The path of Goodman's stepwise search is schematically represented by Table 2.

Table 2 shows that we come to the same conclusions as those obtained by Goodman with the choice of 5% as the critical level, simply by choosing models with lower values of AIC. The fact that AIC does not require the table lookup of the chi-squares with different degrees of freedom adds to the significance of this result. Since AIC is defined with a unique scaling unit, it allows easy extraction of useful information from a collection of fitted models. For example, by comparing the difference of AIC's of H_7 and H_5 with that of H_8 and H_{11}, we can clearly see the deteriorating effect of including S in the model. Also the direct comparison of H_6 and H_7, not possible by the log likelihood-ratio test, is now possible by AIC, and the inferiority of H_6, which contains S, is clearly recognizable. The ability of AIC to allow the researcher to extract global information from the result of fitting a large number of models is a characteristic that is not shared by the conventional model selection procedure realized by some ad hoc application of significance tests.

To avoid possible misconceptions some precautions are in order. The fact that AIC allows simple comparison of models does not justify the mechanistic enumeration of all possible models. The selection of the basic set of models must represent the particular way of looking at the data by the researcher. This point is discussed extensively, in relation to the categorical data analysis,

Table 2. The Path of Goodman's Stepwise Search and the Corresponding AIC's[a]

72.6	6.9				−8.2	−3.3	0
H_1	H_2				H_3	H_4	H_∞
None	Singles				Pairs	Triplets	Saturated
		59.6	−11.7				
		H_6	H_5				
		PU, S	PU, S, T				
		−15.2	−17.5		−16.1		
		H_7	H_8		H_{10}		
		PU, T	PU, PT		PU, PT, ST		
		1.1		−14.0			
		H_9		H_{11}			
		PT, U		PU, PT, S			

[a] The number above each hypothesis denotes the AIC relative to that of H_∞.

by Fienberg (1980), who also treats the detergent-user data. In particular, without proper restriction of the basic set of models, the growing number of possible models easily makes the model selection by minimum AIC quite unreliable. The situation is worse with the selection by repeated applications of tests, as the procedure cannot provide the global view of the models equivalent to the one given by the distribution of AIC over the models.

Another possible confusion is to equate AIC to a test statistic. The choice of one and the same model by the testing procedure defined with the critical level 5% and by the minimum AIC procedure in the above example is merely a coincidence. There are situations where judgements resulting from the use of AIC differ drastically from those based on the sequential application of testing procedures. One typical example of this difference is discussed in Akaike (1983b), where good agreement between the judgement by AIC and that by an expert is observed, while the test-based procedure fails to provide a reasonable explanation of the situation.

It seems that AIC has attracted the attention of people in various fields of application of statistics. This can be seen by the fact that the Institute for Scientific Information denoted the 1974 paper (Akaike, 1974) as one of the most frequently cited papers in the area of engineering, technology, and applied sciences (Akaike, 1981). However, there are rather limited number of theoretical works related to AIC. These include the discussion of the asymptotic equivalence of the minimum-AIC procedure to cross-validation, by M. Stone (1977b); modifications of the criterion by Schwarz (1978) and by Hannan and Quinn (1979); discussions of the relation to the Bayes procedure by Zellner (1978), Atkinson (1980), and Smith and Spiegelhalter (1980); and discussions of the optimality of the MAICE procedure by Akaike (1978a), Shibata (1980), and C. J. Stone (1982). Evidence of the inherent relation between the magic number 2 and the predictive point of view can be found in works by Geisser and Eddy (1979) and Leonard (1977).

When the number of possible alternatives is increased, the MAICE procedure may tend to be sensitive to sampling fluctuations. One solution to this problem is to use some averaging procedure, as is discussed in Akaike (1979). However, this brings us closer to the Bayesian modeling approach, which is discussed in the next section.

9. Entropy-Maximization Principle and the Bayes Procedure

The discussion of the concept of true model and its relation to entropy shows that there is no end to the process of statistical model building. All we can do is attempt to produce better models. When we admit this, then it is easy to accept the following very modest, yet very productive view of statistics: *all statistical activities are directed to maximizing the expected entropy of the predictive distribution in each particular application.* We call this the entropy-maximization principle (Akaike, 1977). The minimum-AIC procedure may be considered as a realization of this principle. The generality of this principle can be seen by the following discussion of the Bayesian approach to modeling.

Consider the set of models given by $\{g_k(\cdot);\ k = 1, 2, \ldots, K\}$, where $g_k(y)$ denotes a predictive distribution specified by the parameter k. Assume that we consider the use of a random mechanism for the selection of the predictive distribution. Our preferences with respect to the models is represented by the distribution of probabilities $w_k(x)$ of selecting the kth model, where $w_k(x)$ is specified by combining our knowledge of the problem and the data x. However, irrespective of the form of the true distribution of y, the following relation holds:

$$E_y \log\left\{\sum_{k=1}^{K} g_k(y)w_k(x)\right\} \geqslant \sum_{k=1}^{K} w_k(x)E_y \log g_k(y),$$

where E_y denotes the expectation with respect to the true distribution of y. This means that the entropy of the true distribution with respect to the averaged distribution $\sum g_k(y)w_k(x)$ is always greater than or equal to that with respect to the distribution chosen by the random mechanism. The entropy-maximization principle suggests that we should consider the use of the averaged distribution $\sum g_k(y)w_k(x)$ as our predictive distribution, rather than a distribution chosen by a random mechanism. Taking into account the fact that a conventional model selection procedure corresponds to a particular choice of $w_k(x)$ which takes either the value 0 or 1, the present result suggests the possibility of improved modeling for the purpose of prediction by extending the basic set of models from $\{g_k(\cdot); k = 1, 2, \ldots, K\}$ to $\{\sum g_k(\cdot)w_k;\ w_k \geqslant 0, \sum w_k = 1\}$.

The problem now is how to define $w_k(x)$. Since the distribution $w_k(x)$, which we will call the inferential distribution, is introduced to define a predictive distribution, we consider the more general problem of the selection of a predictive distribution. Assume that the variable x takes a finite number of discrete values $x = 1, 2, \ldots, I$. Before observing the value of x, we consider the selection of the predictive distribution of x, where the possible predictive distributions of x are given by $f_k(x)$. Since x is not available yet, we consider the use of a probability distribution w_k over k, defined independently of x. Thus we are specifying a probability distribution $w_k f_k(x)$ over (k, x).

When the observation produces $x = x_0$, a Bayesian will say that we should follow the Bayes procedure and replace the distribution $w_k f_k(x)$ by the distribution $w(k, x)$ which is defined by

$$w(k,x) = \begin{cases} \dfrac{w_k f_k(x_0)}{\sum_k w_k f_k(x_0)} & \text{for } x = x_0, \\ 0 & \text{otherwise.} \end{cases}$$

The common counsel of the subjectivist that a probabilistic structure must be based on the whole set of available information is definitely correct, but it does not imply the use of the Bayes procedure. De Finetti (1972, p. 150) mentions that "according to a criterion of temporal coherency" the posterior probability must be the new probability after the person has observed the data. However, no explanation is given about the criterion. In fact Bayes' theorem does not contain any element of time, and thus its temporal interpretation is arbitrary.

There is an essential analogy between Boltzmann's derivation of the exponential family of distributions for energy and the use of the Bayes procedure. To see this we consider more generally an arbitrary distribution $\pi(k,x)$ over (k,x) and try to find a distribution $w(k,x)$ concentrated on $\{(k,x_0)\}$ and such that the Boltzman entropy with respect to the original $\pi(k,x)$ is maximum. This leads to the maximization of

$$\sum_x \sum_k w(k,x)\{\log \pi(k,x) - \log w(k,x)\} + \lambda\left\{\sum_k w(k,x_0) - 1\right\},$$

where λ is the Lagrange multiplier. The solution is given by

$$w(k,x) = \begin{cases} \dfrac{\pi(k,x_0)}{\sum_k \pi(k,x_0)} & \text{for } x = x_0, \\ 0 & \text{otherwise.} \end{cases}$$

This result characterizes the transition from the original distribution to the conditional distribution as the most conservative action that conforms to the observation of the data x_0 yet otherwise maximally retains the structure of the originally assumed distribution. We refer to this particular application of the maximum-entropy method of probability distribution generation as the conditioning principle. That the Bayesian rule of conditionalization is a special case of the principle of minimum information, or of maximum entropy, was also noticed by Williams (1980).

Coming back to Bayesian modeling, we can now see that the assumption of the original distribution $\pi(k, x)$ and the conditioning principle leads to the use of the "posterior distribution" $w(k, x)$ as the inferential distribution $w_k(x)$. That such a definition of the inferential distribution is a reasonable one can be shown as follows. First we assume that when k is given, y and x are independent and the distribution is given by $g_k(y)f_k(x)$. The expected performance of a predictive distribution $h(y|x)$ is then evaluated by $E_k E_{x|k} E_{y|k} \log h(y|x)$, where E_k denotes the expectation with respect to the distribution w_k, and $E_{x|k}$ and $E_{y|k}$ denote the expectations with respect to $f_k(x)$ and $g_k(y)$, respectively. We have

$$E_k E_{x|k} E_{y|k} \log h(y|x) = \sum_x f(x) \sum_y \sum_k g_k(y) w(k|x) \log h(y|x),$$

where $f(x) = \sum f_k(x) w_k$ and $w(k|x) = f_k(x) w_k / f(x)$. This quantity is maximized by putting

$$h(y \mid x) = \sum_k g_k(y) w(k \mid x),$$

which means that, *as long as we assume the validity of the original probabilistic setup*, the use of the posterior distribution $w(k \mid x)$ as the inferential distribution is the best choice. This result was recognized earlier by Kerridge (1961) and Aitchison (1975).

10. Statistical Inference and Bayesian Modeling

What the result of the preceding section has shown is that the conditioning principle leads to the best choice of the inferential distribution *under the assumption of the validity of the Bayesian model defined by* $f_k(y)f_k(x)w_k$. What will happen when we are uncertain about the choice of the "prior distribution" w_k?

Here we recall our basic observation that statistical model building is an unending process. This means that the validity of a model can only be established by a careful analysis of other possibilities. This leads to the situation where we have several alternative prior distributions $w_k^{(i)}$ ($i = 1, 2, \ldots, I$). Here we have to assume a (hyper) prior distribution $\pi(i)$ over these alternatives. When the data x are observed the posterior probability $p(i \mid x)$ of the ith model is given by the relation

$$p(i \mid x) \propto f^{(i)}(x)\pi(i),$$

where $f^{(i)}(x)$ is the likelihood of the ith Bayesian model, defined by

$$f^{(i)}(x) = \sum_k f_k(x) w_k^{(i)}.$$

Thus, even when we do not know how to specify $\pi(i)$, we can see how much relative support is given to each model by the observation x.

Based on the concept of entropy we have demonstrated the objectivity of the log likelihood as a criterion of fit of a probabilistic structure to a set of data. Since each Bayesian model specifies a probabilistic structure for the data, the objectivity holds for the log likelihood $\log f^{(i)}(x)$ as a measure of the goodness of the model. Accordingly, in spite of the firm belief of some strict Bayesians that a probabilistic structure must be constructed whenever there are several possibilities, we may safely insist that the goodness of one Bayesian model relative to another can be evaluated by the difference of the log likelihoods.

Good (1965) calls the procedure of hyperparameter estimation by maximizing the likelihood of a Bayesian model "type II maximum likelihood." The use of the likelihood for the assessment of a Bayesian model is demonstrated in an illuminating paper by Box (1980). The application to the very practical problem of seasonal adjustment is discussed by the present author (Akaike, 1980a).

The discussion of Bayesian modeling will never be complete unless we provide a procedure for the modeling of the situation where no further prior information is available for the modeling. The concept of entropy again finds an interesting application in this type of situation. It has been shown that the well-known Jeffreys ignorance prior distribution (Jeffreys, 1946) can be given an interpretation as the locally or globally impartial prior distribution (Akaike, 1978b). However, this concept is essentially dependent on the continuity of the parameter involved. Recently the present author applied the predictive point of view and the concept of entropy to define a prior distribution that will retain its impartiality for a discrete set of alternatives. For the Bayesian model discussed in the preceding section a minimax-type prior distribution is defined by minimizing

$$\max_k \sum_x f_k(x) \sum_y g_k(y) \log\left(\frac{g_k(y)}{p(y \mid x)}\right),$$

where $p(y \mid x) = \sum g_k(y) f_k(x) w_k / f(x)$ and $f(x) = \sum f_k(x) w_k$. The strict predictive point of view requires us to put $g_k(y) = f_k(y)$. It has been observed by numerical investigation that this definition leads to interesting nontrivial specifications of the prior distribution that exhibits a local uniformity (Akaike, 1983b). Related works in this area are those by Zellner (1977) and Bernardo (1979), based on the earlier work of Lindley (1956), who discussed the use of the Shannon entropy in statistics.

Do these formal procedures for generating prior distributions produce useful results? The answer can be obtained only through the detailed analysis of the final output of each Bayesian model thus obtained. An example of such an analysis is given by Akaike (1980b), where admissibility is proved for a James–Stein type of estimator of a multivariate normal distribution obtained by applying the ignorance prior to the hyperparameter of a prior distribution.

Here again we are reminded of the attitude of Boltzmann, who considered that the justification of the primary distribution used in the derivation of the distribution of the energy could only be obtained through the observation of the validity of the final result. It is the author's view that the use of a Bayesian procedure can only be justified when the procedure produces good results for those data which are "similar" to the present one and for which unequivocal judgment of the results is possible.

11. Conclusion

The predictive point of view, particularly in its strict form, and the concept of entropy can produce a unifying view of statistics. This view is not only conceptually simple and unifying, but also practical and very productive. It leads, for example, to a practical solution to the notoriously difficult problems associated with significance tests involving multiple hypotheses.

The entropy-maximization principle which is obtained by combining the predictive point of view with the concept of entropy clearly states that the search for better models is the purpose of statistical data analysis. From this perspective Bayesian modeling will often be an improvement over non-Bayesian approaches. Nevertheless, the objectivity of the log likelihood, established with the aid of entropy, as an evaluation of a stochastic model provides us with a firm basis for the selection of a Bayesian model when further Bayesian modeling of the situation is impractical.

Statistical models are formulations of our past experiences, and only new interesting problems can stimulate the development of useful models. The fundamental contribution by Boltzmann came from the deep study of one particular real problem. Thus we can see that for the development of statistics the main emphasis should be placed on the search for important practical problems.

Acknowledgements

The author is grateful to A. P. Dawid and T. Leonard for helpful comments. The reference to the work by Williams on Bayesian conditionalization was made possible by the comment of Dawid. The presentation of the paper has been significantly improved by the comments of the editors and reviewers. This work was partly supported by the United States Army under Contract No. DAAG29-80-C-0041 at the Mathematics Research Center, University of Wisconsin—Madison, and by the Ministry of Education, Science and Culture, Grant-in-Aid No.58450058 at the Institute of Statistical Mathematics.

Bibliography

Aitchison, J. (1975). "Goodness of prediction fit." *Biometrika*, **62**, 547–554.

Akaike, H. (1969). "Fitting autoregressive models for prediction." *Ann. Inst. Statist. Math.*, **21**, 243–247.

Akaike, H. (1970). "Statistical predictor identification." *Ann. Inst. Statist. Math.*, **22**, 203–217.

Akaike, H. (1973). "Information theory and an extension of the maximum likelihood principle." In B. N. Petrov and F. Csaki (eds.), *Second International Symposium on Information Theory*. Budapest: Akademiai Kiado, 267–281.

Akaike, H. (1974). "A new look at the statistical model identification." *IEEE Trans. Automat. Control*, **AC-19**, 716–723.

Akaike, H. (1977). On entropy maximization principle. In P. R. Krishnaiah, (ed.), *Applications of Statistics*. Amsterdam: North-Holland, 27–41.

Akaike, H. (1978a). "A Bayesian analysis of the minimum AIC procedure". *Ann. Inst. Statist. Math.*, **30A**, 9–14.

Akaike, H. (1978b). "A new look at the Bayes procedure". *Biometrika*, **65**, 53–59.

Akaike, H. (1979). "A Bayesian extension of the minimum AIC procedure of autoregressive model fitting." *Biometrika*, **66**, 237–242.

Akaike, H. (1980a). "Seasonal adjustment by a Bayesian modeling." *J. Time Series Anal.*, **1**, 1–13.

Akaike, H. (1980b). "Ignorance prior distribution of a hyperparameter and Stein's estimator." *Ann. Inst. Statist. Math.*, **33A**, 171–179.

Akaike, H. (1981). "Abstract and commentary on 'A new look at the statistical model identification'." *Current Contents*, Engineering, Technology and Applied Sciences, **12**, No. 51, 22.

Akaike, H. (1983a). "On minimum information prior distributions." *Ann. Inst. Statist. Math.*, **34A**, 139–149.

Akaike, H. (1983b). "Information measures and model selection." In *Proceedings of the 44th Session of ISI*, **1**, 277–291.

Atkinson, A. C. (1980). "A note on the generalized information criterion for choice of a model." *Biometrika*, **67**, 413–418.

Bahadur, R. R. (1967). An optimal property of the likelihood ratio statistic. In L. M. LeCam and J. Neyman (eds.), *Proc. 5th Berkeley Symp. Math. Statist. and Probab.*, **1**. Berkeley: Univ. of California Press, 13–26.

Bernardo, J. M. (1979). "Reference posterior distributions for Bayesian inference (with discussion)." *J. Roy. Statist. Soc. Ser. B*, **41**, 113–147.

Boltzman, L. (1872). "Weitere Studien über das Wärmegleichgewicht unter Gasmolekülen." *Wiener Berichte*, **66**, 275–370.

Boltzman, L. (1877a). "Bemerkungen über einige Probleme der mechanischen Wärmetheorie." *Wiener Berichte*, **75**, 62–100.

Boltzman, L. (1877b). "Über die Beziehung zwischen dem zweiten Hauptsatze der mechanischen Wärmetheorie und der Wahrscheinlichkeitsrechnung respective den Sätzen über das Wärmegleichgewicht." *Wiener Berichte*, **76**, 373–435.

Boltzmann, L. (1878). "Weitere Bemerkungen über einige Plobleme der mechanischen Wärmetheorie." *Wiener Berichte*, **78**, 7–46.

Box, G. E. P. (1980). "Sampling and Bayes' inference in scientific modelling and robustness." *J. Roy. Statist. Soc. Ser. A*, **143**, 383–430.

Chernoff, H. (1956). "Large sample theory—parametric case." *Ann. Math. Statist.*, **27**, 1–22.

Csiszar, I. (1975). "*I*-divergence geometry of probability distributions and minimization problems." *Ann. Probab.*, **3**, 146–158.

Fienberg, S. E. (1980). *Analysis of Cross-classified Categorical Data* (2nd ed.). Cambridge, MA: M.I.T. Press.

de Finetti, B. (1972). *Probability, Induction and Statistics*. London: Wiley.

Fisher, R. A. (1935). "The fiducial argument in statistical inference." *Ann. Eugenics*, **6**, 391–398. Paper 25 in *Contributions to Mathematical Statistics* (1950). New York: Wiley.

Fisher, R. A. (1936). "Uncertain inference." *Proc. Amer. Acad. Arts and Sciences*, **71**, 245–258.

Geisser, S. and Eddy, W. F. (1979). "A predictive approach to model selection." *J. Amer. Statist. Assoc.*, **74**, 153–160.

Good, I. J. (1965). *The Estimation of Probabilities*. Cambridge, MA: M.I.T. Press.

Goodman, L. A. (1971). "The analysis of multidimensional contingency tables: Stepwise procedures and direct estimation methods for building models for multiple classifications." *Technometrics*, **13**, 33–61.

Guttman, I. (1967). "The use of the concept of a future observation in goodness-of-fit problems." *J. Roy. Statist. Soc. Ser. B*, **29**, 83–100.

Hannan, E. J. and Quinn, B. G. (1979). "The determination of the order of an autoregression." *J. Roy. Statist. Soc. Ser. B*, **41**, 190–195.

Jaynes, E. T. (1957). "Information theory and statistical mechanics." *Phys. Rev.*, **106**, 620–630; **108**, 171–182.

Jeffreys, H. (1946). "An invariant form for the prior probability in estimation problems." *Proc. Roy. Soc. London Ser. A*, **186**, 453–461.

Kerridge, D. F. (1961). "Inaccuracy and inference." *J. Roy. Statist. Soc. Ser. B*, **23**, 184–194.

Kullback, S. (1959). *Information Theory and Statistics*. New York: Wiley.

Kullback, S. and Leibler, R. A. (1951). "On information and sufficiency." *Ann. Math. Statist.*, **22**, 79–86.

Leonard, T. (1977). "A Bayesian approach to some multinomial estimation and pretesting problems." *J. Amer. Statist. Assoc.*, **72**, 869–876.

Leonard, T. and Ord, K. (1976). "An investigation of the *F*-test procedure as an estimation short-cut." *J. Roy. Statist. Soc. Ser. B*, **38**, 95–98.

Lindley, D. V. (1956). "On a measure of the information provided by an experiment." *Ann. Math. Statist.*, **27**, 986–1005.

Mallows, C. L. (1973). "Some comments on C_p." *Technometrics*, **15**, 661–675.

Pearson, K. (1929). "Laplace, being extracts from lectures delivered by Karl Pearson." *Biometrika*, **21**, 202–216.

Rao, C. R. (1961). "Asymptotic efficiency and limiting information." In J. Neyman, (ed.), *Proc. 4th Berkeley Symp. Math. Statist. and Probab.*, **1**. Berkeley: Univ. of California Press, 531–548.

Rao, C. R. (1962). "Efficient estimates and optimum inference procedures in large samples." *J. Roy. Statist. Soc. Ser. B*, **24**, 46–72.

Sanov, I. N. (1957). "On the probability of large deviations of random variables." (in Russian). *Mat. Sbornik N.S.*, **42**, No. 84, 11–44. English transl., *Selected Transl. Math. Statist. Probab.*, **1** (1961), 213–244.

Schwarz, G. (1978). "Estimating the dimension of a model." *Ann. Statist.*, **6**, 461–464.

Shannon, C. E. and Weaver, W. (1949). *The Mathematical Theory of Communication*. Urbana: Univ. of Illinois Press.

Shibata, R. (1980). "Asymptotically efficient selection of the order of the model for estimating parameter of a linear process." *Ann. Statist.*, **8**, 147–164.

Smith, A. F. M. and Spiegelhalter, D. J. (1980). "Bayes factors and choice criteria for linear models." *J. Roy. Statist. Soc. Ser. B*, **42**, 213–220.

Stigler, S. M. (1975). "The transition from point to distribution estimation." In *Proceedings of the 40th ISI Meeting*, **2**, 332–340.

Stone, C. J. (1982). "Local asymptotic admissibility of a generalization of Akaike's model selection rule." *Ann. Inst. Statist. Math.*, **34A**, 123–133.

Stone, M. (1974). "Large deviations of empirical probability measures." *Ann. Statist.*, **2**, 362–366.

Stone, M. (1977a). "Asymptotics for and against cross-validation." *Biometrika*, **64**, 29–35.

Stone, M. (1977b). "Asymptotics equivalence of choice of models by cross-validation and Akaike's criterion." *J. Roy. Statist. Soc. Ser. B*, **39**, 44–47.

Wald, A. (1943). "Tests of statistical hypotheses concerning several parameters when the number of observations is large." *Trans. Amer. Math. Soc.*, **54**, 426–482.

Williams, P. M. (1980). "Bayesian conditionalization and the principle of minimum information." *Brit. J. Philos. Sci.*, **31**, 131–144.

Zellner, A. (1977). "Maximal data information prior distributions." In A. Aykac and C. Brumat (eds.), *New Developments in the Applications of Bayesian Methods*. Amsterdam: North-Holland, 211–232.

Zellner, A. (1978). "Jeffreys–Bayes posterior odds ratio and the Akaike information criterion for discriminating between models." *Economic Letters*, **1**, 337–342.

EXPERIENCES ON THE DEVELOPMENT OF TIME SERIES MODELS

H. AKAIKE
The Institute of Statistical Mathematics
4-6-7 Minami-Azabu, Minato-ku
Tokyo 106, Japan

0. Introduction

The development of statistical models is realized through the accumulation of successful experiences of the analysis of real data. As is discussed in Akaike (1992) statistical modeling activity contains highly subjective or personal aspect which Polanyi (1962) related to the scientific talent of the researcher. However, scientific activity is never isolated from the society and Turing (1969) explicitly characterized the search for new techniques as the" cultural search '. This paper is intended to provide the background information of some of the experiences of the author on time series modeling. It is hoped that the description of the interaction between the author and his environment will provide some suggestion for those who intend to organize effective statistical modeling activity in the future.

The interest in time series of the present author was motivated by the desire to see more structures in observational data than those given by ordinary static statistical models. However, conventional theory of time series at that time was mainly concerned with the analysis of stationary time series and the subject was essentially related to the estimation of autocorrelograms or power spectra and there was not much space for a statistician to incorporate his own idea into the modeling of a time series.

The first experience of time series modeling with a flexible structure was related to the stationary 0-1 process with discrete time parameter defined by the sequence of independently identically distributed intervals between the 1's. The success of the application of this model to the silk production process confirmed the author's belief that proper modeling of the process was the key to the successful implementation of statistical procedure in a practical application.

The theory of feedback control of a random process had been developed under the assumption of a known structure of the process and produced an opportunity for statisticians to contribute to the development of new engineering practice through the development of proper identification procedures of system characteristics. The analysis of

H. Bozdogan (ed.), Proceedings of the First US/Japan Conference on the Frontiers of Statistical Modeling: An Informational Approach, 33–42.

a random process with inner feedback loop posed a difficulty for the direct application of the frequency domain approach and the approach through the fitting of multivariate autoregressive (AR) model was eventually developed. This development was motivated by the implementation of the computer control of a cement kiln process. It also clarified the importance of the problem of order determination of an autoregressive model and formed a starting point for the later introduction of an information criterion AIC.

The introduction of AIC emphasized the necessity of parsimonious parametrization of a statistical model and the use of autoregressive moving average (ARMA) model of multivariate time series was considered to avoid possibly inefficient modeling by AR model. However, there was the problem of identifiability of multivariate ARMA model. A solution to this problem was obtained by the analysis of the structure of the state space representation of the time series. This was realized through the interpretation of the result on system realization in mathematical system theory.

The inflexibility of modeling caused by the requirement of parsimony of parametrization can be eliminated by assuming a prior distribution over the parameter space which is parametrized by a small number of hyperparameters. A seasonal adjustment procedure BAYSEA was realized by using a Bayesian model with a very simple structure. The elimination of the psychological or philosophical barrier against the use of Bayesian models was realized by the informational approach. Consequently a wide area of new modeling activity of time series was opened up.

These are some of the experiences of the author on time series modeling which will be discussed in more detail in the subsequent sections.

1. Gap process modeling vs. control chart approach

At the time of the beginning of the career of the present author as a statistician a common subject of research in time series was the theory of stationary time series as represented by the work of Wold (1937). It did not require much time to learn from the experience of the analysis of the time series of stock market behavior that more detailed structural information on the generating mechanism was necessary than the mere assumption of stationarity to develop an effective analysis of a time series for practical application. The queueing process theory presented an example of time series modeling with more structure. In this theory a significant role was played by the simple random Poisson process used for the representation of the input sequence. The simplicity of the mathematical structure of the process was quite attractive, but the uncritical use of the process as the standard model of the input was not quite acceptable.

It was expected that a simple example which would not allow the approximation by a simple Poisson process would be given by the flow of cars on a road with single lane on each side, as the distribution of the length of the interval between consecutive car arrivals would never show maximum concentration at zero. Based on this idea an actual flow of cars was recorded and the frequencies of the number of cars within the unit interval of 5 seconds were computed. Somewhat disappointingly the resulting frequency distribution was very closely approximated by a Poisson distribution. However, the frequency distribution of the interval length as measured by the unit of 0.5 second, which allowed the existence of arrival of at most one car within the interval, did show a clear dip at zero. The autocorrelogram of the corresponding 0-1 sequence of the car arrivals was approximated very well by the theoretical correlogram obtained from the observed interval length distribution with the assumption of the independence between the intervals. This model of stationary 0-1 process X_n (X_n = 0 or 1) which was defined by a succession of independent intervals between 1's was called the gap process and the distribution of intervals was called the gap distribution (Akaike, 1956).

A visitor to a colleague of the present author was experiencing a difficulty in the trial of the implementation of statistical control of a silk production process. The difficulty was represented by the fact that the frequency distribution of the number of dropping ends, the number of the ends of cocoon filaments being reeled into a single silk thread observed within a unit time interval, did show good approximation to a Poisson distribution but the standard control chart strategy based on the assumption of the simple Poisson process was a failure. The time series of actual dropping ends did show a particular wavy pattern which was unexpected from the assumption of a simple Poisson process.

Obviously, the sequence of dropping ends could have been better represented by the sum D_n of the gap processes $X_n(i)$ of a fixed number k,

$$D_n = X_n(1) + X_n(2) + \ldots\ldots + X_n(k)\ ,$$

where k is equal to the number of cocoons used for the reeling of silk thread. The gap distribution in this case was defined by the length distribution of cocoon filaments and could easily be estimated by using the result of the test reeling for the determination of the boiling condition of cocoons.

The validity of the approximation by a gap process was checked by experiment and a systematic approach to the production process control of silk reeling process was established by Akinori Shimazaki of the Sericultural Experiment Station of the Ministry of Agriculture, now professor of Shinshu University (Akaike, 1959, Shimazaki, 1961). This was historically the first theoretical result on the silk reeling process effectively applied to the control of real silk production process.

The instructive point of this example is that it clearly demonstrates the importance of developing a proper modeling for particular application. In the case of this example the application of the conventional approach by statistical control chart was doomed to fail. The size of each lot of cocoons was quite limited and by the time a record of observations of sufficient length for reliable computation of required statistics was obtained the lot was almost finished. Moreover, such conventional approach would not have been able to detect the existence of structural abnormality, such as the one identified as the effect of the uneven temporal supply of new cocoons by the habitual behavior of a carrier. The modeling approach succeeded in providing a reliable objectively defined reference for the actual production process by using the rather limited number of observations provided as the by-product of the test reeling. This example provides a typical case of successful application of statistical modeling.

From the point of view of modeling the most significant characteristic of this example is that it was based on a direct modeling of the basic probability distribution by the histogram of observations. The crucial point of the modeling was the choice of the time interval used for the definition of the histogram. Although this was the first experience of modeling which found successful application in a production process, the leading idea of the whole process of modeling and application was so clear that the author had a feeling of definite familiarity with the final outcome when it was observed. However, actually it was through this cooperative work on the application that the author really developed the confidence in the practical use of statistical modeling.

2. Frequency response function estimation and feedback system analysis

In spite of the author's conviction that the development of a new model for each particular application was the only way to successful development of time series analysis, there was a vast area of application of a particular type of model, namely the constant linear system model. This was due to the fact that there were already well developed theory and practice for the application of linear dynamic system model in science and engineering.

The first example of power spectrum estimation brought to the attention of the author was that of stationary random vibration of a car tried by Ichiro Kaneshige of Isuzu Motor Company, now director of the Japan Automobile Research Institute. The successful application of the Fourier analytic procedure to the power spectrum estimation developed by J. W. Tukey (Tukey and Hamming, 1949, Blackman and Tukey, 1959) suggested the possibility of extension of the procedure to the estimation of cross-spectrum and frequency response function of two-variate time series. One particular difficulty with the estimation of cross-spectrum was the existence of possibly a quick change of the phase lag at a particular frequency. A practical procedure of estimation of the frequency response function was developed by Akaike and Yamanouchi (1962) by introducing a compensation procedure for the phase lag. The procedure was applied to real data supplied by researchers from various fields and the results were published as a report entitled " Studies on the Statistical Estimation of Frequency Response Functions " in the Annals of the Institute of Statistical Mathematics, Supplement III, 1964.

The basic condition for the application of the procedure is the independence or orthogonality of the input and the noise which contaminates the linear response of the system. Obviously this condition does not hold in the practically important case where feedback exists from the output to the input. A typical example is provided by the control action of a pilot and the response of an airplane in the air.

The condition seriously limited the applicability of the method to the industrial production processes which were usually under the control of human operators. A typical example was the record of the kiln process of the cement production which was shown to the author by Toichiro Nakagawa of the Chichibu Cement Company, now the chairman of the board of directors of the SSK (System Sougoh Kaihatsu) Company. A direct application of the Fourier analytic method of frequency response function estimation to the record of two variables of a cement kiln produced an estimate of which inverse Fourier transform showed significant responses on both positive and negative side of the time axis (Akaike, 1967). This result suggested the necessity of considering the condition of physical realizability to restrict the response to positive time axis.

Since the incorporation of the physical realizability condition into the frequency domain analysis was not easy, attention was turned to the analysis in the time domain. This eventually lead to the application of multivariate AR model for the analysis and control of dynamic systems (Akaike, 1968, 1971, Otomo, et al., 1972, Akaike and Nakagawa, 1972).

When the time domain modeling was tried for the analysis of feedback systems it became quickly clear that the proper choice of the order, the number of past observations used for the representation of the present observation, was quite difficult. The difficulty was particularly significant when the dimensionality of the observation was high. However, even the handling of the simplest case of scalar observation was not easy. The solution to this problem was obtained through the introduction of the concept of final prediction error (FPE) and this eventually lead to the introduction of the information criterion AIC. This development is discussed in Akaike (1992).

The procedure of the analysis and control of a dynamic system through multivariate AR model fitting found interesting applications in various fields, such as engineering, medicine and economics. The computer program package TIMSAC included in Akaike and Nakagawa (1972) was particularly useful as it could be run on a small-sized computer and the decision on the order of the model was almost automatic. One particularly successful application was developed by Hideo Nakamura of the Central Research Laboratory of the Kyushu Electric Power Company, now chief engineer at Bailey Japan Company, for the steam temperature control of thermal electric power plants (Nakamura and Akaike, 1981). This example clearly demonstrated the potential of statistical approach to the identification of system characteristics for the purpose of implementation of optimal computer control. Further development of TIMSAC is described in Akaike (1987).

The realization of the application of the optimal control theory to industrial processes was certainly the most significant contribution of TIMSAC program package. However, from the point of view of the analysis of time series, the decomposition of the power spectrum of each component variable into the contributions from the components of the innovation, the prediction error of the AR model, is most useful. This decomposition is realized by assuming the independence or orthogonality of the component processes of the innovation and is given, for a d-dimensional observation, in the from

$$P_i(f) = P_{i1}(f) + P_{i2}(f) + \ldots\ldots + P_{id}(f) \quad (i=1, 2, \ldots\ldots , d) ,$$

were $P_i(f)$ denotes the power spectrum density at frequency f of the i-th component and $P_{ij}(f)$ the part originating in the j-th component of the innovation. This decomposition reveals the dominant source of variation of the i-th component variable at frequency f.

The importance of such analysis is shown by the application to the analysis of the abnormal behavior of a nuclear power plant developed by Fukunishi (1977). He produced a convincing evidence to support a conjecture about the origin of an abnormal behavior through the comparison of the results of the analyses of the records in normal and abnormal condition. With a proper combination of the analysis of the diagonality of the variance matrix of the innovation this decomposition procedure, which is realized by the program MULNOS in TIMSAC, will continue to be a standard procedure for the basic analysis of feedback systems and will find many more practically important applications.

3. Identifiability of multivariate ARMA model and the state space representation

AIC has produced a formal proof of the necessity of parsimonious parametrization in modeling. A typical example of non-parsimonious parametrization is given by the AR modeling when the power spectrum shows a dip that requires a moving average (MA) representation of the time series. Thus the introduction of AIC emphasized the use of ARMA models. Although the discussion of the distribution of the initial value was required for the definition of the exact likelihood, a reasonable approximation was obtained by fitting the model to the positive definite sequence of sample autocovariances for sufficiently long series of observations. With this definition of the likelihood an almost automatic procedure for the fitting of an ARMA model could be realized by assuming a parsimonious parametrization for scalar time series.

The case of multivariate ARMA model was not simple. In the case of scalar time series an ARMA model could be uniquely specified by assuming the minimality of the orders of the AR and MA part. However, in the case of a multivariate time series, the series could contain a component with lower orders of both AR and MA part than those of another component. In this situation, by adding the representation of the former component with proper time shift to that of the latter, another representation could be obtained without violating the minimality of the orders. This simple observation showed the necessity of careful analysis of the structure of multivariate ARMA model for the maximum likelihood estimation. This is the problem of identifiability, the condition for the uniqueness of the representation of a model.

In 1971 R. E. Kalman decided to visit Japan and sent to the present author a bundle of copies of his papers related to the mathematical system theory. Within the papers there was one on the multivariate system realization that treated the derivation of the state space representation of a multivariate system from the matrix sequence of its impulse response, the sequence $Y_0, Y_1, Y_2, \ldots$ of the matrices of the multiple outputs to the impulsive input of unit intensity applied at each input at the origin of time (Ho and Kalman, 1966). The realization procedure was algorithmically defined by using the singular value decomposition of the generalized Hankel matrix, defined by augmenting the original sequence of the first r impulse response matrices by the sequences obtained by successively shifting up the time parameter,

$$S_r = \begin{bmatrix} Y_0 & Y_1 & \cdots & Y_{r-1} \\ Y_1 & Y_2 & \cdots & Y_r \\ \vdots & \vdots & \cdots & \vdots \\ Y_{r-1} & Y_r & \cdots & Y_{2r-2} \end{bmatrix}.$$

It was obvious that this Hankel matrix could be interpreted as a covariance matrix between the past white noise input vectors and the corresponding present and future output vectors. This interpretation suggested that the algorithm of Ho and Kalman could be interpreted as the procedure for the canonical correlation analysis between the past input vectors and that of the present and future output vectors. This interpretation suggested that a state space representation could be obtained by doing the canonical correlation analysis between the vector of the present and past observations and that of the present and future observations. Finally it was observed that the basis of the linear space spanned by the projections of the present and future observations onto the space spanned by the linear combinations of the present and past observations provided a representation of the state. This observation further lead to the interpretation of the state as the package of the information to be transmitted from the present and past to the present and future. The result of these analyses are summarized in Akaike (1974, 1977) and the related computer programs are included in TIMSAC-74 by Akaike et al. (1976). This example provided a proof of the importance of another type of cross-disciplinary contact for the development of statistics.

The fitting procedure of the multivariate ARMA model given in TIMSAC-74 has not been very successful in finding examples of real applications. This is due to the fact that by properly incorporating related variables into the models the multivariate AR models with rather lower orders have been producing good results in various applications. Thus it seems that the contribution of this analysis of multivariate ARMA model has been mainly the clarification of the statistical meaning of the state of a time series.

4. Seasonal adjustment by a Bayesian modeling

In Japan, a systematic effort for the development of proper prediction of earthquakes started in 1969. This lead to the production of a huge collection of data of related measurements. The present author had an opportunity to join the discussion of the quality of these measurements and found that they usually contained important information in the form of the trend components and that they were often contaminated by significant quasi-periodic components representing the behavior of the earth and by some missing values. This experience renewed the author's interest in the seasonal adjustment procedure developed for the trend analysis of economic time series analysis.

Conventional procedures of seasonal adjustment were mainly realized by successive applications of linear filters to realize the proper extraction of required frequency components, as in the case of the Census Method II X-11 variant. These procedures usually underwent successive modifications through the accumulation of the experience of application. Thus the structures of the procedures tended to be extremely complicated and the possibility of further improvement seemed rather limited.

The application of informational point of view to the analysis of the Bayes procedure produced a new approach that prepared a technical and philosophical basis for the systematic development of practical use of Bayesian modeling (Akaike, 1978). To produce a proof of the practical applicability of the new approach, the author tried a tentative computer program for seasonal adjustment. When the program was applied to an artificially generated time series, it produced an extremely good reproduction of the trend component. This result was reported with other examples at the International Meeting on Bayesian Statistics in Valencia in 1979 (Akaike, 1980a).

This procedure was obtained simply by assuming the representation of the observation y_n as the sum of the trend, seasonal and irregular component

$$y_n = T_n + S_n + I_n ,$$

and a prior distribution that represented the necessary smoothness of the trend and seasonal component by assuming a Gaussian distribution of some proper linear combinations of T_n and S_n. The adjustment of the prior distribution was realized by maximizing the likelihood of the model with respect to the hyperparameters that controlled the smoothness of the trend and seasonal components (Akaike, 1980b). The leading idea of the construction of the Bayesian model was the generation of a prior distribution that was technically understandable and manipulable by the available knowledge of the subject. A practically useful procedure was realized by the computer program BAYSEA (Akaike and Ishiguro, 1980).

It is interesting to note that missing observations did not produce any problem by this approach and the Bayesian modeling was eventually applied to the record of geophysical observations related to the original problem of earthquake prediction by Ishiguro et.al. (1983). The computer program BAYTAP-G written by M. Ishiguro and Y. Tamura is included in TIMSAC-84 (Akaike et al., 1985). A slight modification of BAYSEA found an interesting application in the analysis of circadian rhythm of human body (Akaike, 1983). From the point of view of statistical model building BAYSEA provides a typical example of the use of subjective or psychological information.

The success obtained by BAYSEA produced confidence in the application of models with structures physically not confirmable but were appealing to common sense. Examples of this type of applications are given in Kitagawa and Gersch (1984), and Kitagawa (1987). The latter provides a particularly clear explanation of the role played by AIC in relation to the application of these models. Applications to the spatial data developed by Ogata and Katsura (1988) and Ogata, et al. (1991) provide interesting examples of scientific applications on the analysis of the invisible structure of the earth. These examples clearly show that with the help of the information criterion the activity of statistical modeling has now entered into an entirely new stage.

5. Concluding remarks

The experiences on time series modeling described in this paper are all related to the realization of practically useful application of statistical time series analysis. They suggest that the author had a particular idea on how to handle real problems that this prompted the response of the author to stimulus from the outside. This point could be confirmed by the existence of the déjà vu feeling which the author experienced when the successful application of the gap process model to the silk production process was completed. Nevertheless, it was only through the contact with real problems that the author really learned about what to study in statistics.

The arrivals of the stimuli were certainly quite random and often unexpected. Fortunately, the environment for the reception of the stimuli was always properly maintained. This was made possible by the organization of the Institute of Statistical Mathematics which traditionally emphasized the importance of conducting research keeping contact with real problems.

The combination of interest in practical problems and intention to develop useful application of time series models formed a basis for the realization of author's experiences described in this paper. However, the existence of the environment which allowed the establishment of necessary cross-disciplinary contacts was quite instrumental for this realization. If the modeling activity described in this paper could be considered to have been on a right track, then it would be safe to conclude that only through the effective realization of cross-disciplinary cooperative research one can expect continuous development of time series modeling in the future.

REFERENCES

Akaike, H. (1956). On a zero-one process and some of its applications. *Annals of the Institute of Statistical Mathematics,* Vol. 8, 87-94.

Akaike, H. (1959). On the statistical control of the gap process. *Annals of the Institute of Statistical Mathematics,* Vol. 10, 233-259.

Akaike, H. (1967). Some problems in the application of the cross spectral method. In *Spectral Analysis of Time Series,* (ed. B. Harris), 81-107, John Wiley, New York.

Akaike, H. (1968). On the use of a linear model for the identification of feedback systems. *Annals of the Institute of Statistical Mathematics,* Vol. 20, 425-439.

Akaike, H. (1971). Autoregressive model fitting for control. *Annals of the Institute of Statistical Mathematics,* Vol. 23, 163-180.

Akaike, H. (1974). Markovian representation of stochastic processes and its application to the analysis of autoregressive moving average processes. *Annals of the Institute of Statistical Mathematics,* Vol. 26, 363-387.

Akaike, H. (1977). Canonical correlation analysis of time series and the use of an information criterion. *System Identification : Advances and Case Case Studies,* (eds. R. K. Mehra and D. G. Lainiotis), 27-96, Academic Press, New York.

Akaike, H. (1978). A new look at the Bayes procedure. *Biometrika*, Vol. 65, 53-59.

Akaike, H. (1980). Likelihood and the Bayes procedure. *Bayesian Statistics*, (eds. J.M. Bernardo, M.H. De Groot, D.V. Lindley and A.F.M. Smith), 143-166, University press, Valencia, Spain.

Akaike, H. (1983). Statistical inference and measurement of entropy. *Scientific Inference, Data Analysis, and Robustness*, (eds. G.E.P. Box, J.F.Wu and T. Leonard), 165-189, Academic Press, New York.

Akaike, H. (1987). On the development of TIMSAC program packages. Proceedings of the 46th Session of the ISI.

Akaike, H. (1992). Implications of informational point of view on the development of statistical science. Presented at the First US/JAPAN Conference on The Frontiers of Statistical Modeling, An Informational Approach, May 24-29, 1992, Knoxville, Tennessee, U.S.A.

Akaike, H., Arahata, E. and Ozaki, T. (1976). TIMSAC-74, A time series analysis and control program package (2). *Computer Science Monographs*, No.6, The Institute of Statistical Mathematics, Tokyo.

Akaike, H. and Ishiguro, M. (1980). BAYSEA, A Bayesian seasonal adjustment program. *Computer Science Monographs*, No.13, The Institute of Statistical Mathematics, Tokyo.

Akaike, H. and Nakagawa, H. (1972). *Statistical Analysis and Control of Dynamic Systems*, Saiensu-sha, Tokyo (In Japanese). English version published by Kluwer Academic Publishers, Dordrecht (1988).

Akaike, H., Ozaki, T., Ishiguro, M., Ogata, Y., Kitagawa, G., Tamura, Y.H., Arahata, E., Katsura, K. and Tamura, Y. (1985). TIMSAC-84 Part 1, *Computer Science Monographs*, No.22, The Institute of Statistical Mathematics, Tokyo.

Blackman R. B. and Tukey, J. W. (1959). *The Measurement of Power Spectra From the Point of View of Communications Engineering*. Dover, New York.

Fukunishi, K. (1977). Diagnostic analyses of a nuclear power plant using multivariate autoregressive processes. *Nuclear Science and Engineering*, Vol. 62, 215-225.

Ho, B. L. and Kalman, R. E. (1966). Effective construction of linear state-variable models from input / output functions. *Proceedings of the Third Allerton Conference*, 449-459.

Ishiguro, M., Akaike, H., Ooe, M. and Nakai, S. (1983). A Bayesian approach to the analysis of earth tides. *Proceedings of the Ninth International Symposium on Earth Tides*, (ed. J.T.Kuo), 283-292, E. Schweizerbart'sche Verlagsbuchhandlung, Stuttgart.

Kitagawa, G. (1987). Non-Gaussian state-space modeling of nonstationary time series. (with discussions) *Journal of the American Statistical Association,* Vol. 82, 1032-1063.

Kitagawa, G. and Gersch, W. (1984). A smoothness priors-state space modeling of time series with trend and seasonality. *Journal of the American Statistical Association,* Vol. 79, 378-389.

Nakamura, H. and Akaike, H. (1981). Statistical identification for optimal control of supercritical thermal power plants. *Automatica,* Vol. 17, 143-155.

Ogata, Y. and Katsura, K. (1988). Likelihood analysis of spacial inhomogeneity for marked point patterns. *Annals of the Institute of Statistical Mathematics,* Vol. 40, 29-39.

Ogata, Y., Imoto, M. and Katsura, K. (1991). 3-D spacial variation of b-values of magnitude-frequency distribution beneath the Kanto district, Japan. *Geophysical Journal International,* Vol. 104, 135-146.

Otomo, T., Nakagawa, T. and Akaike, H. (1972). Statistical approach to computer control of cement kilns. *Automatica,* Vol. 8, 35-48.

Polanyi, M. (1962). *Personal Knowledge,* Corrected edition. University of Chicago Press.

Shimazaki, A. (1961). Studies on the statistical control of the raw silk production process I. On the process control in the reeling with fixed number of cocoons. *Bulletin of The Sericultural Experiment Station,* Vol. 16, 403-529.

Tukey, J. W. and Hamming, R. W. (1949). *Measuring Noise Color.* Unpublished memorandum.

Turing, A. M. (1969). Intelligent machinery. *Machine Intelligence 5,* (eds. B. Meltzer and D. Michie), 3-23, Edinburgh University Press.

Wold, H. O. A. (1937). *A study in the Analysis of Stationary Time Series.* Almquist & Wiksell, Uppsala.

IMPLICATIONS OF INFORMATIONAL POINT OF VIEW ON THE DEVELOPMENT OF STATISTICAL SCIENCE

H. AKAIKE
The Institute of Statistical Mathematics
4-6-7 Minami-Azabu, Minato-ku
Tokyo 106, Japan

0. Introduction

Informational approach represents a new trend in the development of statistical science. This paper is intended for the discussion of the informational outlook in relation to the development of statistics or statistical science.

R. A. Fisher's great success in laying down the foundation of mathematical statistics was largely based on his use of mathematical likelihood as the basic concept for the construction of the theory of estimation. However, from the point of view of the informational approach to be discussed in this paper, it appears that the assumption of a known functional form of the distribution as the starting point of the theory blocked the use of log likelihood as a general criterion of fit of models with possibly different functional forms of the distribution.

The informational approach to statistics was realized by the recognition of the relation between the Kullback-Leibler information and the expected log likelihood of a model. The information criterion AIC was developed based on this observation and found various practical applications before its general acceptance by theoretical statisticians.

In this paper an explanation of this popularity is traced back to the philosophical work of C. S. Peirce on the logic of abduction which is concerned with the first starting of a hypothesis and entertaining it (Peirce, 1955, p.151). Fisher emphasized the use of likelihood in the inductive phase of inference under the assumption of a hypothesis, while Peirce insisted that the most original part of scientific work was related to the abductive phase, or the phase of the selection of proper hypotheses. The existence of examples of successful use of AIC in practical applications shows that the informational approach opened up the possibility of developing a systematic procedure for abductive inference with the aid of the informational interpretation of log likelihood.

Since the development of hypotheses depends on the available knowledge and purpose of application, cross-disciplinary cooperative work is required for successful development

H. Bozdogan (ed.), Proceedings of the First US/Japan Conference on the Frontiers of Statistical Modeling: An Informational Approach, 27–38.

of the science of statistics in the future. The informational approach has prepared a basis for the successful development of such cross-disciplinary activity.

1. Review of Fisherian statistics

In this section the contribution of R. A. Fisher to statistics will be reviewed briefly to clarify its relation with the informational approach to statistics. The distinctive characteristic of Fisher's contribution is the reflection of strong influence of the experience of handling statistical problems in real world. However, his first paper on statistics was not directly related to any particular practical application (Fisher, 1912). In the paper he introduced a criterion of fit of a statistical model which was later to be called likelihood.

Fisher did not describe how he was lead to the use of the criterion. However, taking into account the fact that just before that time the method of fitting frequency curves and the test of the goodness of fit by chi-square statistic had been developed by K. Pearson, one possible guess is that the introduction of the criterion was motivated through the need of calculating theoretical frequencies for the chi-square test. It is known that there is a close relation between the chi-square test and the probabilistic interpretation of entropy as developed by L. Boltzmann and that Boltzmann's final work on the subject is based on a representation identical to negative Kullback-Leibler information (Akaike, 1985). Thus it seems that there is an inherent historical connection of Fisher's contribution to statistics with the probabilistic concept of entropy, or information.

As is observed by Stigler(1976) the 1922 paper of Fisher (Fisher, 1922) on the mathematical foundations of theoretical statistics is characterized by its heavy dependence on the concept of parameter. Fisher defined the purpose of statistical methods as the reduction of data and divided the related problems into three types : problems of specification, estimation and distribution. The importance of the problem of specification or the proper choice of the functional form of the distribution for each particular application was clearly recognized by Fisher. He highly evaluated the contribution of K. Pearson on this aspect of statistics by introducing the Pearsonian system of frequency curves and an objective criterion of goodness of fit, the chi-square test statistic. However, in spite of the insightful observation of the importance of the problems of specification, Fisher considered these as entirely a matter for practical statistician and regarded the discussions of theoretical statistics as alternating between problems of estimation and problems of distribution (Fisher, 1922, p.315).

Fisher was keen to clarify the contribution of mathematical theory of statistics to the logic of inductive reasoning. In an expository paper on the advances of statistical methods he tried to demonstrate the adequacy of the concept of likelihood for inductive reasoning in the particular logical situation for which it had been introduced (Fisher, 1935). This was the use of likelihood in the problem of estimation. He considered that the concept of likelihood had been introduced for the handling of the problem of parameter estimation and that the justification for its use was provided by the successful development of the theory of estimation.

In his presidential address to the First Indian Statistical Conference (Fisher, 1938), Fisher pointed out the importance of sufficiently prolonged experience of practical research and of responsibility for drawing conclusions from actual data for teachers of statistics at universities. However, of the three types of problems mentioned in the 1922 paper, only the importance of the problems of distribution was mentioned in relation to the development of the theory of estimation. Further, stressing the successful development of the estimation theory, he even mentioned that statistician's job was only to produce what the given data set contained. This statement makes a peculiar contrast with the statement made in an earlier paragraph of the same paper. It tells that all extensive bodies of data were liable to contain information on points which were not in view when they were collected, and that to recognize such information, and to find the means of eliciting it, was most stimulating part of statistician's task. It seems that this peculiarity reflects the desire to confirm the contribution of mathematical statistics within the framework of inductive logic.

In spite of the heavy reliance of Fisher's theory of estimation on the concept of likelihood his understanding of the nature of the likelihood remained somewhat rudimentary. This can be seen by the explanation of the concept of mathematical likelihood given in his last book as follows :

> *It is, like Mathematical Probability, a well-defined quantitative feature of the logical situations in which it occurs, and like Mathematical Probability can serve in a well-defined sense as a " measure of rational belief ";* .... (Fisher, 1973, p.72).

Similar explanation appears in several other writings. However, no convincing argument was developed for the characterization of the likelihood as a "measure of rational belief ".

The informational point of view to be discussed in the next section shows that this lack of the characterization restricted Fisher's use of likelihood to the local theory of parameter estimation under the assumption of a given form of the distribution. A framework was then established to view the test of significance as the basic procedure for the solution of the problems of specification and restrict the estimation to the parameter of a given model. Thus the test and estimation formed a paradigm to make statistics into what was called a normal science by Kuhn (1970). However, it seems that the use of test procedures advocated by this paradigm eventually produced a very restricted image of statistics in applications which was conditioned by the availability of proper test procedures.

2. Informational view of statistics

In spite of the suggestive title of the 1912 paper to view likelihood as an absolute criterion of fit of frequency curves Fisher actually defined it as a rational measure for the comparison of parameters within a model. This eliminated the possibility of recognizing the potential of log likelihood as a general criterion of fit of a model and restricted its use to the estimation of parameters within a fixed model. Thus the later development of the theory of estimation was concerned with the refinement under the assumption of a known form of the true distribution.

In a real situation, it is certainly quite rare that the exact form of the distribution is known. The success of the analysis of real data thus depends essentially on the choice of the basic model, or the form of the distribution. Only through proper modeling of the structure of the distribution one can get useful information out of data by statistical procedures. A poor model cannot produce useful information with the refinement of the parameter estimation procedure.

According to the present author's personal experience almost each new problem demanded significant effort for the choice of the basic model. One trivial example which clearly explains this was obtained in the analysis of data for the experimental comparison of chemotherapeutics for the treatment of tuberculosis. The medical doctors conducting the experimental study were bothered by the smallness of the number of complete time series of observations of patients and asked the author to provide a proper procedure to fill in the missing observations. A careful examination of the data showed that the clustering of the time series by the initial condition of each patient, represented by the number of cavities in the x-ray chest image, the time series could produce sufficient information to differentiate the effects of the medicines. The choice of the number of cavities as the concomitant variable was the key to the success of the analysis and the simple recovery rate at each month demonstrated sufficient accuracy in each cluster. However, the detection of this concomitant variable required much work. This simple example sufficiently shows that it is the solution of the problem of specification that is really decisive for the successful handling of a real problem.

If the attention had been limited to the empirical choice of the basic model there would have been no chance to get the idea of the use of information criterion. The analysis of engineering problems showed the existence of a vast area of application of a particular type of models, the constant, or time invariant, linear system models. Their frequent use suggested that the identification of the structure of a constant linear system model, based on the time series of observations of a real system, was a practically very important problem. The limitation of the empirical approach to the handling of time series for the identification became clear when the fitting of a constant linear system model, the autoregressive (AR) model, was considered. Conventional procedure of power spectrum estimation was realized by smoothing the periodgram and this was highly empirical and required much experience for proper choice of the final estimate. The estimation could be realized through AR model fitting and thus eliminate the difficulty if a proper procedure for the determination of the order of an autoregression was known. However, when the choice of AR order was contemplated, it became quickly clear that curiously enough this problem had been handled not as a problem of estimation but of a test.

It is interesting to see that Fisher also faced with the problem of order determination in his first work with the long sequence of yields of wheat from Broadbalk when he tried to extract time trends by fitting orthogonal polynomials (Fisher, 1921). Incidentally, in this connection, J. F. Box (1978, p.104) mentions that, fifty years later, the problem of how best to determine the adequacy of fit of a polynomial of any degree has not been satisfactorily solved. Apparently Fisher tried to confirm the adequacy of the chosen order by checking the autocorrelation coefficients of the residuals.

In the case of the order determination of AR models, after the trial of various criteria, the concept of final prediction error (FPE) was introduced which was defined by the expected mean-square one-step ahead prediction error when the estimated parameters were used for prediction (Akaike, 1969, 1970). However, for the extension of the concept of FPE to the case of multivariate AR model, a measure for the evaluation of the multivariate prediction error was required. A seemingly natural choice was the generalized variance, or the determinant of the variance-covariance matrix of the innovation, or the one-step ahead prediction error. It was then noticed that the log of the sample generalized variance of the residuals was asymptotically related to the maximum log likelihood of the stationary Gaussian multivariate AR model (Whittle, 1953). Thus the attention was turned to the method of maximum likelihood.

Certain similarity was observed between the choice of the order of an autoregression and that of the number of factors in the factor analysis, but, in the case of the factor analysis, it was not easy to identify the equivalent of the prediction error of time series. The recognition of the fact that the fitting of the factor analysis model was realized by maximizing the log likelihood (Jöreskog, 1967) lead to the recognition that the negative log likelihood was playing the role of the mean square prediction error. In the case of AR model fitting the estimated parameters were used to compute the predicted value and the analysis of the expected performance of such a predictor lead to the concept of FPE. By analogy, it was clear that the probability distribution defined by the maximum likelihood estimate of the parameter was providing a predictor of which performance was being evaluated by the expected log likelihood. When the use of log likelihood was considered as the basic criterion it was noticed that the expectation of this quantity was related to the Kullback-Leibler information defined by

$$I(f;g) = E_x \log f(x) - E_x \log g(x) ,$$

where $g(x)$ is an approximation to $f(x)$, the true distribution, and E_x denotes the expectation with respect to $f(x)$. The non-negativity of the K-L information showed that the maximum of the expected log likelihood was attained when the model was true.

The information criterion AIC was derived by considering the typical situation of estimation treated by Fisher. When the parametric model $f(x \mid \theta)$ is fitted to the vector of observations $x(N) = (x_1, x_2, \ldots, x_N)$ under the assumption of independent and identical distribution of observations the asymptotic theory of the maximum likelihood estimate suggests that the difference of the twice the expected log likelihoods

$$2E_x \log f(x \mid \theta_0) - 2E_x \log f(x \mid \theta(x(N)))$$

is, under certain regularity condition, asymptotically distributed as a chi-square statistic with the degrees of freedom k, the dimension of θ, where E_x denotes the expectation with respect to the true distribution $f(x \mid \theta_0)$ of $x(N)$ and $\theta(x(N))$ the maximum likelihood estimate. This quantity is asymptotically approximated by

$$2 \log f(x(N) \mid \theta(x(N))) - 2 \log f(x(N) \mid \theta_0)$$

which is a familiar Neyman-Pearson likelihood ratio test statistic to test the hypothesis $\theta = \theta_0$. This shows that the increase of the log likelihood attained by replacing the true parameter θ_0 by its maximum likelihood estimate $\theta(x(N))$ is asymptotically equivalent to the decrease of the expected log likelihood of the predictive distribution $f(\bullet \mid \theta(x(N)))$ from that of the true distribution $f(\bullet \mid \theta_0)$.

Thus to recover the relative relation between the twice expected log likelihoods from that of the twice log likelihoods it is necessary to reduce the value of twice the maximum log likelihood by twice the corresponding chi-square statistic. However, since this chi-square is unobservable, it is replaced by its expectation k. To establish the direct correspondence between the log likelihood and mean square error the negative of the log likelihood is adopted as the criterion to measure the badness of the model and AIC is defined by

$$\text{AIC} = - 2 \log (\text{maximum likelihood}) + 2 (\text{number of parameters}),$$

where log denotes natural logarithm.

This derivation of AIC is based on the assumption that the parametric model contains the true distribution as its member. This represents the situation where the models being considered are just-or over-parametrized. In actual comparison of models the crucial situation is where some of the models are under-parametrized but the corresponding increase of the badness of fit, as measured by the criterion, is of comparable order to the reduction of the number of parameters. This situation is discussed in detail in the original paper of Akaike (1973) by using the result of the large sample theory of the test of hypotheses of multiple parameters developed by Wald (1943).

3. Contribution of informational approach to the generation of hypotheses

By reading ordinary introductory text books of mathematical statistics one often gets an impression that the main use of the theory of statistics is the evaluation of the level of significance of some test statistics or of the accuracy of estimated parameters. Since the level of mathematics required for the solution of the problems of distribution related to the test and estimation procedures demands professional skill, people who are mainly interested in the application of statistical procedures usually have to seek for a model of which handling procedure is well-developed. Thus it was often felt impossible for the people in applied area to develop the use of new models for their own problems. However, from the point of view of application, it is the production and confirmation of interesting and useful hypotheses through proper use of observational data that is really expected of statistical methods. Thus the emphasis of the mathematical aspect of conventional statistical procedures unexpectedly produced an undesirable effect to impede the participation in the development of statistics of the people with ample knowledge of particular area of application but with less professional skill of mathematics.

In this connection it is interesting to refer to the philosophical works of C. S. Peirce on abduction and induction around 1890 (Peirce, 1955). Peirce discusses the process of making use of observed facts, or data, to extract practical knowledge for application to other circumstances than those under which they were observed. For this the necessary additive ingredients are hypotheses and he calls the inferential step of starting and entertaining a hypothesis abduction, or retroduction, and calls the operation of testing a hypothesis by experiment induction. The testing in this case is performed by checking the validity of a prediction based on the hypotheses by observations and the result is used for the forming of confidence on the hypothesis. It can be seen that the statistical process of the test of significance falls into this category of induction. Fisher (1935) considered the essential effect of the general body of researches in mathematical statistics at his time to be a reconstruction of logical rather than mathematical ideas and the work of making sense of figures was considered to be an attempt of inductive logical process for the use of data in making an inference from the particular to the general. It seems that what Fisher highly evaluated of mathematical statistics was the contribution to the logic of induction as defined by Peirce.

It is interesting to note that Peirce placed more importance on the inferential step of abduction or retroduction than on induction. He considered that originality lies in the suggestion of hypotheses and that typical induction only tests a suggestion already made. He also exemplified the process of abduction by Kepler's work on the analysis of the motion of the planet Mars which was realized through a succession of refinements of hypotheses until a complete fit to observations was attained. By this example the importance of the existence of sound and rational motive for the modification of hypothesis is stressed instead of the use of capricious or random modification. The existence of an inherent relation of this concept of abduction with the scientific treatment of the problems of knowledge can be observed by the recently revived interest in the abductive inference in the area of artificial intelligence; see, for example, Josephson et al. (1987).

Within the context of mathematical statistics the abductive reasoning is realized by the refinement of a hypothesis through successive process of modification and testing. Fisher also considered that the free use of empirical frequency formulae was justified by the use of the tests of goodness of fit. However, this process is essentially concerned with the problem of specification which was considered by Fisher to be outside of the concern of theoretical statisticians. Thus the view of statistics developed by Fisher placed the abductive aspect of inference outside the attention of theoretical statisticians. This was a situation which was not quite favorable for the development of statistics as a science of knowledge.

The introduction of AIC significantly changed the situation. The example of Kepler's work shows that the process of abduction as contemplated by Peirce is realized by successive test or evaluation of hypotheses. Thus for the implementation of an abductive inference procedure it is necessary to label the result of the test or evaluation of each hypothesis by a common measure useful for the purpose of comparison. In this connection it must be remembered that Akaike (1985, p.11) has pointed out the importance of the fact that the log likelihood log $f(x \mid \theta)$ provides an unbiased estimate of $E_y \log f(y \mid \theta)$, irrespective of the form of the true distribution f(y) of x, where E_y denotes the expectation with respect to the distribution f(y). This fact represents the most significant aspect of the informational approach in relation to the generation of hypotheses. It provides a justification for the choice of a model with larger log likelihood as a better model, whatever it may be the form of the true distribution. This observation essentially frees the statistical modeling activity from the unproductive concept of " fixed but unknown true distribution " and encourages the proposal of new models.

When the use of a statistical model for the handling of the information supplied by data is considered, it becomes obvious that the definition of true model depends on the objective of the modeling. In particular, the concept of true model depends on the available resources, such as tools and knowledge, for the construction and application of the model. In this sense the concept of true model is highly dependent on the circumstances. Further, the basic choice of a model is realized only through the mental activity of the researcher. In this sense, the concept could be quite subjective or personal.

The emphasis of the subjective nature of statistical modeling activity seemingly makes a significant contrast to the emphasis of the objectivity of statistical procedures by Fisher as represented by his careful handling of the Bayes procedure or the inverse probability. However, Fisher (1973, p.81) also clearly recognized the role of statistician's imagination in a test of significance. Since all the great scientific discoveries or inventions are always quite unexpected from the point of view of contemporary observers and the generation of an innovative hypothesis is always highly dependent on personal activity, it is obvious that blind adherence to the concept of objectivity must be eliminated to regain the creative power of statistical methods in scientific activities. M. Polanyi who stresses the personal aspect of science also points out the importance of the talent to select good hypotheses for investigation and mentions that any theory of inductive inference in which this talent plays no role is a *Hamlet* without prince (Polanyi, 1962, p.30). The importance of the role of initiative in producing human intelligence has been noticed earlier by A. M. Turing in a report written in 1947 (Turing, 1969).

Fisher (1973, p.46) considered that the objectivity of a statistical procedure is represented by the communicability and verifiability of the the result. The above mentioned informational interpretation provides a justification of log likelihood as an objective or inter-subjective criterion of the goodness of statistical models. Thus log likelihood attains the position of a general criterion that can lead the unending search for "truth" through successive improvement of hypotheses, represented by statistical models, within a scientific community. AIC is a typical example that has shown the general applicability of this criterion for the realization of an abductive inference procedure, and encouraged the activity related to the production and comparison of hypotheses by both theoretical statisticians and researchers in the area of application. Since the method of maximum likelihood provides a definite procedure for the production of the estimates of unknown parameters and since the computers allow easy implementation of necessary numerical procedures an efficient procedure for the development of a new hypothesis is realized through successive proposal and comparison of statistical models with the aid of AIC.

This procedure provides a typical example of the realization of the abductive inference procedure discussed by Josephson et al. (1987) for the search of a best hypothesis.

However, the limitation of the above procedure becomes quickly apparent. This is the difficulty of handling the situation where extremely large number of hypotheses can be generated mechanically. Typical example is the polynomial fitting to a temporal series of observations. By equating some of the coefficients equal to zero the number of possible models becomes equal to 2^{k+1} for the k-th degree polynomial. The number of models can easily exceed the number of observations and there is no hope of the simplistic minimization procedure of AIC to lead to any reasonable choice of a best hypothesis.

Again it is quite remarkable that Peirce already considered this type of problem in relation to the search for better hypotheses. In his work Peirce remarks, based on his own experience, that it is not the logical simplicity but the simplicity in terms of facility and naturalness that should be preferred. This remark is based on the conviction that human mind is with a natural bent in accordance with nature's. It seems that the notion of simplicity implies the possession of wider applicability for the generation and explanation of other hypotheses. These observations suggest the need of introducing a hypothesis of higher level that properly represents the natural bent of human mind and governs the collection of original hypotheses. It is interesting to note that, in the case of statistical hypotheses, the role of the governing hypothesis is played exactly by the prior distribution over the models that represent the original hypotheses. Thus Peirce's observation provides an idea on the use of Bayesian models for the elimination of the difficulty of handling a large number of models. However, the observation does not directly lend support to the personalistic view of subjective Bayesians. Any behavior that is only based on rigid subjectivism cannot be differentiated from the behavior of a lunatic person and the result can not be communicated as useful information to others.

The observations made in this section suggest the necessity of clarifying the relation between data and hypotheses. In the inductive logic data are used to check or evaluate hypotheses. In the abductive phase of inference the data prompt the generation of hypotheses. It can be seen that inductive logic is concerned with the decisional use of data, while the abduction is concerned with the analysis or interpretation of data. If the process of the analysis of data is viewed as a process of abduction then it becomes natural and necessary that the hypotheses being generated are data dependent, as the mental search for possible hypotheses will be conditioned by the knowledge of the data. This is the actual state of the analysis of data being practiced in scientific activities and the implicit dependence of the hypotheses on the data poses a problem for the strict application of inductive reasoning. However, if the dependence is realized only through the rejection of definitely inadequate models with very low likelihood this would not seriously invalidate the reasoning. The final validation of the hypotheses will only be realized through the application to real problems.

A drastic view of data is obtained by viewing the knowledge represented by, or used in the forming of, the hypothesis also as a kind of data. Actually the prior knowledge used in the representation of a hypothesis may be considered as a reduced representation of the information supplied by a particular set of past observational data. By this view statistical model construction is concerned with the proper use of the information provided by the whole set of data of the present and past. The past and present data are uniquely determined and in that sense certain, but the knowledge based on the past data is not necessarily perfect or certain. Its validity must always be questioned. This simple observation shows that the rigid Bayesian framework which assumes the prior knowledge represented by the prior distribution as certain is useful only in the inductive or decisional framework. In the abductive phase of the analysis of data the validity of the assumption of the prior distribution must be discussed.

The informational approach suggests that if the process of generating a hypothesis is viewed as the process of forming an expectation represented by a probability distribution then a generally acceptable criterion for the comparison of models could be defined by proper definition of the of log likelihood. Thus a Bayesian model is viewed only as a particular type of statistical model and its relative goodness can be evaluated by the likelihood which is defined by the integral of the likelihood of the data distribution with respect to the prior distribution of the parameter. Some early examples given by the present author on the use of Bayesian models based on this observation are the seasonal adjustment procedure BAYSEA, and the control of improper solutions in the factor analysis (Akaike 1980, 1987).

4. Enhancement of the importance of statistics in society

The history of statistics shows the rises and falls of various schools each focused on particular aspect of statistics and demonstrates the difficulty inherent in the definition of the word statistics. K. Pearson (1973) describes vividly how the word "statistics" was first introduced into English. The word has been traditionally used to denote numerical characteristics of a mass of observations, in particular of a society. Another and more recent use is to denote the science and technology of the effective use of information provided by observational data for the construction and use of hypotheses. The use of the word statistical science for this latter purpose would be more appropriate to eliminate the common confusion of the concept of statistics.

Recent development of computer technology is rapidly increasing the number and variety of observations recorded in media directly accessible from computers. The huge accumulation of data necessitates the use of statistical procedures for the extraction of useful information for particular application. Thus this is the time that society is really in need of statistical science. The discussions in the preceding sections show that the contribution of a statistician would be highly evaluated only when he had contributed significantly to the generation of a new effective hypothesis. For the construction and formulation of such a hypothesis, besides the skill of handling statistical models, deep background knowledge of the area is indispensable. It is the ability to provide a reasonable representation of the background knowledge in the form of a statistical model that is really required of a statistician. Thus, basically it is only through cross-disciplinary cooperative researches that statisticians can make significant direct contributions to society.

The present author found several researchers in unexpected areas of application who used AIC, or its variant for the evaluation of Bayesian models, to produce very interesting results. This shows that the informational approach certainly contributed significantly to the enhancement of abductive use of statistical inference and increased the number of successful examples of application of statistical methods. However, when the complexity of the required model is further increased, the cooperation of professional statisticians will be necessary to make fuller use of the information supplied by data. It is also certain that only through this type of contact with real problems statisticians can gain the idea of the future direction of the development of statistics, as statistics is the science of a particular type of activity of human brain and the function of human brain develops with the accumulation of experience and knowledge in human society. Without the contact with outside world, statisticians' research effort will inevitably be restricted to the analysis of those problems that were modeled by the brain activity based only on the remote past experience.

These observations show the need of formulating statistical science as a science for the development of the concept of statistics through the contact with problems and researchers of various disciplines. The reason of existence of the science can be found in the fact that a statistical model developed in a particular area of application can often be profitably transplanted to other areas. Thus the accumulation of statistical models and experiences of their use is particularly important. Also, as the development of the informational approach shows, it would be safe to say that only the theoretical or logical analyses of the accumulated experiences of application can lead to the development of new concepts. These observations show that only the proper implementation of cross-disciplinary activities can enhance the importance of statistics in society. The informational approach prepared a technical and philosophical basis for the development of such cross-disciplinary activities.

5. Conclusion

The informational view of statistics has clarified that the lack of the characterization of log likelihood as a general measure of rational belief limited Fisher's attention to the use of log likelihood in the inductive logic of estimation and this lead to the unlucky dichotomy of statistical methodology into test and estimation. In contrast with this, the informational point of view has been developed based on the recognition that the choice of the basic model or the solution of the problem of specification constitutes the crucial phase of statistical handling of observational data and log likelihood is useful as a general criterion for the evaluation and comparison of models with different distributional structures.

The review of the philosophical work of Peirce on abduction and induction has shown that the abductive inference, or the generation of hypotheses, plays more significant role in the science of knowledge than the inductive inference. It has been confirmed that AIC produced a realization of the abductive inference procedure. It has further been observed that the informational point of view provides a basis for the free use of the originality based on personal knowledge for the construction of hypotheses represented by statistical models, including Bayesian models.

Thus the informational view allows the development of statistical methods for the important scientific activity of hypothesis generation with proper use of observational data. It may be concluded that informational view has prepared a basis for the enhancement of the recognition of importance of statistical science in society.

References

Akaike, H. (1973). Information theory and an extension of the maximum likelihood principle. *2nd International Symposium on Information Theory.* (eds. B.N.Petrov and F.Csaki), 268-281, Akademiai Kiado, Budapest. Reproduced in *Breakthroughs in Statistics*,Vol. 1, (eds. S.Kotz and N.L.Johnson), Springer-Verlag, New York, (1992), 610-624.

Akaike, H. (1980). Seasonal adjustment by a Bayesian modeling. *Journal of Time Series Analysis,* Vol. 1, 1-13.

Akaike, H. (1985). Prediction and entropy. *A celebration of Statistics,* (eds. A.C.Atkinson and S.E.Fienberg), 1-16, Springer-Verlag, New York.

Akaike, H. (1987). Factor analysis and AIC. *Psychometrika,* Vol. 52, 317-332.

Box, J. F. (1978). *R. A. Fisher, the life of a scientist.* Wiley, New York.

Fisher, R. A. (1912). On an absolute criterion for fitting frequency curves. *Messenger of Mathematics*, Vol. 41,155-160.

Fisher, R. A. (1915). Frequency distribution of the values of the correlation coefficient in samples from an infinitely large population. *Biometrika*, Vol. 10, 507-521.

Fisher, R. A. (1921). Studies in crop variation. I. An examination of dressed grain from Broadbalk. *Journal of Agricultural Science*, Vol.11, 107-135.

Fisher, R. A. (1922). On the mathematical foundations of theoretical statistics. *Philosophical Transactions of the Royal Society of London*, A, Vol.222, 309-368.

Fisher, R. A. (1935). The logic of inductive inference. *Journal of the Royal Statistical Society*, Vol. 98, 39-54.

Fisher, R. A. (1938). Presidential address, First Indian Statistical Conference, 1938. *Sankhyá*, Vol.4, 14-17.

Fisher, R. A. (1973). *Statistical Methods and Scientific Inference.* Third edition. Hafner Press, New York.

Jöreskog, K. G. (1967). Some contributions to maximum likelihood factor analysis. *Psychometrika*, Vol.32, 443-482.

Josephson, J. R., Chandrasekarn, B., Smith Jr, J. W. and Tanner, M. (1987). A mechanism for forming composite explanatory hypotheses. *IEEE Transactions on Systems, Man, and Cybernetics*, Vol.17, 445-454.

Kuhn, T. S. (1970). *The Structure of Scientific Revolutions* Second Edition, enlarged. The University of Chicago.

Pearson, K. (1973). The early history of statistics. *The History of Statistics in the 17th & 18 Centuries*, (ed. E.S.Pearson), 1-9, Charles Griffin, London.

Peirce, C. S. (1955). Abduction and induction. *Philosophical Writings of Peirce*, (ed. J. Buchler), 150-156, Dover, New York.

Polanyi, M. (1962). *Personal Knowledge*, Corrected edition. University of Chicago Press.

Stigler, S. M. (1976). Discussion of a paper by L. J. Savage. *The Annals of Statistics*, Vol. 4, 498-500.

Turing, A. M. (1969). Intelligent machinery. *Machine Intelligence 5*, (eds. B. Meltzer and D. Michie), 3-23, Edinburgh University Press.

Wald, A. (1943). Tests of statistical hypotheses concerning several parameters when the number of observations is large. *Transactions of the American Mathematical Society*, Vol.
54, 426-482.

Whittle, P. (1953). The analysis of multiple stationary time series. *Journal of Royal Statistical Society, Series B*, Vol.15,125-139.

Index

Editors' Note

This volume was planed as a part of the activity to celebrate the 50th anniversary of the establishment of the Institute of Statistical Mathematics, Tokyo. We wish to acknowledge the support provided by the Institute of Statistical Mathematics.

We thank the author of the interview article, Dr. David F. Findley, and the co-authors of the papers included in this volume, Prof. Y. Yamanouchi, Dr. T. Otomo, Dr. T. Nakagawa, Dr. H. Nakamura, Prof. Y. Sakamoto, Prof. Y. Ogata, Prof. M. Ishiguro, Prof. M. Ooe and Prof. S. Nakai, for their understanding on and the permission of the publication of this volume. The photograph of Professor Akaike was provided by Prof. N. Iba.

We are also indebted to the Institute of Statistical Mathematics, Institute of Mathematical Statistics, Elsevier Science Ltd., The Institute of Electrical and Electronics Engineers, Inc., Blackwell Publishers, Biometrika Trustees, Professor J. M. Bernardo, E. Schweizerbart'sche Verlagsbuchhandlung, Psychometric Society, Springer-Verlag, and Kluwer Academic Publishers for their kind permission to reproduce the papers included herein.

Editors
May 28, 1997

Springer Series in Statistics

(continued from p. ii)

Le Cam/Yang: Asymptotics in Statistics: Some Basic Concepts.
Longford: Models for Uncertainty in Educational Testing.
Manoukian: Modern Concepts and Theorems of Mathematical Statistics.
Miller, Jr.: Simultaneous Statistical Inference, 2nd edition.
Mosteller/Wallace: Applied Bayesian and Classical Inference: The Case of *The Federalist Papers.*
Parzen/Tanabe/Kitagawa: Selected Papers of Hirotugu Akaike.
Pollard: Convergence of Stochastic Processes.
Pratt/Gibbons: Concepts of Nonparametric Theory.
Ramsay/Silverman: Functional Data Analysis.
Read/Cressie: Goodness-of-Fit Statistics for Discrete Multivariate Data.
Reinsel: Elements of Multivariate Time Series Analysis, 2nd edition.
Reiss: A Course on Point Processes.
Reiss: Approximate Distributions of Order Statistics: With Applications to Non-parametric Statistics.
Rieder: Robust Asymptotic Statistics.
Rosenbaum: Observational Studies.
Ross: Nonlinear Estimation.
Sachs: Applied Statistics: A Handbook of Techniques, 2nd edition.
Särndal/Swensson/Wretman: Model Assisted Survey Sampling.
Schervish: Theory of Statistics.
Seneta: Non-Negative Matrices and Markov Chains, 2nd edition.
Shao/Tu: The Jackknife and Bootstrap.
Siegmund: Sequential Analysis: Tests and Confidence Intervals.
Simonoff: Smoothing Methods in Statistics.
Small: The Statistical Theory of Shape.
Tanner: Tools for Statistical Inference: Methods for the Exploration of Posterior Distributions and Likelihood Functions, 3rd edition.
Tong: The Multivariate Normal Distribution.
van der Vaart/Wellner: Weak Convergence and Empirical Processes: With Applications to Statistics.
Vapnik: Estimation of Dependences Based on Empirical Data.
Weerahandi: Exact Statistical Methods for Data Analysis.
West/Harrison: Bayesian Forecasting and Dynamic Models, 2nd edition.
Wolter: Introduction to Variance Estimation.
Yaglom: Correlation Theory of Stationary and Related Random Functions I: Basic Results.
Yaglom: Correlation Theory of Stationary and Related Random Functions II: Supplementary Notes and References.

Printed in the United Kingdom
by Lightning Source UK Ltd.
118173UK00001B/12